NANOTECHNOLOGY SCIENCE AND TECHNOLOGY

APPLICATIONS OF CARBON NANOTUBES

NANOTECHNOLOGY SCIENCE AND TECHNOLOGY

Additional books in this series can be found on Nova's website under the Series tab.

Additional e-books in this series can be found on Nova's website under the e-book tab.

MATERIALS SCIENCE AND TECHNOLOGIES

Additional books in this series can be found on Nova's website under the Series tab.

Additional e-books in this series can be found on Nova's website under the e-book tab.

NANOTECHNOLOGY SCIENCE AND TECHNOLOGY

APPLICATIONS OF CARBON NANOTUBES

AJAY KUMAR MISHRA
EDITOR

nova publishers
New York

For permission to use material from this book please contact us:
Telephone 631-231-7269; Fax 631-231-8175
Web Site: http://www.novapublishers.com

Library of Congress Cataloging-in-Publication Data

ISBN: 978-1-62081-918-0

Published by Nova Science Publishers, Inc. † New York

CONTENTS

PREFACE

The discovery of carbon nanotubes has opened new windows for numerous applications in many disciplines of nanoscience and nanotechnology. Carbon nanotubes consist of graphene sheets in form of sp^2 hybridized carbon atoms. These materials have gained the interest of researchers from various disciplines due to their fascinating physico-chemical properties. The small dimensions, strength and the fascinating physical properties of the carbon nanotubes make them a very unique material which shows a variety of promising applications.

Carbon nanotubes possess unusual electronic and electrochemical properties that are applied as mechanical reinforcements in high performance composites, field emitters, nanoprobes in metrology and as templates for the creation of other nanostructures. Carbon nanotubes show unusual properties and thus highly applied in the field of nanotechnology, electronics, materials and polymer science. Fundamental and practical nanotube research has shown possible applications in the fields of energy storage, molecular electronics, nano-mechanic devices, composite materials.

Carbon nanotubes usually play a key role in the development of smart nanomaterials in the area of nanotechnology. A variety of applications are due to carbon nanotubes for their wide applications in the area of sensors, waste water treatment, defense, electronics, and medical sciences. Researchers across the world are still trying to bring newer applications of carbon nanotubes in many areas.

The book describes the applications of carbon nanotubes in various areas including environmental science such as remediation and speciation, water research, medicine including sensors, targeted drug delivery and therapeutics. The application of carbon nanotubes in waste water research which includes organic, inorganic and microbial pollutants has also gained more attention in research. The present book contains a comprehensive discussion about the waste water research as well. Some of the other applications of carbon nanotubes in the field of material science and polymer science have also been discussed.

The researchers and scientists working in the similar domain of research will benefit from the fundamental concepts and advanced approaches described in the book. Researchers involved in multidisciplinary research on carbon nanotubes and their applications will be major beneficiaries of the content of the book. The book is going to also be beneficial to the researchers who are working for their graduate and postgraduate degrees in the area of nanoscience and nanotechnology.

The book also provides a platform for all academics and researchers as it covers a vast background for the recent literature, abbreviations, and summaries. The book will be worth reading for the researchers who are more interested in the general overview of carbon nanotubes and variety of applications in the area on nanoscience, nanomedicine, engineered materials and developments.

The book covers a broader research area that integrates biology, chemistry, physics, electronics, sensors, electronics, medicine, materials science, medical science, polymer science engineering and nanotechnology to present an interdisciplinary approach. The book contains the fundamental knowledge with the recent advancements for the research and development in the field of nanomaterials and nanotechnology. The book also covers recent approach and prospects about the current research and development in nanomaterials. Versatile application of carbon nanotubes in various disciplines of nanoscience and nanotechnology is the backbone of this book.

In: Applications of Carbon Nanotubes
Editor: Ajay Kumar Mishra
ISBN: 978-1-62081-918-0

Chapter 1

CARBON NANOTUBES: MULTIFUNCTIONAL AND MULTIDIMENSIONAL NANOBIOSYSTEMS FOR DIAGNOSTICS AND THERAPEUTICS

S. B. Mishra
Department of Chemical Technology, University of Johannesburg,
Johannesburg, South Africa

ABSTRACT

With unique and fascinating properties, carbon nanotubes have gained a worldwide attention to be used for various applications. Due to toxic behaviour, its application in the field of biotechnology, biomedical science and health science has not been well investigated. However, carbon nanotubes, functionalized carbon nanotubes and their bionancomposite systems are being well exploited for diagnostics and therapeutics applications. Some of the biomedical applications where carbon nanotubes have found it supreme niche area include drug delivery systems, sensors and other therapeutics. This chapter deals with the overview of toxic behaviour and recent advances in formation of nanobiosystems for biomedical applications.

INTRODUCTION

Since the discovery of carbon nanotubes, these have attracted a worldwide attention due to unique and fascinating properties. Carbon nanotubes (CNT) have been exploited in almost every field of science and technology from physics, chemistry, and material science to engineering and technology. The investigation of carbon nanotubes and related nanobiosystems in the field of medicine is always a thrust area as it finds it utility from diagnostics to therapeutics. This chapter deals with the toxic behaviour, biocompatibility and with the some recent developments where carbon nanotubes have been investigated for medical applications.

2. Toxicity of Carbon Nanotubes

The high aspect ratio of diameter 10^6 with an average length of 1mm and diameter 1nm is one of the most promising properties of carbon nanotube to be found its advantage in biomedical field. The surface functionalization of carbon nanotubes allows modulating its in-vivo and in-vitro behavior and toxicity [1]. There is many other reason that are considered as responsible factors inducing toxicity in carbon nanotubes. Some of these factors are the length of the CNT, concentration, the type of modification on the surface of CNT, solvent / dispersant used to solublise it. Structural dissimilarities of each carbon tube with respect to the length, local and over all charges and the type of catalyst used to synthesize CNT affects the potential use of it in the biomedical science.

The type of charge and the charge density on the carbon nanotubes pose huge limitations towards the binding of biological molecules such as DNA thereby affecting the cellular interactions. Catalysts use to fabricate CNT constitutes 20-40% of weight that is expected to catalyze the oxidative species of cells and tissues by generating free radicals that eventually cause the oxidative tissue damage [2-9]. The other important aspect regarding toxicity of carbon nanotubes is that it has been a well known fact for now that pure carbon nanotubes are rendered more toxic when compared to functionalized CNTs. But in few cases it has been reported that even some of the functionalized CNTS have been identified to be toxic in nature where funtionalised groups were larger [10] and also due to nature of the solvent used such as tetrahydrofuran [11]. Jain et al. reported extensively about the pulmonary toxicity of carbon nanotubes in their review article [12].

In another study [13], scientists investigated the toxicological behaviour in rat lungs and in vitro and concluded that these CNTs were able to trigger the lung cell for the release of TNF$\propto$ inducing lung inflammation and fibrosis. When used in lower concentration i.e. < 25 mg/l , multiwalled carbon nanotubes (MWCNT) were not able to destabilize the erythrocyte to hemolysis whereas when a concentration of MWCNT was increased by double the amount, it was able to accelerate the process [14].

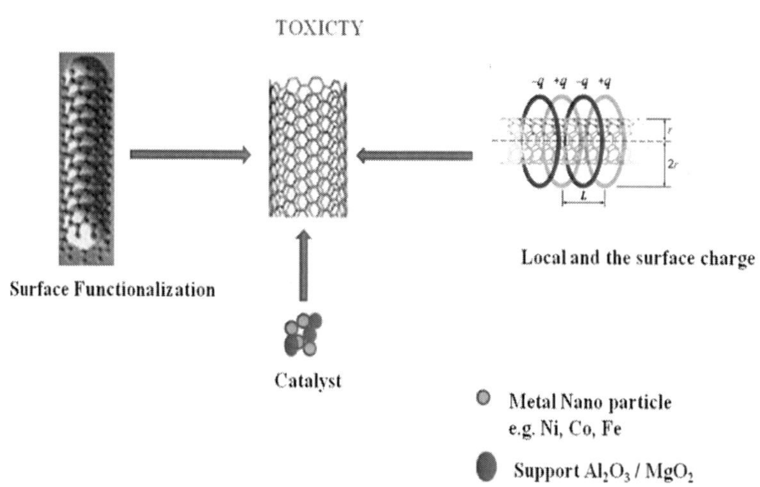

Scheme 1. Represents the major factors as limitations for the potential use of carbon nanotube in biomedical field.

Wick et al. studied the degree and agglomeration effect of CNTs and made their recommendations that the carbonaceous material and the degree of agglomeration were the prime most reasons behind the toxicity [15]. There are many other reports available on different aspects regarding the toxicity and cytotoxicity of carbon nanotubes but it seems that there is still a need to work on toxic behaviour at the cellular level. The toxic behaviour differs with many parameters which include the factors stated above, the conditions and biological environment.

3. Biocompatibility of Carbon Nanotubes

In this section, the biocompatibility studies of CNTs in biological systems will be discussed. In a recent review by Smart et al, the authors reported [16] about need to investigate a long term biocompatibility tests for CNT based biomedical devices such as biomedical implants, drug and vaccine delivery systems and other biomedical diagnostic appliances. Chlopek et al [17] studied the biocompatibility of pristine MWCNT on the viability of fibroblasts and osteaoblasts. They concluded that except the increased level of collagen formation on CNTs was the found and no other detrimental effect was observed. CNT based bioceramic composites with hydroxyapatite were tested for biocompatibility. CNT is this case was found to be bioactive booster [18]. Binder less CNT was examined for biocompatibility and was reported to have strength as compared to that of cancellous bone. Further no toxicity towards bone regeneration or any inflammatory effect was observed.

4. Carbon Nanotubes Based Nanocomposites for Sensor Applications

Biosensors are the sensors or detectors that are capable of identifying the specific chemical moiety by detecting its presence, activity or its concentration via a chemical change. When a biosensor interacts with an analyte contains a specific biomolecule, there occurs a chemical change at the surface of the biosensor which is captured by transducer as "Chemical signal" and biosensor converts it to "Quantifiable signal" in the form of colour or intensity. The following Figure 1 shows a sketch of a typical biosensor. The carbon nanotubes have been widely investigated to develop many different types of biosensor. The main component where carbon nanotubes play its role is the signal transducer part. Different types of nanocomposites have been prepared as per the requirement of a typical biosensor. The other common names of biosensors include immunosensor, obtrudes, chemical canneries, resonant mirrors, glucometers, biochips, bio computers etc.

In this section we will focus on the carbon nanotubes as a multifunctional nanobio-systems that have been successfully used to form biosensors. Wang et al. developed a glucose sensor using carbon nanotube modified electrode with a stability of 86.7% after a shelf life of four months [19]. Authors suggested that the use of open ended CNT is beneficial as it allows the enzyme into CNT those results in high sensitivity. In a recent study, carbon nanotubes along with silica gel was used as a filler in polyacrylonitrile matrix to detect glucose even at 1μM [20].They concluded that the use of carbon nanotube in the composite material provided the superior conductivity to the biosensor.

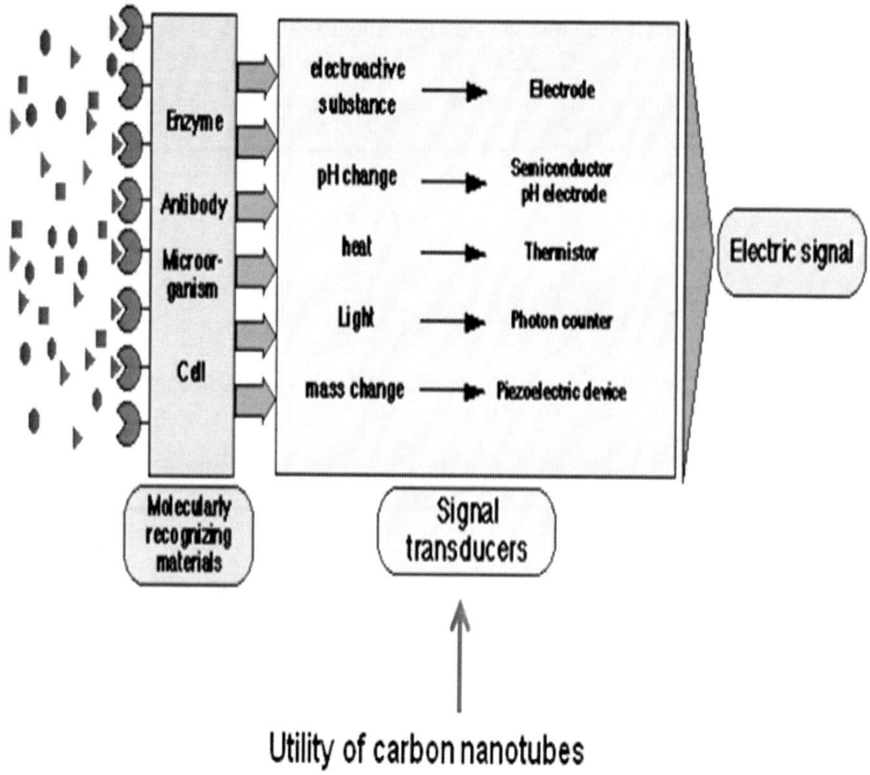

In a similar study, the researchers fabricated an amperometric biosensor using a composite material of carbon nanotubes, phenathiozine and enzyme. Phenathiozine was an effective electron transfer mediator that lowered the potential substantially [21]. Functionalized graphene carbon nanotube hybrid material was chosen to develop hydrogen peroxide biosensor. The biosensor so formed were able to response faster with lower detection limits when compared to earlier reports [22]. The synergistic effect of the hybrid along with the bimetallic gold/platinum nanoparticles was the proposed reason behind this efficient biosensor. Laschi et al, prepared the disposable electrodes using carbon nanotubes for biosensor applicatons. The researchers used carbon nanotube dispersed in polyetheneimine and modified with it , the screen printed carbon electrode to detect hydrogen peroxide [23]. A composite of CNT with gold nanoparticles were used to fabricate an amperometric biosensor for oxalate sensing that showed a response time of 7 s and a sensitivity of 0.50mAmM−1 in 1–800_M. This biosensor was stable upto five months and showed reusability for more than 100 times [24]. An electrochemical sensor for detecting nucleic acid was developed using a composite material of single walled CNTs with methyl blue as an electrochemical label and modified single stranded DNA [25]. The biosensor so formed was having an improved sensitivity due to surface to bulk ratio and an excellent electrical conductivity. Researchers recommended this type of biosensor to detect gene related diseases. In a unique study a layer of entangled CNTs was prepared using alumina substrate to immobilize glucose oxidase. The biosensor so formed showed a linear response for 12-62 μM of glucose concentration. The authors recommended the electrode for various oxidase based biosensor [26]. CNT-chitosan composites were synthesized in order to incorporate enzymes without the use of crosslinking agents. The biosensor was developed using the above composite material to detect laccase. The researchers concluded that with the help of this biosensor, it is possible to identify different substrate with high specificity and also can be applied to fabricate biofuel cells and other bioelectronics devices [27]. A glassy carbon electrode was upgraded using CNT and gold nanoparticles embedded in poly (allylamine hydrochloride) film to develop polyphenol oxidase biosensor. This was used to identify the sulphite in red and white wine in the range of 0.5-22μM/L [28]. Hoerse radish peroxidase was immobilized onto the CNT/ nano platinum-poly(neutral red) composite membrane to be used as a biosensor for hydrogen peroxide. This biosensor had a detection limit of 1.1×10^{-6}. This type of biosensor was recommended for direct electrochemistry and biocatalysis [29]. In a study, platinum nanoparticles were electrodeposited onto the CNT and polyaniline nanocomposites for glucose detection [30]. With a response time of 5 sec, the biosensor was able to detect upto 1μM with a sensitivity of 16.1 μAmM^{-1}. Layer by layer technique was used to develop glucose biosensor using by alternate layer of CNT and the glucose oxidase onto the surface of the graphite electrode. The authors reported that set of different layers was able to change the amperometric response and a linear range with low detectatble limit was observed [31].In another research, platinum nanospheres were used to modify the surface of CNTs and the nanocomposites was eventually applied onto to the glassy carbon electrode. Further gold nanoparticles encapsulated by L cystine and poly (diallyldimethylammonium) chloride were deposited onto the composite material along with the glucose oxidase. The authors reported a freedom of interference from co existing electro active species that resulted in high sensitivity and a better response from the immobilized enzyme [32]. To detect Hepatitis B virus, a biosensor was fabricated using CNT with a field effect transistor structure. The authors observed a linear relationship conductance and concentration of Hepattitis B antigen [33].

4. Carbon Nanotubes for Drug Delivery Systems

Carbon nanotubes have been an interesting material for biologists for developing drug delivery systems. The surface of the carbon nanotube allows covalent and non covalent linking of biomolecules that can be used for drug delivery, silencing RNAs and helps in gene transfer as shown in scheme below.

In this section we will focus on the various recent researches done using carbon nanotubes used in drug delivery devices. Prakash et al have published a review article on the polymeric nanohybrids and functionalized carbon nanotubes which were developed to deliver anticancerous drugs [34]. They have discussed extensively on the potential, limitations and the current scenario of carbon nanotubes as drug delivery system for cancer therapy. In another study, the researchers investigated the mechanism of transport of gemcitabine though CNTs. They made use of molecular dynamics and steered molecular dynamics simulation to study the objective [35]. Cobalt ferrite nanoparticles were synthesized by solvothermal technique and impregnated into CNTs for magnetic resonance imaging and to deliver doxorubicin, an anticancerous drug [36]. The drug was effectively released intracellularly that showed cytoxicity against HeLa cancer cells. The systems were successfully applied for diagnositics and therupatics of cancer cells. The “Magic bullet” concept was relieved when

CNT based nanocapsules were prepared to study the suction behavior of cis platin based on acceptance condition and suction energy [37]. Folate decorated CNTs were used as drug delivery systems for the control release of doxorubicin.

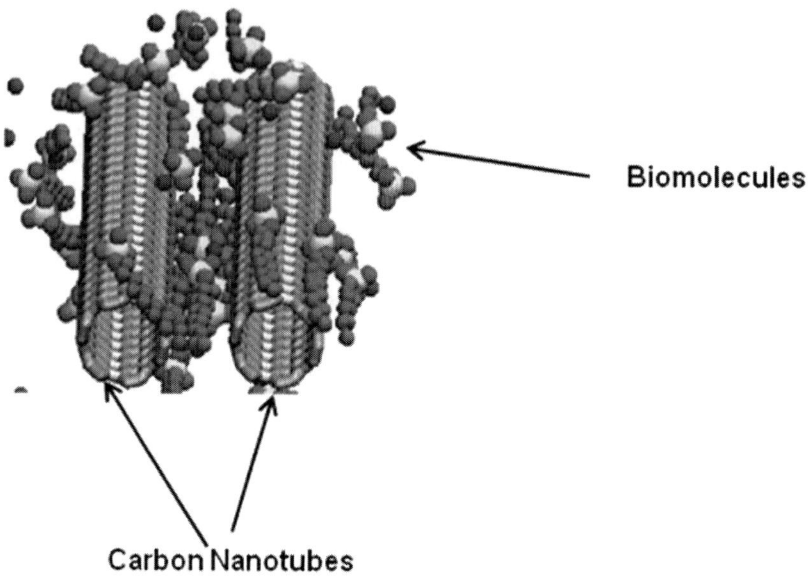

The anticaerocus drug was attached to the CNTs through π-π stacking interaction followed by encapusulation using folic acid conjugated chitosan [38].In another attempt , CNTs were modified by folic acid –chitosan composite to be used as targeteted in vivo and in vitro release of doxorubicin to SMMC 7221 liver cancer cells [39]. The drug delivery system killed not only the cancer cells but also subsided the growth of new cells simultaneously with low side effects. Lee et al studies the effect of CNT on drug delivery in electro sensitive transdermal drug delivery systems [40]. The system was prepared using electrospinning method with the help of polyethylene oxide and pentaerytheritol triacrylate polymers. It was reported that with the increase in applied electric voltage the release of the drug was increased. Cellulose acetate was used to functionalize the CNT for the release of indomethacin. The use of this natural biopolymer improved the controlled release of the drug when compared to the system where virgin CNT were used in coating solutions. The authors recommended the use of acid oxidized CNT in coating solution [41]. Magnetic dual targeted nanocarriers were developed using iron oxide nanoparticles and the CNT for the delivery of doxorubicin with a high drug loading and pH dependent controlled release. The system prepared with magnetic targeting and ligand receptor interactions effectively killed U87 cells [42]. In another study, the researchers successfully reported the delivery of SiRNA using cationically functionalized CNT based drug delivery system. The system show 10-30% silencing activity with 10-60% cytotoxicity [43]. Magnetite nanoparticles were used to functionalise the CNTs for magnetic lymphatic targeting drug delivery systems. The CNTs were loaded with chemotherapeutic agents into the pores to be delivered to regional targeted lymph nodes [44]. In another very recent study, the researchers investigated ultra short CNT to be used for encapsulating cisplatin. The delivery of this drug was confined to the cancer cells only [45].

SUMMARY

The toxicity and biocompatibility of carbon nanotubes play a little or negligible role in the biomedical applications such as biosensor and drug delivery systems. The CNT based composites play a crucial role in achieving the sensor with finest parameters. On the other hand CNT also promotes the targeted delivery with low side effects.

REFERENCES

[1] C. P. Firme III, P.R. Bandaru, Nanomedicine: Nanotechnology, Biology, and Medicine 6, 245 (2010).
[2] L. Lacerda, A. Bianco, M. Prato, K. Kostarelos, Adv. Drug Deliv. Rev. 58, 1460, (2006).
[3] H. Hu, P. Bhowmik, B. Zhao, M.A. Hamon,M.E Itkis, R.C Haddon, Chem. Phys. Lett. 25, 345 (2001).
[4] P.R.Bandaru, C. Daraio, K. Yang, A.M. Rao, *J. Appl. Phys.* 101, 094307(2007).
[5] P.R. Bandaru, C. Daraio, S. Kin, A.M. Rao, Nat. Mater. 4, 663 (2005).
[6] R.Singh, D. Pantarotto, D. McCarthy, O. Chaloin, J.Hoebeke, C.D. Partidos, *J. Am. Chem. Soc.*,127, 4388 (2005).
[7] G. Pagona, N. Tagmatarchis, Curr Med.Chem. 13, 1789 (2006).
[8] A.A.Shvedova, V.Castranova, E.R. Kisin, D. Schwegler-Berry, A.R. Murray, V.Z. Gandelsman VZ, J Toxicol. Environ. Health, 66, 1909 (2003).
[9] V.E. Kagan, H. Bayir, A.A. Shvedova, Biol Med. 4, 313 (2005).
[10] N.W.S Kam, H. Dai, J. Am. Chem. Soc. 127, 6201 (2005).
[11] S.Zhu, F. Obersdorster, M.L Haasch, Marine Environ. Res. 62, S5-9.u (2006)
[12] J. Kayat, V. Gajbhiye, R.K. Tekade, N. K. Jain, Nanomedicine: Nanotechnology, Biology, and Medicine, 7 , 40 (2011).
[13] J. Muller, F. Huaux, N.Moreau, P. Misson, J-F. Heilier, M.Delos, M. Arras, A. Fonseca, J.B. Nagy, D. Lison, Toxicology and Applied Pharmacology, 207, 221 (2005).
[14] S.V. Prylutska, I.I. Grynyuk, O.P. Matyshevska, V.M. Yashchuk,Yu, I. Prylutskyy, U. Ritterd, P. Scharffd, Physica E. 40, 2565, (2008).
[15] P. Wick, P. Manser, L.K. Limbach, U. Dettlaff-Weglikowska, F.Krumeich, S.Roth, W. J. Stark, A. Bruinink, Toxicology Letters, 168, 121 (2007).
[16] S.K. Smart, A.I. Cassady, G.Q. Lu, D.J. Martin, Carbon ,44, 1034 (2006).
[17] J. Chłopek, B. Czajkowska, B. Szaraniec, E. Frackowiak, K. Szostak, F. Be´guin , Carbon, 44, 1106 (2006).
[18] E.J. Tercero, S.Namin, D. Lahiri, K. Balani, N.Tsoukias, A. Agarwal, Materials Science and Engineering, C 29, 2195 (2009).
[19] S.G. Wang, Q.Zhang, R. Wang, S.F. Yoon, Biochemical and Biophysical Research Communications, 311, 572 (2003).
[20] R. Nenkova, D. Ivanova, J. Vladimirova, T. Godjevargova, Sensors and Actuators B 148, 59 (2010).
[21] T.Hoshino, S. Sekiguchi, H. Muguruma, Bioelectrochemistry, (2011) [In press].
[22] Q.Sheng, M.Wang, J. Zheng, Sensors and Actuators B, 160, 1070 (2011).

[23] S. Laschi, E. Bulukin, I. Palchetti, C. Cristea, M. Mascini, ITBM-RBM, 29, 202 (2008).

[24] C.S. Pundir, N. Chauhan, Rajneesh, M. Verma, Ravi, Sensors and Actuators B, 155, 796 (2011).

[25] K.Yang and C. Zhang, Biosensors and Bioelectronics, 28, 257 (2011).

[26] S.Maghsoodi, Z. Gholami, H. Chourchian, Y. Mortazavi, A.A. Khodadadi, Sensors and Actuators B, 141, 526 (2009).

[27] Y.Liu, X. Qu, H. Guo, H.Chen, B. Liu, S. Dong, Biosensors and Bioelectronics, 2195 (2006).

[28] E.R. Sartori, F.C. Vicentinia, O. F.Filho, Talanta, (2011) [In Press].

[29] Y. Zhang, R. Yuan, Y. Chai, Y.Xiang, C. Hong, X. Ran, *Biochemical Engineering Journal*, 51, 102 (2010).

[30] H.Zhong, R.Yuan, Y. Chai, W. Li, X. Zhong, Y.Zhang, Talanta, 85, 104 (2011).

[31] J. Huang, Y. Yang, H. Shi, Z. Song, Z. Zhao, J. Anzai, T. Osa, Q. Chen, Materials Science and Engineering, C 26, 113 (2006).

[32] Y.Wang, R.Yuan, Y. Chaia, W. Li, Y. Zhuo, Y.Yuan, J.Li, *Journal of Molecular Catalysis B: Enzymatic*, 71, 146 (2011).

[33] J.Oh, S. Yoo, Y.W.Chang, K.Lim K. Yoo, Current Applied Physics, 9, e229 (2009).

[34] [34] S.Prakash, M. Malhotra, W.Shao, C. Tomaro-Duchesneau, S.Abbasi, Advanced Drug Delivery Reviews, (2011), [In Press].

[35] U. Arsawang, O. Saengsawang, T. Rungrotmongkol, P.Sornmee, K. Wittayanarakulb, T.Remsungnen, S. Hannongbua, *Journal of Molecular Graphics and Modelling*, 29, 591 (2011).

[36] H. Wu, G.Liu, X.Wang, J.Zhang, Y.Chen, J. Shi, H.Yang, H. Hu, S.Yang, Acta Biomaterialia, 7, 3496 (2011).

[37] T.A. Hilder and J.M. Hill, Current Applied Physics, 8, 258 (2008).

[38] H. Huang, Q. Yuan, J.S. Shah, R.D.K. Misra, Advanced Drug Delivery Reviews, (2011) [In Press].

[39] Z. Ji G. Lin, Q. Lu , L. Meng, X. Shen, L.Dong, C. Fu, X. Zhang, Journal of Colloid and Interface Science, 365, 143(2012).

[40] J. S. Im, B.C. Bai, Y.S. Lee, Biomaterials, 31, 1414 (2010).

[41] S.S. Madaeni, K. Derakhshandeh, S.Ahmadi, V. Vatanpour, S. Zinadini, Journal of Membrane Science, (2011), [In Press].

[42] Y.J. Lua, K.C. Wei, C.C.M.Mac, S.Y. Yang, J.P. Chen, Colloids and Surfaces B: Biointerfaces, 89,1 (2012).

[43] K. Varkouhi, S. Foillard, T. Lammersa, R. M. Schiffelers E. Doris, W. E. Hennink, Gert Storm, *International Journal of Pharmaceutics*, 416, 419 (2011).

[44] F.Yang, D. Fu, J. Long, Q.X. Ni, Medical Hypotheses, 70, 765 (2008).

[45] A.Guven , I. A. Rusakova , M. T. Lewis, L.J. Wilson, Biomaterials, 1 (2011).

In: Applications of Carbon Nanotubes
Editor: Ajay Kumar Mishra
ISBN: 978-1-62081-918-0

Chapter 2

COUPLING CARBON NANOTUBE WITH PHOTOSENSITIZER FOR CANCER PHOTOTHERAPY

***Chung-Hao Wang*[1] *and Ching-An Peng*[2]**
[1]Department of Chemical Engineering, National Taiwan University, Taipei, Taiwan, Republic of China
[2]Department of Chemical Engineering, Michigan Technological University, Houghton, Michigan, US

ABSTRACT

Carbon nanotube has been used as a thermal-coupling agent for hyperthermic treatment of cancer cells under near-infrared (NIR) laser irradiation. To further enhance the efficacy of phototherapy, multi-walled carbon nanotubes (MWNTs) can be functionalized with photosensitizers which produce singlet oxygen to kill cancer cells by exposure to light emitted diode (LED) light. In this study, photosensitizer rose bengal (RB) and target ligand folic acid (FA) were coupled with MWNTs and examined by UV-Vis, Fourier transform infrared (FT-IR), photoluminescence (PL) and X-ray photoelectron spectroscopy (XPS) spectral analysis to confirm the chemical binding of RB and FA on MWNTs. RB-tagged MWNTs functionalized with FA (RB-MWNT-FA) were incubated with MCF-7 breast cancer cells for 24 hr. After replaced with fresh culture media, cell culture plates were first exposed with 540-nm LED light and then followed with 808-nm NIR laser. The amount of singlet oxygen triggered by RB detected by a sensing probe was increased significantly for MCF-7 cells first challenged with RB-MWNT-FA for 24 hr and then sequentially exposed with LED light and NIR laser. The cell viability determined by MTT assay revealed close to 90% eradiation rate for cells treated with RB-MWNT-FA which is much better than cells treated separately with MWNT-FA (40% killing rate) and RB-MWNT (60% killing rate). In conclusion, functionalizing carbon nanotube with photosensitizer (i.e., RB) and targeting moiety (i.e., FA) can significantly increase the death of malignant cells by the combined irradiation of LED light and NIR laser.

INTRODUCTION

Carbon nanotube (CNT) has very interesting physicochemical properties such as ordered structure with high aspect ratio, ultra-light weight, high mechanical strength, high electrical conductivity, high thermal conductivity, metallic or semi-metallic behavior and high surface area. The combination of these characteristics makes CNT a unique material with the potential for diverse applications including biomedicine. For photothermolysis of malignant cells, Kam et al. [1] showed CNTs functionalized with folic acid internalized by HeLa cells overexpressed with folate receptors led to cell death under 808-nm NIR laser light exposure. CNTs were also functionalized with anti-GD2 and anti-CD133 monoclonal antibodies to specifically target neuroblastoma and stem-like glioblastoma cells, respectively and thermally ablated malignant cells by NIR laser irradiation [2-3].

Photodynamic therapy (PDT) has been widely used in both oncological (e.g., tumors and dysplasias) and non-oncological (e.g., age-related macular degeneration, localized infection, and nonmalignant skin conditions) applications [4-7]. Rose bengal (RB) is a water-soluble xanthene dye that can convert oxygen molecule to singlet oxygen upon green light irradiation. The singlet oxygen ($^1O_2$) is a cytotoxic agent and reacts rapidly with cellular components to cause damage that ultimately leads to cell death and tumor destruction [8]. Although RB has the potential as a photosensitizer for photodynamic therapy (PDT), the application of RB in PDT has been limited due to insufficient lipophilicity and lack of tumor-accumulating property.

Folic acid is an attractive target ligand due to its high binding affinity for the folate receptors (K_d~10^{-10} M). Folic acid is efficiently internalized into the cells through the receptor-mediated endocytosis even when conjugated with a wide variety of molecules [9,10]. Folic acid and its conjugate were widely used for selective delivery of anticancer agents to cells with folate receptor, which are overexpressed in many kinds of human cancer cells such as ovarian, breast and prostate cancers while only minimally distributed in normal tissues [11,12].

Our results showed that combination of PDT and hyperthermic therapy could enhance anticancer effectiveness of CNTs conjugated with RB and FA. According to the MTT assay, MCF-7 cells left only 13% viable cells after pre-treated with RB-MWNT-FA for 24 hr followed with LED light and NIR laser exposure; while the control group using NIH 3T3 still remained 85% cell viability with the same treatment conditions.

EXPERIMENTAL

Materials

Multi-walled carbon nanotube (MWNT) was purchased from CNT Co., Ltd. (Incheon, Korea). HNO_3 and H_2SO_4 were obtained from Showa Chemicals (Japan). $SOCl_2$, dimethyl formamide (DMF), Sb_2O_3, 4',6-diamidino-2-phenylindole (DAPI), rose bengal and folic acid were purchased from Sigma (St Louis, MO). All chemicals were received and used without further purification. Dulbecco's modified Eagle's medium (DMEM), trypsin, fetal bovine serum (FBS), and 24-well tissue cultured plate were all purchased from HyClone (Logan,

UT). 3-(4,5-cimethylthiazol-2-yl)-2,5-diphenyl tetrazolium bromide (MTT) was purchased from Sigma (97.5% TLC, Sigma). The singlet oxygen sensor green reagent was purchased from Invitrogen (Carlsbad, CA).

Functionalization of MWNT by RB and FA

In a typical experiment, 10 mg of MWNTs were added to a 20 mL of a mixture of sulfuric acid and nitric acid in a composition of 3/1 by volume. The mixture was sonicated in a bath for 10 min at room temperature. The mixture was then refluxed at 80°C for 48 h. At the end of the reaction, the mixture was washed with deionized water for 5 times. The obtained carboxyl edge-functionalized MWNT was dispersed in 15 mL of DMF and 5 mL of $SOCl_2$ was added slowly under stirring at room temperature.

After the addition was completed, the temperature was increased to 85°C and maintained over night under nitrogen atmosphere. The products were filtered, washed with DI water and dried under vacuum freeze dryer. The dried chlorinated MWNT product was collected, weighed and stored in a vial. 50 mg of chlorinated MWNT, 1 mg of Sb_2O_3, and 20 mg of RB or FA or RB+FA mixture (1:1 weight ratio) were charged in a two-necked round bottom flask containing 15-mL DI water, and stirred at 85°C for 24 hr. NaCl was formed as a by-product, which can be removed from the reaction medium by repeated washing with excess of DI water. The product was washed and filtered until the filtrate turned into colorless. The product was dried under vacuum freezer for 24 hr. The samples were collected, weighed and stored in a vial. The mechanism and reaction is shown in Figure 1. The black product obtained was the RB functionalized MWNT.

Characterizations of RB-MWNT-FA

All samples were analyzed by Fourier transform infrared spectroscopy (FT-IR, Thermo Nicolet NEXUS470, Waltham, MA), UV-Vis spectrophotometer (Jasco Model V-570, Japan), photoluminescence spectrophotometer (Jasco Model FP-6000, Japan), **X-ray photoelectron spectroscopy (XPS,** Thermo Scientific, Theta Probe, UK**).**

Endocytosis of RB-MWNT-FA

1×10^5 cells of MCF-7 and NIH 3T3 cells were inoculated in coverslip-bottomed Petri dishes (MatTek, MA) containing 2 mL of DMEM supplemented with 10% FBS. After 24 hr incubation, MCF-7 and NIH 3T3 cells were treated separately with 5 and 25 μg/mL of RB-MWNT-FA for 24 hr.

Then, 20 μg DAPI were added to each dish and allowed to incubate for 1 h at room temperature for nucleus staining. Cell images were taken by laser confocal microscopy (Leica TCS SP5, Wetzlar, Germany).

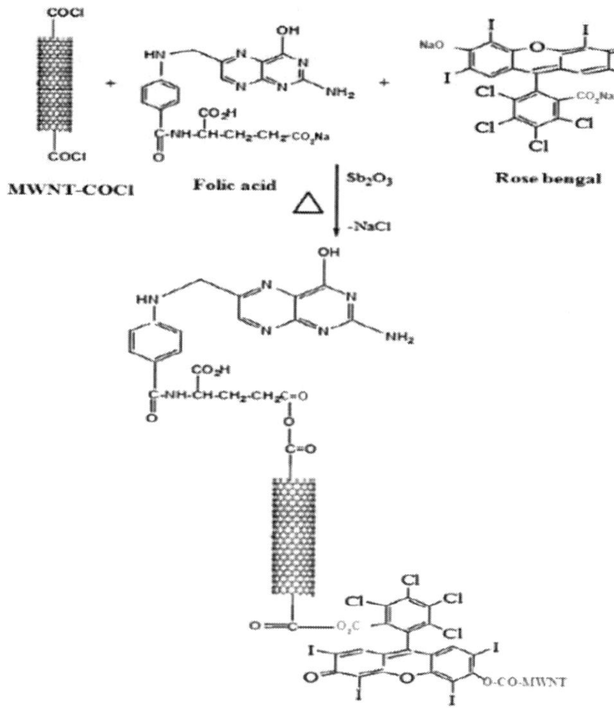

Figure 1. Synthesis of RB/FA functionalized MWNT.

Photolysis Effects of MWNT Derivatives

First, 1 $\times 10^5$ MCF-7 and NIH 3T3 cells were cultured in 24-well tissue cultured plates for 24 hr. The MCF-7 and NIH 3T3 cells were treated separately with MWNT, RB-MWNT, MWNT-FA, and RB-MWNT-FA for 24 hr, replaced with fresh medium and then exposed to 540 ± 10-nm LED light (ITRI, Taiwan) only for 4 J, 808-nm NIR laser light only with 600 mW/cm^2 for 3 min (Opto Power Corporation, Tucson, AZ), or LED light followed with NIR laser. The 808-nm NIR laser beam was delivered to the irradiation zone through a 1.5-m long, 600-μm single core fiber with a numerical aperture of 0.37, followed by a 25-mm focal length fused-silica biconvex lens. After three different treatments, cell viabilities were analyzed by MTT assay.

Singlet Oxygen Production

To measure the production of $^1O_2$ by MWNT derivative, the sensing probe Singlet Oxygen Sensor Green, which is highly selective for singlet oxygen without appreciable response to hydroxyl radical or superoxide, was used for this study. In brief, after MCF-7 and NIH 3T3 cells were challenged with MWNT derivatives, fresh culture media containing 1 μM sensing probe were used to replace the treated medium. After LED light and NIR laser exposure, the fluorescence intensity was measured by a microplate reader (SpectraMax M2e, Molecular Devices, Sunnyvale, CA) with λ_{ex} of 488 nm and λ_{em} of 525 nm.

RESULTS AND DISCUSSION

FT-IR Analysis

Figure 2A shows the FT-IR spectrum of pristine MWNT. It shows one broad peak around 3443 cm^{-1} due to the OH-stretching vibration. Trace amounts of water molecules are attached with the MWNT during the synthesis itself. Further, it can be confirmed by a peak at 1640 cm^{-1} occurring due to the bending vibration of water molecules. A broad peak at 592 cm^{-1} confirms the M–O (metal-oxide) stretching vibration (metal oxides used for the synthesis of MWNT). Figure 2B indicates the FT-IR spectrum of MWNT-CO-RB conjugate. A broad peak at 3443 cm^{-1} describes the OH stretching of carboxyl group of unreacted oxidized MWNT. The C–H symmetric (2863 cm^{-1}) and asymmetric (2924 cm^{-1}) stretching were also seen. The carboxyl carbonyl stretching appears at 1640 cm^{-1}. Appearance of a broad peak at 1070 cm^{-1} explains the ester C–O–C linkage in the conjugate. The iodine (attached with RB) peak appears at 734 cm^{-1}. The appearance of new peaks due to I and C–O–C ester linkage confirms the binding of RB with MWNT through diester linkage. Figure 2C and 2D indicate the FT-IR spectra of MWNT functionalized with FA and RB + FA mixture, respectively. The new peaks that appear, particularly, the peaks at 1605 and 1326 cm^{-1} are corresponding to the amide C=O linkage and C–N stretching of FA-functionalized MWNT. Peaks corresponding to halogen are appearing around 570 cm^{-1}. This observation confirms the chemical modification of MWNT with RB + FA mixture. For the sake of comparison, the FT-IR spectrum of pristine RB is also indicated in Figure 2 as Figure 2E. A broad peak at 3393 cm^{-1} is due to the OH stretching of water molecules associated with RB. The carbonyl stretching appears at 1620 cm^{-1}. The aromatic C–H stretching appears at 958 cm^{-1}. Peaks at 662 and 530 cm^{-1} are corresponding to Cl and I stretching vibrations.

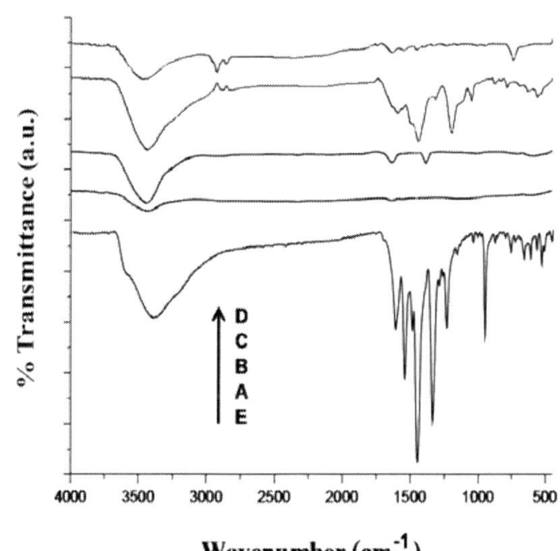

Figure 2. FT-IR spectra of (A) MWNT, (B) RB-MWNT, (C) MWNT-FA, (D) RB-MWNT-FA, (E) RB.

UV-Vis Spectrum

Absorbance spectrum of pristine RB is shown in red line of Figure 3. The spectrum ranging from 250 to 460 nm is smooth, and then a sharp peak at 558.8 nm confirmed the monomeric form of pristine RB. A small hump at 521.5 nm inferred the dimeric nature of RB. The absorbance spectrum of RB-conjugated MWNT is shown in green line of Figure 3. As a result, one can observe the same pattern of peaks but with slight shifting. The monomeric form of RB is seen at 552.8 nm, whereas the dimeric form is broadened without a significant peak. During the functionalization reaction, RB can interact with MWNT through alkoxide or ester linkage. The shifting of monomeric form of RB confirmed the chemical binding of RB with MWNT. The black line of Figure 3 exhibits the electronic absorption spectrum of simultaneous binding of RB (disodium salt) and FA (monosodium salt) with MWNT. The monomeric and dimeric forms of RB were observed at 574.3 and 537.3 nm, respectively. In this case, due to the competitive bindings of RB and FA with MWNT, the peak was red shifted. Appearance of these two peaks along with FA peaked around 300 nm confirmed the simultaneous chemical binding of both RB and FA with MWNT preferably through their ester linkage.

Photoluminescence Spectrum

The red line of Figure 4 indicates the photoluminescence (PL) spectrum of pristine RB with an emission peak at 569 nm. MWNT after functionalization with RB (the green line of Figure 4) showed an emission peak at 554 nm. From the black line of Figure 4, one can observe a PL emission peak at 580 nm. This red shift confirmed the role of FA on the functionalization with MWNT in the presence of RB. In addition, FA binding on MWNT revealed a peak around 512 nm. The emission spectra confirm the functionalization of MWNT with both FA and RB.

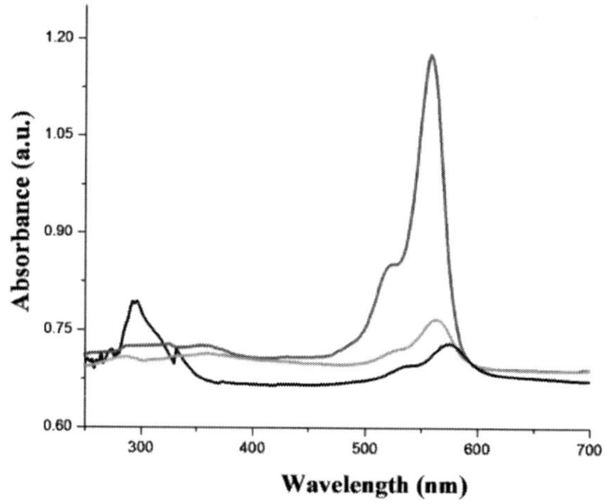

Figure 3. UV-Vis spectra of RB (red line), RB-MWNT (green line), and RB-MWNT-FA (black line).

XPS Analysis

Figure 5A indicates the XPS spectrum of MWNT. It shows the C1s peak at 283.3 eV. The XPS of MWNT-CO-RB conjugate is shown in Figure 5B. It shows C1s, O1s, Cl2p, and I3d peaks at 283.3, 533.2, 197.26, and 633.9 eV, respectively. The important point here is that a peak corresponding to Na1s at 1071 eV has completely disappeared. This confirms the existence of chemical binding between MWNT and RB. During the conjugation reaction, the NaCl were eliminated as a by-product. Figure 5C indicates the XPS of FA-functionalized MWNT. The N1s peak appears at 397.88 eV at the same time when the Na1s peak has disappeared. This confirms the chemical binding of FA with MWNT. Figure 5D represents the XPS spectrum of MWNT functionalized with RB and FA mixture (1:1 ratio) simultaneously. Peaks due to N1s, Cls, O1s, and I3d are appeared with the disappearance of Na1s peak. This evidences the chemical interactions of both RB and FA with MWNT. In both cases, the %N was found to be 1.65% (MWNT–FA) and 0.33% (RB–MWNT–FA), respectively. For the sake of comparison, the XPS of pristine RB is given in Figure 5E. The Cl2p, C1s, O1s, I3d, and Na1s peaks were observed at 200.09, 283.95, 530.83, 619.86, and 1071.36 eV, respectively. The XPS confirms the structure of RB-bound MWNT. Yong et al. [13] reported about the XPS of functionalized MWNT. The binding energy of C1s of MWNT was observed at 284.6 eV, and the O1s of oxidized MWNT was seen at 532.3 eV [14]. Our investigations are in accordance with these two references [13,14].

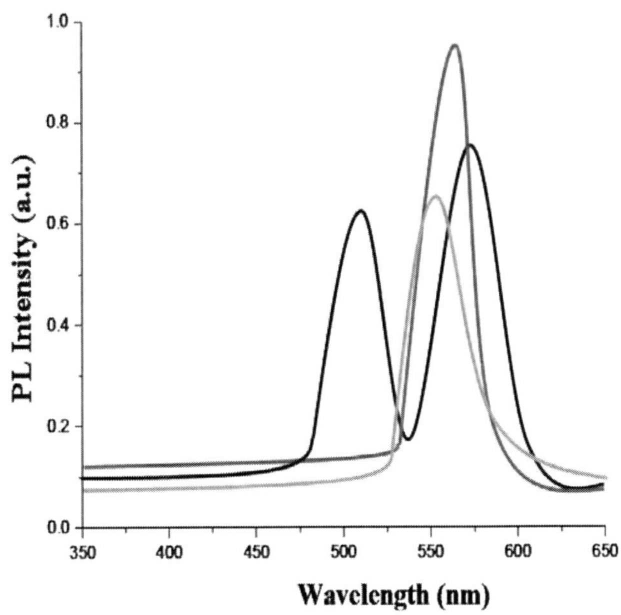

Figure 4. PL spectra of RB (red line), RB-MWNT (green line), and RB-MWNT-FA (black line).

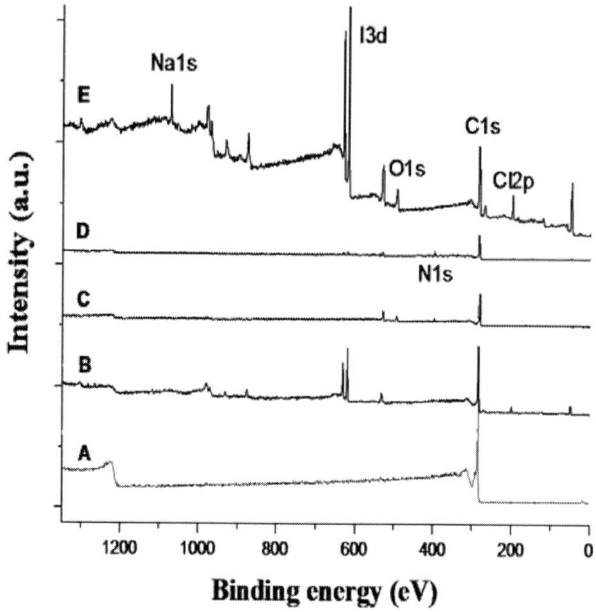

Figure 5. XPS spectra of (A) MWNT, (B) RB-MWNT, (C) MWNT-FA, (D) RB-MWNT-FA, (E) RB.

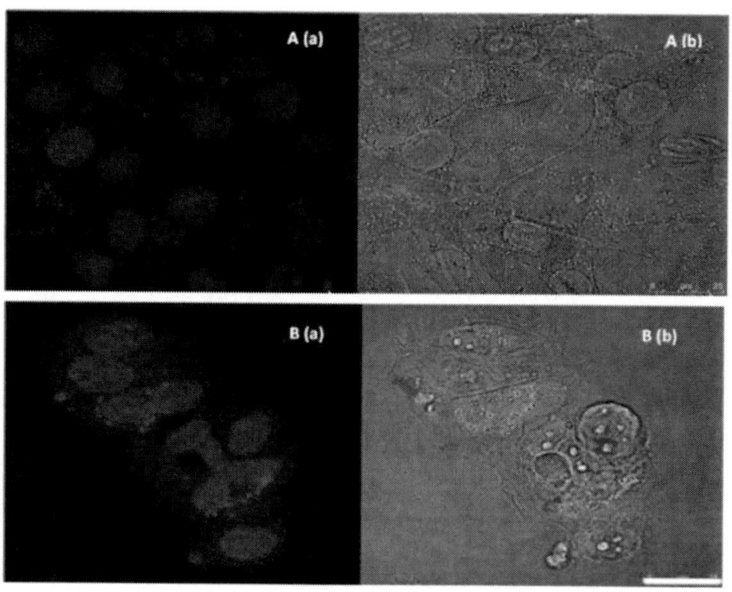

Figure 6. Confocal images of (A) NIH 3T3 and (B) MCF-7 cells co-cultured with the RB- MWNT-FA for 24 hr illustrated as (a) fluorescent image and (b) bright-field overlayed with fluorescent image. Red fluorescence reveals RB-MWNT-FA located in cytoplasm. The nucleus is stained with blue DAPI dye. Scale bar = 25 μm.

Endocytosis of RB-MWNT-FA

After 24-hr incubation with pristine MWNT and different types of functionalized MWNT, both NIH 3T3 and MCF-7 cells ingested only a small amount of MWNT and RB-MWNT due to their nonspecific adsorption onto cell surface. For both cell types treated with MWNT-FA or RB-MWNT-FA, MCF-7 cells with abound folate receptors expressed on cell surface endocytosed large amounts of folate-tagged MWNT, while much less were taken up by NIH 3T3 cells lacking of folate receptor expression. The images illustrated in Figure 6 are typical ones demonstrated that MCF-7 cells could internalize more RB-MWNT-FA into cytoplasm than NIH 3T3 cells could.

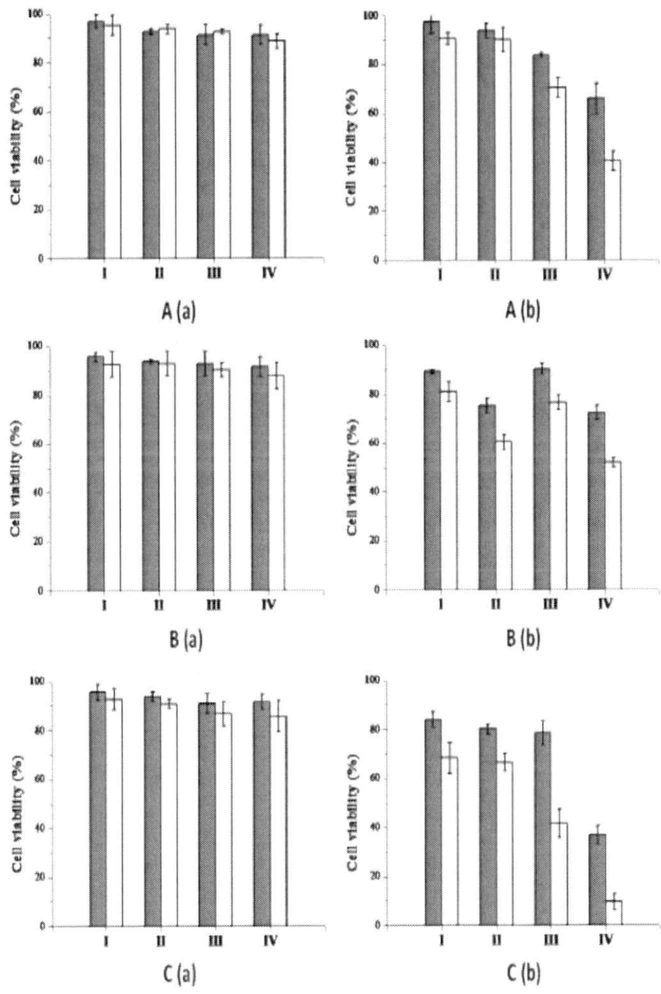

Figure 7. MTT assay of (a) NIH 3T3 and (b) MCF-7 cells co-cultured with the MWNT (I), MWNT-FA (II), RB-MWNT (III) and RB-MWNT-FA (IV) by (A) 4 J/cm2 of 540±10 nm LED light only, (B) 1 W/cm2 of 808-nm Near-IR laser for 3 min only, (C) 4 J/cm2 of 540±10 nm LED light and 1 W/cm2 of 808-nm Near-IR laser for 3 min after cell cultured for 24 hr. The gray bar chart represents the cells treated with 5 μg/mL MWNT and its derivatives for 24 hr. The white bar chart represents the cells treated with 25 μg/mL MWNT and its derivative for 24 hr.

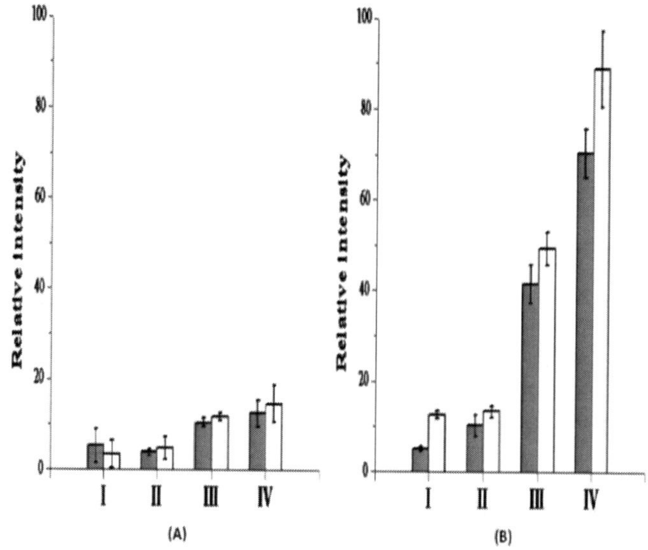

Figure 8. Singlet oxygen detection of (A) NIH 3T3 and (B) MCF-7 cells co-cultured with the MWNT (I), MWNT-FA (II), RB-MWNT (III) and RB-MWNT-FA (IV). The gray bar chart represents the cells treated with 5 μg/mL MWNT and its derivatives for 24 hr. The white bar chart represents the cells treated with 25 μg/mL MWNT and its derivatives for 24 hr.

Dual Effects of RB-MWNT-FA With Photo-Treatment

As shown in Figure 7C(b), after challenged with MWNT derivatives for 24 hr, MCF-7 cells irradiated with green LED light and NIR laser, dropped cell viability drastically to 13% with 25 μg/mL RB-MWNT-FA treatment. On the contrary, the control group NIH 3T3 cells still exhibited up to 85% cell viability. This result clearly demonstrated that the targeting moiety folic acid labeled MWNT can specifically bind on the surface of MCF-7 cells overexpressed with folate receptors and lead to significant internalization of RB-MWNT-FA nanomaterials which trigger RB-related singlet oxygen production by green LED light and generate MWNT-associated thermal energy by NIR laser light. Figure 7 also showed that the RB-MWNT-FA resulted in lower cell viability compared with MWNT, MWNT-FA and RB-MWNT for two different dosages 5 and 25 μg/mL used to challenge cells for 24 hr. The MWNT-FA (with 25 μg/mL) without RB can spatially binding on the cells of MCF-7 cells, increase the cellular temperature with 808-nm laser exposure and result in around 60% cell viability as shown in Figure 7B(b)-II and 7C(b)-II. For RB-MWNT without FA, the amounts of nanomaterials internalized by MCF-7 cells were less than the ones labeled with FA and therefore led to 80% viability (shown in Figure 7A(b)-III and 7B(b)-III) and 40% cell viability (shown in Figure 7C(b)-III).

Furthermore, Figure 8(B)-IV showed that MCF-7 cells taken up more RB-MWNT-FA released highest amount of singlet oxygen (90% intensity) which resulted in nearly 90% killing rate as shown in Figure 7C(b)-IV. For the 25 μg/mL RB-MWNT, the singlet oxygen production rate was approximately 50% which is almost half less than RB-MWNT-FA due to the lack of targeting moiety. For the control group NIH 3T3 cells, the detected level of singlet oxygen was less than 20% which is consistent with the cell viability results given in Figure 7.

REFERENCES

[1] Z. Liu, H. Song, L. Yu and L. Yang, *Appl. Phys. Lett. 86*, 113109 (2005).
[2] C-H Wang, Y-J Huang, C-W Chang, W-M Hsu and Peng, C-A, *Nanotechnology* 20, 315101 (2009).
[3] C-H Wang, S-H Chiou, C-P Chou, Y-C Chen, Y-J and Peng, C-A, *Nanomedicine: Nanotechnology, Biology, and Medicine* 7, 69 (2011).
[4] T. N. Demidova and M. R. Hamblin, *Int. J. Immunopathol. Pharmacol.* 17, 245 (2004).
[5] S. B. Brown, E. A. Brown and I. Walker, *Lancet. Oncol.* 5, 497 (2004).
[6] C. A. Kendall and C. A. Morton, *Technol. Cancer Res. Treat.* 2, 283 (2003).
[7] T. J. Dougherty, *J. Clin. Laser Med. Surg.* 20, 3 (2002).
[8] M. T. Jarvi, M. J. Niedre, M. S. Patterson and B. C. Wilson, *Photochem. Photobiol.* 82,1198 (2006).
[9] N. Karousis, H. A. Boucetta and N. Tegmatarchis, *Mater. Sci. Engg. B.* 152, 8 (2008).
[10]Y. Zhang, M. C. A. Stuart and F. Picchioni, *Macromolecules.* 41, 6141 (2008).
[11]K. Balasubramanian and M. J. Burghard, *Mater. Chem.* 18, 3071 (2008).
[12]G. Xu, W. T. Wu, Y. Wang and Y. You, *Polymer* 47, 5909 (2006).
[13]X. G. Yong, X. Ru, W. Hu and Z. Q. Ren, *Chin. Sci. Bull.* 53, 2297 (2008).
[14]G. Xu, W. T. Wu, P. Wanf and F. Lu, *Nanotechnology* 17, 2458 (2006).

In: Applications of Carbon Nanotubes
Editor: Ajay Kumar Mishra
ISBN: 978-1-62081-918-0

Chapter 3

SYNTHESIS OF CARBON NANOTUBES AND ITS APPLICATION OF POLYMER ENGRAFTED IN CELLULAR UP-TAKE

B. Kothandapani and Ajay Kumar Mishra*
Department of Chemical Technology, University of Johannesburg,
Johannesburg South Africa

ABSTRACT

Carbon nanotubes are internationally renowned nanomaterials for their technological applications in the molecular devices. In this book chapter, the importance of functionalization of these nanomaterials has been discussed as it plays a vital role in tailoring their properties and the engineering of nanotube devices. Carbon nanotubes can be converted into environmentally responsive materials by grafting polymers onto the surface of the wall of the tubes which can be achieved either by covalent attachment or by non-covalent absorption.

For this purpose, suitably controlled radical initiating groups are anchored to the carbon nanotubes, from which polymerization is achieved via "grafting from" technique. These non covalent/covalent attachments of the polymer chains onto the surface of carbon nanotubes have attracted the scientist because they endow the surfaces of the carbon nanotubes with novel structures and properties.

These organic/inorganic hybrid materials after preparation would show excellent stabilities in the solvents of the interest in contrast to pristine nanotubes. In addition, synthesis of carbon nanotubes by various methods will be discussed in order to compare and relate with the conventional techniques used for industrial purposes. Finally, the applications of carbon nanotubes in the biological field especially for cellular uptake have also been mentioned in detail.

* Corresponding author: Email - amishra@uj.ac.za, Tel - 0027115596180; Fax – 0027115596425.

Introduction

Carbon is the sixth element of the periodic table and each carbon atom has six electrons occupying $1s^2$, $2s^2$ and $2p^2$ atomic orbitals. It has an ability to hybridize in sp (e.g. C_2H_2), sp^2 (e.g. graphite) or sp^3 (e.g. CH_4) forms. Discoveries of very stable nanometer size sp^2 hybrid carbon materials such as fullerenes [1], carbon nanotubes [2] and graphene [3] have stimulated the research in this field. Among these different forms of carbon nanomaterials, the present chapter is focused on carbon nanotubes. Since carbon nanotubes derive most of the physical properties from graphene, it is therefore useful to understand this material first before we proceed to discuss carbon nanotubes.

Sp2 Hybridization

Graphene is a one-atom-thick planar sheet of sp^2 bonded carbon atoms that are densely packed in a honeycomb crystal lattice [4]. A schematic diagram representing the 1-D graphene lattice is shown in Figure 1. Several graphene layers are stacked over each other to form bulk graphite. These layers are loosely bonded to each other and hence can slip freely, making the extraction of graphene possible from graphite crystal through mechanical/chemical exfoliation [5]. Due to the presence of sp^2 hybridized carbon atoms in one plane, graphene is atomically two dimensional.

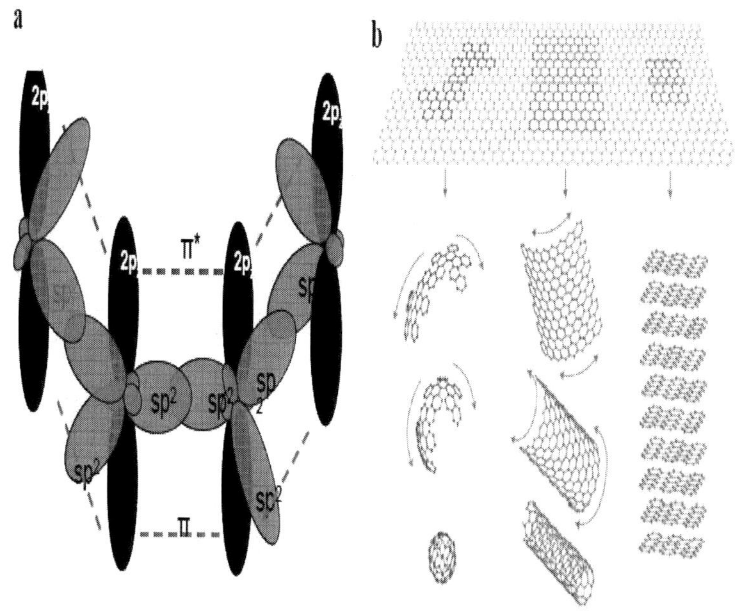

Figure 1. sp2 hybridization of carbon and its derived materials a) The three sp2 hybridized orbital are in-plane, with 2p orbital orthogonal to the plane, π and π* denotes the bonding and anti bonding orbital and b) Graphene as the source of three different materials- Fullerene (left), carbon nanotube (Center) and bulk graphite (right).

Carbon Nanotubes

In 1991, a Japanese scientist Iijima [6] discovered new form of carbon-graphitic tubules. He also suggested the possible structure of these tubes, which was later proven right. The first carbon nanotubes (CNT), which were observed and described by Iijima, had walls built from two to fifty graphene sheets, and so they became known as multiwalled carbon nanotubes (MWNT). Later, single walled nanotubes (SWNT) were also identified.

A single walled carbon nanotube is technically defined as a cylinder made up of rolled up sheets of graphene as described artistically in Figure 1. However, rolling up graphene is not the actual way in which a nanotube forms; the diameter of carbon nanotubes typically varies from 0.7-3 nm. Due to such small diameters, nanotubes become quasi one dimensional. They can posses a single shell or multiple shells, as depicted in Figure 2. Carbon nanotubes with single graphene layers are called single walled carbon nanotubes (SWNT) while those with more than one graphene layer are multiwalled carbon nanotubes (MWNT). The length of nanotubes can be up to centimeters, giving them an astonishing length/diameter ratio of 10^7. Carbon nanotubes can be metallic and semi conducting due to the presence of a hollow inside the nanotubes, in addition to unique hardness and mechanical strength. These metallic /semiconducting properties make carbon nanotubes especially attractive to scientists. Following this, the chemical structure of carbon nanotubes does not allow them to react or make complexes with any other elements of the periodic table [7]. This peculiar feature of carbon nanotubes renders it to be interesting nanomaterials for functionalization and modification to suit to the applications of carbon nanotubes [8, 9].

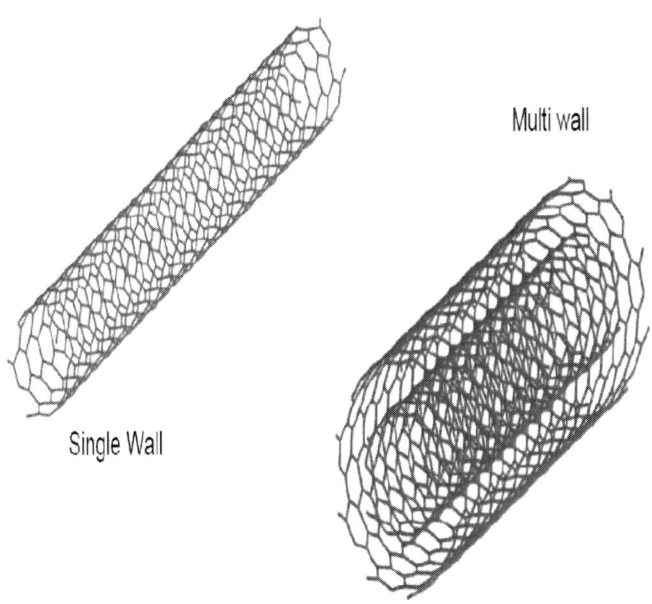

Figure 2. Single and multiwalled carbon nanotubes as hypothetically one can visualize the formation of single walled carbon nanotube through rolling single graphene sheet into a cylinder. For multiwalled nanotubes bi-layer graphene sheet will be the starting material.

Properties of Carbon Nanotubes

Carbon nanotubes are chemically inert by nature and the presence of an empty space inside them in addition to their high mechanical strength and electrical conductivity make them attractive for use in structural applications, medicine and optoelectronics [10]. In these applications, CNT can be sketched as sheets of graphite rolled into seamless cylinders. Depending on the number of sheets, these nanotubes are called single- or multiwalled carbon nanotubes, SWNT or MWNT, respectively. The distance between two graphite sheets in MWNT is typically 0.340 nm, slightly greater than the distance between two consecutive sheets in single crystal graphite, 0.335 nm, which may be explained by specific cylindrical geometry [11].

CNT can be metallic or semiconducting, depending on the crystal structure of the graphene sheets. The electronic transport occurs ballistically (i.e. almost without scattering) in metallic nanotubes. This allows the nanotubes to conduct electricity with minimal resistance, thus diminishes heating [12]. The maximum current conductivity, for example, in copper, the second after silver, best conductive material, is 59.6×10^6 S/m (pure copper). Bulk CNTs produced by arc discharge method have the conductivity ~ 100 S/cm, or 10^4 S/m, [13] annealed nanotubes, produced by chemical vapor deposition method have the conductivity 2×10^5 S/m [14]. Some of the fascinating properties of the carbon nanotubes are given below.

Mechanical properties of carbon nanotubes are outstanding as the mechanical strength is nearly 200 times stronger than steel. Density normalized Young's modulus and ultimate tensile strength of typical CNT are 19 and 56 times that of a steel wire and 2.4 and 1.7 times that of silicon carbide nanorods, respectively in addition to their high elasticity [15].

It has high thermal conductivity as it is dependent on the temperature and the large phonon mean free paths. It has an unusually high value of thermal conductivity up to 6,600 W/m-k at room temperature which is comparable to the thermal conductivity of a diamond or a layer of graphite.

It has an excellent chemical and thermal stability which is estimated to be 2800 °C in vacuum and 750 °C in air.

It is one dimensional and hence has high length /diameter (aspect) ratio i.e., 132,000,000:1 significantly larger than any other material. Because of this nanoscale dimension, electron propagates only along the tube's axis and electron transport involves many quantum effects. Because of this property, carbon nanotubes are frequently referred to as one-dimensional.

It has a high chemical storage capacity and hence represents a new direction for solid-state hydrogen storage especially to store large amounts of hydrogen at room temperature. Relying on physio-sorption alone will lead to a bond that is too weak, and chemisorption will lead to a bond that is too strong. Thus, a weak covalent bond between carbon and hydrogen exist where the electron donation from the (s) orbital of hydrogen to the carbon tubes is established but does not break the hydrogen-hydrogen bond.

Applications of Carbon Nanotubes

Due to the unique structural and electronic properties of carbon nanotubes (CNTs). The surface modification of carbon nanotubes with polymeric structures improves the solubility of the nanotube-polymer conjugates, even with low degree of functionalization of the conjugate unlike pristine CNT which doesn't dissolve in the solvent of interest. Furthermore, the versatility of polymer chemistry allows for control over the final properties of the nanotube polymer conjugate, which are dictated by the chemical and physical characteristics of the grafted polymer [16] as shown in Figure 3.

Synthesis of Carbon Nanotubes

Production methods for carbon nanotubes (CNTs) can be broadly divided into two categories namely chemical and physical depending upon the process used to extract atomic carbon from the carbon carrying precursor. Chemical methods rely upon the extraction of carbon solely through catalytic decomposition of precursors on the transition metal nanoparticles, whereas physical methods also use high energy sources, such as plasma or laser ablation to extract the atomic carbon as they give bulk quantities of CNTs which could then be treated chemically to remove any carbon soot or nanoparticles present in the mixture. Three different methods are generally used to synthesis carbon nanotube: arc-discharge, laser-ablation and chemical vapour deposition.

* Telecommunication, cell phones
* Fuel cell
* Medical equipment
* Composites for aircraft
* Carbon Nanotube computer

Figure 3. Photograph of CNT samples in tetrahydofuran solvent (a) CNTs, (b) Initiator -functionalized CNTs and (c) polymer-functionalized CNTs.

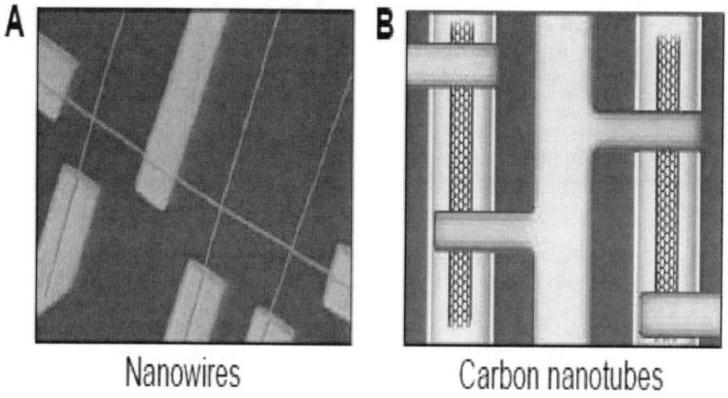

Figure 4. shows an approach towards nanotube computer in which (a) Diodes and transistors based on semiconductor nanowires are assembled with micro fluidics [17] and (b) Carbon nanotube transistors are connected by gold interconnects to construct logic circuits [18].

Arc Discharge Method

Iijima discovered nanotubes in 1991 while experimenting with the synthesis of fullerenes in which he discovered that the needle-like carbon filaments are grown during the arc discharge evaporation of carbon in argon at 100 Torr pressure. Iijima described these filaments, which are grown on the negative end of the graphite electrode, as rolled graphite sheets inserted into each other. The number of these sheets varied from 2 to 50. In 1992 Ebbesen and Ajayan optimized the conditions for the nanotube growth to obtain a larger quantity ~ 1 gram of nanotubes by the arc discharge evaporation method [19]. The inert gas atmosphere (flow of Ar or He) of ~ 500 Torr pressure is created in the reaction vessel. Two graphite rods under DC current (~ 100A) and with applied potential of ~ 18 V are brought together at approximately 1 mm distance. As a result, the discharge occurred and plasma formed. Carbon nanotubes along with some carbon nanoparticles are deposited on the larger rod, finally it results in nanotubes of 75% yield. These nanotubes had the same structure as those described by Iijima, and consisted of 2 to 50 concentric graphitic sheets. The minimum diameter of the tubes is 2 nm and maximum is 20 nm [20]. In 1993, two papers describing similar syntheses of single walled carbon nanotubes are published in the same issue of Nature [21] in which two groups of researchers independently found that by co-evaporation of carbon and iron or cobalt, the single walled nanotubes could be grown. For this synthesis the reaction vessel is filled with a mixture of argon and methane (40 Torr argon and 10 Torr methane) in which there were two graphitic electrodes, an anode of 10 mm in diameter and a twice bigger cathode containing traces of iron as shown in Figure 5, following the DC current of 200 A at 20 V applied to the rods generated heat. The iron under these conditions formed iron carbide particles with the carbon from the cathode and/or methane in which the carbon nanotubes with only one shell are grown on these nanoparticles. Smallest tubes are 0.75 nm in diameter, the biggest are 1.6 nm. With cobalt as a catalyst for the growth of the single walled nanotubes the reaction conditions are slightly different. The reaction vessel is instead filled with helium at 100-500 Torr, and the applied current is ~ 100 A. The nanotubes produced with cobalt as a catalyst also had single walls and the average diameter is 1.2 nm.

Laser Ablation Method

The second, most popular, early method of single-walled nanotube synthesis is laser ablation - laser vaporization of a graphite rod in an oven with Co and Ni as catalysts [22] as shown in Figure 6. This mechanism of the nanotube growth is similar to the arc discharge method. The laser is pointed at the carbon-metal composite in a chamber heated up to 1200°C, where argon flowed at 500 Torr with the flow rate of 50 sccm (standard cubic centimeters per minute).

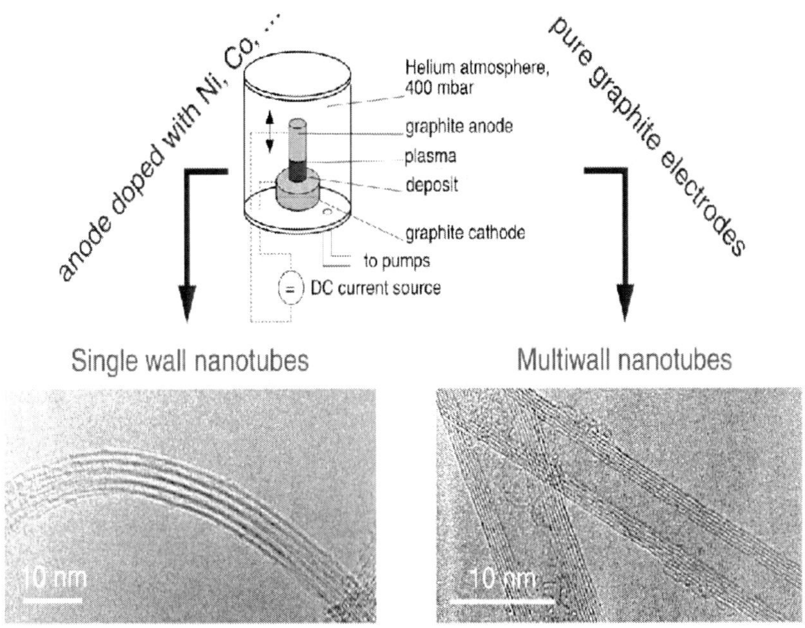

Figure 5. Arc-discharge apparatus (along with electron microscopy pictures of the products obtained with doped and pure anodes).

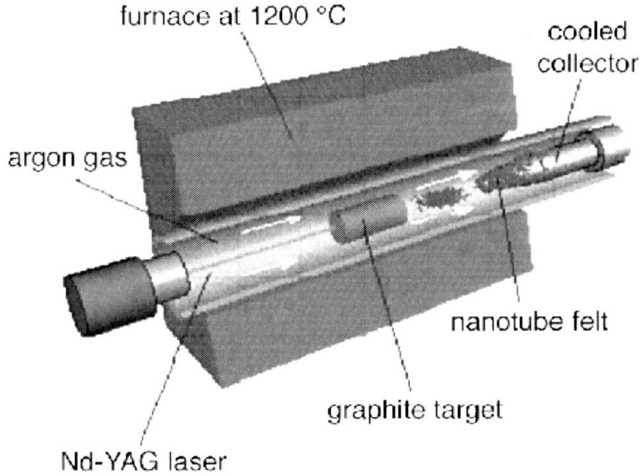

Figure 6. Laser-Ablation Method.

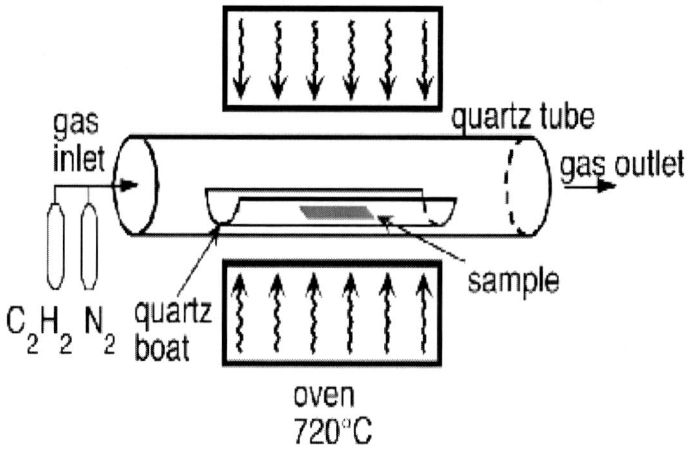

Figure 7. shows chemical vapour deposition method.

The soot produced by the laser is deposited onto a cooled copper collector [23] and this method is optimized to give a higher than 70% yield of nanotubes. The advantages of this method in comparison to the arc discharge method are that the resulting nanotubes are "clean", with less amorphous carbon and no graphite particles on the outer walls.

Chemical Vapour Deposition Method

Chemical vapour deposition CVD is one of the main interesting synthetic procedures [24] because it employs hydrocarbon decomposition over substrate where metal nanoparticles have been placed as shown in Figure 7. Ni, Fe or Co or alloys of these three elements are used. Impurities due to pyrolytic graphite are few, but metal nanoparticles are present, especially at the periphery of CNTs. MWCNTs are grown by decomposition of methane. Further SWCNTs are produced by floating CO with $Fe(CO)_5$ with a disproportionation process.

The substrate is placed in an oven which is maintained at 800 °C after which carbon – bearing gas such as methane is slowly added. The energy source is used to crack the molecule into atomic carbon, follows by combination to form nanotube.

$$\text{Methane} \xrightarrow{800\ ^{o}\text{C}} \text{MWNT}$$

$$\text{Hydrocarbon} + \text{Fe/Co/Ni catalyst} \xrightarrow{1000\ ^{o}\text{C}} \text{SWNT}$$

Implementation of the different techniques for the preparation carbon nanotubes, it needs purification and processing after synthesis to be used for any application. Usually metal nanoparticles and amorphous carbon are present as a synthetic residue. Purification concerns acidic treatments that removes amorphous particles and oxidizes metals that were used as catalysts.

In the same context, more severe purification procedure can shorten CNTs and produce oxidized nanotubes. In general, CNTs are a fluffy powder difficult to manage, while chemical functionalization contributes to prepare more homogenous and soluble powder.

Surface Modification on Carbon Nanotubes

The unique properties of carbon nanotubes make it desirable for different applications. For most of these applications nanotubes require functionalization, such as changing some of the graphite properties to make nanotubes soluble in different media, or attaching different groups or even inorganic particles for future utilization of modified nanotubes.

Hence the stabilization of nanotubes can be divided into three categories.

1. Surfactant stabilization.
2. Covalent attachment of polymers.
3. Polymer wrapping of the molecules.

Surfactant Stabilization

The surfactant stabilization of carbon nanotubes is described in this sub-division. Pyrene can be non-covalently attached on the carbon nanotube surface in which a carboxylic derivative of pyrene is used as interlinker for the binding of capped magnetic nanoparticles on the carbon nanotubes as shown in Figure 8. This is expected to increase organophilic character of the capped nanoparticles that induces high solubility in organic media for the modified carbon nanotubes as shown in Figure 9.

Hence, a new strategy of CN linkage to magnetic nanoparticles is described in order to show how carbon nanotubes can enhance its solubility. In other words magnetic nanoparticles are attached to carbon nanotubes by means of an interlinker molecule, a carboxylic derivative of pyrene as shown in Figure 10. The procedure relies on the known property of pyrene to interact with CN through π-π stacking interactions resulting in its attachment to the nanotube surface [25], in which the available carboxylic groups of pyrene derivatives at the terminal can be further linked to metal or metal oxide nanoparticles.

Finally, the inorganic nanoclusters are well dispersed on the nanotube surface, while at the same time the organophilic sheath surrounding the inorganic partner offers the ensuing composites high solubility in organic solvents [26], ferrofluids based on CN/magnetic oxides can find specific application in biology and biomedicine [27]. In addition, facile recovery of the starting CNs is successful because of the use of pyrene as interlinker. CNs in the magnetic composite materials remains intact since the nanoparticles are not directly attached to the tubes. This is important for the study of interactions between magnetic and electronic properties of the obtained composites. Hence a big limitation in using CNs in nanoelectronics is handling them without destroying or altering their properties is overcome by this method. On the contrary, the as-synthesized magnetically modified carbon nanotubes could be exploited in this way since they could be handled by magnetic forces without altering their electronic or mechanical properties.

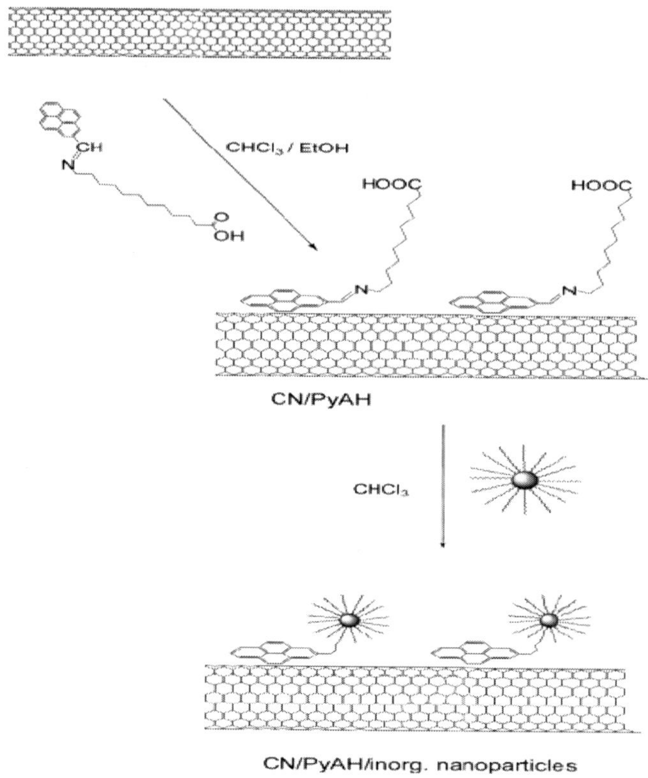

Figure 8. Scheme shows the modification of carbon nanotubes by capped magnetic nanoparticles via "pyrene intermediate".

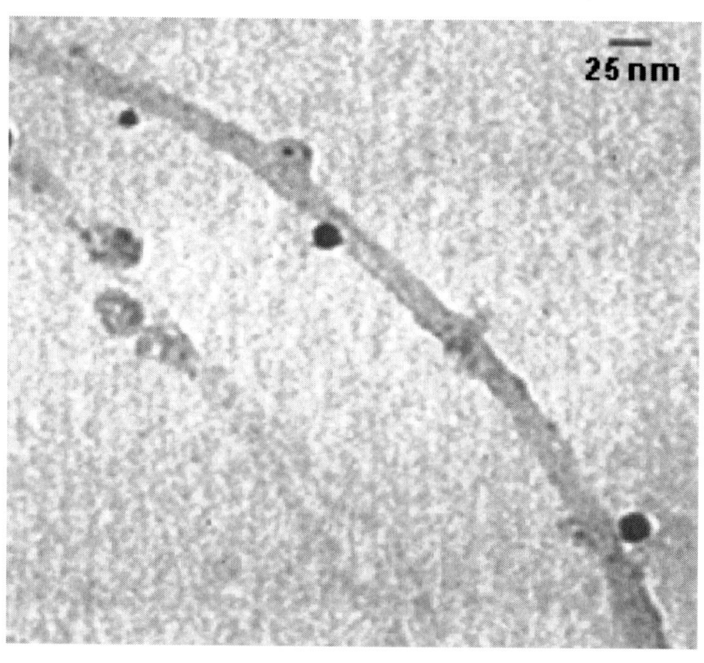

Figure 9. Transmission electron microscopy image showing iron oxide nanoparticles bound on the carbon nanotubes surface.

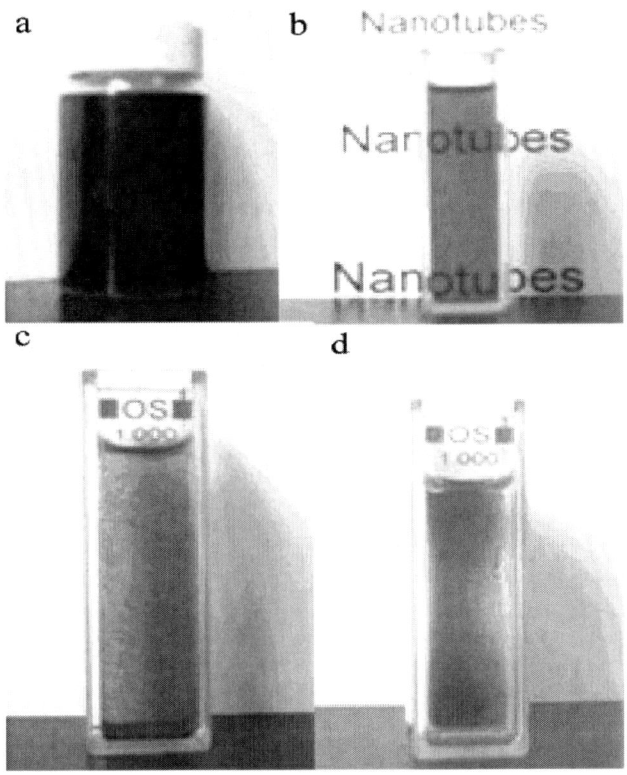

Figure 10. Photographs of (a,b) composites in chloroform in different concentration, (c) dispersion of composite in acetone and (d) its response to a permanent magnet.

Covalent Attachment of Polymers to Carbon Nanotubes

Another fabulous polymer modification of carbon nanotubes is by "grafting from" technique to develop composition hybrids for fabricating novel materials with desirable properties. For this purpose, modification of carbon nanotubes (CNT) is reported widely in which synthetic polymers or bio-macromolecules is grafted or assembled onto the convex surface of CNT via covalent bonds [28] or "chemisorption". This method involves the immobilization of an initiator through end-groups that react with the active sites on the CNT surface. These sites are capable of initiating the polymerization of a monomer leading to the synthesis of polymer encapsulated CNTs in other words the primary polymer layer can be prepared from this active initiator immobilized on the CNT surface In this way, the "grafting from" technique [29] of growing the polymer brush is more effective which allows for the synthesis of brushes with a wide range of chemical nature and high grafting densities via covalent bond formation.

To control the grafting polymerization, especially the length of the grafted polymer chains, controlled radical polymerization is one of the most promising polymerization approaches to controlling molecular weight and its distribution within the resulting polymers [30]. A core-shell hybrid nanostructure, possessing a hard backbone of carbon nanotube CNTs and a soft shell of brush like polymer is prepared via controlled radical polymerization

(CRP) at elevated temperature. The molecular weight of polymers is well controlled as it is big advantage of CRP, which in turn helps to control the thickness of the shell layer. TEM images of the samples provided direct evidence for the formation of a core-shell structure, i.e., the CNT coated with polymer layer as shown in Figure 11. Moreover, this method can be extended to the copolymerization system, which helps to open a new dimension for preparing novel CNT-based nanomaterials and molecular devices with tailor made structures and fine properties.

Transmission electron microscopy (TEM) of carbon nanotube and its polymer grafting is shown in Figure 11 in which picture (a) shows the image of a part of crude CNT. Picture (b-e) displays the images of modified CNT with various thicknesses of the polymer layers having somewhat shallower and more transparent color. The average thicknesses of the enwrapped polymer layers are 3.8, 5.3, 8, and 14 nm, respectively. It is found that the average thickness of the polymer layer increases with an increase in time. Therefore, the thickness of a polymer layer can be well controlled through the CRP method, also 7.8 nm thickness polymer layer could be achieved for successful growth of block copolymers in figure 11.

Figure 12 explains solubility of polymer grafted CNT in the solvent of the interest: The crude CNT is insoluble in any solvents tested. After oxidation of CNT, CNT-COOH is partially soluble in water. If the CNT-COOH was placed firstly in $CHCl_3$, followed by adding water to the mixture, CNT-COOH would move to the water phase and aggregated on the interface of water and $CHCl_3$. CNT-OH showed some extent of solubility in $CHCl_3$ and CNT-Br showed relatively good solubility in polar solvents. The CNT-polymer samples showed good solubility in non-polar solvents and poor solubility in polar solvents. Compared with the CNT-polymer, CNT-block co-polymer is only soluble in strong polar solvents like DMF because of its strong polar hydrophilic chains.

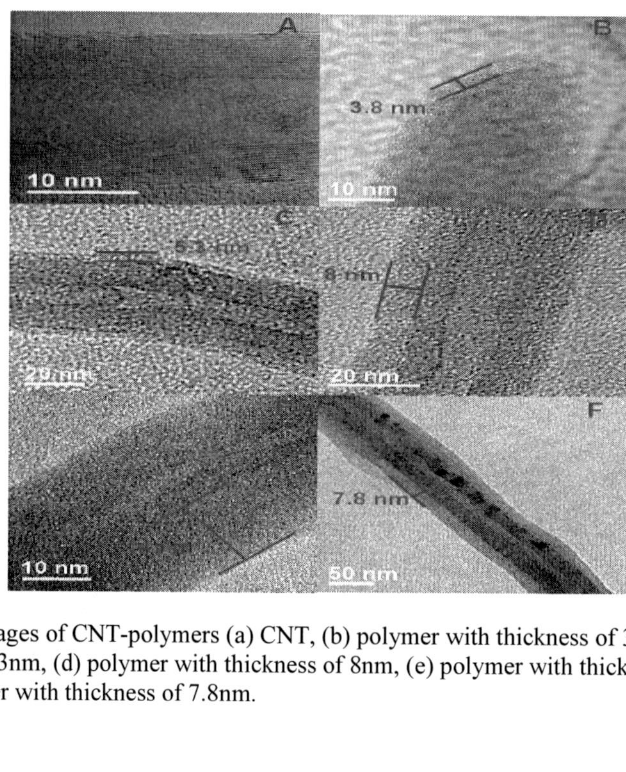

Figure 11. TEM images of CNT-polymers (a) CNT, (b) polymer with thickness of 3.8nm, (c) polymer with thickness of 5.3nm, (d) polymer with thickness of 8nm, (e) polymer with thickness of 14nm and (f) block co-polymer with thickness of 7.8nm.

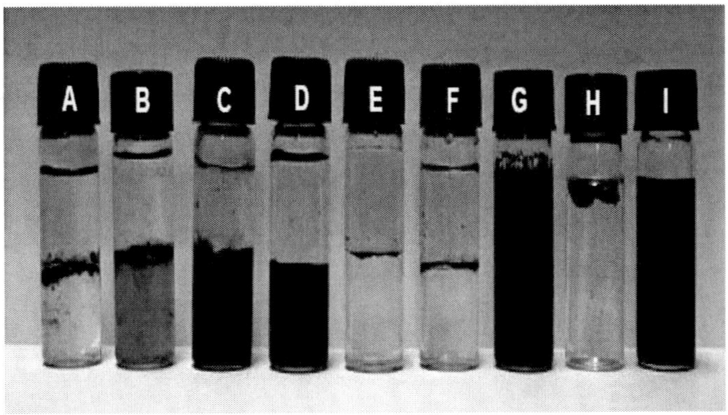

Figure 12 Photo Image of CNT - polymer (a) CNT-COOH in CHCl3/H2O, (b) CNT-OH in CHCl3/H2O, (c) CNT-Br in CHCl3/H2O, (d) CNT-Polymers in CHCl3/H2O, (e) polymers physically coated on CNT in CHCl3/H2O, (f) CRP on MWNT-COOH in CHCl3/H2O, (g) CNT-polymers in CHCl3, (h) CNT-block-copolymers in CHCl3 and (i) CNT-block-co-polymers in DMF.

Polymer Wrapping of the Carbon Nanotubes

Non-covalent chemical modifications of the nanotube surface are essential for the creation of functional device architectures. So in this sub-division, a new general procedure for carbon nanotube modification based on polyelectrolyte layer-by-layer assembly will be discussed. Noncovalent modification has emerged as a leading synthetic strategy for nanotube modification in which multilayer structures around individual carbon nanotube bridges is built by first modifying the nanotube surface with a pyrene derivative followed by layer-by-layer deposition of polyelectrolyte macro-ions on the nanotubes. This is confirmed using transmission electron microscopy images that show the formation of nanometer-thick polymer nanoshells around the nanotubes. Unlike covalent modification, noncovalent modification of the CNT surface typically preserves sp^2 hybridization of the carbon atoms hence its inherent electronic properties of the nanotube is retained [31]. This method involves building of polymer nanoshells using the layer-by-layer deposition on carbon nanotube templates [32]. For this purpose, ionic pyrene derivatives is first adsorb on suspended carbon nanotube which serves as "intermediate" to create a charged template and this helps to facilitate subsequent polymer deposition of poly(diallyldimethylammonium chloride) [PDDA] and sodium salt of polystyrene sulfonate [PSS] polyelectrolytes by layer-by-layer assembly on pyrene-primed carbon nanotubes as shown in Figure 13.

Figure 14 shows the transmission electron microscope images of nanotubes is shown in picture (a) as well as the polymer coated is shown in picture (b) confirms that nanotubes are coated with polymer layer with reasonable few nanometer thickness. To confirm the presence of both polyelectrolytes used in layer-by-layer deposition on the nanotube, elemental mapping using energy-filtered TEM is performed. As the molecular structures of the polymers indicate, PSS is the only source of sulfur in the system, while nitrogen appears only in PDDA and $PyrNH_3$. Moreover, nitrogen from $PyrNH_s$ should be confined to the nanotube surface because the pyrene moiety lays flat on the nanotube. On the other hand, sulfur from PSS and nitrogen from PDDA should be distributed uniformly throughout the coating, which is exactly what the elemental maps show (picture c, d). Therefore, TEM imaging provides strong evidence for the formation of our layer-by-layer composite. These multilayer polyelectrolyte shells on individual carbon nanotubes introduce nearly unlimited opportunities for the incorporation of various functionalities into nanotube devices, which, in turn, opens up the possibility of building more complex multicomponent structures.

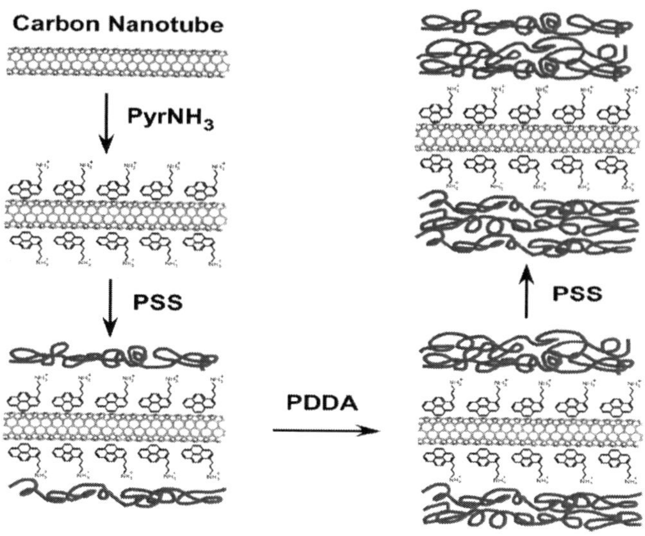

Figure 13. Carbon nanotube modification using layer-by-layer electrostatic self-assembly of PSS and PDDA.

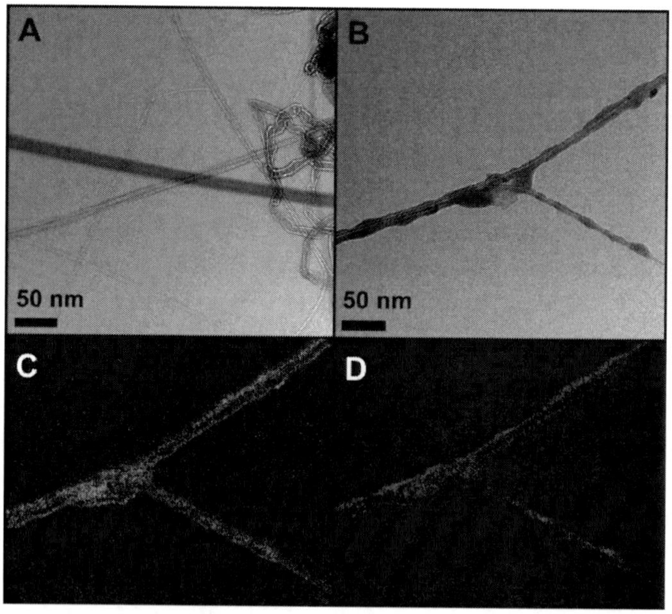

Figure 14 Transmission electron microscopy images of (a) CNTs, (b) nanotubes coated with multi-alternating PSS/PDDA polyelectrolyte layers, (c) Sulfur and (d) nitrogen elemental maps of the polyelectrolyte layers.

Application in Cellular Uptake

Carbon nanotubes (CNTs) are well known for its application as a novel carrier system for both small and large therapeutic molecules. CNTs can be functionalized (i.e. surface-engineered) with certain functional groups in order to manipulate their physical or biological properties. The cellular uptake of CNTs has been confirmed in a range of studies but the mechanism of CNT penetration into cells is still not well-understood. Because of their needle-like shapes, CNTs might be able to perforate cellular membrane and pass into the cellular components without causing apparent cell damage [33]. The nano-injector is designed using an atomic force microscope (AFM) tip and functionalized CNTs attached to a model cargo compound via a disulfide linker. The CNTs nano-injector successfully transported into the cell where the disulfide bond is broken, resulting in the release of the cargo compound within the cytosol as shown in Figure 15. This is followed by successful retraction of the nano-injector with no apparent cell damage being produced.

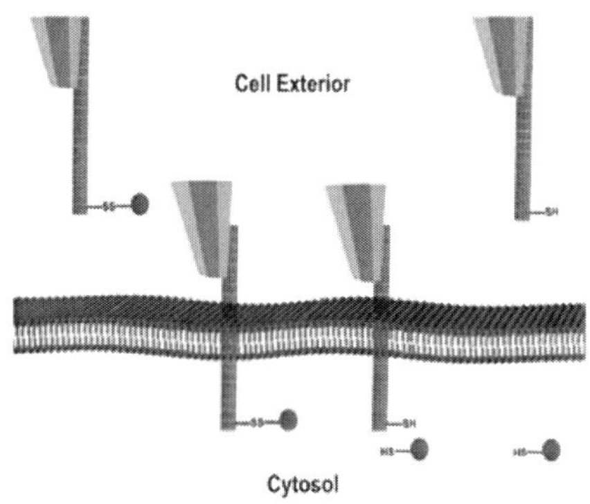

Figure 15. A schematic diagrams showing that an atomic force microscopy-controlled CNT-based nano-injector was able to penetrate into a cell and release the attached cargo compound after the breakage of the disulfide bond.

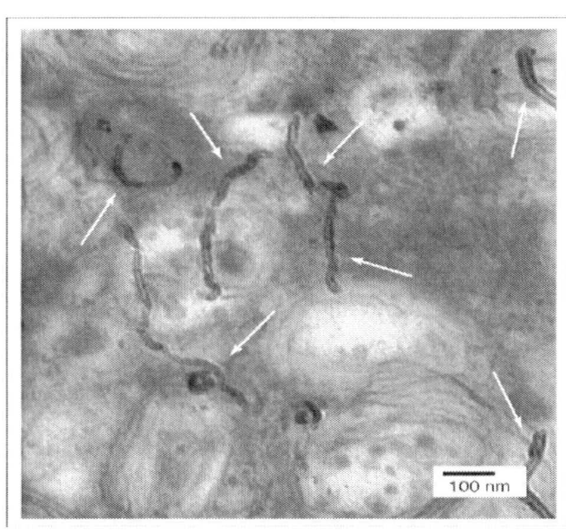

Figure 16. The perpendicular positioning of CNTs (pointed at by the white arrows) during internalization into HeLa cells suggests that cellular uptake of CNTs by the cells.

The perpendicular positioning of the nanotubes to the cell membranes suggests that uptake of CNTs is similar to that of nano-needles which diffuse through cell membranes without causing cell death. The protein attached to CNTs-biotin is detected in the endosomes, suggesting that uptake of the nanotubes occurred via endocytosis. By contrast, no protein is internalized in the absence of the nanotubes. Using confocal microscopy, functionalized CNTs labeled with a fluorescent agent have been shown to penetrate through the cell to the cytoplasm or the nucleus of fibroblasts as shown in Figure 16.

Conclusion

CNTs can be functionalized so that certain molecules are attached to their surfaces via covalent or non-covalent bonding. This covalent attachment of the polymer with CNT result in the loss of the electronic properties of nanotubes in contrast non-covalent attachment which help to retain the electronic properties of the nanotube by preserving sp^2 hybridization. The needle-like shape of the CNTs enables them to perforate cellular membranes and transport the carried therapeutic molecules to the cellular components. This process is thought to take place via endocytosis. CNTs have exclusive properties that would make them appropriate not only in the medical field but also in the field of electronic device as well as in various other fields. CNTs have been introduced to drug delivery research in the recent years and therefore extensive research is expected to be conducted in the forthcoming years in order to explore their potential.

References

[1] T. Guo, C.M. Jin and R.E. Smalley, *Journal of Physical Chemistry* 95, 4948 (1991).
[2] S. Iijima and T. Ichihashi, *Nature* 363, 603 (1993).
[3] K. S. Novoselov, A. K. Geim, S. V. Morozov, D. Jiang, Y. Zhang, S. V. Dubonos, I. V. Grigorieva and A. A. Firsov, *Science* 306, 666 (2004).
[4] K. S. Novoselov, A. K. Geim, S. V. Morozov, D. Jiang, M. I. Katsnelson, I. V. Grigorieva, S. V. Dubonos and A. A. Firsov, *Nature* 438, 197 (2005).
[5] Y. B. Zhang, Y. W. Tan, H. L. Stormer and P. Kim, *Nature* 438, 201 (2005).
[6] S. Iijima, *Nature* 354, 56 (1991).
[7] S. J. Son, X. Bai and S. Lee, *Drug Discovery Today* 12, 657 (2007).
[8] G. Pagona and N. Tagmatarchis, *Current Medicinal Chemistry* 13, 1789 (2006).
[9] K. Balasubramanian and M. Burghard, *Small* 1, 180 (2005).
[10] R. H. Baughman, A. A. Zakhidov and W. A. de Heer, *Science* 297, 787 (2002).
[11] P. M. Ajayan, *Chem. Rev.* 99, 1787 (1999).
[12] W. J. Liang, M. Bockrath, D. Bozovic, J. H. Hafner, M. Tinkham and H. Park, *Nature* 411, 665 (2001).
[13] T. W. Ebbesen and P. M. Ajayan, *Nature* 358, 220 (1992).
[14] D. Mattia, M. P. Rossi, B. M. Kim, G. Korneva, H. H. Bau and Y. Gogotsi, *Journal of Physical Chemistry B* 110, 9850 (2006).
[15] E. W. Wong, P. E. Sheehan and C. M. Lieber, *Science* 271, 1971 (1997).

[16] H. Li, F. Cheng, A. M. Duft and A. Adronov *J. Am. Chem. Soc.* 127, 14518 (2005).
[17] Y. Huang, X. Duan, Y. Cui, L. Lauhon, K. Kim and C. M. Lieber, Science 294, 1313 (2001).
[18] S. J. Tans, A. R. M. Vershueren and C. Dekker, Nature 393, 49 (1998).
[19] K. Hernadi, A. Fonseca, J. B. Nagy, D. Bernaerts, J. Riga and A. Lucas, *Synthetic Metals* 77, 31 (1996).
[20] T. W. Ebbesen and P. M. Ajayan, *Nature* 358, 220 (1992).
[21] D. S. Bethune, C. H. Kiang, M. S. Devries, G. Gorman, R. Savoy, J. Vazquez and R. Beyers, *Nature* 363, 605 (1993).
[22] A. Thess, R. Lee, P. Nikolaev, H. J. Dai, P. Petit, J. Robert, C. Xu, Y. H. Lee, S. G. Kim, A. G. Rinzler, D. T. Colbert, G. E. Scuseria, D. Tomanek, J. E. Fischer and R. E. Smalley, *Science* 273, 483 (1996).
[23] T. Guo, P. Nikolaev, A. Thess, D. T. Colbert and R. E. Smalley, *Chemical Physics Letters* 243, 49 (1995).
[24] A. Jorio, D. Gene and D. S. Mildred, *Carbon Nanotubes: Advanced topics in the synthesis, structure, properties and application*, 1^0 ed Springer (2008).
[25] L. Qu, B. Martin, W. Huang, F. Fu, D. Zweifel, Y. Lin, Y. P. Sun, C. E. Bunker, B. A. Harruff, J. R. Gord and L. F. Allard, *J. Chem. Phys.* 117, 8089 (2002).
[26] N. J. Turro, P. H. Lakshminarasimhan, S. Jockusch, S. P. O'Brien, S. G. Grancharov and F. X. Redl, *Nano Lett*er 2, 325 (2002).
[27] M. Shim, N. W. S. Kam, R. J. Chen, Y. Li and H. Dai, *Nano Lett*er 2, 285 (2002).
[28] Y. Lin, A. M. Rao, B. Sadanadan, E. A. Kenik and Y. P. Sun, *J. Phys. Chem. B* 106, 1294 (2002).
[29] M. Sano, A. Kamino, J. Okamura and S. Shinkai, *Langmuir* 17, 5125 (2001).
[30] J. Qiu, B. Charleux and K. Matyjaszewski, *Prog. Polym. Sci.* 26, 2083 (2001).
[31] S. Niyogi, M. A. Hamon, H. Hu, B. Zhao, P. Bhowmik, R. Sen, M. E. Itkis and R. C. Haddon, *Acc. Chem. Res.* 35, 1105 (2002).
[32] G. Decher, *Science 277*, 1232 (1997).
[33] A. Bianco, K. Kostarelos and M. Prato, *Current opinion in biotechnology* 9, 674 (2005).

In: Applications of Carbon Nanotubes
Editor: Ajay Kumar Mishra
ISBN: 978-1-62081-918-0

Chapter 4

CARBON NANOTUBES AS MATERIAL SUPPORT FOR PT-BASED CATALYSTS AND THEIR USE IN DIRECT METHANOL FUEL CELLS

Juan Manuel Sieben*[*1,2] and Marta M. E. Duarte[1,3]
[1]Instituto de Ingeniería Electroquímica y Corrosión (INIEC). Departamento de Ingeniería Química. Universidad Nacional del Sur, Bahía Blanca - Argentina
[2]Consejo Nacional Investigaciones Científicas y Tecnológicas, Argentina
[3] Comisión de Investigaciones Científicas, Argentina

ABSTRACT

In the last decade carbon nanotubes have turned out to be an important industrial material, being produced hundreds of tons each year. The unique structure of carbon nanotubes makes them an alternative material for catalyst support in fuel cells devices due to their unique one-dimensional nanostructure, high surface area, excellent electronic conductivity, and high chemical and mechanical stability. However, raw nanotubes do not have sufficient binding sites for anchoring metals nanoparticles, which frequently lead to poor dispersion and agglomeration of the nanoparticles. Hence, functionalization of carbon nanotubes is generally an essential issue to further applications. The presence of these surface functional groups modifies the chemical and physical properties of the carbon, improving wettability, adsorption, and cation exchange capacity. Furthermore, these groups act as nucleation centers or anchoring sites, limiting the particle growth and improving the dispersion of metallic crystallites and the stability of the supported catalysts.

This chapter is devoted to review the available scientific literature, motivated by the enormous world-wide demand of increasing the production of energy from green renewable technologies. In the need of developing new, better-performing catalysts for direct methanol fuel cells applications, the present survey is focused on synthesis of catalytically active metal nanoparticles, such as Pt, and Pt-Me (Me: Ru, Sn, Mo, Os, etc.) decorating the external walls or the interior of the carbon nanotubes and carbon nanotubes/polymer composites.

[*] Tel/fax: +54-291-4595182.

1. INTRODUCTION

Carbon nanotubes (CNTs) were first observed in the latest seventies by Endo et al. [1] but it was not until the Iijima's publication in 1991 [2] that these nano-archetypical materials sparked off a real interest in the scientific community. The tremendous impact of Iijima's paper stimulated intensive research from a fundamental point as well as from an advanced point of view. Since then, the yearly number of publications related to carbon nanotubes has been increasing exponentially to over 22,000 papers in 2010 [3]. The success of CNTs can be associated with their novel and unique structure and remarkable mechanical, thermal, and electrical properties (can be either metallic or semiconducting depending on the tube diameter and helicity) along with low density, being single-walled carbon nanotube (SWCNT) [4] and multi-walled carbon nanotube (MWCNT) [2] the principal forms.

Nowadays, CNTs are produced industrially by using the catalyst-assisted chemical vapor deposition (CVD) process developed by Endo et al. in the 70's [1], although a few companies provide SWCNTs synthesized by pulsed laser vaporization, high pressure electric arch discharge and radio frequency (RF) plasma. In addition, it should be remarked that the capacity of annual production of MWCNT and SWCNT are 300 and 7 tons/year respectively, currently dominated by Japanese manufacturers [6]. The growing demand of the nanotubes (a global market on the order of thousands of tons/year is predicted to be necessary over the next decades) can be associated with their many potential applications such as power sources for implantable medical devices (glucose fuel cell) [7,8], nanoelectronic and photovoltaic devices [9-11], supports for metal catalysts [12-17], superconductors [18], electromechanical actuators [19], electrochemical capacitors [20,21], nanowires [22,23], conductive and high strength composites [24-26], field emission displays and radiation sources [27-29], batteries [30], chemical and biological sensors [31-35], field-effect transistor [36,37], environmental remediation technology and pollutant identification [38-40], electroanalysis of biologically and environmentally significant species including proteins [41], NADH [42], neuro-transmitters [31], cytochrome c [43], sulphide [44], epinephrine [45], hydrazine compounds [46], and artificial muscles [47].

The use of CNT-supported Pt-based catalysts in low temperature fuel cells results very promising owing to the unique electrical and structural properties of CNTs and to the high total surface area of SWNT and MWNT (BET areas between 400-900 m^2 g^{-1} and 200-400 m^2 g^{-1} are often reported for the as-grown materials [48], respectively). More specifically, the tubular morphology of carbon nanotubes, the presence of surface defects (edge sites, tips, etc.) and highly mesoporous 3-D structure can provide specific active sites for anchoring metal crystallites [49], although functionalization of CNTs is generally prerequisite to further applications due to their high hydrophobicity. Besides, the high conductivity of CNTs, which implies high electrochemically accessible area and easier charge transfer at the electrode /electrolyte interface, is considered to contribute to the high performance of the CNT-supported metal electrodes. In addition, the tubular three-dimensional morphology of the carbon nanotubes makes the fuel diffusion easier [50], increasing the percentage of catalyst utilization.

This chapter offers an overview of the available bibliography dealing with carbon nanotubes, single- and multi-walled, as supports for Pt-based catalysts, with particular attention on the effect of substrate characteristics on catalysts properties, stability and performance, especially for low-temperature fuel cell.

2. Carbon Nanotubes Functionalization and Composites

2.1. Outline

Despite the fact that carbon nanotubes have remarkable electrical and structural properties, the dispersion of this material in aqueous media is very problematical because the nanotubes are highly entangled (they have a tendency to aggregate together), due to their very high, long-range van der Waals forces of attraction. In order to use the carbon nanotubes in practical applications it is essential to disperse them in a solvent. To overcome this problem, the functionalization (non-covalent or covalent) has received much attention recently, since it decreases the long-range van der Waals forces of attraction and increases the CNT-matrix /electrolyte interaction achieving a homogeneous dispersion of CNTs. Therefore, surface functionalization can improve wettability for polar solvents [51], and compatibility with polymers in composites [52]. Non-covalent functionalization is carried out by nonchemical-bond interactions through π-π electronic and hydrophobic interactions, etc. which are done by adsorption of charged surfactants and polyelectrolytes [53-55], non-wrapping adsorption of partially collapsed block copolymers [56] and complexation by π-π interactions [57,58]. On the other hand, the covalent functionalization is achieved by the covalently bonded functional groups like -COOH, -OH, halogen, etc. [59-61]. Typically, the oxidation of CNT is usually carried out by treatment with aqueous solutions of acids or oxidizing agents, such as H_2SO_4/HNO_3 [62], H_2O_2 [63], $KMnO_4$ [64], and $(NH_4)_2S_2O_8$ [65], etc. In addition a milder gas oxidation treatment with ozone [58,66] or O_3/UV [67] emerges as an interesting alternative, since the surface of the nanotubes is decorated with active moieties without several degradation of the mechanical properties of CNTs.

2.2. Covalent Functionalization of CNTs

2.2.1. Oxidative "Purification"

Oxidative purification treatment is the most common method to improve the dispersion of the CNTs in water. The number of surface oxygenated groups depends on the nature of the oxidant and the conditions applied during the reaction (temperature, pH, pressure, etc.). Generally, surface groups such as carboxylic, quinone and phenolic functionalities (–COOH, -C=O and –OH) are created at nanotubes end-tips or defective sites (edge sites) of the CNTs during the oxidation process [62-67]. This process takes place at the nanotube tips and defect sites and moves gradually from the outermost graphitic layer toward the nanotube core in MWCNTs [6,68], while the oxidation process occurs at nanotube end-tips and the outer layer of the bundle in SWCNTs [69].

The first work dealing with this issue was published by Iijima's group in 1993 [70]. The authors modified the defective sites on the carbon nanotubes surface by an oxidative treatment in air at above 700 °C, creating hemispherical end-caps opening in CNTs and defects in their sidewalls. In 1995, Esumi et al. [71] carried out the chemical oxidation of CNTs with nitric acid and sulfonitric mixed acid refluxed at 140 °C, resulting in the formation of a high concentration of acidic groups on the nanotubes surface. This functionalization process was more effective with the acid mixture and more pronounced at longer pretreatment times. The oxidation of carbon nanotubes has been frequently carried out by aqueous solutions of oxidizing agents and strong mineral acids such as $KMnO_4$ [64,72,73], H_2O_2 [63,65], $KMnO_4/H_2SO_4$ [73,74], HNO_3 [15, 75-77], H_2SO_4 [78], H_2SO_4/HNO_3 [62,74, 79], etc. In this process, the oxygenated functionalities are covalently bond to the carbon nanotubes due to the interactions between the surface defects or tips in the CNTs and oxidants or strong acids, changing the sp^2 hybridization of CNTs to sp^3 by the formation of carboxylic and phenolic groups, while gas phase oxidation (e.g., oxygen plasma treatment) produces mainly hydroxyl and carbonyl groups. Li et al. [73] observed that the oxidation of MWCNTs with $KMnO_4$, H_2O_2 and HNO_3 introduced different functionalities such as carboxyl, lactone and phenolic surface group, being the amount of carboxylic and lactone groups formed on the MWCNTs treated with HNO_3 higher than on those oxidized with the other two reactive, whereas the phenolic groups predominated on the CNTs treated with $KMnO_4$. On the other hand, Ma et al. [67] produced functionalized MWCNTs through the combination of ozone and UV light with phenolic surface groups as the dominant functionality.

These oxidative procedures open the possibility to obtain nanotube surfaces modified with both small molecules and macromolecules due to the replacement of the carboxyl and hydroxyl surface groups by chemical attachment.

2.2.2. Coupling Replacement Reactions

In the most common coupling process the molecules of long chain with amine group moieties can be attached to the nanotube surface by amidation reaction, in which a first stage of chlorination with $SOCl_2$ to form the acyl chloride intermediate is usually required [80]. For instance, ethylenediamine [15], diphenylamine [81], cysteamine ($NH_2CH_2CH_2SH$) [82] or polymers with amino side groups such as poly-(propionylethylenimine-co-ethylenimine) (PPEI-EI) [83,84] poly-(diphenylamine) PDPA [81], can be attached onto the oxidized nanotubes through the amide formation, improving the dispersion of CNTs in water. On the other hand, direct amide linkages between the carboxylic groups and poly-(amidoamine) (PAMAM) [85] or octadecylamine (ODA) [86] onto oxidized CNTs can be obtained by direct amidation reaction. In addition, Hsieh et al. [79] treated carboxylic functionalized MWCNTs used acyl chlorination ($SOCl_2$) and then amidation process with a NH_3 solution to enable the formation of the terminal amino groups on the sidewall of the CNTs. Furthermore, acyl chloride intermediate can be used to ester-functionalize carbon nanotubes with straight-chain alkyl alcohols such as octadecylalcohol (ODO) [87], dendrons (methyl-3,5-dihexadeco-xybenzoate and methyl-3,5-dihydroxybenzoate [88]).

Apart from that, other reactions can be used to attach molecules to the surface of the carbon nanotubes. However, they are not especially useful to the preparation of materials supports for the deposition of catalysts in fuel cell applications [89,90], although between them, fluorination may have promising applications on the preparation of highly functionalized and hydrophilic supports since the addition of fluorine drastically raises the reactivity of the nanotube side walls. Therefore, the fluorine sites can be replaced by alkoxides, organic peroxides (such as lauroyl, benzoyl and tertbutyl), etc. [89] through a specific chemical reaction, for instance, alkylation with Grignard reagent.

Plasma activated aligned CNTs were used for the chemical immobilization of polysaccharides [91]. For example, amino-dextran chains were immobilized onto plasma-treated aligned carbon nanotubes through the formation of Schiff-base linkages, which were further stabilized by reduction with sodium cyanoborohydride in aqueous solution. The resultant materials exhibited a highly hydrophilic surface with the integrity of the nanotube structure being largely maintained.

2.3. Non-Covalent Functionalization of CNTs

2.3.1. Physical Wrapping

Non-covalent functionalization of CNTs is of particular interest because it does not modify the physical properties of CNTs (i.e., conserves the nanotubes electronic structures by preventing rupture of nanotube sp^2 conjugated structure), but improves solubility and processability. This type of functionalization mainly involves surfactants, biomacromolecules or wrapping with polymers. The carbon nanotubes are transferred to the aqueous phase in the presence of surfactants [92] and then are surrounded by the hydrophobic components of the corresponding micelles, becoming the interaction stronger when the hydrophobic part of the amphiphilic contains an aromatic group [93].

It is well known that conjugated polymers can wrap CNTs and reduce van der Waals interaction between CNTs, resulting in increased dispersibility of SWNTs [94,95]. The driving force to wrap the carbon nanotubes sidewall with polymer chains is strong π-π interaction between the carbon nanotubes and the conjugated polymer, forming supra-molecular complexes [95,96]; for example, poly-(vinylpyrrolidone) (PVP) or poly-(styrene sulfonate) (PSS), poly-(phenylacetylene) (PA) can be wrapped onto the CNTs [97,98]. In addition, CNTs have been wrapped with a copolymer (e.g. poly-(p-phenylvinylene-co-2,5-dioctylloxy-m-phenylenevinylene) without changing the electrical property of the nanotubes and improving their dispersion in organic solvents [99,100]. Besides, CNTs have been encapsulated inside poly-(styrene)-block-poly-(acrylic acid) micelles, improving the dispersion of CNTs in several polar and non-polar solvents [101].

2.4. Composites of CNTs

2.4.1. Chemical and Electrochemical Polymerization

Conducting polymers such as polyaniline (PANI), polypyrrole (PPy), polythiophene (PTh) and their derivatives can be attached to CNTs surfaces by two-main routes, i.e. the chemical and electrochemical oxidation in various organic solvents and in aqueous media. Therefore, the in-situ polymerization improves the processability, and electrical, magnetic and optical properties of CNTs, being the formation of the composites environmentally friendly and facile. In addition, the properties of polymer-CNTs composites depend on a multitude of factors that include the type (SWNT, MWNT), chirality, purity, defect density, and dimensions (length and diameter) of the nanotubes, nanotubes loading, dispersion state and alignment of nanotubes in the polymer matrix, and the interfacial adhesion between the nanotubes and the polymer matrix [102].

The typical chemical synthetic route involves the oxidative polymerization in the presence of the carbon nanotubes by using ammonium persulfate (APS: $(NH_4)_2S_2O_8$), $K_2S_2O_8$ or $FeCl_3.6H_2O$ in some cases. Qu et al. [103] and Zhao et al. [104] synthesized polypyrrole-MWNTs nanocomposites with APS oxidant and used it as a Pt and Pt-based catalyst support, studying the influence of feeding ratio of Py/MWNTs on the properties of PPy-MWNTs nanocomposites. On the other hand, Cho et al. [105] described a simple approach to the synthesis of MWNT/PPy nanotubes by the in-situ chemical polymerization of pyrrole on the CNTs using ferric chloride as an oxidant, and investigated the effect of the monomer concentration on the coating and properties of the resulting complex nanotubes. In addition, Wu et al. [106] prepared highly conductive PPy-MWCNTs composites via in-situ chemical oxidation polymerization using various concentrations of cationic polyelectrolyte poly-(styrene sulfonate) (PSS) and APS.

Notwithstanding, polyaniline (PANI)-CNT composites are the most investigated hybrid supports for fuel cell catalysts. Konyushenko et al. [107] prepared uniform coatings of PANI-MWCNTs with CNTs contents between 0 and 80 wt.% by using APS as oxidant. The authors evaluated the conductivity and hydrophility of the different coating films. In addition, Qiao and coworkers [108] prepared a PANI-MWCNTs composite of 20 wt.% CNTs to use as anode materials for microbial fuel cells. Guo at al. [109] used a water/oil microemulsion system for synthesizing PANI nanoparticles, which were employed for preparing PANI-MWCNTs nanocomposites via in-situ chemical oxidative polymerization, producing nanocables of high conductivity. Canobre et al. [110] coated MWCNTs with protonated PANI in situ during the chemical polymerization of aniline, and improvement in the covering of CNT composites was attained by the association of poly(2,5-dimercapto-1,3,4-thiadiazole) (PDMcT). Whereas, Jiang et al. [111,112] prepared poly-(N-acetylaniline) (PAANI) composites for fuel cell catalysts. Reddy and coworkers [113] prepared MWCNTs-f-PANI as catalyst support. The authors initially functionalized the acyl chloride terminated CNTs (MWCNTs-COCl) with 2,5-diaminobenzenesulphonic acid (DABSA) via amide bond formation, followed by surface initiated in situ chemical oxidative graft polymerization of aniline in the presence of APS as an oxidizing agent. Apart from that, polypyrrole oligomers were deposited on MWCNTs (PPO-MWCNTs), and then used as support for Pt-based catalyst [114].

Alternatively, electrochemical polymerization has several advantages, particularly, there is no need for added oxidants and electrodeposited conducting polymers are naturally integrated as a continuous uniform film on the electrode, saving the use of a binder. So, the conductivity of the composite is expected to be higher than that of chemically prepared ones, which results for practical uses in electrocatalysis [115]. Wu et al. [116] prepared nanoporous composite films of multi-walled carbon nanotubes and polyaniline by using galvanostatic deposition from acidic aqueous solutions, whereas Li et al. [117,118] and Bavio et al. [119] used cyclic voltammetry to form composite films of MWCNTs or SWCNTs and polyaniline with good uniformity and dispersion. In addition, Jiang et al. [120] synthesized a 3D nanocomposite of poly-(N-acetylaniline) (PAANI) and MWNTs by cyclic voltammetry, in which the inner layer was comprised of MWNTs and the outer layer was PAANI (core-shell structure), and after that, used as support of Pt nanoparticles. Furthermore, Peng et al. [115] growed polyaniline (PANI), polypyrrole (PPY) and poly-(3,4-ethylenedioxythiophene) (PEDOT) via cyclic voltammetry from solutions containing acid treated CNTs and the corresponding monomer. These composites exhibited good mechanical integrity and higher electronic and ionic conductivity. Besides, PEDOT-MWCNT composite films were fabricated by a simple potentiostatic step [121,122], the PEDOT polymer chains enclosed the nanotubes and formed a porous structure; while Lota et al. [123] growed PEDOT on MWCNTs by both chemical and electrochemical oxidative polymerization of EDOT monomer, obtaining better properties with that prepared by electrochemical polymerization.

3. The Use of CNTs as Support of Catalysts

Starting from the work of Che et al. [17] that reported the behavior of carbon nanotube membranes catalyzed with Pt-Ru particles for methanol oxidation, there have been a great number of publications that describe the use of CNTs as support of catalysts for direct methanol fuel cells. More works report an outstanding performance of MWCNT supported Pt and Pt based catalysts for methanol oxidation. However, Antolini [48] in his comprehensive review on carbon supports for low temperature fuel cells advised prudence when comparing the electrochemical activity of CNTs supported catalysts with carbon blacks catalysts. The exceptional results appear to have several origins, such as the high surface area to support a high dispersion of nanoparticle catalysts, high conductivity due to its crystalline nature, a pore structure suitable for reactant transport, the absence of poison species and a favorable interaction between the catalyst particles and the carbon nanotubes [124]. For example, CNTs are frequently compared with the most widely used Vulcan XC-72R carbon support [125]. Vulcan has an electronic conductivity of 4.0 S cm^{-1} and a specific surface area of 237 m^2 g^{-1} while CNTs have significantly higher electronic conductivities of 104 S cm^{-1} and specific surface areas of 200-900 m^2 g^{-1}. Furthermore, CNTs have no micropores smaller than 2 nm, whereas Vulcan XC-72R has a large ratio of micropores which are smaller than 2 nm. Therefore, the platinum nanoparticles may sink into the micropores of the porous carbon particles, and will reduce the number of three-phase boundary reactive sites, thus reducing the Pt utilization.

Several issues are involved in the attainment of a highly active electrode. Although catalysts with high utilization of the metal, generally platinum, are widely used, it is difficult to obtain nanoparticles with homogeneous size using conventional methods because the pristine surfaces of carbon nanotubes are relatively inert, and they tend to form large bundles, which results in the agglomeration of the nanoparticles [126]. Therefore, the pretreatment and functionalization of CNTs surface are fundamental aspects of the preparation of the catalyst.

The method of preparation and deposition of the nanoparticles is another important question. Lee et al. [125] reviewed the literature on the synthesis of CNTs and CNFs Pt electrocatalysts and remarked that the synthesis method decidedly influences the morphology, utilization and activity of the supported catalyst. The long term stability of the nanoparticles under electrochemical operation is yet another aspect that must be considered [48].

3.1 Influence of the Pretreatment and Functionalization of CNTs

Most authors describe a substantial enhancement in the dispersion of the catalysts particles on oxidized CNTs when compared to the pristine material. Tian et al. [127] reported the improvement of Pt nanoparticles dispersion and distribution after the oxidation treatment of MWCNTs with both H_2SO_4/H_2O_2 and H_2SO_4/HNO_3 solutions in comparison to that prepared with as-received MWCNTs, in which the distribution of Pt nanoparticles was not uniform and large clusters were formed. The average size of Pt particles was 4.2 nm, 3.5 nm and 2.9 nm for the Pt/MWCNTs prepared on as-received, H_2SO_4/H_2O_2 treated, and H_2SO_4/HNO_3 treated MWCNTs, respectively. Another example is given by Halder et al. [128] that prepared Pt-CNTs composites using the controlled attachment of a precursor phase on the walls of the nanotubes. They found that the formation of uniform coating of nanoparticles on the nanotubes is primarily due to the attachment of the intermediate on the tube walls. The non-functionalized CNTs do not show appreciable attachment of platinum intermediate, and hence, the number of platinum nanoparticles is found to be lower. These authors remarked that not only is the surface functionalization important for the attachment of nanoparticles, but also the catalytic activity is related to the active sites on the carbon surfaces. By increasing the number of oxygen containing functional groups (defects) on the carbon surface by several different ways of oxidative treatment, the properties of carbon-supported metal catalysts can be tuned.

Chetty et al. [129] used plasma treatment to prepare nitrogen-doped carbon nanotubes as support to disperse Pt-Ru nanoparticles. They found that the Pt-Ru particles supported on the plasma treated CNTs shows higher catalytic activity for methanol oxidation in comparison to the conventional dual acid treatment method, and also to the commercial E-TEK catalysts. It is likely that the enhanced activity of nitrogen-doped electrode may be a consequence of the composition of the nanoparticles themselves, which is influenced by the support material. The presence of nitrogen can have some influence on the surface electron density, which in turn can influence the deposition kinetics of Pt and Ru. The authors proposed that nitrogen plasma treatment produces pyridinic and pyrrolic nitrogen species on the CNTs surface. During impregnation, the metal ions are more likely to interact with and attach to the graphene sheets containing these surface functional groups, which serve as the anchoring sites, by ion exchange or coordination reaction. During reduction with hydrogen, the formation of highly dispersed Pt-Ru particles on the N-CNTs surface occurs.

Maiyalagan et al. [50] studied the same system and commented that functional groups of the carbon support might influence the electronic structure of Pt nanoparticles. The nitrogen functional group on the carbon nanotubes surface intensifies the electron withdrawing effect against Pt and the decreased electron density of platinum facilitates the oxidation of methanol. The onset potential of methanol oxidation at nitrogen containing carbon nanotubes supported catalysts occurs at 0.22 V, which is relatively more negative to that of the other electrodes. This may be attributed to the high dispersion of platinum catalysts and the nitrogen functional groups on its surface. This author [130] compared sulphur and nitrogen functionalized CNTs and found that the suphur containing electrodes showed the lowest activity towards methanol oxidation, suggesting the existence of specific interaction between Pt and sulphur on the carbon support which inhibited the rate of the reaction.

Nitrogen functionalized carbons were also used by Sun [131], that obtained CN_xNT with the bamboo structure of interesting characteristics, and Chen [132] that synthesized nitrogen doped carbon nanotubes (CN_x) of a high nitrogen concentration. Jiang et al. [133] obtained Pt-Ru nanoparticles with a high dispersion, a narrow size distribution of 2.5-3.5nm and tunable chemical composition when nitrogen doped CNTs were used. In general, the nitrogen functionalities in CNTs act as the defect sites to enhance nucleation of Pt particles [134].

Sulfur containing CNTs has also been studied as Pt catalysts support. Dispersion and size of Pt nanoparticles in Pt/CNTs nanocomposites are remarkably affected by the S/CNT ratio [135]. Besides, sulfonated MWCNTs have been used as support of Pt nanoparticles [136] and palladium catalysts [137], showing improved properties with respect to the unsulfonated counterpart.

Non-covalent functionalization of CNTs has attracted particular attention because it enables the properties of the hybrids of nanoparticles and CNTs to be tailored while still preserving nearly all the intrinsic properties of CNTs. For example, Wang et al. [138] studied the functionalization of MWCNTs by 1-aminopyrene (1-AP) and the synthesis of bimetallic Pt-Ru electrocatalysts on the functionalized support. The authors indicated that the most important advantage is that, in contrast to the acid-treated MWCNTs, where the harsh chemical acids are used to produce carboxylic acid sites on the surface, the 1-AP-functionalization treatment preserves the integrity and the electronic structure of CNTs without affecting the catalyst dispersion, electrocatalytic activity and also enhancing stability. Another example is presented in [139], where 8-hydroxyquinoline (HQ) functionalized multi-walled carbon nanotubes (MWCNTs) were used as catalyst support.

Although carbon functionalization affects the dispersion of the catalysts nanoparticles, other effects have been described. For example, Chen and coworkers reported the spontaneous deposition of Pt nanoparticles [140] by spontaneous reduction of $PtCl_6^{2-}$ ions via the oxygen-containing functional groups introduced on CNTs defect sites by oxidative treatments. The driving force for Pt deposition could be the difference between the potential of $PtCl_6^{2-}$ reduction and the potential of CNTs oxidation. This implies that a galvanic cell between functional CNTs defects and $PtCl_6^{2-}$ ion exists, i.e., a spontaneous electron transfer from the nanotube defects (oxidation) to the metal ions (reduction) is produced.

3.2 Methods of Preparation Used in the Synthesis of CNTs Supported Catalysts

Although it is a well known fact that Pt-Ru bimetallic systems are better electrocatalysts for methanol oxidation, most of the works refer to the preparation of Pt/CNTs electrodes because their use permits to evaluate the influence of substrate, structure and morphology on the electrochemical behavior. For different causes, ethylene glycol is the most used reducing agent in the synthesis of Pt nanoparticles, so the information will be divided between methods of impregnation and reduction where ethylene glycol is not used, methods where EG is used, electrochemical methods and miscellaneous.

3.2.1. Impregnation and Reduction

Table 1 collects the information available of the catalysts prepared by impregnation and reduction or reduction in aqueous solution (excluding EG) for methanol oxidation. Most refers to Pt and Pt-Ru catalysts, although some other metals such as Pd and oxides have been essayed.

Impregnation with a precursor followed by reduction with hydrogen gas is a well known technique. Most differences between the existent works refer to the type of precursor or the manner of impregnation. Some precursor are for example H_2PtCl_6 and/or $RuCl_3$ in aqueous solution [131,141,142] or in acetone solution [137], $Pt(NO_2)_2(NH3)_2$ and $RuNO(NO_3)_3$ [143]

and carbonyl compounds [144,145]. Bambagioni et al. [146] used organometallic precursors such as dimethylcyclooctadieneplatinum, cyclooctadiene cyclooctatriene ruthenium and tris-(dibenzylideneacetone)-dipalladium.

An interesting variant is given by Cao et al. [147] that synthesized Pt/SWCNTs from platinum carbonyl complexes. Purified SWCNTs, SDBS, and water were mixed ultrasonically to accomplish a stable dispersion. The dispersed CNTs were added to the Pt-carbonyl complexes mixture at about 0 ºC under nitrogen and stirred for more than 3 h. In this step, the green color of Pt carbonyl mixture completely disappeared in a few minutes. This could result in the bonding of Pt clusters onto the SWCNTs via a surface reaction and may favor a stronger interaction with CNTs. The resultant solution was homogeneous, visually black, and could be stored at room temperature under air. Then the solvent was removed from the prepared homogeneous Pt-SWCNT mixture at ca. 55 ºC by purging with highly pure nitrogen. Subsequent to the above steps, the powder was subjected to hydrogen treatment at room temperature. The sample was washed and dried under nitrogen. Nearly monodispersed Pt nanoparticles with a diameter of ca. 2.2 nm are formed and broadly dispersed on the surface of individual and/or small bundles of SWCNTs, even at a very high metal loading.

Another method involves the suspension of CNTs in a solution that contains a metal precursor and the posterior addition of some reducing agent such as $NaBH_4$, formaldehyde or hydrazine. Figuereido et al. [148] compared the effect of different methods to prepare Pt and Pt-Ru catalysts. Pt was loaded on CNTs by incipient wetness impregnation, wherein a calculated amount of hexachloroplatinic acid dissolved in a minimum amount of water was added dropwise under vacuum and ultrasonic mixing. The catalysts were then dried under vacuum. On the other hand, Pt/Ru (1:1) catalysts with 10 % (total) metal load were prepared by sequential impregnation on the carbon supports. Thus, Ru was initially impregnated from a $RuCl_3$ solution. The catalyst was then dried under vacuum and kept overnight in an oven. Subsequently, Pt was added from a calculated amount of the hexachloroplatinic acid precursor. Reduction of the metal precursors was achieved in the liquid phase, by slow addition of sodium formate, or hydrazine hydrate or sodium borohydride, or in the gas phase, under hydrogen flow at 350 °C. For the monometallic samples, the dispersion of Pt is bimodal, i.e., there are small particles anchored on the support (diameter 6-15 nm), and very large particles (generally around 100 nm), although it seems that the $NaBH_4$ treatment in solution prevents the formation of large particles. Concerning the activation of platinum, the best results were obtained with $NaBH_4$ and sodium formate. For the bimetallic sample there was a pronounced effect of ruthenium pre-deposition on platinum dispersion, no bimodal particle size distribution was observed, and only small particles were obtained. The authors concluded that in general, reduction of the catalyst with sodium borohydride leads to higher dispersion and more homogeneous distribution of the metal particles on the supports.

One effective variation was given by Kosaka et al. [144]. Single-walled nanohorns (SWNHs) powder was put into a stirred solution containing $NaHSO_3$ and H_2PtCl_6, and then dispersed with a homogenizer. H_2O_2 was added to the mixture to form Pt oxide colloids that were immediately adsorbed on the surface of the SWNHs. The excess of H_2O_2 was then removed by boiling the solution. The obtained Pt oxide particles supported on the SWNHs were reduced in hydrazine solution, dried and further reduced in H_2 gas. For the best catalyst, a surface area of 93 $m^2\ g^{-1}$ was achieved.

Furthermore, a typical procedure was given by Huang et al. [149]. Catalysts were prepared by an HCHO reduction method, in which surface-oxidized SWCNTs were suspended in water by sonication and then mechanically stirred for 4 h, during which time H_2PtCl_6 was added to the solution dropwise and NaOH was added to adjust the pH of the solution to about 11. Pt was attached to the surface-oxidized SWNTs by adding formaldehyde to the solution.

Gattia et al. [150] used formaldehyde as reducing agent to prepare Pt supported on different substrates. They observed that aggregates formed on the CNTs and Vulcan carbon black and described how the formation of aggregates of Pt particles depended on the carbon support. A larger amount of aggregates of Pt particles was found in SWNHs and MWNTs samples than in Vulcan sample. For the Pt/SWNHs sample, the aggregate diameter was in the 50–100nm range, while for the Pt/MWNT sample the diameter ranged from 20 to 50 nm. The aggregates observed were formed by small Pt particles (<4 nm). On the other hand, Pt deposited on Vulcan was not well dispersed: in fact, the sample exhibited some carbon primary particles not covered by Pt crystals. These authors also compared the reduction of the catalyst surface area during a simulated test and observed the complete atomic redistribution of Pt particles, followed by the formation of large Pt particles of 10–30nm in diameter. The particles shape was also changed. However, while in the Pt/SWNHs and Pt/VU electrodes the redistribution of Pt particles seemed to give the same results (similar particles in shape, smoothed surface, and diameter), in the case of the Pt/MWNTs electrode it is possible to observe the presence of some larger particles, but with a nanostructured surface. These results could explain the better performance toward methanol oxidation reaction of the Pt/MWNTs electrodes. Diffusion of Pt atoms on carbon surface was used to explain the formation of Pt aggregates.

Table 1. Catalysts for methanol oxidation prepared by impregnation and reduction or reduction in aqueous solution, excluding EG

Catalyst	Method	Substrate	Particle size	Ref.
Pt	Mixing of surfactant-assisted solubilized SWCNTs and Pt carbonyl complexes, adsorption of Pt and H_2 reduction gas	SWCNTs	2.2 nm	[144]
Pt	Wet chemical hydrogen reduction	MWCNTs	80 nm	[154]
Pt	Impregnation with acetone H_2PtCl_6 solution, reduction with H_2 gas at 350 °C	Sulfur-modified CNTs	2.8 to 3.6 nm	[137]
Pt	Impregnation with H_2PtCl_6 aqueous solution and H_2 gas reduction	CNTs with controlled diameters	1-2 nm	[143]
Pt	Impregnation with aqueous H_2PtCl_6 solution and reduction with hydrogen gas	Nitrogen containing CNTs	3 nm	[131] [132]
Pt, Pt-Ru	Impregnation with the precursors and reduction with H_2 gas at 580 °C	Aligned graphitic CNTs	Pt 7.1 ± 0.4 nm. Pt-Ru 1.59 ± 0.03 nm	[145]
Pt	Controlled attachment of a precursor phase on the walls of the CNTs and reduction with $NaBH_4$	MWCNTs	2-3 nm	[50]
Pt	Borohydride reduction method	Photochemically oxidized MWCNTs		[155]

Table 1. (Continued).

Catalyst	Method	Substrate	Particle size	Ref.
Pt	Synthesis of colloidal Pt oxide particles and reduction with hydrazine	SWCNHs	2.9 nm	[151]
Pt	Reduced by 1-dodecanol	SDS micelle functionalized SWCNTs and MWCNTs	1.60 to 1.87 nm	[156]
Pt	Impregnation process and reduction with formaldehyde	MWNTs, SWNHs	Agglomerates	[150]
Pt	Reduction of H_2PtCl_6 with $NaBH_4$ on thiolated MWNTs and elimination of the thiol groups by a heat-treatment in H_2	thiolated MWNTs	2 nm	[157]
Pt	Glacial acetic acid as a reducing agent.	MWCNTs grown directly on carbon paper	2-4 nm	[158]
Pt	Aqueous solution reduction of H_2PtCl_6 with HCHO	SWCNTs	8.7 nm	[154]
Pt-Ru	Reduction with $NaBH_4$	3D nanostructured CNTs array	2-5 nm	[159]
Pt-Ru	$NaBH_4$ as a reducing agent with the surfactant TOAB	MWCNTs	3-4 nm	[160]
Pt-Ru	Deposition and reduction processes of Pt and Ru precursors	cup-stacked-type carbon nanotubes (CSCNTs)	5 nm	[161]
Pt, Pt-Ru	Reduction with sodium formate, hydrazine hydrate and sodium borohydride solutions, hydrogen gas	MWCNTs	3.5 a 5 nm	[149]
Pt	Reduction with formate in tetrahydrofuran/water/ethanol mixture solution	SWNTs and MWNTs	3.5 nm	[153]
Pd, Pt-Ru	Impregnation with organometallic precursors and reduction with hydrogen	MWCNTs	4.3 nm (Pd) 5.9 nm (Pt-Ru)	[148]
Pt-CeO_2 composites	Co-precipitation method.	MWCNTs	3.4 nm	[162]
Pt-Ru	Modified impregnation method and reduction with $NaBH_4$	MWCNTs	2.7 nm	[163]
Pt-Ru and Pt	Reduction with NaBH4	Functionalized grapheme on MWCNTs	5-10 nm	[164]
Pt, Pt-Ru	Reduction with $NaBH_4$	MWCNTs	3-4 nm	[165]
Pt-Ru	Reduction with hydrogen gas and formaldehyde	Carbon nanotube-vapor grown carbon fiber		[146]
Pt-Ru	Reduction with $NaBH_4$	MWCNTs	2-3 nm	[155]
Pt-Ru/CeO_2	Reduction with $NaBH_4$	MWNTs	~2.4 nm	[166]
Pd	Reduction with $NaBH_4$	Sulfonated MWNTs	4 nm	[139]
Pt-Ru	$NaBH_4$ as a reducing agent with the surfactant TOAB	MWCNTs	2.6-2.8 nm	[160]
Pt-Ru	Chemical impregnation	SWCNTs		[167]
Pt-Ru	Reduction of carbonyl precursor with hydrogen gas			[147]
Pt-Ru	Impregnation-reduction method from chloride precursors	Nitrogen-doped carbon nanotubes (N-CNTs)	10 nm	[130]
Pd-Ag	Reduction with $NaBH_4$	CNTs activated by alkali treatment		[168]
Pd, Pd-Co	Reduction with $NaBH_4$	CNTs		[153]

Other variants include the use of organic solvents to facilitate the dispersion of CNT and the impregnation with the precursors. Prabhuram et al. [151] described the procedure to obtain Pt-Ru nanoparticles with atomic composition 1:1 supported on the MWCNTs. The Pt-Ru/MWCNTs nanocatalyst was synthesized by taking appropriate amount of MWCNTs and precursors H_2PtCl_6 and $RuCl_3 \cdot 6H_2O$ in a tetrahydrofuran (THF) solvent. The resulting suspensions were stirred for 2 h and then, the solvent THF was evaporated. After complete evaporation of the solvent, the dry mixtures were dispersed in diluted methanol and ultrasonicated. Subsequently, an excess quantity of $NaBH_4$ solution was added to the mixtures with vigorous stirring for the complete reduction of Pt and Ru from their respective metal salts. After that, the mixtures were stirred for 2 h at room temperature, filtered and washed. In some cases surfactants [152,153] and stabilizers were added to prevent the coalescence of the metal particles, the formation of CNT bundles or to help in the dispersion and attachment of nanoparticles to the CNT surface.

3.2.2. Electrodeposition

Electrodeposition has been used by many research groups to improve Pt and Pt-Ru utilization because the process requires electronic contact and the metal particles must attach to the CNTs surface. Moreover, electrochemical techniques offer a certain degree of control over the catalyst physico-chemical properties. Catalyst particle size and composition can be controlled by the experiment conditions. However electrodeposition has a drawback. After initial formation of a platinum particle, further deposition continues on the same particle resulting in increase of the size of the particle [169]. In Table 2 some details of the works that have used this technique with CNTs are shown. In most cases the reported particle sizes are larger than those obtained with chemical reduction methods.

Girishkumar et al. [170] used potentiostatic pulses to deposit platinum nanoparticles on OTE/SWCNT films by electrochemical reduction of $PtCl_6^{2-}$. The electrode was immersed in an aqueous solution of hexachloroplatinic acid, and reduction was carried out by applying a sequence of pulses at a potential of -350 mV (SCE). It is known that pulsing can give rise to uniform particle size and a good distribution. By varying the duration of the pulse, particles of different diameters were obtained.

Guo et al. [171] developed an all electrochemical process to oxidize the CNT surfaces and disperse Pt particles. The CNT paste electrode was subject to a potential cycling from 1.8 to -0.4 V for 10 min in order to produce oxide functional groups at the defect sites located at the ends and/or the sidewalls of the carbon nanotubes. Then octahedral complexes of Pt(IV) were formed on the SWNT surface from K_2PtCl_4 aqueous solutions by CV; finally the surface complexes on the SWNTs were converted to platinum nanoparticles through cycling from 1.6 to -0.25 V in 0.1 M H_2SO_4 solutions.

Variations appear in the type of potential function utilized: potentiostatic [172], repetitive potentiostatic steps [173], cyclic potential and single and multiple pulse galvanostatic polarizations [174]. Kim and Park [175] used a step-potential plating method and investigated the influence of applied potential and interval time in the characteristic of the deposits and the particle sizes.

Nagle and Rohan [176] used electrochemical techniques and solution phase deposition to obtain aligned arrays of carbon nanotubes in a platinum matrix composite. The catalyst material was electrodeposited in an oriented manner on the nanoscale using anodized aluminum oxide (AAO) templates.

Tsai et al. [177] electrodeposited platinum and platinum-ruthenium nanoparticles on dense carbon nanotubes directly grown on carbon cloths in ethylene glycol (EG) containing H_2SO_4 aqueous solutions. It was found that EG enhanced the dechlorination of Pt and Ru precursor salts and led to the formation of nanoparticles. The authors compared the Pt-Ru/CNT catalysts obtained by electrodeposition with a commercial catalyst from Johnson Matthey Co. and reported mass activities of 547.2 A g^{-1} and 249.1 A g^{-1} respectively.

Addition of EG during the electrodeposition process had a distinctly positive effect on dispersion of the Pt or Pt-Ru catalysts and reduction of their particle sizes. That is, the addition of EG not only had a negligible impact on the ionic conductivity of the deposition electrolyte but also favorably reduced the particle sizes of Pt and Pt-Ru catalysts and prevented them from serious agglomerations.

3.2.3. Polyol Methods

The polyol process has been used for preparing colloidal metal particles. In this process a polyol, most commonly ethylene glycol, solution containing the metal precursor salts is refluxed in a range of 393 to 443 K, where the polyol decomposes homogeneously to release the reduction of the metal ion precursor. A support material may be optionally present to capture the depositing metal particles [182]. Although conventional conductive heating, with its inherent heterogeneous temperature distribution, was used to thermally activate the polyol in many cases [183], microwave heating has been increasingly used as an attractive alternative for nanosized materials synthesis. Microwave radiations can increase the reaction kinetics by 1-2 orders of magnitude, and promote the formation of uniform metal nanoparticles. The advantage of the microwave irradiation is that it transfers heat to the substance uniformly through the microwave-transparent reaction container (i.e. glass or plastic), leading to a more homogeneous nucleation and shorter crystallization time compared to conventional heating during which unavoidable temperature gradients occur, in particular when large volumes of solutions are used. This gradient of temperature may adversely affect the particle size distribution and yield [184]. Works that have used the polyol method for synthesis of Pt, Pt-Ru, and other catalysts supported on CNTs are listed in Table 3.

Table 2. Preparation of electrocatalysts by electrodeposition

Catalyst	Method	Substrate	Particle size	Ref.
Pt	Electrodeposition by potentiostatic pulses	SWCNTs	20 nm	[172]
Pt	Electro-reduction of the Pt(IV) complex formed on the SWNTs surface	SWCNTs	4-6 nm	[173]
Pt	Electrodeposition by potentiostatic method	MWCNTs	60-80 nm	[174]
Pt	Co-electrodeposition of Pt and CNTs	MWNTs		[178]
Pt	Multiple potential-step method	well-aligned CNTs arrays	30-70 nm	[175]
Pt	Single and multiple pulse galvanostatic polarizations	MWCNTs		[176]
Pt, Pt-Ru	Potentiostatic electrodeposition in ethylene glycol solution	CNTs	4.5-9.5 nm (Pt) 4.8-5.2 nm (Pt-Ru)	[179]
Pt-Ru	Potentiostatic electrodeposition	MWCNTs	60-80 nm	[178]
Pt-MoO_x	Electrochemical codeposition	CNTs/graphite	50 nm	[179]
Pt-Ru	Step potential deposition	CNTs	4.8 nm	[177]
Au	Two-step electrochemical process.	(PANI) grafted MWCNTs	8-10 nm	[180]
Ultrathin Pt films	Potentiostatic electrodeposition	Aligned array of CNTs		[181]

Guha et al. [183] made use of ethylene glycol both as the reducing medium as well as surfactant for the synthesis of platinum colloidal particles. The procedure involved first the sonication of o.5 g of the functionalized carbons in 50 ml EG. An aqueous solution of chloroplatinic acid was then added to the carbon slurry very slowly over the course of 4 h. The solution mixture was kept stirring and NaOH in EG solution was added in sufficient quantities required to raise the mixture pH to about 11-12. The reaction mixture was then transferred to a three–necked flask and refluxed for 3 h at 140 ºC under an argon blanket. The flask was allowed to cool and the supported catalyst filtered and washed and allowed to dry for 24 h. Agglomeration was minimal and the deposited platinum particles were found to lie in the size range of 2-5 nm. This homogeneous distribution and small particle size has been attributed to effect of the glycol which is denser compared to aqueous media. Higher viscosity of the EG media ensures that once the colloidal platinum particles have been precipitated and anchored on the carbon surface, diffusion to other sites on carbon is inhibited. Such diffusion is usually the rate limiting step in supported catalyst synthesis, hence its retardation would lead to the platinum particle nuclei growing at a uniform rate and as a result largely eliminating agglomeration. Ethylene glycol also acts as mild reducing agent, itself getting oxidized to acetaldehydes and subsequently carbon dioxide. The mild reducing agent favors smaller sized particles.

Chen and coworkers [184] describe a typical process for the synthesis of Pt/CNTs using microwave assisted polyol method. 1.0 ml of an aqueous solution of $H_2PtCl_6{\cdot}6H_2O$ was mixed with 25 ml of EG and KOH. After that, 0.040 g of CNTs were uniformly dispersed in the mixed solution by ultrasound. The beaker was placed in the center of a household microwave oven and heated for 50 s under microwave powder of 700W. The resulting suspension was filtered and the residue was washed with acetone. The solid product recovered as such was dried at 393 K over night in a vacuum oven. The authors concluded that the rapid and uniform microwave irradiation heating provided a more homogeneous environment for the quick nucleation and growth of Pt particles. The function groups of the surface oxidized CNTs were suitable for heterogeneous nucleation and interrupted particles agglomeration. Therefore, the microwave synthesized Pt nanoparticles were very uniform in size, and well dispersed on CNTs surface.

Jeng et al. [185,186] developed a modified polyol method to prepare CNTs supported Pt-Ru catalysts. As an example, a known amount of acid-oxidized CNTs was added to 50 ml EG aqueous solution. The above mixture was ultrasound sonicated, followed by high-speed stirring for 30 min, to form a homogeneous paste so that EG was able to completely cover the carbon nanotube surfaces. Then, $H_2PtCl_6{\cdot}6H_2O$ and $RuCl_3$ as Pt-Ru electrocatalyst precursors were dissolved in 10 ml EG, together with addition of 1 ml of 1 M $NaHSO_3$ aqueous solution as a modifier, forming complexed salts or ions. After that, the catalyst precursor salt/EG/$NaHSO_3$ solution was slowly added to the prepared CNTs/EG paste and the pH was adjusted to about 4. The solution was subjected to high-speed stirring for the complexed metal salts or ions to adhere to the surfaces of carbon nanotubes. In the meantime, the resultant mixture was heated at 130 °C using a microwave heater for 60 min to form Pt-Ru/CNTs with the pH monitored. The reaction solution was refluxed during reaction. The authors compared the electrocatalytic activity on methanol electro-oxidation of a commercial Pt-Ru/C, and the prepared Pt-Ru/CNT, and found a better performance for Pt-Ru/CNTs than for the commercial Pt-Ru/C electrocatalyst obtained from Johnson-Matthey. This can be attributed to several factors, including (i) the prepared Pt-Ru/CNTs electrocatalyst possesses perfect nanoparticle sizes and ideal compositions that enhance its electrocatalytic activity, (ii) the functional groups formed on the CNTs surface result in high hydrophilicity that creates better electrochemical reaction environments on the electrode, and (iii) the high electronic conductivity of the CNTs lowers the resistance in methanol electro-oxidation.

Other modifications are found for example in the work by Liu et al. [187] that synthesized stable thiol-encapsulated Pt and Pt-Ru colloids by the polyol method and then incorporated the carbon support by combining a toluene dispersion of Pt or Pt-Ru colloid with a suspension of SWCNTs.

The effects of pH on the size and uniformity of Pt particles and their electrocatalytic activity were investigated by Li and coworkers [188]. The preparation of Pt/CNTs was carried out in ethylene glycol solutions of H_2PtCl_6 precursor salts with different pH by microwave heating. TEM images of Pt/CNTs showed that Pt particles size supported on CNTs decreases with the increase of EG solution pH. The average sizes of Pt particles ranged from 5.8 to 2.7 nm, respectively, for pH between 3.6 and 9.2. In the case of pH 3.6 and 5.8, Pt nanoparticles are apparently agglomerated and not well dispersed on CNTs, exhibiting size ranges from 2.0 to 9.0 nm. In the case of pH 7.4 and 9.2, Pt nanoparticles are hardly agglomerated and appear well dispersed on the surface of CNTs. The majority of the Pt particles vary from 2.5 to 4.5 nm for pH 7.4, and from 1.5 to 3.5 nm for pH 9.2. The effects of pH may be explained considering that ethylene glycol could be oxidized to acetate, that would act as a stabilizer for metal colloidal particles. But acetic acid under the lower solution pH is a poor stabilizer. Therefore, when the pH of ethylene glycol solution increases, Pt particles become smaller and more uniform, and are better dispersed on CNTs surface. So the size of Pt particles could be selected by adjusting the solution pH.

Further adjustment in the technique can be achieved by non-covalent functionalization of the CNTs. Wang et al. [189] assembled platinum nanoparticles (NPs) on non-covalently poly-(diallyldimethylammonium chloride) (PDDA) functionalized MWCNTs via microwave-assisted polyol reduction and seed-mediated growth methods. The non-covalent functionalization by PDDA not only leads to a high density and homogeneous distribution of surface functional groups but also preserves the intrinsic properties of the CNTs without any chemical oxidation treatment. Pt NPs were uniformly deposited on MWCNTs support with Pt loading of 10-93 wt.%. The electrochemical characterizations show that the catalytic activity of Pt/MWCNTs depends strongly on the loading of Pt NPs on MWCNTs and the catalytic activity of Pt/MWCNTs for methanol electro-oxidation, CO electro-oxidation, and oxygen electro-reduction increases with the Pt loading. The authors claimed that the electrocatalytic activities of Pt/MWCNTs catalysts are fundamentally correlated to the interconnectivity of Pt NPs on CNTs. The magnitude of the interconnectivity of Pt NPs is a critical factor influencing their electrocatalytic activity, and the interconnected Pt NPs are more active than the isolated Pt NPs. The high electrocatalytic activity of highly interconnected Pt NPs is considered to be related to the increased active intergrain boundaries. On the other hand, the interconnected Pt NPs would significantly weaken their chemical adsorption with oxygen-containing species (i.e., CO_{ads} and OH_{ads}), resulting in the promoted electrocatalytic activity for CO and methanol oxidation. The increase of interconnectivity of Pt NPs also reduces the interface resistance among particles for electron charge transfer.

Tri-metallic catalysts have been synthesized with the polyol method such as Pt-Ru-Ir [190] or Pt-Ru-Mo [191]. For example, Liao et al. [192] used acetone as the solvent to form highly active Pt-Ru-Ir/CNTs catalysts. The procedure used for the synthesis of the catalysts involved the following steps: Specific ratios of hexachloroplatinic acid, dihydrogen hexachloroiridate, ruthenium(III) chloride hydrate, and sodium citrate were dissolved in a mixture of acetone and ethylene glycol and then stirred for 30 min to entirely dissolve the citrate. Carbon powder (XC-72R) or carbon nanotubes were then added to the mixture, followed by pH adjustment to >10 by the dropwise addition of a KOH/ethylene glycol solution with vigorous stirring.

The solution was then transferred into an autoclave and left to react at 160 °C for 8 h, followed by filtering, washing, and then drying in air at 90 °C. The particle size was found to be ~1 nm, and the catalyst showed excellent activity toward methanol oxidation. It was found that the addition of Ir significantly improved the dispersion of the metallic components, thus enhancing the activity of the catalyst.

Table 3. Catalysts for methanol oxidation prepared by polyol methods

Catalyst	Method	Substrate	Crystallite size	Ref
Pt	Polyol method	Fluorine tin oxide (FTO) and MWCNTs composites	4-8 nm	[193]
Pt	Synthesis of Pt NPs in alkaline EG solution	Nitrogen-doped CNTs	3 nm	[136]
Pt/MWCNTs + HPMo composite catalyst	EG reduction method	MWCNTs	1.8 nm	[190]
Pt-Ru	$NaHSO_3$ and calcium hydroxide aqueous solutions were employed as additives to EG forming a modified reducing agent.	CNTs	3.3 nm	[187]
Pt	Nanoparticles synthesized by the polyol processes and then dispersed on MWCNTs/CC sample	MWCNTs synthesized on carbon cloth modified by microwave digestion		[194]
Pt	Microwave heating ethylene glycol solutions of platinum salt with different pH	CNTs	5.8-2.7 nm	[191]
Pt	Microwave heated polyol process	MWCNTs	3.1 nm	[184]
Pt	EG reduction method	Sulfonated CNTs		[138]
Pt	Chemical reduction of H_2PtCl_6 in EG-water solution	MWCNTs	3.5 nm	[195]
Pt	Pt nanoparticles (reduced by EG), modified by the organic molecule triphenylphosphine	MWCNTs	1.9 ± 0.5 nm	[196]
Pt	Reduction with ethylene glycol	CNTs grown on carbon cloth (CNTs/CC)	3.77 nm	[197]
Pt	Microwave-assisted polyol reduction	Non-covalently poly-(diallyldimethylammonium chloride) functionalized MWCNTs	1.5 nm	[192]
Pt	Rapid intermittent microwave irradiation (IMI) technique	MWCNTs	3 nm	[129]
Pt, Pt-Ru	Spontaneous reduction of $PtCl_6^{2-}$ ion as a result of direct redox reactions with oxygen-containing functional groups	CNTs	10-20 nm	[142]

Table 3. (Continued).

Catalyst	Method	Substrate	Crystallite size	Ref
Pt	Intermittent microwave heating	HF treated MWCNTs	2.1-4 nm	[198]
Pt	Reduction in glycol	Activated carbon nanotubes	4.5-5.0 nm	[199]
Pt-Ru	Ethylene glycol reduction,	Bamboo-shaped carbon nanotubes (BCNTs)	2-3 nm	[200]
Pt_x-Ru_y x = 25-100 % y = 0-75%	Reduction with ethylene glycol	MWCNTs	3-5 nm	[201]
Catalyst	Method	Substrate	Crystallite size	Ref
Pt-Ru	Microwave-assisted polyol process.	MWCNTs	2-6 nm	[202]
Pt-Ru	Modified polyol method	CNTs	2-5 nm	[203]
Pt-Ru	Synthesis of colloidal particles using EG solution as solvent and reducing agent and posterior electrophoretic deposition technique	CNTs	2.08 nm	[204]
Pt-Ru	Ethylene glycol (EG) as reducing agent under hydrothermal condition	Carbon nanotube and tin oxide (CNTs/SnO_2)	2.5 nm	[205]
40 wt.% Pt-20 wt.% Ru	Modified polyol synthesis method	MWCNTs	3.7 nm	[206]
Pt-Ru	Modified polyol synthesis method	MWCNTs	3.5-4.0 nm	[187]
Pt-Ru	Microwave-assisted polyol process	MWCNTs with different lengths and diameters	1.8-3.1 nm	[128]
Pt-Ru	Microwave-polyol method	Nitrogen-doped CNTs	2.5-3.5 nm	[135]
Pt-Ru	Ethylene glycol reduction method.	SWCNTs, double-walled nanotubes (DWNTs), and MWCNTs	2-3 nm	[207]
Pt-Ru-Ir	Reduction with ethylene glycol in acetone solution with citrate as a complexing agent and stabilizer	CNTs	~1 nm	[190]
Pt-Ru-Mo	Ethylene glycol reduction method	MWCNTs	2.06 nm	[189]
Pt-Ru-PMo12	Microwave-assisted polyol process	MWCNTs	1-4 nm	[208]
Pt-PMo12	Microwave heated H_2PtCl_6 and $H_3PMo_{12}O_{40}$ (PMo12) ethylene glycol (EG) solution	MWCNTs	2.6-4.7 nm	[209]
Pt-SnO_2	Ethylene glycol reduction method.	Nitrogen doped carbon nanotubes (CNx)	1.5-4 nm	[134]
Pt-Zn	Microwave-assisted polyol method	MWCNTs	3-5 nm	[210]
Pt-Ru	Microwave-assisted polyol process.	1-aminopyrene functionalized MWCNTs	2 nm	[140]

3.2.4. Supercritical Fluid Synthesis

The use of supercritical fluids (SCF) for the synthesis and processing of nanomaterials has gained considerable interest in recent years. SCF exhibits an attractive combination of the solvent properties of a gas and a liquid. It can dissolve solutes such as a liquid and yet possess low viscosity, high diffusivity, and zero surface extension like a gas. Furthermore, the solvent strength of SCF can be varied by manipulating fluid temperature and pressure, thus allowing a degree of control and rapid separation of products, which is impossible using conventional solvents [211]. It provides a rapid, direct, and clean approach to preparing nanomaterials and nanocomposites. These special and unique features make SCF an attractive medium for delivering reactant molecules to areas with high aspect ratios, complicated surfaces, and poorly wettable substrates. Supercritical carbon dioxide ($scCO_2$) allows reactive components to penetrate inside the porous materials themselves, partitioning into the inner regions of the porous supports. Lin and coworkers [212,213] deposited platinum and platinum/ ruthenium nanoparticles on CNTs surface in $scCO_2$. A better electrocatalytic performance of bimetallic alloy Pt-Ru on the surface of CNTs from $scCO_2$ was observed for the methanol oxidation.

In the procedure given by An et al. [213] H_2PtCl_6 or the mixture of H_2PtCl_6 and $RuCl_3$ were used as precursors, and reduced to Pt or Pt-Ru by methanol and deposited on MWCNTs with the aid of supercritical carbon dioxide (scCO2), resulting in Pt/CNT or Pt-Ru/CNTs nanocomposites. In a typical experiment, a known amount of MWCNTs were ultrasonically dispersed in methanol aqueous solution (4:1, v/v) of metal precursors. Then the suspension was loaded in a stainless autoclave. After flushing with nitrogen gas, the autoclave was sealed and CO_2 was charged into the vessel up to the desired pressure. The vessel was moved to an oven of 120 ºC and maintained at this temperature for 2 h, and then cooled to room temperature naturally. After the vessel was depressurized, the dark precipitate was separated from the solution by centrifugation and washed repeatedly. The composites prepared at higher precursor:CNTs ratio were almost completely wrapped with the nanoparticle layer while the CNTs were decorated with separate nanoparticles for the composites prepared at lower precursor:CNTs ratio. In absence of CO_2, the produced nanoparticles were free of CNTs and aggregated into large particles. This is understandable because the CNTs were not modified by any functional groups, therefore, the surfaces of the MWCNTs were very inert and the metal precursor could not be absorbed on MWCNTs readily from methanol aqueous solution. When CO_2 was charged into the methanol aqueous solution of metal precursor, the volume of the liquid expanded due to the dissolution of CO_2, and the density, viscosity, surface tension and the solvent power of the methanol-rich phase decreased gradually. This was favorable to the metal precursor to be adsorbed on the surfaces of MWCNTs from the CO_2-expanded methanol solution. Under the reaction conditions (e.g., 120 ºC), the precursors adsorbed on MWCNTs converted to Pt-Ru nanoparticles, and the nanoparticles generated from the fluid phase preferentially deposited on MWCNTs due to the low density and viscosity of the fluid.

Taylor et al. [214] investigated how high-temperature liquid and supercritical fluid synthesis conditions could influence the Pt particle size and fuel cell performance of these nanocomposite catalyst materials. High-temperature methanol is a suitable medium for Pt particle size deposition, generating smaller particles (90 % measuring ~5 nm). The addition of a surfactant, such as sodium dodecyl sulfate, to the fluid medium improved the fuel cell performance and Pt utilization by 27 to 63 %, with the best catalyst material, Pt/SWCNTs, having a peak power density of 449 mW cm^{-2}. This synthesis technique was a very fast (<5 min) process with no reactor sidewall effects on Pt particle size or yield.

3.2.4. Ion-Exchange Methods

A platinum cation complex can exchange with H^+ of the acid functional groups on the surface of SWNTs. The sites of -COOH are proton and electron accessible. All the Pt nanoparticles loaded on these sites are, therefore, expected to be electrochemically active. So, the higher Pt utilization in Pt/SWNTs catalysts can be obtained. Wang et al. [5] prepared Pt/SWNTs catalysts by an ion-exchange method. The detailed preparation process was described as follows: The oxidized carbon nanotubes were suspended in ultra pure water. Then a solution of the platinum cation-complex salt (Pt $(NH_3)_2(NO_2)_2$), was added to the solution containing carbon nanotubes drop by drop. As a consequence of this immersion, the hydrogen ions of the functional group present on the surface of the carbon nanotubes underwent an ion exchange reaction with the platinum cation complex. The immersed carbon nanotubes were then filtrated to ensure that only the cations $Pt(NH_3)_2^{2+}$ bounded to the support were kept, the excess being washed away. The obtained catalyst was dried in a vacuum oven and then reduced to its metallic form at 200 °C in a flowing mixture of Ar–H_2. The authors claimed that these catalysts showed the higher catalytic activities for methanol oxidation when compared with a commercial catalyst. The results can be linked to the smaller fine platinum particles and their higher Pt utilization of the catalyst. Another version of this technique was developed by Waje and coworkers [215] that functionalized the surface of CNTs with sulfonic acid groups which acted as sites for the adsorption of Pt(IV) ions.

3.2.4. Other Methods

Sputtering is a procedure widely used to obtain model catalysts. Sun et al. [133] used DC sputtering to prepare Pt nanoparticles on CN_xNT/Si. Furthermore, Wang et al. [216] deposited platinum and ruthenium on CN_xNTs-carbon cloth composite electrodes by physical vapour deposition. γ Radiolysis is another method, which has been used to produce nanoparticles. The main advantage of this method is the controlled reduction of metal ions without any external reducing agents. Samant et al. [217] used γ-radiolysis assisted preparation of a composite of functionalized multiwalled carbon nanotubes (f-MWCNTs) with Pt (Pt-f-MWCNTs). γ-Radiolysis of water leads to the formation of a hydrated electron, hydroxyl radicals and an H-atom (H_2O + γ-radiation → e^-_{aq} + $H^{\cdot}$ + $OH^{\cdot}$). $PtCl_6^{2-}$ ions are preferentially reduced to Pt^0 by hydrated electrons, ($E^0_{(H2O/e-\ aq.)}$ = -2.87 V). Several miscellaneous methods are summarized in Table 4.

There are several interpretations about the influence of CNT properties on the supported Pt/CNT catalysts characteristics. Carmo et al. [223] compared the particle size obtained on CNTs and Vulcan XC-72 and concluded that Pt particle size supported on Vulcan XC-72 was significantly higher. The authors suggest that the nanotube morphological structure favors the dispersion of the electrocatalysts on the support by preventing the coalescence of the growing nuclei. The large differences in particle size cannot be explained by the presence of the other metals in the nanotubes. Similar differences were observed for Pt-Ru. The crystallite size of Pt-Ru supported on the nanotubes was between 4.06 to 4.62 nm, while that of the Pt-Ru on Vulcan XC-72 was much higher (6.39 nm). It has been suggested that Ru promotes the dispersion of the electrocatalyst; the unalloyed amorphous material that may reside on or near the surface of the Pt-Ru alloy particles may help prevent sintering during the deposition or during the thermal reduction processes.

Some work has been done on the influence of the carbon substrate on the Pt nanoparticles (Pt-NPs) morphology. Chen and coworkers [163] prepared Pt/XC-72, Pt/MWNTs and Pt/SWNTs catalysts and observed that the type of carbon support material affects significantly the morphology of Pt-NPs. TEM images show that spherical Pt-NPs are uniformly dispersed on XC-72 while the Pt-NPs in Pt/MWNTs catalyst are also spherical accompanied by some slight agglomeration This slight discrepancy may be correlated with the higher surface area of XC-72, which facilitates better metal dispersion. On the other hand the nanoparticles of Pt/SWNTs catalyst present "needle" shape and some Pt-NPs are aggregated. This particular needle-shape morphology of Pt-NPs likely originated from the unique nature of SWNTs with very large length/ diameter ratio. Alternative bamboo like structures of CNTs [202] presents a large amount of pentagon defects in the walls. It has been suggested that these pentagon defects have a strong interaction with metal particles. The facile deposition of high loading and high dispersion of Pt-Ru nanoparticles on the BCNTs suggests a strong interaction of the bimetallic nanoparticles with the support.

With respect to fuel cell applications, it is not only important how much the surface heterogeneities can influence the dispersion of Pt particles on the CNTs support surface, but how their electrochemical properties will be affected by aggregate size and surface morphology [158]. Some particular aspects of the influence of CNTs in methanol electro-oxidation have been discussed by different authors. Probably the bifunctional effect between Pt nanoparticles and functionalized CNTs defects is the most important.

Understanding the kinetics of methanol oxidation on SWCNTs and comparing the results with pure platinum is necessary to prove the electrocatalytic activity of CNTs toward methanol oxidation. Girishkumar et al. [170] compared SWCNTs/Pt electrode with unsupported platinum and found that the SWCNTs/Pt electrode was more active.

Table 4. Catalysts for methanol oxidation prepared by miscellaneous methods

Catalyst	Method	Substrate	Crystallite size	Ref
Pt-Ru	Microwave-assisted imidazolium ionic liquid method	MWCNTs	2-5 nm	[218]
Pt-Ru	Decorated in supercritical carbon dioxide	CNTs	5-10 nm	[213]
Pt, Pt-Ru	Reduction with methanol in supercritical CO_2	MWCNTs	5 nm	[215]
Pt	Synthesized in supercritical carbon dioxide ($scCO_2$) using platinum(II) acetylacetonate as metal precursor	CNTs	5-10 nm	[214]
Pt	Supercritical fluid synthesis	SWCNTs functionalization with sodium dodecyl sulfate	2.8 nm	[216]
Pt	Ion exchange method.	SWCNTs		[126]
Pt	Nanoparticles prepared by a ligand-exchange process	Curled CNTs onto carbon black	~2.3 nm	[219]
Pt	Exchange with functional groups and reduction in H_2	CNTs grown directly on carbon paper	2-2.5 nm	[217]
Pt	Sputtering	Arrayed nitrogen-containing carbon nanotube	2 nm	[133]
Pt-Ru	Sputtering	CNTs grown directly on carbon cloth		[218]
Pt-Ru	Ru is spontaneously deposited on CNT, and the product is used as a substrate for Pt deposition.	MWCNTs		[220
Pt-Ru	γ- Irradiation	Functional polymer grafted MWNTs	2.5 ± 0.7 nm	[221]
Pt	γ-radiolysis of K_2PtCl_6	HOOC-functionalized MWCNTs		[218]
$Pt-SnO_x$		8-Hydroxyquinoline (HQ) functionalized MWCNTs	2.2 nm	[141]
Pd	Raspberry hollow Pd nanosphere	CNTs functionalized by polymer wrapping	32 nm	[222]

The onset potential for the methanol oxidation is at a much lower potential for the CNTs/Pt electrode (200 mV vs. SCE) than the unsupported Pt electrode (400 mV). The authors concluded that the potential shift along with the enhancement of the anodic current using CNTs electrodes arises from a catalytic role of CNTs in promoting the methanol oxidation process.

Pt nanoparticles are surrounded by the adjacent oxygen-containing groups such as carboxylic acid and hydroxyl groups [142]. A possible bifunctional effect between Pt nanoparticles and functional CNTs defects, which is similar with the well-known bifunctional mechanism between Pt and Ru, occurs during methanol electro-oxidation. In other words, Pt is responsible for catalyzing the dehydrogenation of methanol, while functional CNTs defects provide the oxygen-containing groups

Similar explanation is given by Samant et al. [218] when reporting the electro-oxidation of methanol oxidation on Pt-multiwalled carbon nanotube composites (Pt-f-MWCNTs), prepared by γ-radiolysis of K_2PtCl_6 in the presence of HOOC-functionalized multiwalled carbon nanotubes. They found that unlike the polycrystalline Pt electrode, voltammograms of Pt-f-MWCNTs in acid methanol solution did not show any CO desorption peak, and attributed this fact to the 'active oxy groups' on the oxidized MWCNTs, which perhaps facilitate the formation of formate as an intermediate over CO.

Poh et al. [224] modified CNTs surfaces by a simple method using citric acid and claimed that this method produces more functional groups such as carboxyl and hydroxide than the HNO_3/H_2SO_4 treatment, permitting to obtain more Pt nanoparticles with smaller average particle size. Amongst other causes, they attributed the high electrocatalytic activity of the Pt/CA-CNTs to the increase in the number of surface hydroxyl groups that might facilitate the removal of the CO intermediate that is adsorbed on Pt surface during the electro-oxidation of methanol.

Other aspects that enhance the electrocatalytic activity of Pt-Ru/CNT are related to the existence of the distinctive Pt crystallite phases, i.e., Pt(110), on the Pt-Ru particles supported on the MWCNTs is likely to be the reason of the enhanced activity of the methanol oxidation reaction [153].

With respect to the preference of SWNTs or MWCNTs as catalyst support, there are conflicting opinions. Chen et al. [163] have found that Pt/SWNTs catalyst exhibits better electro-catalytic activity than Pt/MWNTs and Pt/XC-72 catalyst and connect their results with a change of the electronic structure of Pt-NPs (i.e., ligand effect) due to strong Pt-SWNTs interaction which results in the weakening of the CO adsorption on the Pt-NPs, thereby reducing the over-potential of methanol oxidation. On the other hand, Gattia et al. [150] found a better performance of the Pt/MWNTs electrodes toward methanol oxidation reaction.

Li et al. [124] indicated that a suitable diameter and length of MWCNTs is crucial for the electrocatalytic activity of the Pt-Ru/MWCNTs. The MWCNTs sample with a smaller diameter and similar tube length suffers more structural damages during the oxidative pretreatment, and thus contains a higher amount of amorphous carbon impurities and more oxygen containing functional groups on their surface compared with those with larger diameter. The co-existence of amorphous carbon impurities on the support lowered the accessible surface area of the Pt-Ru nanoparticles, hampered the dispersion of the Pt-Ru nanoparticles and induced the formation of a low degree of Pt-Ru alloy. This result suggests that the length of MWCNTs support has little effect on the specific activity per unit surface area of the Pt-Ru/MWCNTs. It is also demonstrated that the Pt-Ru nanoparticles supported on the MWCNTs with large tube diameter tends to exhibits a higher specific activity per unit surface of methanol oxidation reaction. This may be mainly attributed to both high degree of Pt-Ru alloy on the MWCNTs samples with large tube diameter and high electronic conductivity of these MWCNTs samples. Other authors have remarked that CNTs with larger diameters are not only more favorable for dispersion of Pt nanoparticles with more uniform sizes, but also less vulnerable to deactivation and more tolerable towards CO poisoning during MOR [143].

CONCLUSION

Carbon nanotubes are promising catalyst supports for methanol oxidation in fuel cells. The different variants that can be achieved with their structural characteristics, the extended surface area, and the many alternatives that can be introduced by functionalization and formation of composites and their electronic properties may be helpful in the development of customized catalysts for fuel cells.

This chapter presented a summary of the treatments needed to improve the CNTs properties for their use as catalyst support, i.e., functionalization of the nanotubes by oxidative treatment or formation of composite materials. In addition, it has also compiled a variety of methods for preparing nanostructured Pt and Pt-based catalysts. We hope that this material will be useful to the reader, allowing them to select the best method in order to produce high-performance catalysts for the electro-oxidation of methanol in DMFCs.

ACKNOWLEDGMENTS

We are very grateful for the financial support from the Agencia Nacional de Promoción Científica y Tecnológica (ANPCYT, Argentina) and PGI Universidad Nacional del Sur.

REFERENCES

[1] Oberlin, M. Endo, T. Koyama, *J. Cryst. Growth* 32, 335 (1976).

[2] S. Iijima, *Nature 354*, 56 (1991).

[3] Source: ISI Web of Science, http://www. thomsonreuters.com.

[4] D. S. Bethune, C. H. Kiang, M. D. de Vries, G. Gorman, R. Savoy, J. Vazquez, R. Beyers, *Nature* 363, 605 (1993).

[5] T. Belin, F. Epron, *Mater. Sci. Eng. B* 119, 105 (2005).

[6] P. Eklund, P. Ajayan, R. Blackmon, A.J. Hart, J. Kong, et al. WTEC International Assessment of Research and Development of Carbon Nanotube Manufacturing and Applications. Technical Report. *World Technology Evaluation Center Inc.* (2007).

[7] L. Hussein, Y. J. Feng, N. Alonso-Vante, G. Urbana, M. Krüger, *Electrochim. Acta* 56, 7659 (2011).

[8] S. Kerzenmacher, J. Ducrée, R. Zengerle, F. von Stetten, *J. Power Sources* 182, 1 (2008).

[9] Bachtold, P. Hadley, T. Nakanishi, C. Dekker, *Science* 294, 1317 (2001).
[10] H. Ago, K. Petritsch, M. S. P. Shaffer, A. H. Windle, R. H. Friend, *Adv. Mater.* 11, 1281 (1999).
[11] S. Peng, Y. Wu, P. Zhu, V. Thavasi, S. G. Mhaisalkar, S. Ramakrishna, *J. Photochem. Photobiol. A* 223, 97 (2011).
[12] Y. T. Kim, K. Ohshima, K. Higashimine, T. Uruga, M. Takata, et al., *Angew. Chem. Int. Ed.* 45, 407 (2006).
[13] J. Wang, G. Yin, Y. Chen, R. Li, X. Sun, *Int. J. Hydrogen Energy* 34, 8270 (2009).
[14] D. He, S. Mu, M. Pan, *Carbon* 49, 82 (2011).
[15] L. Q. Hoa, M. C. Vestergaard, H. Yoshikawa, M. Saito, E. Tamiya, *Electrochem. Commun.* 13, 746 (2011).
[16] V. Selvaraj, M. Alagar, *Electrochem. Commun.* 9, 1145 (2007).
[17] G. Che, B. B. Lakshmi, E. R. Fisher, C. R. Martin, *Nature* 393, 346 (1998).
[18] Y. Kasumov, R. Deblock, M. Kociak, B. Reulet, H. Bouchiat, et al., *Science* 284, 1508 (1999).
[19] R. H. Baughman, C. Cui, A. A. Zakhidov, Z. Iqbal, J. N. Barisci, et al., *Science* 284, 1340 (1999).
[20] Niu, E. K. Sichel, R. Hoch, D. Moy, H. Tennet, *Appl. Phys. Lett.* 70, 1480 (1997).
[21] V. Khomenko, E. Frackowiak, F. Béguin, *Electrochim. Acta* 50, 2499 (2005).
[22] P. M. Ajayan, S. Iijima, *Nature* 361, 333 (1993).
[23] Y. H. Lee, Y. T. Jang, B. K. Ju, *Appl. Phys. Lett.* 86, 173103 (2005).
[24] X. L. Xie, Y. W. Mai, X. Ping, *Mater. Sci. Eng. Rep.* 49, 89 (2005).
[25] R. Andrews, M. C.Weisenberger, *Curr. Opin. Solid State Mater. Sci.* 8, 31 (2004).
[26] Qian, E. C. Dickey, R. Andrews, T. Rantell, *Appl. Phys. Lett.* 76, 2868 (2000).
[27] W. A. de Heer, A. Chatelain, D. Ugarte, *Science* 270, 1179 (1995).
[28] Bachtold, P. Hadley, T. Nakanishi, C. Dekker, *Science* 294, 1317 (2001).
[29] W. B. Choi, Y. W. Jin, H. Y. Kim, S. J. Lee, M. J. Yun, et al., *Appl. Phys. Lett.* 78, 1547 (2001).
[30] G. Che, B. B. Lakshmi, C. R. Martin, E. R. Fisher, *Langmuir* 15, 750 (1999).
[31] P. J. Britto, K. S. V. Santhanam, P. M. Ajayan, *Bioelectrochem. Bioenerg.* 41, 121 (1996).
[32] J. J. Davis, R. J. Coles, H. A. O. Hill, *J. Electroanal. Chem.* 440, 279 (1997).
[33] Y. T. Jang, S. I. Moona, J. H. Ahnb, Y. H. Lee, B. K. Ju, Sens. *Actuators B* 99, 118 (2004).
[34] R. Ionescu, E. H. Espinosa, E. Sotter, E. Llobet, X. Vilanova, et al., *Sens. Actuators B* 113, 36 (2006).
[35] U. Yogeswaran, S. Thiagarajan, S.-M. Chen, *Anal. Biochem.* 365, 122 (2007).
[36] B. H. Chen, H. C. Lin, T. Y. Huang, J. H. Wei, H. H. Wang, et al., *Appl. Phys. Lett.* 88, 93502 (2006).
[37] Y. Huang, X. Duan, Y. Cui, L. J. Lauhon, K. H. Kim, C. M. Lieber, *Science* 294, 1313 (2001).
[38] X. Zhang, J. Fu, Y. Zhang, L. Lei, *Sep. Purif. Technol.* 64, 116 (2008).
[39] Z.-Q. Liu, J. Maa, Y.-H. Cui, B.-P. Zhang, *Appl. Catal. B* 92, 301 (2009).
[40] J.Wang, X. Ma, G. Fang, M. Pan, X. Ye, S. Wang, *J. Hazard. Mater.* 186, 1985 (2011).
[41] J. J. Davis, R. J. Coles, H. A. O. Hill, *J. Electroanal. Chem.* 440, 279 (1997).
[42] M. Musameh, J. Wang, A. Merkoci, Y. Lin, *Electrochem.Commun.* 4, 743 (2002).

[43] Z. Wang, J. Liu, Q. Liang, Y. Wang, G. Lou, *Analyst* 127, 653 (2002).
[44] N. S. Lawrence, R. P. Deo, J. Wang, *Anal. Chim. Acta* 517, 131 (2004).
[45] Salimi, C. E. Banks, R. G. Compton, *Analyst* 129, 225 (2004).
[46] Y. Zhao, W. D. Zheng, H. Chen, Q. M. Lou, *Talanta* 58, 529 (2002).
[47] U. Vohrer, I. Kolaric, M.H. Haque, S. Roth, U. Detlaff-Weglikowska, *Carbon* 42, 1159 (2004).
[48] Antolini, *Appl. Catal. B* 88, 1 (2009).
[49] B. Rajesh, K. Ravindranathan Thampi, J.-M. Bonard, N. Xanthopoulos, et al., *J. Phys. Chem. B* 107, 2701 (2003).
[50] T. Maiyalagan, B. Viswanathan, U. V. Varadaraju, *Electrochem. Commun.* 7, 905 (2005).
[51] J. Chen, Y. Liu, W. A. Weimer, M. D. Halls, D. H. Waldeck, G. C.Walker, *J. Am. Chem. Soc.* 124, 9034 (2002).
[52] L. Zhao, L. Gao, *Colloid. Surface. A* 224, 127 (2003).
[53] Star, T.R. Han, J. Christophe, P. Gabriel, K. Bradley, G. Gruner, *Nano Lett.* 3, 1421 (2003).
[54] Szleifer, R. Yerushalmi-Rozen. *Polymer* 46, 7803 (2005).
[55] H. Wang, W. Zhou, D. L. Ho, K.I. Winey, J.E. Fischer, C.J. Glinka, et al. *Nano Lett.* 4, 1789 (2004).
[56] Nativ-Roth, R. Shvartzman-Cohen, C. Bounioux, M. Florent, D. Zhang, et al., *Macromolecules* 40, 3676 (2007).
[57] R. J. Chen, Y. Zhang, D. Wang, H. Dai, *J. Am. Chem. Soc.* 123, 3838 (2001).
[58] X. Lu, L. Zhang, X. Xu, N. Wang, Q. Zhang, *J. Phys. Chem. B 106*, 2136 (2002).
[59] Y. C. Chiang, W. H. Lin, Y. C. Chang, *Appl. Surf. Sci.* 257, 2401 (2011).
[60] S. Murugesan, K. Myers, V. R. Subramanian, *Appl. Catal. B* 103, 266 (2011).
[61] Yang, Z. Zhang, X. Men, X. Xu, X. Zhu, *J. Colloid Interface Sci.* 346, 241 (2010).
[62] Y. Xing, L. Li, C. C. Chusuei, R.V. Hull, *Langmuir* 21, 4185 (2005).
[63] Y. Liang, H. Zhang, B. Yi, Z. Zhang, Z. Tan, *Carbon* 43, 3144 (2005).
[64] C. Lu, H. Chiu, *Chem. Eng. J.* 139, 462 (2008).
[65] S. Haydar, C. Moreno-Castilla, M. A. Ferro-García, F. Carrasco-Marín, J. Rivera-Utrilla, et al., J. P. Joly, *Carbon* 38, 1297 (2000).
[66] D.B. Mawhinney, V. Naumenko, A. Kuznetsova, J. T. Yates, J. Liu, R. E. Smalley, *J. Am. Chem. Soc.* 122, 2383 (2000).
[67] P. C. Ma, J. K. Kim, B. Z. Tang, *Carbon* 44, 3232 (2006).
[68] S. C. Tsang, Y. K. Chen, P. J. F. Harris, M. L. H. Green, *Nature* 372 (1994) 159.
[69] Dujardin, T. W. Ebbesen, A. Krishnan, M. M. Treacy, *Adv. Mater.* 10, 611 (1998).
[70] P. M. Ajayan, T. Ichihashi, S. lijima, *Chem. Phys. Lett.* 202, 384 (1993).
[71] Esumi, M. Ishigama, I. Nakajima, K. S. Awada, H. Honda, *Carbon* 34, 279 (1996).
[72] D. Dutta, R. Dubey, J. Yadav, T. C. Shami, K. U. Bhasker Rao, *New Carbon Mater.* 26, 98 (2011).
[73] Y.-H. Li, S. Wang, Z. Luan, J. Ding, Cailu Xu, Dehai Wu, *Carbon* 41, 1057 (2003).
[74] N. Zhang, J. Xie, V. K. Varadan, *Smart Mater. Struct.* 11, 962 (2002).
[75] C. Y. Du, T. S. Zhao, Z. X. Liang, *J. Power Sources* 176, 9 (2008).
[76] W. Xia, Y. Wang, R. Bergstraßer, S. Kundu, M. Muhler, *Appl. Surf. Sci.* 254, 247 (2007).

[77] R. Cuervo, E. Asedegbega-Nieto, E. Díaz, S. Ordóñez, A. Vega, A. B. Dongil, I. Rodríguez-Ramos, *Carbon* 46, 2096 (2008).
[78] K. Cuentas-Gallegos, R. Martínez-Rosales, M. E. Rincón, G. A. Hirata, G. Orozco, *Opt. Mater.* 29, 126 (2006).
[79] C.-T. Hsieh, H. Teng, W.-Y. Chen, Y.-S. Cheng, *Carbon* 48, 4219 (2010).
[80] Agüí, P. Yáñez-Sedeño, J. M. Pingarrón, *Anal. Chim. Acta* 622, 11 (2008).
[81] P. Santhosh, K.M. Manesh, A. Gopalan, K.-P. Lee, Sens. *Actuators B* 125, 92 (2007).
[82] Z. Liu, Z. Shen, T. Zhu, S. Hou, L. Ying, Z. Shi, Z. Gu, *Langmuir* 16, 3569 (2000).
[83] R. Czerw, Z. Guo, P. M. Ajayan, Y.-P Sun, D. L. Carroll, *Nano Lett.* 1, 423 (2001).
[84] J. E. Riggs, Z. Guo, D. L. Carro, Y. P. Sun, *J. Am. Chem. Soc.* 122, 5879 (2000).
[85] L. Tang, Y. Zhu, L. Xu, X. Yang, C. Li, *Talanta* 73, 438 (2007).
[86] J. Chen, A. M. Rao, S. Lyuksyutov, M. E. Itkis, M. A. Hamon, et al., *J. Phys. Chem. B* 105, 2525 (2001).
[87] A. Hamon, H. Hui, P. Bhowmik, H. M. E. Itkis, R. C. Haddon, *Appl. Phys. A* 74, 333 (2002).
[88] K. Fu, W. Huang, Y. Lin, L. A. Riddle, D. L. Carroll, Y. P. Sun, *Nano Lett.* 1, 439 (2001).
[89] V. Khabashesku, W. E. Billups, J. L. Margrave, *Acc. Chem. Res.* 35, 1087 (2002).
[90] J. Chen, M. A. Hamon, H. Hu, Y. Chen, A. M. Rao, P. C. Eklund, R. C. Haddon, *Science* 282, 95 (1998).
[91] Q. Chen, L. Dai, M. Gao, S. H., A. Mau, *J. Phys. Chem. B* 105, 618 (2001).
[92] S. Bandow, A. M. Rao, K. A. Williams, A. Thess, R. E. Smalley, P. C. Eklund, *J. Phys. Chem. B* 101, 8839 (1997).
[93] G. Sahoo, S. Rana, J. W. Cho, L. Li, S. H. Chan, *Progress Polym. Sci.* 35, 837 (2010).
[94] Hirsh, *Angew. Chem. Int. Ed.* 41, 1853 (2002).
[95] T. Ogoshi, T. A. Yamagishi, Y.Nakamoto, *Chem. Commun.* 4776 (2007).
[96] Hiura, T. W. Ebbesen, K. Tanigaki, *Adv. Mater.* 7, 275 (1995).
[97] M. J. O'Connell, P. Boul, L. M. Ericson, C. Huffman, Y. Wang, et al., *Chem. Phys. Lett.* 342, 265 (2001).
[98] Xu and B. Z. Tang, *Polym. Mater. Sci. Eng.* 80, 408 (1999).
[99] S. A. Curran, P. M. Ajayan, W. J. Blau, D. L. Carroll, J. N. Coleman, et al., *Adv. Mater.* 10, 1091 (1998).
[100] B. Dalton, C. Stephan, C. S. J. N. Coleman, B. McCarthy, P. M. Ajayan, et al., *J. Phys. Chem. B* 104, 10012 (2000).
[101] Y. Kang, T. A. Taton, *J. Am. Chem. Soc.* 125, 5650 (2003).
[102] E. Antolini, *Appl. Catal. B* 100, 413 (2010).
[103] B. Qu, Y.-T. Xu, S. Lin, Y.-F. Zheng, L.-Z. Dai, *Synth. Met.* 160, 732 (2010).
[104] Zhao, J. Yang, L. Li, H. Li, J. Wang, Y. Zhang, *Int. J. Hydrogen Energy* 34, 3908 (2009).
[105] N. G. Sahoo, Y. C. Jung, H. H. So, J. W. Cho, *Synth. Met.* 157, 179 (2007).
[106] T.-M. Wu, H.-L. Chang, Y.-W. Lin, *Compos. Sci. Tech.* 69, 639 (2009).
[107] E. N. Konyushenko, J. Stejskal, M. Trchová, J. Hradil, J. Kovářová, et al., *Polymer* 47, 5715 (2006).
[108] Y. Qiao, C. M. Li, S.-J. Bao, Q.-L. Bao, *J. Power Sources* 170 79 (2007).
[109] H. Guo, H. Zhu, H. Lin, J. Zhang, *Mater. Lett.* 62, 3919 (2008).

[110] S. C. Canobre, D. A.L. Almeida, C. P. Fonseca, S. Neves, Electrochim. Acta 54, 6383 (2009).
[111] C. Jiang, X. Lin, *J. Power Sources* 164, 49 (2007).
[112] C. Jiang, H. Chen, C. Yu, S. Zhang, B. Liu, J. Kong, *Electrochim. Acta* 54, 1134 (2009).
[113] R. Reddy, B. C. Sin, K. S. Ryu, J.-C. Kim, H. Chung, Y. Lee, *Synth. Met.* 159, 595 (2009).
[114] Y. Zhao, X. Yang, J. Tian, F. Wang, L. Zhan, *J. Power Sources* 195, 4634 (2010).
[115] C. Peng, J. Jin, G. Z. Chen, *Electrochim. Acta* 53, 525 (2007).
[116] M. Wu, G. A. Snook, V. Gupta, M. Shaffer, D. J. Fray, G. Z. Chen, *J. Mater. Chem.* 15, 2297 (2005).
[117] Z. Wang, Z.-Z. Zhu, J. Shi, H.-L. Li, *Appl. Surf. Sci.* 253, 8811 (2007).
[118] Z.-Z. Zhu, Z. Wang, H.-L. Li, *Appl. Surf. Sci.* 254, 2834 (2008).
[119] M. A. Bavio, T. Kessler, A. M. Castro Luna, *Int. J. Hydrogen Energy* 35, 5930 (2010).
[120] C. Jiang, H. Chen, C. Yu, S. Zhang, B. Liu, J. Kong, *Electrochim. Acta* 54, 1134 (2009).
[121] S. Bhandari, M. Deepa, A. K. Srivastava, C. Lal, R. Kant, Macromol. Rapid Commun.**29,** 1959 (2008).
[122] Zhang, X. Li, W. Guo, T. Hreid, J. Hou, et al., *Electrochim. Acta* 56, 3147 (2011).
[123] Lota, V. Khomenko, E. Frackowiak, *J. Phys. Chem. Sol.* 65, 295 (2004).
[124] J. Li, Y. Liang, Q. Liao, X. Zhu, X. Tian, *Electrochim. Acta* 54, 1277 (2009).
[125] K. Lee J. Zhang, H. Wang, D. P. Wilkinson, *J. Appl. Electrochem.* 36, 507 (2006).
[126] J. J. Wang, G. P. Yin, J. Zhang, Z. B. Wang, Y. Z. Gao, *Electrochim. Acta* 52, 7042 (2007).
[127] Z. Q. Tian, S. P. Jiang, Y. M. Liang, P. K. Shen, *J. Phys. Chem. B* 110, 5343 (2006).
[128] Halder, S. Sharma, M. S. Hegde, N. Ravishankar, *J. Phys. Chem. C* 113, 1466 (2009).
[129] R. Chetty, S. Kundu, W. Xia, M. Bron, W. Schuhmann, et al., *Electrochim. Acta* 54, 4208 (2009).
[130] T. Maiyalagan, *Appl. Catal. B* 80, 286 (2008).
[131] C.-L. Sun, L.-C. Chen, M.-C. Su, L.-S. Hong, O. Chyan, et al., *Chem. Mater.* 17, 3749 (2005).
[132] Y. Chen, J. Wang, X. Meng, Y. Zhong, R. Li, et al., *Int. J. Hydrogen Energy* 36, 11085 (2011).
[133] S. Jiang, L. Zhu, Y. Ma, X. Wang, J. Liu, et al., *J. Power Sources* 195, 7578 (2010).
[134] H.-Y. Du, C.-H. Wang, H.-C. Hsu, S.-T. Chang, U.-S. Chen, et al., *Diamond Relat. Mater.* 17, 535 (2008).
[135] R. Ahmadi, M. K. Amini, *Int. J. Hydrogen Energy* 36, 7275 (2011).
[136] H. J. Wang, H. Yu, F. Peng, P. Lv, *Electrochem. Commun.* 8, 499 (2006).
[137] Z.-P. Sun, X.-G. Zhang, R.-L. Liu, Y.-Y. Liang, H.-L. Li, *J. Power Sources* 185, 801 (2008).
[138] S. Wang, X. Wang, S. P. Jiang, *Langmuir* 24, 10505 (2008).
[139] C. Hu, Y. Cao, L. Yang, Z. Bai, Y. Guo, et al., *Appl. Surf. Sci.* 257, 7968 (2011).
[140] J. Chen, M. Wang, B. Liu, Z. Fan, K. Cui, Y. Kuang, *J. Phys. Chem. B* 110, 11775 (2006).
[141] A.-Y. Lo, N. Yu, S.-J. Huang, C.-T. Hung, S.-H. et al., *Diamond Relat. Mater.* 20, 343 (2011).

[142] G. Che, B. B. Lakshmi, C. R. Martin, E. R. Fisher, *Langmuir* 15, 750 (1999).
[143] M.-Y. Lo, I.-H. Liao, C.-C. Huang, *Int. J. Hydrogen Energy* 32, 731 (2007).
[144] Kosaka, S. Kuroshima, K. Kobayashi, S. Sekino, T. Ichihashi, et al., *J. Phys. Chem. C* 113, 8660 (2009).
[145] L. Ocampo, M. Miranda-Hernández, J. Morgado, J. A. Montoya, P. J. Sebastian, *J. Power Sources* 160, 915 (2006).
[146] V. Bambagioni, C. Bianchini, A. Marchionni, J. Filippi , F. Vizza, et al., *J. Power Sources* 190, 241 (2009).
[147] J. Cao, C. Du, S. C. Wang, P. Mercier, X. Zhang, et al., *Electrochem. Commun.* 9, 735 (2007).
[148] J. L. Figueiredo, M. F. R. Pereira, P. Serp, P. Kalck, P. V. Samant, J. B. Fernandes, *Carbon* 44, 2516 (2006).
[149] J.-E. Huang, D.-J. Guo, Y.-G. Yao, H.-L. Li, *J. Electroanal. Chem.* 577, (2005) 93.
[150] D. M. Gattia, M. V. Antisari, L. Giorgi, R. Marazzi, E. Piscopiello, et al., *J. Power Sources* 194, 243 (2009).
[151] J. Prabhuram, T. S. Zhao, Z. X. Liang, R. Chen, *Electrochim. Acta* 52, 2649 (2007).
[152] Y. Chen, G. Zhang, J. Ma, Y. Zhou, Y. Tang, T. Lu, Int. *J. Hydrogen Energy 35*, 10109 (2010).
[153] Y. Wang, X. Wang, C. M. Li, *Appl. Catal. B* 99, 229 (2010).
[154] S. Ghosh, C. R. Raj, *J. Phys. Chem. C* 114, 10843 (2010).
[155] Y. Jang, S. H. Lee, K. C. Park, W. Wongwiriyapan, C. Kim, et al., *Electrochem. Commun.* 11, 1472 (2009).
[156] C.-L. Lee, Y.-C. Ju, P.-T. Chou, Y-C. Huang, L.-C. Kuo, J.-C. Oung, *Electrochem. Commun.* 7, (2005) 453.
[157] S. W. Lee, S. C. W. Sheng, N. Yabuuchi, Y.-T. Kim, T. Mitani, et al., *J. Am. Chem. Soc.* 131, 15669 (2009).
[158] S. Saha, R. Li, X. Sun, *J. Power Sources* 177, 314 (2008).
[159] H.-J. Ahn, W. J. Moon, T.-Y. Seong, D. Wang, *Electrochem. Commun.* 11, 635 (2009).
[160] J. Prabhuram, T. S. Zhao, Z. K. Tang, R. Chen, Z. X. Liang, *J. Phys. Chem. B* 110, 5245 (2006).
[161] C. Kim, Y. J. Kim, Y. A. Kim, T. Yanagisawa, K. C. Park, et al., *J. Appl. Phys.* 96, 5903 (2004).
[162] D.-J. Guo, Z.-H. Jing, *J. Power Sources* 195, 3802 (2010).
[163] Z. Cui, C. Liu, J. Liao, W. Xing, *Electrochim. Acta* 53, 7807 (2008).
[164] Jha, R. I. Jafri, N. Rajalakshmi, S. Ramaprabhu, *Int. J. Hydrogen Energy* 36, 7284 (2011).
[165] N. Jha, A. L. M. Reddy, M. M. Shaijumon, N. Rajalakshmi, S. Ramaprabhu, *Int. J. Hydrogen Energy* 33, 427 (2008).
[166] Z. Sun, X. Wang, Z. Liu, H. Zhang, P. Yu, L. Mao, *Langmuir* 26, 12383 (2010).
[167] J.-F. Drillet, H. Bueba, U. Dettlaff-Weglikowskab, R. Dittmeyera, S. Roth, *J. Power Sources* 195, 8084 (2010).
[168] Y. Wang, Z. M. Sheng, H. Yang, S. P. Jiang, C. M. Li, *Int. J. Hydrogen Energy* 35, 10087 (2010).
[169] C. R. K. Rao, D. C. Trivedi, Coord. Chem. Rev. 249, 613 (2005).
[170] G. Girishkumar, K. Vinodgopal, P. V. Kamat, *J. Phys. Chem. B* 108, 19960 (2004).
[171] D.-J. Guo, H.-L. Li, *J. Electroanal. Chem.* 573, 197 (2004).

[172] Z. He, J. Chen, D. Liu, H. Tang, W. Deng, Y. Kuang, *Mater. Chem. Phys.* 85, 396 (2004).

[173] H. Tang, J. H. Chen, Z. P. Huang, D. Z. Wang, Z. F. Ren, et al., *Carbon* 42, 191 (2004).

[174] C. Paoletti, A. Cemmi, L. Giorgi, R. Giorgi, L. Pilloni, et al., *J. Power Sources* 183, 84 (2008).

[175] S. Kim, S.-J. Park, *Electrochim. Acta* 53, 4082 (2008).

[176] L. C. Nagle, J. F. Rohan, *J. Power Sources* 185, 411 (2008).

[177] M.-C. Tsai, T.-K. Yeh, C.-H. Tsai, *Electrochem. Commun.* 8, 1445 (2006).

[178] Z. He, J. Chen, D. Liu, H. Zhou, Y. Kuang, *Diamond Relat. Mater.* 13, 1764 (2004).

[179] Y.-P. Gan, H. Huang, W.-K. Zhang, *Trans. Nonferrous Met. Soc. China* 17, 214 (2007).

[180] Santhosh, A. Gopalan, K.-P. Lee, *J. Catal.* 238, 177 (2006).

[181] X. Su, J. Wu, B. J. Hinds, *Carbon* 49, 1145 (2011).

[182] W. Chen, J. Zhao, J. Y. Lee, Z. Liu, *Mater. Chem. Phys.* 91, 124 (2005).

[183] Guha, W. Lu, T. A. Zawodzinski Jr., D. A. Schiraldi, *Carbon* 45, 1506 (2007).

[184] M. Sakthivel, A. Schlange, U. Kunz, T. Turek, *J. Power Source*s 195, 7083 (2010).

[185] C.-C. Chien, K.-T. Jeng, *Mater. Chem. Phys.* 99, 80 (2006).

[186] K.-T. Jeng, C.-C. Chien, N.-Y. Hsu, S.-C. Yen, S.-Du Chiou, et al., *J. Power Sources* 160, 97 (2006).

[187] Z. Liu, X. Y. Ling, B. Guo, L. Hong, J. Y. Lee, *J. Power Sources* 167, 272 (2007).

[188] X. Li, W.-X. Chen, J. Zhao, W. Xing, Z.-D. Xu, *Carbon* 43, 2168 (2005).

[189] S. Wang, S. P. Jiang, T. J. White, J. Guo, and X. Wang, *J. Phys. Chem. C* 113, 18935 (2009).

[190] X. Guo, D.-J. Guo, J.-S. Wang, X.-P. Qiu, L.-Q. Chen, W.-T. Zhu, *J. Electroanal. Chem.* 638, 167 (2010).

[191] N. Kakati, J. Maiti, J. Y. Oh, Y. S. Yoon, *Appl. Surf. Sci.* 257, 8433 (2011).

[192] S. Liao, K.-A. Holmes, H. Tsaprailis, V. I. Birss, *J. Am. Chem. Soc.* 128, 3505 (2006).

[193] D.-J. Guo, Z.-H. Jing, *J. Colloid Interface Sci.* 359, 257 (2011).

[194] C.-C. Chen, C.-F. Chen, C.-M. Chen, F.-T. Chuang, *Electrochem. Commun.* 9, 159 (2007).

[195] Y. Li, W. Gao, L. Ci, C. Wang, P. M. Ajayan, *Carbon* 48, 1124 (2010).

[196] Y. Mu, H. Liang, J. Hu, L. Jiang, L. Wan, *J. Phys. Chem. B* 109, 22212 (2005).

[197] S.-L. Chen, C.-T. Lin, C.-C. Chieng, F.-G. Tseng, *J. Power Sources* 195, 1640 (2010).

[198] Y. Li, F. P. Hu, X. Wang, P. K. Shen, *Electrochem. Commun.* 10, 1101 (2008).

[199] J. J. Niu, J. N.Wang, *Electrochim. Acta* 53, 8058 (2008).

[200] L. Gan, R. Lv, H. Du, B. Li, F. Kang, *Electrochem. Commun.* 11, 355 (2009).

[201] C.-T. Hsieh, J.-Y. Lin, S.-Y. Yang, *Physica E* 41, 373 (2009).

[202] Z. Liu, J. Y. Lee, W. Chen, M. Han, L. M. Gan, *Langmuir* 20, 181 (2004).

[203] N.-Y. Hsu, C.-C. Chien, K.-T. Jeng, *Appl. Catal. B* 84, 196 (2008).

[204] Y.-K. Hsu, J.-L. Yang, Y.-G. Lin, S.-Y. Chen, L.-C. Chen, K.-H. Chen, *Diamond Relat. Mater.* 18, 557 (2009).

[205] N. Kakati, J. Maiti, S. H. Jee, S. H. Lee, Y. S. Yoon, *J. Alloys Compd.* 509, 5617 (2011).

[206] K.-T. Jeng, C.-C. Chien, N.-Y. Hsu, W.-M. Huang, S.-D. Chiou, S.-H. Lin, *J. Power Sources* 164, 33 (2007).

[207] W. Li, X. Wang, Z. Chen, M. Waje, Y. Yan, *J. Phys. Chem. B* 110, 15353 (2006).

[208] D. M. Han, Z. P. Guo, Z. W. Zhao, R. Zeng, Y. Z. Meng, et al., *J. Power Sources* 184, 361 (2008).

[209] Z. P. Guo , D. M. Han, D.Wexler, R. Zeng, H. K. Liu, *Electrochim. Acta* 53, 6410 (2008).

[210] C.-T. Hsieh, W.-M. Hung, W.-Y. Chen, J.-Y. Lin, *Int. J. Hydrogen Energy* 36, 2765 (2011).

[211] Y. Lin, X. Cui, C. H. Yen, C. M. Wai, *Langmuir* 21, 11474 (2005).

[212] Y. Lin, X. Cui, C. Yen and C. M. Wai, *J. Phys. Chem. B* 109, 14410 (2005).

[213] G. An, P. Yu, L. Mao, Z. Sun, Z. Liu, et al., *Carbon* 45, 536 (2007).

[214] D. Taylor, R. C. Sekol, J. M. Kizuka, S. D'Cunha, C. M. Comisar, *J. Catal.* 259, 5 (2008).

[215] M. M Waje, X. Wang, W. Li, Y. Yan, *Nanotechnology* 16, S395 (2005).

[216] C.-H. Wang, H.-Y. Du, Y.-T. Tsai, C.-P. Chen, C.-J. Huang, et al., *J. Power Sources* 171, 55 (2007).

[217] K. M. Samant, V. S. Joshi, G. Sharma, S. Kapoor, S. K. Haram, *Electrochim. Acta* 56, 2081 (2011).

[218] D.-J. Guo, *J. Power Sources* 195, 7234 (2010).

[219] W.-F. Chen, J.-P. Wang, C.-H. Hsu, J.-Y. Jhan, H. Teng, P.-L. Kuo, *J. Phys. Chem. C* 114, 6976 (2010).

[220] W.-C. Chang, M. T. Nguyen, *J. Power Sources* 196, 5811 (2011).

[221] D.-S. Yang, K.-S. Sim, H.-D. Kwen, S.-H. Choi, *J. Nanomater.* 8 (2011), doi:10.1155/2011/13472.

[222] Z. Liu, B. Zhao, C. Guo, Y. Sun, Y. Shi, et al., *J. Colloid Interface Sci.* 351, 233 (2010).

[223] M. Carmo, V. A. Paganin, J. M. Rosolen, E. R. Gonzalez, *J. Power Sources* 142, 169 (2005).

[224] C. K. Poh, S. H. Lim, H. Pan, J. Lin, J. Y. Lee, *J. Power Sources* 176, 70 (2008).

In: Applications of Carbon Nanotubes
Editor: Ajay Kumar Mishra
ISBN: 978-1-62081-918-0

Chapter 5

CARBON NANOTUBES APPLICATIONS IN SOLID-PHASE EXTRACTION

Antonio V. Herrera-Herrera, María Asensio-Ramos, Miguel Ángel González-Curbelo and Javier Hernández-Borges
Department of Analytical Chemistry, Nutrition and Food Science,
University of La Laguna (ULL), La Laguna, Tenerife, Canary Islands, Spain

ABSTRACT

Cylindrical carbon nanostructures known as carbon nanotubes (CNTs), have found a significant place in the Analytical Chemistry field due to their structure and special properties. In particular, their large surface area, their capability of interacting through non-covalent forces and their general stability have consolidated their use in the last decade as sorbents in solid-phase extraction (SPE). Still, the possibility of functionalization and thus, of changing their sorption selectivity, have extended their application in this area. This chapter expects to provide a general overview of the state of the art of the appliance of CNTs for the extraction of inorganic and organic analytes, including tables and general comments on the extraction conditions, method performance and attained results. Solid-phase microextraction (SPME) applications have also been considered for comparison purposes.

1. INTRODUCTION

More than 50 years ago, Roger Bacon discovered a new carbon fiber while he studied carbon on the proximity of its triple point [1]. He observed straight, hollow carbon tubes whose structure seemed to be formed by graphitic layers. Later, in the 1970s, Oberlin et al. [2] and Endo et al. [3, 4] observed again these tubes, generated by a gas phase process. They even talked about a rolled up single wall layer of graphene.

The formal discovery of carbon nanotubes (CNTs) occurred in 1991, when engineer Sumio Iijima employed a previously optimized arc discharge method for the synthesis of fullerenes in multigram amounts [5] and observed the formation of assemblies with a needle-like structure in the electrodes [6]. Transmission electron microscopy (TEM) revealed that these needles were formed by concentric tubes that contained between 2-50 graphene sheets, that is, multi-walled CNTs (MWCNTs). Later in 1993, Iijima and Ichihashi [7] and Bethune et al. [8] simultaneously studied that the addition of some metals, as Fe or Co, to one of the electrodes helped the formation of tubular structures with a single graphite layer, that is, single-walled-CNTs (SWCNTs). Nowadays it is possible to synthesize CNTs of different structures in high quantities with a relatively precise number of layers by either arc-discharge [9], laser ablation [10] and chemical vapor deposition [11], which are the methods most commonly employed. Even though, economically feasible large-scale production and purification techniques are still under development.

CNTs show unique physical, chemical, thermal and mechanical properties, which will be later described, and which make them suitable for a wide range of applications. In this sense, their most important application fields are surely technology/electronics and electrochemistry, in which they are used in the construction and improvement of electrochemical detectors. However, they are currently being intensively used in other fields like biomedicine, i.e. biomedical and tissue engineering, drug delivery, neuroengineering, gene therapy, and biosensor technology, among others. As a result, they are probably one of the most intensively investigated nanostructured materials [12].

The surface of CNTs can be easily modified in order to change/improve their properties. For this purpose, two approaches have been developed: covalent or non covalent functionalization. The first of them frequently includes an acidic and oxidative treatment with a strong acid (i.e. nitric or sulfuric acid) -this is also a way of reducing impurities- in which carboxylic groups are introduced in their structure followed by different substitution reactions in which the desired group is finally linked to the surface. However, the degree of covalent functionalization should not be too high since oxidation creates holes on their sidewalls that may significantly affect the nanotube properties [13]. The second, consists on non covalent modifications by, for example, aromatic compounds, surfactants or polymers, by means of π-π stacking or hydrophobic interactions [14, 15]. In this way, many of their desired properties can be preserved, while improving their solubility quite remarkably. Such modified materials may also have different uses depending of the final required characteristics.

As a result of the latest publications in the field, it is clear that CNTs have also found an important place in Analytical Chemistry; in particular, in their use as solid-phase extraction (SPE) and solid-phase microextraction (SPME) materials. The technique probably comprises nowadays the sample preparation procedure most commonly applied. As a result, this chapter reviews and discusses the different applications of CNTs as SPE and SPME sorbents of both organic and inorganic analytes, with the aim of paying particular attention to those works that have provided relevant and interesting real world applications in the analytical field.

2. Role of Carbon Nanotubes in Analytical Chemistry

Over the last two decades, CNTs have been widely studied since their properties make them suitable for a wide range of applications. In this sense, the impact of this new material in analytical sciences is really worthy.

When talking about the properties of CNTs, authors usually refer to the macroscopic nanotube material [16]. Regarding this, it is difficult to determine the properties of a single tube. Yet, efforts have been made in this sense, especially with more simple SWCNTs, to elucidate some of their properties. For MWCNTs, however, the measurement of physical properties is more difficult, but could be figured out from those of SWCNTs.

Structurally, a SWCNT can be described as a rolled up graphene sheet which is closed in its ends with fulleren-like hemispherical tips. The strength of the C-C bond is graphite-like in which hybridation of carbon is sp^2. However, folding of the sheet can be accomplished in diverse ways along different lattice vectors leading to armchair, zigzag and chiral tubes. Depending on the way of rolling up, the CNT will show different chirality and diameter, and thus, different properties.

SWCNTs diameters range from 0.3 to approximately 5 nm, while MWCNTs can have a layer spacing of 0.3-0.4 nm and diameters between 1.4 and 100 nm [17]. In both of them, lengths could be up to centimeters. Mechanical properties of CNTs depend on the nanotube growth method and processing, as well as on the amount of defects [17], but in general, it can be stated that CNTs are one of the strongest materials in nature, especially on their axial direction. They own a high tensile strength, high resilience and undergo plastic deformation [18].

CNTs possess a π system similar to that of a graphene sheet, but the formation of this system and its behavior depends on the degree of overlapping of the p_z orbitals. This degree of overlapping, and thus, the electrical properties, depends also on the way that the SWCNTs have been rolled up [16]. For this reason, they have found application as metals (armchair tubes) or semiconductors (chiral tubes). Concerning this aspect, their electrical properties such as conductivity, dielectric constant, impedance, refractive index, etc. are sensitive to the interaction with miscellaneous analytes, or even with complex biological molecules [19]. This is the base of their certainly extended use as gas sensors, which have been applied for the detection of polar (NH_3, CO, O_3) and non-polar (He, Ar, N_2, O_2, CO_2) gases [20]; or biosensors, in which CNTs are nanoelectrodes modified with biomolecules, such as proteins, antibody-specific proteins, enzimes or oligonucleotids [21]. Furthermore, CNTs also provide a large active surface to electrodes of small dimensions (nanoelectrodes), electrocatalytic properties and good electron transfer. In this respect, numerous applications describe their use as electrode materials or as modifiers of conventional working electrodes in analytical voltammetry [22]. Besides, they have found a place in anodic striping voltammetry, due to the conjunction of the already mentioned electrical properties together with the highly sorption capacity [23, 24]. On the other hand, it is worth mentioning their use for the construction and improvement of electrochemical detectors in microchip capillary electrophoresis (CE) [25, 26].

This excellent aforementioned adsorption capacity, and the fact that CNTs own greatly thermal and chemical stability (they remain stable up to 1200°C), suggested their use -both SWCNTs and MWCNTs- as stationary phases in gas chromatography (GC) [27-29]. Such stability prevents column bleeding and allows working at higher temperatures than with conventional columns. Normally, a thin layer is deposited on the surface of the chromatographic column by chemical vapor deposition (CVD) [30]. In this way, the structure and surface of CNTs provide superior peak shape. However, this field has not been fully explored and only a small number of applications have been carried out. Even less explored is the use of these materials as stationary phases in liquid chromatography (LC) [27, 30], and they have been normally employed in combination with a second packing material [31] or incorporated in a porous polymeric monolith [32]. Some attention has also been paid to the use of CNTs as pseudostationary phases in CE or capillary electrochromatography (CEC) [27, 30]. For this purpose, CNTs are introduced in the capillary by non-covalent functionalization with surfactants such as sodium dodecyl sulfate (SDS) [33] or adding covalently-functionalized CNTs as additives to the background electrolite (BGE), i.e. oxidized-CNTs, to improve resolution [34]. It is expected that a higher number of works will still come in this sense.

CNTs are also characterized by their high surface area and by possessing hollow cores, especially MWCNTs, which can accommodate foreign molecules. This fact, together with their strong binding capacity for hydrophobic molecules and their ability to establish π-π electrostatic interactions makes them suitable as sorbent materials in SPE and SPME. Furthermore, compared with other carbonaceous material, CNTs are relatively non-porous and more selective. The following section of this chapter will discuss in detail this specific analytical application.

3. CNTs in SPE

Nowadays, SPE is one of the preferred and more exploited extraction techniques in sample pretreatment. Introduced in the mid-1970s and being the first commercial cartridges introduced in 1978 [35], it has experimented in the last decades an important development in terms of format, automation and introduction of new sorbents. SPE is commonly developed by enclosing the stationary phase into small columns, cartridges or disks. Then, sorption of the target analytes and subsequent elution is carried out. In dispersive SPE (dSPE), the extraction is not carried out in a column or in a disk, but in the bulk solution. As one can deduce, its performance is easier, faster and less tedious.

SPME was first developed by Pawliszyn and coworkers in 1989 [36, 37]. This method consists in the use of a fused silica fiber covered by a polymeric sorbent and was initially introduced as a rapid, simple, reliable and flexible procedure with a great capacity of concentration for a wide variety of analytes.

Many of the traditional sorbents for SPE and SPME are limited in terms of selectivity and in many occasions, other sample matrix components can also affect the ability of the sorbent to retain the target analytes [38]. That is why numerous advances have been done in this field with the aim of getting a more selective and higher retention [39]. In this sense, CNTs have emerged as one of the new sorbent materials assayed for the extraction of both inorganic and organic analytes [40].

As already mentioned in the introduction and as it will be described in the following subsections, the possibility of covalently functionalizing the surface of CNTs under strong chemical conditions to introduce particularly hydroxyl, carbonyl and carboxylic groups modifies the selectivity and versatility of the material [41]. Alternatively, non-covalent functionalization which includes ionic interactions, hydrogen bonds, electrostatic forces, π-π stacking, van der Waals forces, etc. is also a source of selectivity. As it will be immediately demonstrated leant on a revision of the literature, this fact has given rise to an enormous field of application.

3.1. Inorganic Analytes

As it is widely known, the presence of inorganic species and the knowledge of their concentration are important in diverse areas. In particular, the direct determination of metal ions from complex matrices is difficult due to the low concentrations found and the presence of possible interferences.

The mechanism of interaction of CNTs with metal ions is believed to fall on the presence of functional groups in the surface of the material [42]. From a revision of the literature, it is clear that for metal extraction it is very common to oxidize CNTs either with concentrated HNO_3 [43-51], $KMnO_4$ [52, 53], H_2SO_4/HNO_3 (3:1 v/v) [54, 55] or even NaClO 70% (w/v) [56]. In this way, complexation is not necessary, contrary to what has been usually done with SPE with other sorbent materials. In fact, when CNTs are not modified, metals complexation is necessary for their later extraction by CNT-SPE [57-61].

In the adsorption of metal ions by CNTs, pH control is of basic importance due to the existence of the so-called point of zero charge [40, 42]. It describes the condition when the electrical charge density on a surface is zero, and it is usually determined in relation to the pH. Thus, at pH higher than the pH at the point of zero charge, the surface of functionalized CNTs (carboxyl and hydroxyl groups) becomes negatively charged and improves the ion-exchange capacity. This is advantageous for CNTs interaction with positive charged ions, because it permits the extraction only by controlling the pH. In general, when the pH increases, the adsorption of metal ions is stronger since the charge of the surface becomes more negative. On the contrary, when the pH is low, protons compete with metal cations for the same sites on the surface of CNTs. When oxidized CNTs are used, pH must be maintained normally in the range 4-8. As an example, Stafiej and coworkers [47] studied the sorption behavior of oxidized-MWCNTs towards some divalent metal ions by systematically changing water pH.

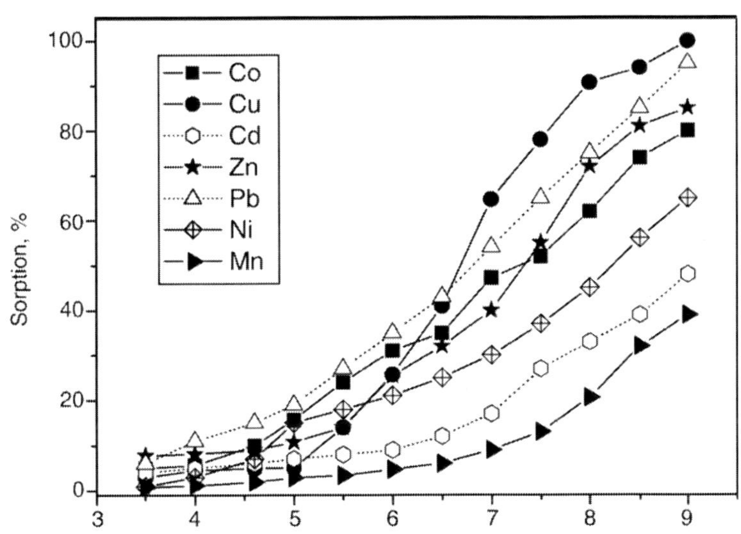

Figure 1. Sorption of metal ions on oxidized-MWCNTs as a function of pH. 50 mg of CNTs; metals concentration 10 μg mL−1, contact time 4 h. Reprinted from [47] with permission from Elsevier.

As can be seen in Figure 1, this affinity increases in the order Cu (II), Pb (II), Zn (II), Co (II), Ni (II), Cd (II) and Mn (II), which agrees with the tendency observed for the electronegativity of these ions and the first stability constant of the associated metal hydroxide.

On the other hand, when metals are being determined as complexes, the effect of the pH is different because it influences the stability of the metal-ligand specie (and also controls possible lateral reactions). As an example, the sorption of Rh (III) as a 1-(2-pyridylazo)-2-naphtol (PAN) complex on oxidized-MWCNTs of outer diameter (o.d.) 3-20 nm is maximum when the pH is around 3.2-4.7 [49]. Also, the recovery of Cu (II), Cd (II), Pb (II), Zn (II), Ni (II) and Co (II) as ammonium pyrrolidine dithiocarbamate (APDC) chelates on raw-MWCNTs of o.d. of 20-50 nm was found to decrease when pH was higher than 7 and was maximum at pH 2 [58].

As it is also well known, depending on the pH value, metal ions will be in a different form. For example, between pH 2 and 4 mercury exists as a mixture of Hg^{2+}, $Hg(OH)^+$ and $Hg(OH)_2$ [51]. However, between pH 4 and 7, $Hg(OH)_2$ is dominant; while at pH higher than 7, the negatively charged species $Hg(OH)_3^-$ and $Hg(OH)_4^{2-}$ are predominant. In the work by El-Sheikh et al. [51], mercury was extracted from natural waters in its colloidal $Hg(OH)_2$ form, using oxidized-MWCNTs at pH 7. For this purpose, 200 mg of sorbent were used and elution was carried out with 5 mL of 1 M HNO_3. Recovery was around 96% and limit of detection (LOD) was 0.0123 μg/L which permitted, after microwave-based digestion, the accurate determination of mercury in phosphate rock. Metals retention onto raw-MWCNTs at pH higher than 7 as their corresponding metal hydroxides has also been described. In this sense, a study of the appropriate dimensions of MWCNTs for the sorption of precipitated hydroxides was also carried out by El-Sheikh [62] for the extraction of Cu (II), Zn (II), Pb (II) and Cd (II) maintaining a pH of 9. It was found that long MWCNTs of 5-15 μm allowed a better preconcentration under these conditions than shorter (1-2 μm) MWCNTs, as well as MWCNTs of external diameter in the range 10-30 nm *versus* those in the ranges 10-20 nm, 20-40 nm, 40-60 and 60-100 nm. LODs achieved with the optimum MWCNTs (5-15 μm x 10-30 nm) were between 8.7 ng/L for Zn (II) and 70.9 ng/L for MnO_4^-.

Once oxidation of CNTs has been carried out, organic species can also be immobilized on their surface, which clearly extends the possibilities of these materials. Regarding the determination of inorganic analytes, this could be especially interesting because cations can interact with organic molecules immobilized on the surface in a more selective way as they do with untreated or oxidized CNTs. Important efforts have been done in this sense. For example, a cationic surfactant, such as cetyltrimethylammonium chloride (CTAC), was added to oxidized-MWCNTs leading to the formation of hemimicelle/admicelle aggregates on the CNTs [63]. This occurs through electric interactions between the cationic alkylchain of CTAC and oxygen-containing functional groups of the CNTs. This material was used for retention of As (V). After elution with 10 mL of 1 M HNO_3, it was later analyzed by hydride generation-atomic fluorescence spectrophotometry (HG-AFS). Recoveries were in the range 94-104%, which were attributed to the interaction between the negatively charged oxyanion $H_2AsO_4^-$ at pH 5-6 and CTAC. MWCNTs have also been functionalized with biomolecules, such as the aminoacids L-Cys [64], L-Tyr [65, 66] and L-Ala [67, 68], proteins [69] or even bacteria like *Pseudomonas aeruginosa* [70]. In this last case, the microorganism was immobilized onto MWCNTs for trace heavy metals extraction, *i.e.* Co (II), Cd (II), Pb (II), Mn (II), Cr (II) and Ni (II). Immobilization was done in 250 mg of MWCNTs by wetting the mixture with distilled water and heating at 105ºC for 1 h. Figure 2 shows the scanning electron microscopy (SEM) photograph of *P. aeruginosa* immobilized MWCNTs. Metal extraction was carried out from samples like boiled wheat, canned fish, black tea and lichen through microwave digestion with HNO_3 and H_2O_2. After dilution with water and pH adjustment to 8.5-9.5, metal ions were preconcentrated with *P. aeruginosa*-modified-MWCNTs. Elution was carried out with 10 mL of 1 M HNO_3 and analyzed by flame atomic absorption spectroscopy (FAAS). The adsorption capacity was found to be between 5.25 and 6.23 mg/g for all ions.

Other compounds have also been immobilized onto MWCNTs, as PAN [71, 72], *N,N*'-bis(2-hydroxybenzylidene)-2,2'(aminophenylthio)ethane (NBHAE) [73], *N*-phenyl-1,2-hydrazinedicarbothioamine [74], 5-(40-dimethylaminobenzyliden)-rhodanine [75], ethylenediamine (ETA) [76], etc. For instance, ETA was immobilized on oxidized-MWCNTs in the presence of *N,N*'-dicyclohexylcarbodiimide for the extraction and preconcentration of Cr (III), Fe (III) and Pb (II) from balsam pear leaves and river water [76]. TEM was used for characterization of raw MWCNTs, oxidized-MWCNTs and modified-MWCNTs with ETA. Figure 3 shows how raw MWCNTs are entwined by means of Van der Waals forces (a).

On the other hand, oxidized-MWCNTs are shorter (b), a fact that is described in the literature [77]. In contrast, modified-MWCNTs showed wider diameters (50-80 nm) in comparison with raw and oxidized-MWCNTs (15-65 nm). In this work, apart from classical on-column SPE, dSPE experiments were carried out determining directly the concentration of the remaining metal in aqueous solution by inductively coupled plasma optical emission spectrometry (ICP-OES). The method showed excellent selectivity towards Cr (III), Fe (III) and Pb (II) over other ions. Maximum adsorption capacity was found to be 39.58 mg/g for Cr (III), 28.69 mg/g for Fe (III) and 54.48 mg/g for Pb (II).

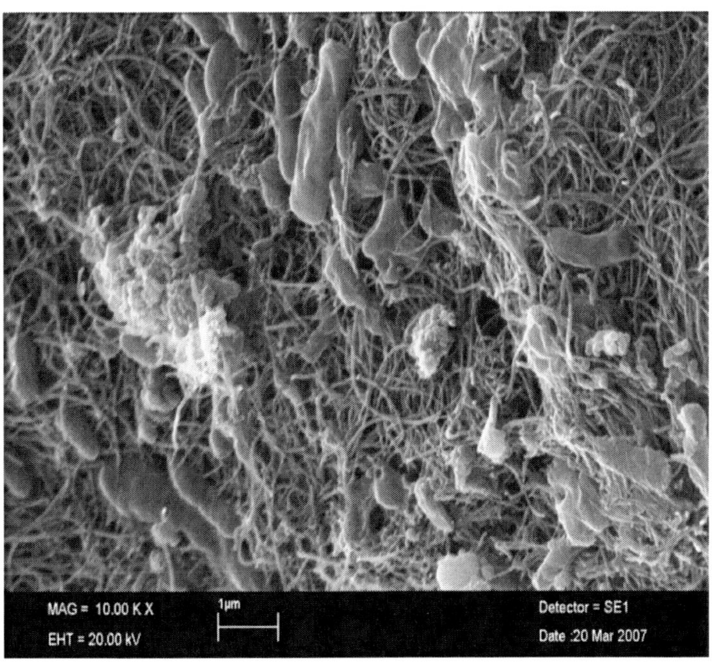

Figure 2.- SEM photograph of Pseudomonas aeruginosa immobilized MWCNTs. Reprinted from [70] with permission from Elsevier.

Table 1. CNTs-SPE applications for the extraction of inorganic analytes

Analyte	Matrix	CNTs type and outer diameter	Sorbent amount (mg)	Elution solvent (mL)	Methodology	Recovery (%)	LOD	Ref.
Au (III)	Lake and sea water	Oxidized-MWCNTs o.d.: 30-60 nm	50	3% thiourea (w/v) in 1 M HCl (2 mL)	SPE-FAAS	95-103	0.15 μg/L	[46]
Cu (II), Co (II), Pb (II)	CRM (mussel), natural water	Oxidized-SWCNTs -	20	0.5 M HNO_3 (2 mL)	SPE-ICP-MS	96-109	1.2-39 ng/L	[48]
Rh (III) (as PAN complex)	Tap, well, river, sea, and waste water and platinum iridium alloy	Oxidized-MWCNTs o.d.: 3-20 nm	20	DMF (5 mL)	FAAS	96-105	0.010 μg/L	[49]
Ag (I)	Tap, river, well, snow and spring water	Oxidized-CNTs/silica gel (1:30 w/w) -	21.7	10% (v/v) HNO_3 (7.5 mL/min for 4 s)	FI-SPE-FAAS	96-108	0.35 μg/L	[50]
Hg (II)	CRM (tobacco leaves), tap, reservoir and water, phosphate rock	Oxidized-MWCNTs o.d.: 40-60 nm	200	1 M HNO_3 (5 mL)	SPE-UV/Vis spectrometry	88-95	0.0123 μg/L	[51]
Cr (III), Cr (VI) (as APDC quelate)	River and waste water	MWCNTs o.d.: 30-50 nm	300	1 M HNO_3 in acetone (10 mL)	SPE-FAAS	95-98	0.90 μg/L	[57]

Analyte	Matrix	CNTs type and outer diameter	Sorbent amount (mg)	Elution solvent (mL)	Methodology	Recovery (%)	LOD	Ref.
Cu (II), Cd (II), Pb (II), Zn (II), Ni (II), Co (II) (as APDC chelates)	CRMs (drinking water, apple leaves, wheat gluten), tap, spring and sea water, tomato sauce, canned fish, tobacco, chickpea	MWCNTs o.d.: 30-50 nm	300	1 M HNO_3 in acetone (10 mL)	SPE-FAAS	95-98	0.30-0.60 µg/L	[58]
As (III), As (V), Sb (III), Sb (V) (as APDC complex)	CRM (natural water), sea, river, lake and ground water	SWCNTs o.d.: 2 nm	10	20% (v/v) HNO_3 (4.8 mL/min for 25 s)	FI-SPE-HG-DC-AFS	92-107	2.1-4.3 ng/L	[60]
As (III), As (V), Sb (III), Sb (V) (as APDC complex)	Mineral, tap, river, spring and sea water	MWCNTs Av. diameter: 9.5 nm	20	1 M ammonia (200 µL)	SPE-ETAAS	94-104	0.02-0.05 µg/L	[61]
Cd (II), Pb (II), Cu (II), Zn (II), MnO_4^-	Tap, reservoir and stream water	MWCNTs o.d.: 10-30 nm	200	1 M HNO_3 (10 mL)	SPE-FAAS	81-100	8.7-378 ng/L	[62]
Cd (II)	Soil, polluted farmland soil, human hair, tea, and rice	Modified-MWCNTs (L-Cys) Av. diameter: 40-60 nm	10	0.5 M HCl (2 mL/min for 30 s)	FI-SPE FAAS	63-98	0.28 µg/L	[64]

Table 1. (Continued).

Analyte	Matrix	CNTs type and outer diameter	Sorbent amount (mg)	Elution solvent (mL)	Methodology	Recovery (%)	LOD	Ref.
Co (II)	Natural water	Modified-MWCNTs (L-Tyr)	10	10% (v/v) HNO_3 (-)	FI-SPE-FAAS	101	50 ng/L	[65]
Total Tl, Tl (III)	Tap water	Modified-MWCNTs (L-Tyr) (-)	10	10% (v/v) HNO_3 (-)	FI-SPE-STPF-ETAAS	>98	150 ng/L	[66]
Ni (II), Pb (II)	CRMs (municipal sludge, lake sediment)	MWCNTs Oxidized-MWCNTs Modified-MWCNTs (L-Ala) (-)	4	5% (v/v) HNO_3 (50 µL)	FI-SPE-SPE-ETAAS	-	10-30 ng/L	[68]
Cu (II), Ni (II)	CRMs (soils, tea leaf, human hair, tea)	Modified-MWCNTs (6His-tagged protein) o.d.: 10-20 nm	15	0.2 M imidazole-HCl (2 mL/min for 20 s)	FI-SPE-FAAS	81-100	0.31-0.63 µg/L	[69]
Co (II), Cd (II), Pb (II), Mn (II), Cr (III), Ni (II)	CMRs (tomato leaves, bovine liver), boiled wheat, canned fish, black tea, lichen, spring and tap water	Modified-MWCNTs (*Pseudomonas aeruginosa)* o.d.: 30-50 nm	250	1 M HNO_3 (8-10 mL)	SPE-FAAS	95-102	0.24-2.60 µg/L	[70]
Co (II)	Tap and river water	Modified-MWCNTs (PAN) Av. diameter: 10-30 nm	40	0.5 M HNO_3 (3 mL)	SPE-FAAS	98-100	0.55 µg/L	[71]

Analyte	Matrix	CNTs type and outer diameter	Sorbent amount (mg)	Elution solvent (mL)	Methodology	Recovery (%)	LOD	Ref.
Zn (II)	CRMs (pepperbush, human hair, tea leaves, chlorella, mussels, pond sediment), well, drinking, and waste water	Modified-MWCNTs (PAN) o.d.: 3-20 nm	30	0.5 M HNO_3 (4 mL)	SPE-FAAS	>98	0.07 pg/mL	[72]
Au (III), Mn (II)	Tap, well and waste water	Modified-MWCNTs (NBHAE) (-)	40	0.1 M $Na_2S_2O_3$ (4 mL)	SPE-FAAS	94-102	0.01-0.03 µg/L	[73]
Cd (II)	Well, river and tap water	Modified-MWCNTs (*N*-phenyl-1,2-hydrazinedicarbothioamide) o.d.: 10-40 nm	50	0.1 M HNO_3 (5 mL)	SPE-FAAS	98	0.22 ng/mL	[74]
Pd (II)	CRM (platinum-iridium alloy), waste water, road dust	Modified-MWCNTs (5-(40-dimethylaminobenzyliden)-rhodanine) o.d.: 3-20 nm	40	0.4 M thiourea (3 mL)	SPE-FAAS	99	0.3 ng/mL	[75]
Cr (III), Fe (III), Pb (II)	CRM (river sediment), balsam pear leaves, river water	Modified-MWCNTs (ETA) o.d.: 15-65 nm	dSPE: 20 SPE: 50	1 M HCl (2 mL)	dSPE-ICP-OES SPE-ICP-OES	99-100	0.19-0.33 µg/L	[76]
Ga (III)	Fly ash	Ga (III)-imprinted/MWCNTs Diameters: 20-40 µm	200	EtOH/acetic acid (9:1 v/v) (3 mL)	SPE-FAAS	87-94	3.03 ng/mL	[78]
V (IV), V (V)	CRM (water), pond and tap water	MWCNTs	CNTs filled a volume equivalent to 25 µL of water	1 M HCl (50 µL, 0.2 mL/min for 0.6 min)	FI-SPE-ETAAS	96-102	19 ng/L	[79]

Table 1. (Continued).

Analyte	Matrix	CNTs type and outer diameter	Sorbent amount (mg)	Elution solvent (mL)	Methodology	Recovery (%)	LOD	Ref.
Eu (III), Gd (III), Ho (III), La (III), Sm (III), Tb (III), Yb (III)	CRM (tea leaves), lake water, synthetic sea water	MWCNTs -	30	1 M HNO_3 (2 mL)	SPE-ICP-AES	94-102	3-57 ng/L	[82]
$H_2AsO_4^-$	River and tap water	Modified-MWCNTs (CTAC) o.d.: 50 nm	200	2 M HNO_3 (8 mL)	FI-SPE-HG-AFS	94-104	2 ng/L	[83]
Cd (II)	Urine	Oxidized-MWCNTs o.d.: 10-20 nm	45	0.5 M HNO_3 (1.5 mL)	FI-SPE-ETA-AAS	97-100	0.010 µg/L	[87]

CRM: certified reference material; HG-DC-AFS: hydride generation-double channel atomic fluorescence spectrometry; ETA-AAS: electrothermal atomization atomic absorption spectrometry; - Not indicated.

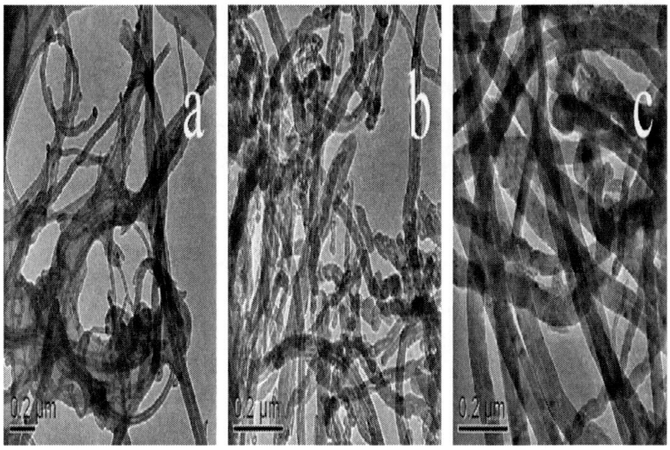

Figure 3. TEM images of MWCNTs (a), MWCNTs-COOH (b) and MWCNTs-EDA (c). Reprinted from [76] with permission from Elsevier.

In the literature, the repairing of the structure of oxidized-CNTs, which is not a usual task, has also been mentioned. For the preparation of a new Ga (III)-imprinted/MWCNT composite material for selective sorption of Ga (III), MWCNTs were first oxidized [78] as shown in Figure 4. *P*-electron conjugated structure is damaged after oxidation. The degree of damage is higher when oxidation conditions are harder. For further repairing of the structure, aniline was used as patch of the conjugate structure. A condensation reaction takes place between the amine group of aniline and the acyl chloride group of CNTs, leading to covalent bonds. Then, polimerization reaction could efficiently take place. Sample preparation consisted in calcination at 550°C, lixiviation of Ga (III) with 5 mL of 6 mM HCl, centrifugation and complexation with 8-hydroxyquinoline (8-HQ). An amount of 200 mg of Ga (III)-imprinted/MWCNT composite was used for determination of Ga (III) in the resulting solution, obtaining recoveries of 87-94% and a LOD of 3.03 ng/mL.

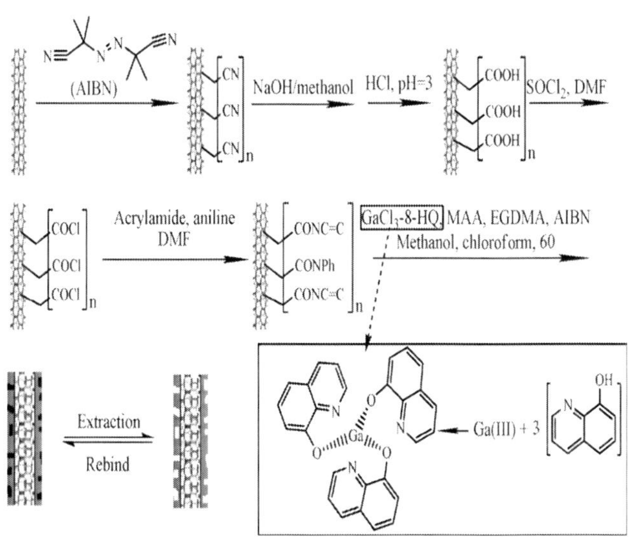

Figure 4. Synthesis route of Ga (III)-imprinted/CNTs sorbent. Reprinted from [78] with permission from Elsevier.

As can be seen in Table 1 -which compiles different works in the literature dealing with CNTs-SPE of inorganic species-, in many works, SPE has been developed online, or by what is called "flow injection" (FI), in combination with the determination technique. However, due to the small size of CNTs, when using them as mini-column packing materials they tend to create flow resistance. To solve this problem, oxidized-MWCNTs have been immobilized on the surface of silica gel in the proportion 1:30 (w/w) [50]. To fabricate the SPE column, 21.7 mg of the CNTs/silica gel composite were introduced in a 10 cm length x 2.8 mm internal diameter polytetrafluoroethylene (PTFE) tube and set online with a flame atomic absorption spectrometry (FAAS) system for the determination of Ag (I) in different types of waters (tap, river, well, snow and spring water). Retention of silver was accomplished by sorption of nascent AgO_2 precipitated with NaOH. However, in complex waters (river and tap waters) other metals could form precipitates at high pH, so to avoid interferences, addition of NaF and ethylenediaminetetraacetic acid (EDTA) was absolutely necessary.

On the other hand, not only simple metal retention has been studied but also metal speciation which is very important in the environmental, toxicological and biological fields, due to the fact that different oxidation states of a same metal can have different properties. In this sense, CNTs have been used for speciation of chromium [57], vanadium [79] thallium [66, 80], arsenic and antimony [60, 61] or even lead, mercury and tin organometallic compounds [81]. Speciation can be developed in diverse ways. For example, speciation of chromium in natural waters [57], which was present as Cr (VI) and Cr (III), has been initially carried out by MWCNT-SPE of Cr (VI) as an APDC complex and then, after oxidation of Cr (III) with H_2O_2, of total chromium also as Cr (VI)-APCD. Cr (III) amount was determined as the difference between both values. Recovery, LOD and absorption capacity reached were 95%, 0.90 μg/L and 9.50 mg/g, respectively, with FAAS determination. On the other hand, one of the oxidation states can be masked to permit the determination of the other. In this sense, in the work by Gil et al. [79], V (IV) was masked with 1,2-cyclohexanediaminetetraacetic acid (CDTA) to selectively determine V (V) with a FI-MWCNT-SPE-electrothermal atomic absorption spectrometry (ETAAS) method (LOD of 19 ng/L). Another approach that has been exploited in this sense is the selectivity of one of the oxidation states towards CNTs at a certain pH. For instance, Tl (III) shows specific interaction with L-Tyr modified MWCNTs at pH 5 [66]. Such property was used by Pacheco et al. [66] to determine Tl (III) in tap water by FI-Tyr-MWCNTs-SPE-stabilized temperature platform furnace (STPF)-ETAAS after elution from the SPE column with 10% (v/v) HNO_3, while total thallium was directly determined by STPF-ETAAS. In this case, it was confirmed that acids can cause a reversible change in the tertiary structure of amino acids providing release of metals from the binding cavity. Also simultaneous determination and speciation of Sb (III), Sb (V), As (III) and As (V) has been carried out determining first Sb (III) and As (III) by complexing them with APDC and then total antimony and arsenic by the reduction of the pentavalent species to the trivalent ones. Adsorption of the trivalent species has been done with SWCNTs [60] and also with MWCNTs [61].

In general, and regarding the SPE protocol, because metal ions are generally poorly absorbed onto oxidized CNTs at low pH, desorption from CNTs has been normally carried out with diluted HNO_3 [43-45, 47, 48, 50-55, 59, 60, 62, 65, 66, 70-72, 72, 74, 80, 82-87] or HCl [64, 76, 79], which are believed to cause competition between protons and metal ions. This can also be considered as an important alternative to avoid the use of highly contaminant solvents. When complexes are involved, elution might be carried out with an organic solvent containing an acid, such as acetone and HNO_3 for Cr (VI)-APDC chelates [57] and Cu (II), Cd (II), Pb (II), Zn (II), Ni (II) and Co (II)-APDC chelates [58]. Also, dimethylformamide (DMF) has been employed for elution of Rh (III)-PAN complex [49] or 1 M ammonia for As (III) and Sb (III)-APDC chelates [61]. In other cases, the addition of a modifier to the eluent resulted favorable. For desorption of Au (III) from 50 g of oxidized-MWCNTs, the addition of 3% (w/v) thiourea to 2 mL of 1 M HCl resulted of crucial importance to obtain quantitative recoveries (98.6%), while 3% (w/v) thiourea or 1 M HCl alone provided lower recoveries [46]. Also for desorption of Au (III) simultaneously with Mn (II) from 40 mg of modified-MWCNTs with NBHAE [73], diverse eluents were assayed (EDTA, HNO_3, HCl, NH_2CSNH_2, KSCN, Na_2SO_3 and $Na_2S_2O_3$), observing that especially for Au (III), recoveries improved markedly when 4 mL of a solution of 0.1 M $Na_2S_2O_3$ was used.

Regarding the type of sample matrices analyzed, the vast majority of the works describes direct SPE from all types of waters. Also in a good number of the works, certified reference materials (CRM) have been used to validate the methodology. However, those that analyze metal ions in complex real samples different than waters are still relatively low. In this sense, organometalic compounds have been determined in coastal sediments [81], Pb (II) in physiological serum, garlic, Ginkgo Biloba [45] and road dust [75], Cu (II), Cd (II), Pb (II), Zn (II), Ni (II) and Co (II) in tomato sauce, canned fish, tobacco and chickpea [58], Cd (II) in soil, human hair, tea, rice [64], urine [87] and cigarettes [44], Co (II), Cd (II), Pb (II), Mn (II), Cr (III), Ni (II) in wheat, canned fish, tea and lichen [70], Cr (III), Fe (III) and Pb (II) in balsam pear leaves [76], Rh (III) in platinum iridium alloy [49], Ga (III) in fly ash [78], Hg (II) in phosphate rock [51] or Pb (II), Cd (II) and Cu (II) in rice [56].

In summary, application of CNTs to the extraction of metal ions from water samples is much extended and introduction of modifications and alternatives have been widely developed. Also, efforts have been and are also being carried out to spread their application for a wide variety of matrices.

3.2. Organic Analytes

3.2.1. Pesticides

Pesticides are widely used in agriculture and therefore they are present in the environment and in food matrices, being highly studied because of their high toxicity and widespread application. For this reason, highly sensitive monitoring methods are necessary to ensure the food supply and for a proper risk assessment on the environment fluxes and on their occurrence in ecosystems. However, pesticide analysis is a laborious mission because of the low concentrations at which these compounds are commonly present in complex matrices. In this sense, SPE is one of the most popular techniques in sample preparation to analyze pesticides due to its high precision and throughput, low solvent consumption and the relative simplicity for carrying out simultaneous extraction and concentration of multiclass pesticides. In fact, the use of SPE using CNTs as sorbent material has found its main field of application in the analysis of pesticides due to their strong affinity towards them. Analytical methods, based on CNTs-SPE and developed for the determination of pesticides, are summarized in Table 2. As can be observed in the literature, CNTs have been mostly used for the extraction of pesticides belonging to the same family such as organophosphorus pesticides (OPPs) [88-94], organochloride pesticides (OCPs) [95], triazines [96-102], sulfonylurea herbicides [103-

109], phenoxyalkanoic acids [110], chloroacetanilide herbicides [111, 112], triazolinone pesticides [113], neonicotinoid insecticides [114] and pyrethroids insecticides [115, 116], and in a lesser extent, for the simultaneous extraction of multiclass pesticides [117-122]. Most of the works have been developed for the analysis of water samples [90, 92, 94-99, 100-107, 110-120, 122-124], whereas only in few cases non aqueous samples were studied, i.e. sewage sludge [88], garlic [89], fruit juices [91], virgin olive oil [121] and soils [93, 99, 108].

The first work in which CNTs were used as sorbents for the extraction of pesticides in conventional SPE was developed by Zhou et al. [96]. A MWCNTs-SPE-high performance liquid chromatography (HPLC)-diode array detection (DAD) method was used to extract two triazine herbicides (atrazine and simazine) from river, reservoir, waste and tap water with recoveries in the range 58-104% and LODs between 9.0 and 33 ng/L. These values are below the maximum residue limits (MRLs) regulated by the European Union (EU) directive on drinking water quality (98/83/CE), established in 0.1 μg/L for individual pesticides and 0.5 μg/L for total pesticides [125]. Initially, the factors that could influence the preconcentration were investigated in purified water including sample volume and its flow rate and pH, as well as the type and volume of the eluent. In the end, the optimized method which consisted on the extraction of 500 mL of water at pH 6-7 with a flow rate of 7 mL/min and 4 mL acetonitrile (ACN) as eluent was applied to real water samples. This scheme is widely used in the literature to optimize CNTs-SPE parameters because good initial conditions can be easily obtained from a simple matrix and later, they can be subsequently extended to a more complex matrix. As another example of the high potential of CNTs as SPE stationary phases for the determination of pesticides from different types of matrices, Min et al. [99] developed a single MWCNTs-SPE procedure to quantify atrazine and its main metabolites (deisopropyl-atrazine and deethyl-atrazine) from two different types of water and an agricultural soil. Again, extraction parameters were optimized using a spiked purified water sample. In addition, for the analysis of soil samples, a previous pretreatment was found necessary, selecting the extraction solvent as well as the amount of sample. High recoveries, above 72%, were obtained in all cases and low LODs for waters (0.02-0.05 μg/L) and soil (0.3-1 μg/kg). In the same way, Ravelo-Pérez et al. [90] developed a MWCNTs-SPE-(GC)-nitrogen phosphorus detection (NPD) method to determine seven OPPs (ethoprofos, diazinon, chlorpyrifos-methyl, fenitrothion, malathion, chlorpyrifos, fenamiphos) and one thiadiazine (buprofezin) in water samples. Different SPE parameters such as the amount of MWCNTs, sample volume, pH and type and amount of the eluent were optimized by means of a step by step approach, using spiked Milli-Q water. The procedure was later extended to the extraction of mineral, run-off and ground waters, obtaining for most of the pesticides recoveries in the range 57-116% and LODs between 18-117 ng/L. Subsequently, the same authors [91] extended the application of this already optimized SPE procedure to the extraction of these pesticides from different commercial fruit juices (apple, grape, orange and pineapple). However, unsatisfactory recoveries were found when the optimum conditions were applied in a very beginning. As a result, some modifications were studied in order to obtain higher recoveries. In this sense, only a 1:1 (v/v) dilution of the sample with Milli-Q water and the introduction of a washing protocol consisting in different volumes of Milli-Q water were sufficient to obtain acceptable recoveries between 73-103% and LODs in the range 1.91-7.32 μg/L.

As can be seen in Table 2, most SPE studies have used MWCNTs [88-106, 108, 110-124] instead of SWCNTs [107, 121, 124] (which have been mostly used in SPME applications) because the presence of concentric layers of graphene offer a better affinity to pesticides. Besides, the functionalization of CNTs changes its properties, and thus, it modifies the interaction between sorbent and analytes. In this sense, an interesting work was carried out by López-Feria et al. [121], in which the sorption capabilities of two types of CNTs (MWCNTs with o.d. 20-30 nm and carboxylated (c)-SWCNTs with o.d. 1-2 nm) were evaluated and compared to determine eight multiclass pesticides in virgin olive oil. c-SWCNTs turned out to be more appropriate to extract all analytes from a spiked diluted virgin olive oil sample which can be ascribed to the additional interaction provided by the carboxylic moiety present in the c-SWCNTs. Furthermore, the cartridge, which was packed with 30 mg of the material, could be reused for at least 100 times without losing retention capacity. Finally, mean recoveries were obtained in the range 79-105% with LODs between 1.5 and 3.0 μg/L. Another study which also includes an important assay about the effect of functionalization was developed by El-Sheikh et al. [104] to extract three pesticides (atrazine, propoxur and methidathion), representing three different chemical families, from environmental water samples. MWCNTs oxidation was studied with various oxidizing agents (HNO_3, H_2O_2 and ammonium persulphate) and also without oxidation. The best recoveries were obtained when MWCNTs oxidized with HNO_3 was used. In this case, the recoveries of methidathion were improved by 17%, while the recoveries of atrazine and propoxur were improved only slightly. In contrast, when MWCNTs oxidized with H_2O_2 or ammonium persulphate, mean recoveries were lower than when oxidation was not carried out. In the same work, authors also studied the influence of the o.d. (i.e. 10-20 nm, 10-30 nm, 20-40 nm, 40-60 nm, 60-100 nm) on the recovery of the pesticides, with MWCNTs of the same length (5–15 μm) previously oxidized with HNO_3. As shown in Figure 5, the highest recovery for atrazine and propoxur was obtained with o.d. 40-60 nm MWCNTs, while for methidathion the highest recovery was almost the same with o.d. 40-60 nm MWCNTs, o.d. 20-40 nm MWCNT and o.d. 60-100 nm MWCNTs. In others words, the best recoveries were obtained using the highest o.d. In this sense, it should be remarked that MWCNTs most commonly used in the literature for pesticide extraction have had high o.d., i.e., 20-40 [104], 30-60 nm [95-98, 103, 105, 106, 114-116, 118], 40-60 nm [99, 100, 102, 104, 108, 112, 117, 119, 123], 40-70 nm [101] or 60-100 nm [104], while fewer authors have used lower o.d. MWCNTs, i.e., 8.6 nm [111; 113], 3-10 nm [92], 5-10 nm [110], 8-10 nm [94], 10-15 nm [90, 91, 93 120, 122], 13-16 nm [109], 10-20 nm [104], 10-30 nm [104] or 20-30 nm [121]. Moreover, in the work by El-Sheikh et al. [104] all experiments involved the use of 200 mg of each MWCNT. In this sense, and as can be seen in Table 2, the use of quantities lower than 200 mg has been very common for the extraction of pesticides residues [88, 90-92, 101, 107, 120, 121].

CNTs performance have not only been compared with themselves, but also with other types of stationary phases like C_{18} [97, 98, 105, 107, 110, 119], activated carbon [97, 107, 119], graphitized carbon black (GCB) [123] and Oasis HLB [124]. In most cases, CNTs generally exhibited higher sorption capability or at least similar effectiveness. One of these works was also developed by El-Sheikh et al. [119] who compared the enrichment efficiency of MWCNTs, C_{18} silica and activated carbon for the extraction of propoxur, atrazine and methidathion from tap, reservoir and stream water. Final extracts were analyzed by HPLC-UV. The SPE procedure was optimized for each sorbent and the only difference between them was the amount of stationary phase: i.e. 200 mg of MWCNTs, 400 mg of C_{18} silica and 300 mg of activated carbon. Subsequently, the optimized SPE procedures were compared in terms of analytical performance, application to environmental waters, cartridge reusability and adsorption capacity. Results showed that precision, linear range and LODs of the optimized procedure using activated carbon were very close to the ones obtained with the other two sorbents. Concerning recovery values, they were in the range 81-108% for MWCNTs, 60-96% for C_{18} silica and 70-98% for activated carbon. However, MWCNTs presented a higher adsorption capacity: almost three times more than that of activated carbon or C_{18}. This explains why 300 mg of activated carbon and 400 mg of C_{18} silica were needed to carry out the same extraction than 200 mg of MWCNTs. In the end, experimental results indicated that the three sorbents may be re-used over 100 times after proper cleaning and re-conditioning.

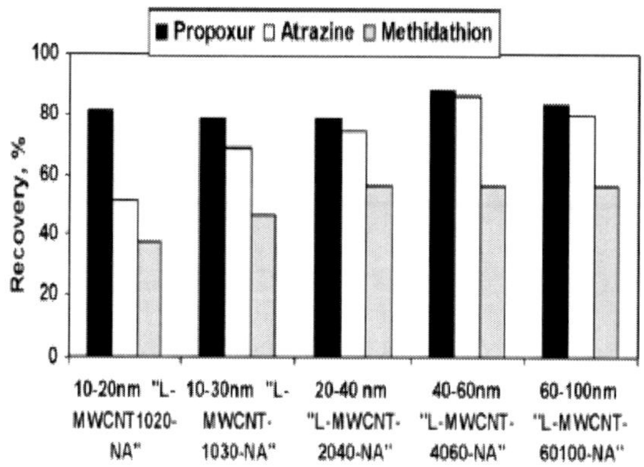

Figure 5. Effect of the o.d. of MWCNT on the recovery of the pesticides: 50 mL of water sample (pH 5) spiked with 50 ng/mL of the three target pesticides; 0.200 g of MWCNTs; washing solvent: 10 mL of 10% (v/v) ACN aqueous solution; eluting solvent: 10 mL of ACN. Reprinted from [104] with permission from Elsevier.

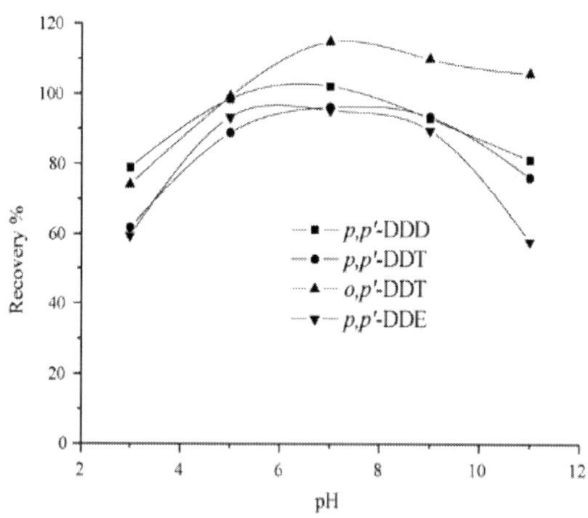

Figure 6. Effect of the pH of water sample on the recovery of the pesticides. Concentration of sample solution: 1 ng/mL of o,p′-DDT, p,p′-DDD, p,p′-DDE and p,p′-DDT; 0.100 g of MWCNTs; washing solvent: 10 mL MeOH and 10 mL water; eluent: 5 mL DCM; flow rate of sample solution: 8 mL/min; volume of sample solution: 200 mL. Reprinted from [95] with permission from Elsevier.

Table 2. CNTs-SPE and SPME applications for the extraction of pesticides

Analyte	Matrix	CNTs type and outer diameter	Sorbent amount (mg)	Elution solvent (mL)	Methodology	Recovery (%)	LOD	Ref.
SPE								
11 OPPs	Sewage sludge	MWCNTs -	6	n-Hexane (100 μL)	μSPE-GC-MS	-	1-7 pg/g	[88]
3 OPPs	Garlic	MWCNTs -	-	-	SPE-SWV	97-104	5.0 μg/L	[89]
7 OPPs and 1 thiadiazine	Mineral, ground and run-off water	MWCNTs o.d.: 10-15 nm	40	DCM (20 mL)	SPE-GC-NPD	40-119	18-117 ng/L	[90]
7 OPPs and 1 thiadiazine	Apple, grape, orange and pineapple juices	MWCNTs o.d.: 10-15 nm	40	DCM (20 mL)	SPE-GC-NPD	73-103	1.91-7.32 μg/L	[91]
7 OPPs and 1 thiadiazine	Forestal, ornamental and agricultural soils	MWCNTs o.d.: 10-15 nm	100	DCM (20 mL)	SPE-GC-NPD	54-91	2.97-31.6 ng/g	[93]
Chlorpyrifos and phosalone	Well water, tap water and river water	MWCNTs o.d.: 8-10 nm	100	DCM (3 mL)	SPE-HPLC-UV	85-100	4.02 and 1.02 ng/L	[94]
4 OCPs	Underground, tap and reservoir water	MWCNTs o.d.: 30-60 nm	100	DCM (5 mL)	SPE-HPLC-UV	90-116	4.0-13 ng/L	[95]
Atrazine and simazine	River, reservoir, waste and tap water	MWCNTs o.d.: 30-60 nm	100	ACN (4 mL)	SPE-HPLC-DAD	58-104	33 and 9.0 ng/L	[96]
Atrazine and simazine	River, reservoir, waste and tap water	MWCNTs o.d.: 30-60 nm	100	ACN (4 mL)	SPE-HPLC-DAD	60-105	33 and 9.0 ng/L	[97]

Table 2. (Continued).

Analyte	Matrix	CNTs type and outer diameter	Sorbent amount (mg)	Elution solvent (mL)	Methodology	Recovery (%)	LOD	Ref.
3 Triazines	Tap, ground, sewage and snow water	MWCNTs o.d.: 30-60 nm	100	DCM (5 mL)	SPE-HLPC-UV	88-110	12-34 ng/L	[98]
3 Ttriazines	Surface and underground water and agricultural soils	MWCNTs -	100	EtOAc (4 mL)	SPE-GC-MS	72-110	0.02-0.05 μg/L (waters), 0.3-1 μg/kg (soil)	[99]
Atrazine and simazine	Tap, river and sea water	MWCNTs o.d.: 40-70 nm	30	Acetone (5 mL)	SPE-GC-MS	87-110	2.5 and 5.0 ng/L	[101]
11 Triazines	River water	MWCNTs -	100	ACN/water (90:10 v/v) (5 mL)	SPE-RRLC-MS	73-98	<0.1 ng/L	[102]
3 Sulfonylurea herbicides	Tap, sea, reservoir and well water	MWCNTs o.d.: 30-60 nm	100	ACN with 1% acetic acid (v/v) (12 mL)	SPE-HPLC-DAD	84-111	5.9-11 ng/L	[103]
2 Sulfonylurea herbicides and 1 triazine	Tap, reservoir and stream water	MWCNTs o.d.: 10-20 nm, 10-30 nm, 20-40 nm, 40-60 nm, 60-100 nm	200	ACN (10 mL)	SPE-HPLC-DAD	81-108	37.5-220 ng/L	[104]
5 Sulfonylurea herbicides	Tap, well, river and canal water	SWCNTs o.d.: 1.5-2.5 nm	30	ACN with 1% acetic acid (v/v) (10 mL)	SPE-HPLC-DAD	79-102	1.1-7.2 ng/L	[107]
4 Sulfonylurea herbicides	Soil	MWCNTs o.d.: 40-60 nm	150	-	SPE-HPLC-DAD	-	-	[108]
Chlorsulfuron and metsulfuron methyl	Reservoir, lake, creek and underground water	MWCNTs o.d.: 13-16 nm	6	MeOH 50% (v/v) with ACN 2% (250 μL)	SPE-CE-UV	86-108	0.36 and 0.40 μg/L	[109]

Analyte	Matrix	CNTs type and outer diameter	Sorbent amount (mg)	Elution solvent (mL)	Methodology	Recovery (%)	LOD	Ref.
3 Chloroacetanilide herbicides	Tap and reservoir water	MWCNTs o.d.: 40-60 nm	100	Acetone (1 mL)	SPE-GC-MS	89-116	2-6 ng/L	[112]
Carfentrazone -ethyl	Tap and river water	MWCNTs o.d.: 8.6 nm	100	EtOAc (4 mL)	SPE-GC-ECD	74-80	10 ng/L	[113]
3 Neonicotinoid insecticides	Tap, ground and reservoir water	MWCNTs o.d.: 30-60 nm	100	MeOH (4 mL)	SPE-HPLC-UV	88-110	5.4-6.7 ng/L	[114]
12 Multiclass pesticides	Milli-Q, tap and river water	MWCNTs o.d.: 40-60 nm	100	Acetone/*n*-hexane (1:1, v/v) (5 mL)	SPE-GC-MS	82-104	10-30 ng/L	[117]
5 Multiclass pesticides	Tap, ground, rain and reservoir water	MWCNTs o.d.: 30-60 nm	100	DCM (6 mL)	SPE-HPLC-UV	79-109	2.99-6.04 ng/L	[118]
1 Triazine, 1 carbamate and 1 OPP	Tap, reservoir and stream water	MWCNTs o.d.: 40-60 nm	200	ACN (5 mL)	SPE-HPLC-DAD	81-108	36.0-220 ng/L	[119]
7 Multiclass pesticides	Milli-Q and mineral water	MWCNTs o.d.: 10-15 nm	40	ACN with 5% formic acid (v/v) and DCM with 5% formic acid (10 mL and 20 mL)	SPE-NACE-UV	47-94	27–58 ng/L	[120]
8 Multiclass pesticides	Virgin olive oil	MWCNTs o.d.: 20-30 nm SWCNTs o.d.: 1-2 nm	30	EtOAc (0.5 mL)	SPE-GC-MS	79-105	1.5-3.0 µg/L	[121]

Table 2. (Continued).

Analyte	Matrix	CNTs type and outer diameter	Sorbent amount (mg)	Elution solvent (mL)	Methodology	Recovery (%)	LOD	Ref.
9 Multiclass pesticides	Milli-Q water	MWCNTs o.d.: 10-15 nm	90	DCM (35 mL)	dSPE-nano-LC-UV	36-101	16–67 ng/L	[122]
SPME								
8 OCPs	Lake and waste water	SWCNTs Diameter <1-2 nm	2000	-	SPME-GC-ECD	45-116	0.19-3.77 ng/L	[126]
3 Benzimidazoles	River and untreated water	Oxidized-MWCNTs o.d.: 10-20 nm	2.0	-	SPME-HPLC-UV	76-105	0.3–1.5 μg/L	[127]
13 Multiclass pesticides	Green tea, oolong tea, white tea and flower tea	SWCNTs o.d.: <2 μm	-	-	SPME-GC-MS	75-118	27–230 ng/L	[128]
4 OCPs	Lake and waste water	Carboxylated-SWCNTs -	50	-	SPME-GC-ECD	90-101	0.07-4.36 ng/L	[129]

SWV: square-wave voltammetry; RRLC: rapid resolution liquid chromatography; - Not indicated.

Other extraction parameters that should be studied in CNTs-SPE are the type and volume of elution solvent in order to completely desorb the analytes. Particularly, when a wide range of multiclass pesticides with different polarity must be simultaneously extracted, solvent selection is critical. For low polarity pesticides desorption may be carried out with non-polar solvents like *n*-hexane [88]. However, and more frequently, for more polar compounds solvents such as ACN [92, 96, 97, 119], acetone [116], methanol (MeOH) [123], ethyl acetate (EtOAc) [99, 111, 113, 121], dichloromethane (DCM) [90, 91, 93-95, 98, 118, 122], or different mixtures of them [100, 102-104, 106, 107, 110, 117, 120] provide a better elution over a wide polarity range.

No less important is the effect exerted by the pH. Enrichment factors are also controlled by the affinity of the analyte for the sorbent phase and its solubility in the liquid phase. For pesticides with acidic and basic functional groups, the pH of the liquid phase determines the molecular state present (ionic or neutral state) and therefore the distribution ratio of the analyte between the two phases. For this reason, a pH in the range 5-7 is usually selected in most of the works. Furthermore, extreme pH values are not suitable since the possible hydrolysis of some of the pesticides under strong acidic or basic conditions can happen. As a clear example of the effect of this parameter on the recoveries of pesticides, Zhou et al. [95] investigated the pH, among other factors, to extract four OCPs (o,p′-dichlorodiphenyl-trichloroethane (DDT), p,p′-DDT, p,p′-1,1-dichloro-2,2-bis-(4′-chlorophenyl)ethane (DDD) and p,p′-1,1-dichloro-2,2-bis-(4′-chlorophenyl)ethane (DDE)) from underground, tap and reservoir water. For this purpose, the pH value was changed between 3 and 11 in purified water sample. As shown in Figure 6, maximum recoveries, in the range 95-102%, were found at pH 7 while the recoveries of the target compounds decreased over and below that pH value. The lowest recoveries were obtained for all pesticides when sample pH was equal to 3 or 11.

Although CNTs have been mostly utilized as SPE cartridges (normally laboratory made), CNTs stationary phases can also be used in the format of a disk or in the dSPE mode. The first of them (SPE disks) have been introduced more recently allowing to use a greater flow without the risk of canalization. This is the case of the work of Niu et al. [107] which used a novel SWCNT disk to extract a group of sulfonylurea herbicides (metsulfuron-methyl, chlor-sulfuron, bensulfuronmethyl, pyrazosulfuron-ethyl and chlorimuron-ethyl) from tap, well, river and canal water. A comparison study between SWCNTs (30 mg), activated carbon (500 mg) and C18 (500 mg) disks showed that the first exhibited similar extraction ability for these pesticides, but unlike for the activated carbon disk, analytes adsorbed onto SWCNTs disks could be easily desorbed with ACN containing 1% (v/v) acetic acid. Moreover, authors also studied the possibility of using a double or triple-SWCNT disk due to the reasonable thickness of the material. Experiments showed the great analytical potential of the procedure for the SPE of large volumes (up to 3.0 L) of real water samples. Katsumata et al. [111] also used a SPE disk to preconcentrate atrazine and simazine from water samples, but using 30 mg of MWCNTs as sorbent. Recoveries of spiked atrazine and simazine in the water samples were in the range of 87–110% with LODs between 2.5 and 5.0 ng/L. Regarding the use of CNTs in the dSPE mode for pesticide extraction, only one work has been developed till now, which was carried out by Asensio-Ramos et al. [122]. Concretely, carbaryl, fensulfothion, fenamiphos, fipronil, profenofos, fonofos, disulfoton, nitrofen and terbufos were determined in Milli-Q water using MWCNTs of o.d. 10-15 nm and 0.1-10 μm length and nano-LC-UV for their later separation and detection. Extraction time compared to conventional SPE was highly reduced as well as the complexity of the procedure. Furthermore, when MWCNTs-dSPE was compared with conventional SPE in terms of extraction efficiency, using the same CNTs and extraction conditions, the same recovery values were practically obtained for all the studied pesticides (in the ranges 33–107% for the dispersive mode vs. 31–95%). It is also worth noting that MWCNTs were reused up to five times, representing an important reduction of the waste of stationary phase. The determination of pesticides residues after CNTs-SPE has been predominantly carried out by HPLC or GC (see Table 2). GC with several detectors like NPD, electron capture detector (ECD) or mass spectrometry (MS) has been used for volatile or thermally stable compounds, being GC-MS the best-established instrumental technique for the analysis of trace pesticides in food and environmental samples. HPLC is also becoming at present an essential analytical tool in the field taking into account the high number of published works. However, the use of MS coupled with HPLC has not been applied after CNTs-SPE of pesticides. Moreover, up to now, there exists few works in the literature concerning the use of CE after the extraction of pesticides by CNTs as stationary phases. In one of them, Asensio-Ramos et al. [120] combined non-aqueous capillary electrophoresis (NACE)-UV with o.d. 10-15 nm MWCNTs-SPE to determine seven multiclass pesticides (pirimicarb, pyrifenox, penconazol, carbendazim, cyromazine, pyrimethanil and cyprodinil) in Milli-Q and mineral water samples. The optimized BGE consisted of a mixture of MeOH and ACN (1:2 v/v) with 90 mM SDS and 20.5 mM $HClO_4$. Electrokinetic injection of the sample dissolved in 78:22 (v/v) MeOH/ACN, 1.11 mM $HClO_4$ was used as on-line preconcentration technique. SPE parameters were also optimized being the following: 750 mL of water at pH 8.0, extraction with 40 mg of MWCNTs as stationary phase, and elution with 10 mL ACN and 20 mL DCM with 5% (v/v) formic acid. Mean recovery values ranged between 53 and 94% for Milli-Q water and between 47 and 93% for mineral waters, with LODs in the range 27–58 ng/L bellow EU MRLs. In another work, Springer and Lista [109] determined chlorsulfuron and metsulfuron methyl in reservoir, lake, creek and underground water using only 6 mg of MWCNTs of o.d. 13-16 nm as sorbent material and capillary zone electrophoresis (CZE)-UV as separation technique. CZE separation time was less than 9 min. A preconcentration step based on a continuous flow system was investigated, minimizing the amount of organic solvents and sample volume. Mean recovery percentages were over the range 86-108% and the LODs were 0.40 and 0.36 μg/L for chlorsulfuron and metsulfuron methyl, respectively. CNTs have also been used as coating materials in SPME fibers to determine pesticides in different water samples [126, 127, 129] and teas [128]. However, the number of studies is still very limited. Even though, the high selectivity and sensitivity of the technique may surely increase its use in the coming years. Generally, the selectivity and sensitivity depends greatly on the right selection of the fiber coating and its thickness with respect to the compounds of interest. In this sense, CNTs have shown to be an alternative to the poly(dimethylsiloxane) (PDMS) or poly(acrylate) (PA) fibers which are commonly used. The particular structures of CNTs as well as their thermal and physical resistance have promoted their use in this aspect. Lu et al. [126] studied for the first time the feasibility of SWCNTs as adsorbent for SPME, which were attached onto a stainless steel wire through an organic binder, using OCPs as model compounds. Compared with commercial PDMS fibers to extract these pesticides from lake and waste water, the SWCNT fiber showed higher extraction efficiency (recoveries between 45 and 116%) because of their larger surface area, micropore and network structure. Moreover, the SWCNT fiber also showed better thermal stability (over 350°C) and longer life span (over 150 times). Wu et al. [128] also compared a novel SWCNTs coated fiber to determine pesticides in tea samples with commercial SPME fibers, i.e., PDMS and PA. Results showed that the SWCNTs fiber also exhibited higher extraction efficiency and equivalent service life (more than 70 times) as well as good linearity range, LODs (27-230 ng/L), recoveries (75-118%), precision and accuracy.

3.2.2. Pharmaceuticals

Regarding pharmaceuticals (see Table 3), CNTs have been used to extract them from pork meat [131, 132, 134], urine [130, 133, 137], eggs [132, 138], water [135, 136, 140], honey [138], milk [138], human plasma [139], chicken [141] and kiwi roots [142]. However, the use of these stationary phases is not as extense as for other organic compounds because the vast majority of drugs are very polar. The first work in which CNTs were used to extract pharmaceuticals was developed by Pang et al. [130]. In this case propranolol, cinchonine, quinine, benzoyl-L-arginine, Z-L-arginine, benzoyl-L-arginine ethyl ester and benzoyl-DL-arginine-4-nitroanilide were extracted from urine using MWCNTs, although the same material was also used as matrix assisted laser desorption/ionization (MALDI)-MS) matrix. For this purpose, 10 mg of CNTs were rinsed with ACN and water, suspended in 1.0 mL 50% v/v MeOH and sonicated for 3 minutes. Later, this suspension was mixed with the selected analytes (or urine samples), sonicated, centrifuged and located onto the target of the MALDI-MS. Since then, MWCNTs have been used to extract benzodiazepines [131], antibiotics of different families [132, 136, 138], barbiturates [134] and antidepressants [137]. SWCNTs have also been used in this field [135, 136]. In few occasions, basically in the works of Suárez et al. [135] and Niu et al. [136], the extraction performance of both type of materials have been compared. Niu et al [136], for example, investigated the applicability of CNTs as packing SPE sorbents for the extraction of sulfonamides, cephalosporines and phenolic compounds from tap and well water. The extraction capability of MWCNTs and SWCNTs was also compared with traditional sorbents like C_{18} and GCB. Since compounds with more than one amino group or hydroxyl group are more difficult to desorb using CNTs or GCB, great efforts were made to optimize desorption conditions. Conventional water miscible solvents (such as MeOH or ACN) were not capable to elute the studied analytes (except the ones belonging to the cephalosporins family). That is why several amino compounds (i.e. ammonium acetate, aminoacetic acid and ammonium acetate) were tested as additives of the eluent. Effective elution was obtained with 30:70 v/v of 0.3M ammonium acetate/MeOH. For phenolic compounds better recoveries were obtained when the carbonaceous sorbents were pretreated with ascorbic acid. Sulfonamide compounds required larger volumes of the eluent (i.e. 5 mL) to achieve higher recoveries while phenolic compounds required the smallest volumes. In general, the antibiotics were more retained onto CNTs (recoveries > 80%) than onto CGB and, between them, MWCNTs had higher extraction ability than SWCNTs. Compared with C_{18}, CNTs were much superior for the extraction of the highly polar compounds. In fact, C_{18} was not suitable to extract trace levels of these polar compounds from environmental waters. MWCNTs were also compared with C_{18} sorbents for the analysis of ten sulfonamides in eggs and pork meat [132] and four benzodiazepines in pork [131], in which their extraction capacity was found clearly superior. On the contrary, C_{18} and MWCNTs extraction capacity was found very similar for the extraction of barbiturates from pork [134].

Table 3. CNTs-SPE and SPME applications for the extraction of pharmaceuticals

Analyte	Matrix	CNTs type and outer diameter	Sorbent amount (mg)	Elution solvent (mL)	Methodology	Recovery (%)	LOD	Ref.
3 Quinolinic alkaloids, arginines and peptides	Urine	MWCNTs	10	-	dSPE-MALDI-TOF-MS	-	-	[130]
4 Benzodiazepines	Pork	MWCNTs o.d.: 3-25 nm	200	Hexane/acetone (5:5 v/v) (5 mL)	SPE-GC-MS	75-104	2-5 μg/kg	[131]
10 Sulfonamides	Eggs and pork	MWCNTs o.d.: 60-100 nm	80	Water/MeOH (78:22 v/v) (1 mL)	On-line SPE-HPLC-UV	66-86	4.1-10.0 ng/L	[132]
3 Non steroidal anti-inflammatory drugs	Urine	Carboxylated-SWCNTs immobilized on inert porous glass -	15	MeOH (200 μL)	On-line SPE-CE-MS	99-102	1.6-2.6 μg/L	[133]
3 Barbiturates	Pork	MWCNTs o.d.: 20-40 nm, 40-60 nm, 60-100 nm	250	Acetone/EtOAc (3:7 v/v) (5 mL)	SPE-GC-MS/MS	76-96	0.2-0.1 μg/kg	[134]
3 Tetracyclines	Surface water	SWCNTs o.d.: 1.2-1.5 nm, 0.7-1.2 nm MWCNTs o.d.: 20-50 nm	6	ACN/acetic acid (90:10 v/v) (200 μL)	On-line SPE-CE-MS	99-103	0.30-0.69 μg/L	[135]
4 Cephalosporines, 4 sulfonamides and 6 phenolic compounds	Tap and well water	MWCNTs o.d.: 30-60 nm SWCNTs o.d.: 0.5-1.6 nm	100	Cephalosporines: ACN/MeOH/water (3 or 4 mL) Sulfonamides: 0.3 M ammonium acetate/MeOH 30:70 (v/v) (5 mL)	SPE-HPLC-UV	55-102	27-38 ng/L	[136]

Analyte	Matrix	CNTs type and outer diameter	Sorbent amount (mg)	Elution solvent (mL)	Methodology	Recovery (%)	LOD	Ref.
9 Antidepressants	Urine	MWCNTs o.d.: 20-50 nm	30	ACN/acetic acid (97:3 v/v) (2 mL)	SPE-HPLC-UV	72-97	12.3-50.0 μg/L	[137]
Chloramphenicol	Egg, honey, milk	MWCNTs o.d.: 40-60 nm	100	ACN (5 mL)	SPE-HPLC-MS/MS	96-102	0.004 μg/kg (egg), 0.003 μg/kg (honey) 0.003 μg/kg (milk)	[138]
10 Quinolones	Human plasma	Magnetic-MWCNTs o.d.: 5-9 μm	1.28	MeOH with 10% NH_4OH (1 mL)	dSPE-UPLC-UV	70-100	5.8-14.5 μg/L	[139]
11 Quinolones	Milli-Q, mineral, tap and waste water	Oxidized-MWCNT o.d.: 110-170 nm	150	Acetone/MeOH (3:1 v/v) (25 mL)	dSPE-CE-DAD	62-114	0.028-0.094 μg/L	[140]
3 Macrolides	Chicken muscle	MIP-MWCNTs o.d.: 20-40 nm	100	EtOH/acetic acid (90:10 v/v) (2 mL)	SPE-HPLC-UV	85-96	-	[141]
Ursolic acid	Root of kiwi	MIP-MWCNTs o.d.: 30-50 nm	300	EtOH/acetic acid 9:1 v/v) (30 mL)	SPE-HPLC-UV	80	-	[142]

- Not indicated.

Like other SPE sorbents, although in a lower number of occasions, CNTs have also formed part of automated and miniaturized procedures as it has been demonstrated in the articles of Fang et al. [132] and Suárez et al. [133, 135]. In the first of them [132], the conventional sample loop of a six port injector valve of a HPLC system was replaced by a preconcentration column packed with CNTs to on-line interface SPE with HPLC. Five grams of sample (pork or egg) were mixed with ferrite potassium cyanide and zinc sulphate. The mixture was heated for 5 min at 75ºC, mixed with ACN and sonicated 5 min. After filtration samples were diluted to 200 mL and introduced onto the SPE microcolumn at a flow rate of 4.5 mL/min while the injector valve was in the load position. Sulfonamides (sulfadiazine, sulfathiazole, sulfamerazine, sulfamoxol, sulfamethizole, sulfadimidine, sulfamethoxy-pyridazine, sulfachlorpyridazine, sulfadoxine, sulfisoxazole) were eluted in the backflush mode by the mobile phase at a flow rate of 1.0 mL/min by switching HPLC from load to inject position. Finally the valve was turned to the load position for next sample preconcentration while analytes were separated in the chromatographic system. Recoveries between 70 and 86% with relative standard deviations (RSDs) in the range 0.9-7.6% as well as clean chromatograms (see Figure 7) were obtained. In the articles of Suárez et al. [135] similar dispositives consisting on three valves were used to make the on-line SPE-CE-MS analysis.

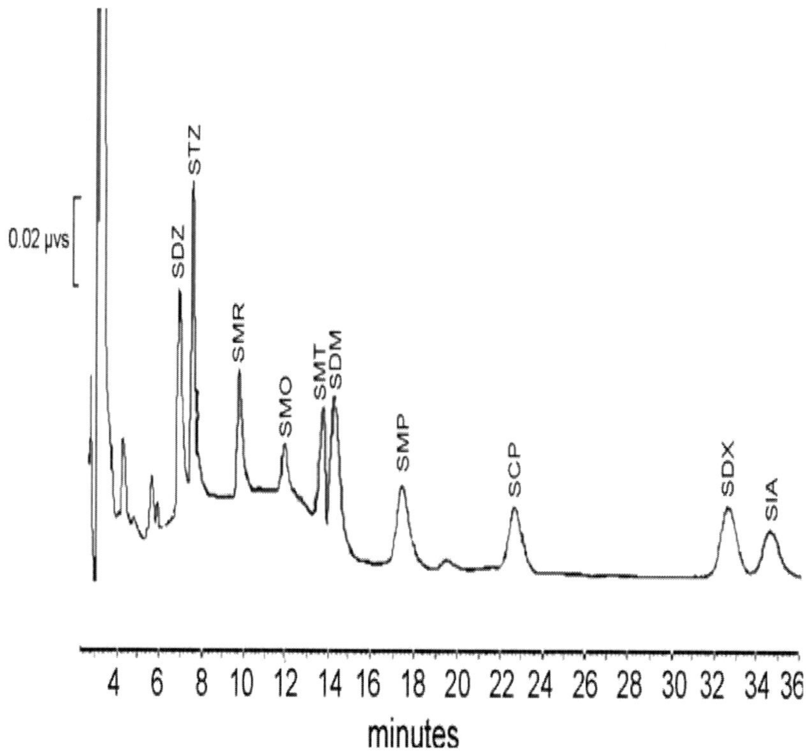

Figure 7. Chromatogram of the on-line SPE preconcentration of 100 mL of the real sample spiked at 4 ng/g of sulphonamides. Mobile phase: water (pH=3.20)/MeOH 78:22 v/v. Flow rate of 4.5 mL/min. UV detection at 270 nm. Peak identification: SDZ, sulfadiazine; STZ, sulfathiazole; SMR, sulfamerazine; SMO, sulfamoxol; SMT, sulfamethizole; SDM, sulfadimidine; SMP, sulfamethoxypyridazine; SCP, sulfachlorpyrydazine; SDX, sulfadoxine; SIA, sulfisoxazole. Reprinted from [132] with permission from Elsevier.

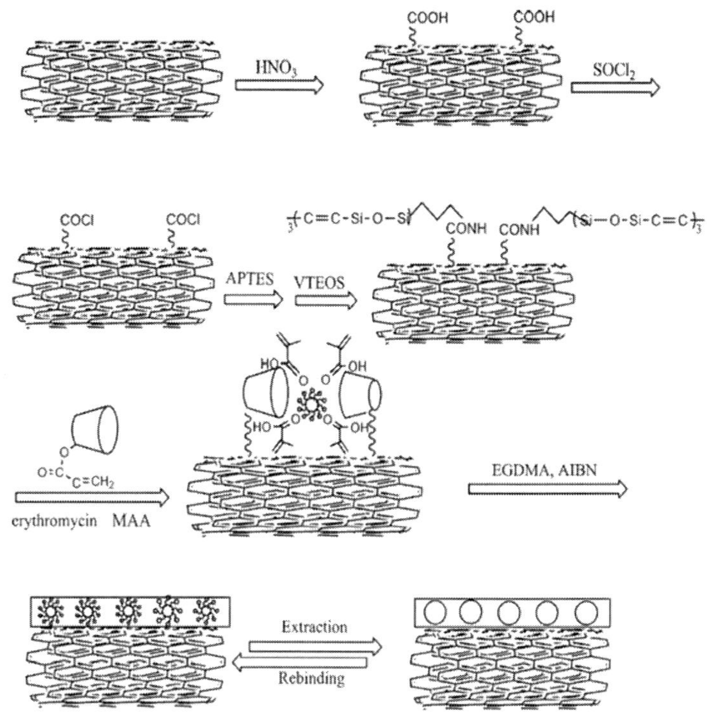

Figure 8. Synthesis protocol of MWCNTs-MIPs. First step: oxidation with HNO_3, second step: reflux in $SOCl_2$, third step: obtaining of vinyl functionalized MWCNTs and last step: preparation of MWCNTs-MIPs. Reprinted from [141] with permission from Elsevier.

Modified CNTs have also been used to extract pharmaceuticals [133, 139-142] using both easy and complex modification methods. As an example of the first of them, the work of Herrera-Herrera et al. [140] described the use of oxidized-MWCNTs to extract eleven quinolones from Milli-Q, mineral, tap and waste waters. For this purpose, 10 g of raw MWCNTs were suspended in 1 L of 3.0 M HNO_3 and stirred at 80ºC and 500 rpm for 24 h. Then, the suspension was diluted with Milli-Q water, filtered and washed until neutral pH. Extraction recoveries were compared with those obtained with the same whole extraction procedure under optimum conditions (250 mL of water at pH=5.0, 150 mg MWCNTs, elution with 25 mL of 3:1 v/v acetone/MeOH) using the same MWCNTs without oxidation. Results obtained, i.e. extraction recoveries in the range 78-116% for oxidized-MWCNTs and between 49 and 74% for MWCNTs, demonstrated how such a simple oxidation enhances the extraction capacity of MWCNTs. It is worth mentioning that this article also represents one of the few works [139, 140] in which dSPE with MWCNTs was used to extract drugs.

As an example of the application of complex modification methods, MWCNTs have been used as support materials for the synthesis of novel molecularly imprinted polymers (MIP) to extract three macrolides (i.e. erythromycin, roxithromycin and spiramycin) [141] and ursolic acid [142]. For this purpose, Zhang et al. [141] oxidized MWCNTs by stirring in HNO_3 at 80ºC for 24 h. The oxidized-MWCNTs were refluxed in $SOCl_2$ for 24 h at 60ºC and the obtained MWCNTs-COCl were dispersed in tetrahydrofurano (THF) and mixed with 3-aminopropyltriethoxysilane (APTES). After the solution was stirred at 50ºC for 12 h, vinyltriethoxysilane (VTEOS) was added and stirred at 50ºC for 12 h more.

Finally, vinyl functionalized MWCNTs were incubated with ACN and acryloyl-β-cyclodextrin, methacrylic acid (MAA) and erythromycin (template) were dissolved in the above solution. The mixture was magnetically stirred at 60ºC for 2h and ethylene glycol dimethacrylate (EGDMA) and azobisisobutyronitrile (AIBN) were added to reflux for 24 h at 60ºC. Figure 8 summarizes this synthesis protocol. Xi et al. [142] developed a similar method in which MWCNTs were oxidized and functionalized with polyethylene glycol (PEG). MWCNT-MIPs were later prepared using ursolic acid as template. In both works, MWCNTs-MIPs were morphological characterized. Zhang et al. [141] studied the obtained CNTs by TEM and SEM, whereas Xi et al. [142] used SEM and infrared (IR) analysis. Adsorption isotherms were also obtained in both works. The obtained MWCNTs-MIPs demonstrated their extraction efficiency with recoveries higher than 80% and RSDs lower than 5%.

3.2.3. Miscellaneous

Apart from pesticides and pharmaceuticals, CNTs have been used to extract other organic compounds such as endocrine disruptors [143, 144], halogenated phenols [144-146], biological compounds [147-150] and contaminants like phthalate esters [144, 151], volatile organic compounds [152], biphenyls [146], nitrophenols [153], polycyclic aromatic hidrocarbons (PAHs) [154], linear alkylbenzene sulfonates [155], cationic surfactants [156], amines [157], parabens [158] and chemical warfare agents [159]. Analyzed samples included water [143, 145, 146, 151-157, 159, 160], air [152], milk [149], cosmetics [158] and biological fluids (i.e. human blood [147, 150] and urine [148]). Table 4 compiles some of these works. As for other SPE applications, several experimental factors affecting the extraction efficiency such as sample pH, sample flow rate and volume and elution solvent type and volume have been optimized following a step by step approach in the majority of occassions. For this purpose either GC [146, 152, 157, 159, 160] or HPLC [143, 145, 148, 149, 151, 153-155, 158] and even UV-Vis spectrometry [147, 150] or MALDI-MS [156] have been coupled off-line to the SPE procedure.

In this case, also different functionalized CNTs, including oxidized-MWCNTs [146, 156], MWCNTs-PEG [146], MWCNTs-octadecylamine functionalized (MWCNTs-ODA) [146], carboxylated-MWCNTs [155], magnetic-MWCNTs [149, 159], magnetic-SWCNTs [149], magnetic-MWCNTs-COOH [149], and magnetic-MWCNTs-NH_2 [149], have been used to extract organic compounds. An example is the work of Salam and Burke [146] in which non-functionalized and chemically modified MWCNTs (oxidized-MWCNTs, MWCNTs-PEG, MWCNTs-ODA) were used to extract polyhalogenated organic pollutants. Results demonstrated that MWCNTs-ODA showed a higher adsorption capacity (recoveries between 93 and 100%) for the extraction of the selected group of compounds (penta-chlorophenol, 2,4,5-trichlorophenol, 3,3′,4,4′-tetrachlorobyphenyl and 2,2′,5,5′-tetra-bromobyphenyl) followed by non-modified MWCNTs (recoveries in the range 81-93%) and MWCNTs-PEG (recoveries from 55 to 93%). Compared with C_{18} (recoveries in the range between 47% and 68%), MWCNTs showed a much higher extraction efficiency. Another interesting article is the work of Ding et al. [149] in which the fabrication of magnetic CNTs based on aggregation wrap was proposed. CNTs (SWCNTs, MWCNTs, MWCNTs-COOH and MWCNTs-NH_2) were vortexed with magnetic nanoparticles in a suitable solvent (DMF) and then magnetic nanoparticles were wrapped into the CNT bundles that formed during the aggregation process. The micromorphologies of MWCNTs, magnetic nanoparticles and magnetic MWCNTs were observed by TEM (see Figure 9).

Table 4. Some applications of CNTs to SPE and SPME for the extraction of miscellaneous compounds

Analyte	Matrix	CNTs type and outer diameter	Sorbent amount (mg)	Elution solvent (mL)	Methodology	Recovery (%)	LOD	Ref.
SPE								
3 Endocrine disruptors	Tap, river, sea and waste water	MWCNTs o.d.: 30-60 nm	500	MeOH (2.5 mL)	SPE-HPLC-FD	90-104	18-83 ng/L	[143]
3 Endocrine disruptors, 4 phthalate esters and 5 chlorophenols	Tap, river and waste water	SWCNTs o.d.: 1.5-2.5 nm	30	MeOH or ACN (8-15 mL)	SPE-HPLC-FD	59-100	7.0-38 ng/L	[144]
5 Chlorophenols	Tap and river water	MWCNTs o.d.: 30-60 nm	300	MeOH at pH 10 (6 mL)	SPE-HPLC-UV	88-109	80-800 ng/L	[145]
2 Polychlorophenols and 2 tetrahalogenated biphenyls	River water	MWCNTs o.d.: - Oxidized-MWCNTs o.d.: - MWCNTs-PEG o.d.: - MWCNTs-ODA o.d.: -	20	Acetone (-)	SPE-GC-ECD	55-100	1.0-6.0 μg/L	[146]
2 Basic proteins (hemoglobin and cytochrome *c*)	Human blood	MWCNTs o.d.: 60-100 nm	8-10	0.5 M NaCl and 0.025 M phosphate buffer at pH 8 (200 μL and 200 μL)	On-line-SPE-UV-VIS spectrophotometry	90-98	60-120 μg/L	[147]
3 Biogenic thiols	Urine and human blood	MWCNTs i.d.: 60-100 nm	80	ACN (300 μL)	SPE-HPLC-FD	92-113	0.004-0.080 nM	[148]

Table 4. (Continued).

Analyte	Matrix	CNTs type and outer diameter	Sorbent amount (mg)	Elution solvent (mL)	Methodology	Recovery (%)	LOD	Ref.
3 Estrogens	Milk	Magnetic-SWCNTs o.d.: 2 nm Magnetic-MWCNTs o.d.: 10-20 nm Magnetic-MWCNTs-COOH o.d.: 10-20 nm Magnetic-MWCNTs-NH_2	2.5	Acetone (1 mL)	dSPE-HPLC-FD	88-112	1.21-2-35 µg/L	[149]
Bovine hemoglobin and serum albumin	Human blood	MWCNTs-PDDA -	15	0.04 M citrate buffer at pH 7 (200 µL)	On-line-SPE-UV-VIS spectrophotometry	90	1.0 mg/L	[150]
4 Phthalate esters	Tap, river and sea water	MWCNTs o.d.: 30-60 nm	500	ACN (5 mL)	SPE-HPLC-DAD	68-105	180-860 ng/L	[151]
16 volatile organic compounds	Air and water	MWCNTs o.d.: 15-30 nm	650		Purge and trap-GC-FID	82-111	-	[152]
2-nitrophenol, 2,6-dichloroaniline and naphtalene	Distilled water	SWCNTs and MWCNTs o.d.: -	0.3	ACN (50-200 µL)	µ-SPE-HPLC-UV	-	0.1-3 µg/L	[153]
10 PAHs	Tap and river water	MWCNTs o.d.: 30-50 nm	500	ACN (4 mL)	SPE-HPLC-UV	79-117	79-117	[154]
4 Linear alkylbenzene sulfonates	River water	Carboxilated-MWCNTs o.d.: 30-60 nm	200	MeOH (3 mL)	SPE-HPLC-UV	85-106	20-30 ng/L	[155]
4 Cationic surfactants	River and waste water	Oxidized-MWCNTs o.d.: 10-20 nm	100	-	dSPE-AP-MALDI-MS	91-98	10-15 ng/L	[156]

Analyte	Matrix	CNTs type and outer diameter	Sorbent amount (mg)	Elution solvent (mL)	Methodology	Recovery (%)	LOD	Ref.
27 amines: 2 anilines, 12 chloroanilines, 10 N-nitrosamines and 3 aliphatic amines	River, pond, tap, well, drinking, swimming pool and waste water	MWCNTs i.d.: 2-15 nm	80	EtOAc:ACN (9:1 v/v) (150 μL)	On-line-SPE-GC-MS	-	-	[157]
4 Parabens	Moisturizing cream, antiwrinkle cream, make-up, lotion, shampoo, hands cream and after sun	MWCNTs o.d.: 20-30 nm	20	Acetone (2 mL)	SPE-HPLC- C-CAD	85-104	0.5-2.1 mg/L	[158]
9 Chemical warfare agents	Distilled, tap and muddy water	Magnetic-MWCNTs o.d.:7-15 nm	40	MeOH (800 μL)	dSPE-GC-FPD	55-96	0.05-1 μg/L	[159]
4 Polychlorobenzenes	River water	MWCNTs o.d.: 30-60 nm	15	-	SPE-GC-MS	80-103	80-600 ng/L	[160]
SPME								
5 Phenols	River and waste water	Oxidized-MWCNTs o.d.: 10-20 nm	2	-	SPME-HPLC-UV	24-119	0.25-3.67 μg/L	[161]

Table 4. (Continued)

Analyte	Matrix	CNTs type and outer diameter	Sorbent amount (mg)	Elution solvent (mL)	Methodology	Recovery (%)	LOD	Ref.
6 Phenols	Sea and tap water	SWCNTs o.d.: -	0.4-0.9	-	SPME-HPLC-UV	88-102	0.9-3.8 μg/L	[162]
6 Phenols	Lake water	MWCNTs o.d.: -	100	-	SPME-GC-FID	88-116	1.05-65.9 ng/L	[163]
5 Polar aromatic compounds	Pond water	MWCNTs/Nafion o.d.: -	-	-	SPME-GC-FID	83-106	0.03-0.57 μg/L	[164]
5 Polybrominated diphenyl ethers	River water, waste water and milk	MWCNTs o.d.: 20-40 nm	-	-	SPME-GC-ECD	90-119	3.6-8.6 ng/L	[165]
4 Anilines	River and waste water	Oxidized-MWCNTs o.d.: 20-40 nm	-	-	In-tube SPME-HPLC-UV	56-110	40-130 ng/L	[166]
7 Polybrominated diphenyl ethers	Reservoir and waste water	SWCNTs o.d.: < 2 nm	90	-	SPME-GC-ECD	74-109	0.08-0.8 ng/L	[167]

C-CAD: corona-charged aerosol detector; PEG: polyethylene glycol, ODA: octadecylamine, PDDA: poly(diallyldimethyl-ammonium chloride), -: Not indicated.

It could be observed that magnetic nanoparticles were nearly monodispersed and spherical with a mean diameter of about 100 nm. It could also be clearly seen that magnetic nanoparticles were wrapped by the CNTs bundles and, as a result, CNTs could be separated from the matrix solutions by an external magnet. The resultant magnetic-CNTs were applied as sorbents for the dSPE of three estrogens (estradiol, ethinyloestradiol and hexestrol) in milk. Extraction efficiencies of magnetic CNTs were investigated under their individual optimal conditions. Estrogens extraction recoveries generally decreased in the following order: magnetic-SWCNTs > magnetic-MWCNTs > magnetic MWCNTs-NH_2 > magnetic-MWCNTs-COOH. Since authors claimed that compared with MWCNTs, SWCNTs are relatively expensive, magnetic-MWCNTs were chosen as the final sorbents. The LODs for three estrogens ranged from 1.21 to 2.35 μg/L. The reproducibility of dSPE with different batches of magnetic-CNTs was acceptable with RSD values lower than 3.6%. Non-covalent wrapping functionalization was also used by Du et al. [150] to fabricate poly(diallyldimethyl-ammonium chloride) (PDDA) wrapped MWCNTs. MWCNTs-PDDA were packed in a micro-column to on-line couple the SPE procedure with UV/Vis spectrometry for determination of bovine hemoglobin and serum albumin in human blood. These proteins adsorbed onto the MWCNTs-PDDA composites were eluted with a citrate buffer as stripping reagent with a recovery of 90% resulting in an enrichment factor of 14.

Both SWCNTs and MWCNTs have been also explored in this field as high performance sorbents for μ-SPE within the needle of a syringe by Sae-Khow and Mitra [153]. The dispositive was fabricated by introducing 300 mg of CNTs into a 100 mm long capillary plugged with glass wool at both ends. Adsorption isotherms were obtained and extraction efficiency and enrichment factor were studied for 2-nitrophenol, 2,6-dichloroaniline and naphthalene. However, no samples were investigated and optimization was done with distilled water. Analytes were concentrated by drawing several mL of water into the syringe through the needle, and then desorbed in few microliters of ACN. This work represents the first μ-SPE application of CNTs with the aim of miniaturizing conventional SPE into smaller beds and minimizing solvent consumption. The technique is comparable to SPME. Miniaturization and automatization can also be done by on line coupling SPE with the determination method. Such approach was also followed by Du et al. [147, 150] and Jurado-Sánchez et al. [157] as on-line CNTs SPE for the extraction of basic [147] and acidic proteins [150] from human blood and 27 amines from water, respectively. In this way, elution solvent was reduced from several mL to very few microliters. Recoveries obtained with this technique were very similar to the ones obtained by off-line SPE (higher than 90% for basic and acidic proteins –recoveries not given for amines-) with LODs values between 60 and 120 μg/L for basic proteins and 1.0 mg/L for acidic proteins.

CNTs have also been disposed in formats different from classical SPE. This is the case of CNTs deposited on fibers which finally result in a SPME approach [161-167]. The methods most commonly used to deposit CNTs include sol-gel fiber coating [167] and heat coating on the surface of fused-silica tube or fiber [165, 166]. CNTs-SPME fibers have been used to extract mainly phenols [161-163] and polybrominated diphenyls ethers [165, 167], although anilines [167] and other polar compounds [164] have been analyzed too. Both MWCNTs (modified [161, 164, 166] and non modified [163, 165]) and SWCNTs [162, 167] have been used as coating materials. Water samples have been in most cases the matrix of choice (i.e. river [161, 165, 166], waste [161, 165, 166, 167], sea [162], tap [162], lake [163], pond [164] and reservoir water [168]) except in the work of Wang et al. [165] in which milk samples were also selected. In this last work, extraction capacity of laboratory made MWCNTs-fibers was compared with that of commercialized SPME fibers of PDMS-divinylbenzene (PDMS-DVB). MWCNTs provided better results than the classical stationary phase. As another clear example, MWCNTs fibers were found superior to PA fibers [164] to extract polar organic compounds like p-nitroaniline, m-cresol, aniline, phenol and o-dinitrobenzene. In the work by Liu et al. [166], in tube-SPME coupled to HPLC-UV was used for the determination of substituted aniline compounds. In this case, oxidized-MWCNTs were coated on the outer surface of the fused-silica tube and inserted in a polyether ether ketone tubing which was fixed on the six-port injection valve of the HPLC instrument. Recoveries higher than 53% what means LODs between 40 and 130 ng/L were obtained.

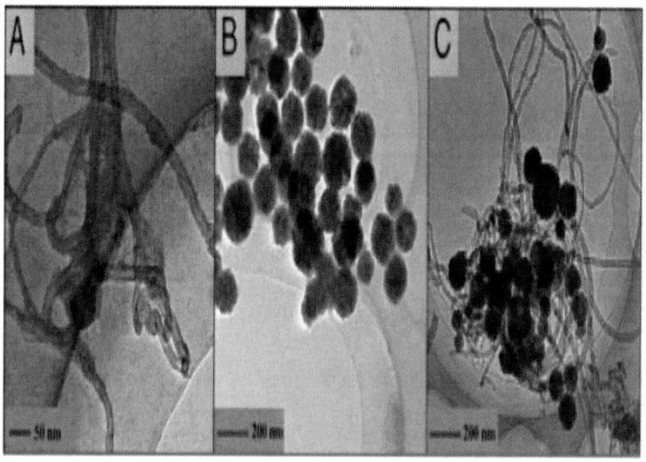

Figure 9. TEM images of the nano-materials: MWCNTs (A), magnetic nanoparticles (B) and magnetic-MWCNTs (C). Reprinted from [149] with permission from Wiley-VCH.

CONCLUSION

Sorptive extraction methods are receiving increasing attention now and in the future. In particular, the introduction of new sorbent materials like CNTs for use in extraction and microextraction techniques has attracted the interest of analytical chemists because of their suitable properties for this purpose: high surface area and internal cavities, chemical, mechanical and thermal stability and their ability to establish π-π interactions as well as excellent Van der Waals interactions with other molecules, especially hydrophobic ones.

In this chapter, the different applications of CNTs for the SPE and SPME of organic and inorganic analytes have been reviewed and summarized. Many interesting results have been reported in the literature, which clearly demonstrate their suitability as sorbent materials. In general, the majority of these works have been focused on the extraction of organic analytes, mainly pesticides. In most cases, water samples have been the matrices of choice also using MWCNTs, since they show better sorption properties than SWCNTs. In many of these applications, small quantities of CNTs are enough to ensure a feasible and repeatable extraction of the required analytes. In some cases it has even been demonstrated that they can be reused several times.

In addition, their different structure and size increase the number of possible applications of the technique. Besides, their functionalization also offers an alternative and challenging application field in SPE, which is still to be fully explored, since their extraction selectivity can be modified/changed according to the analytical problem.

Finally, it should be mentioned that CNTs are nowadays commonly used in laboratories due to their commercial availability in large quantities. They constitute today a multidisciplinary material that is finding an important place in Analytical Chemistry. However, new contributions are still required, especially regarding real world applications with complex matrices.

REFERENCES

[1] K.H. Meyer, R. Bacon, *J. Appl. Phys.* 30, 2035 (1959).
[2] Oberlin, M. Endo, T. Koyama, *J. Cryst. Growth* 32, 335 (1976).
[3] M. Endo, A. Oberlin, T. Koyama, *Jpn. J. Appl. Phys.* 16, 1519 (1977).
[4] M. Endo, A. Tamagawa, T. Koyama, *Jpn. J. Appl. Phys.* 16, 1771 (1977).
[5] W. Krätschmer, L.D. Lamb, K. Fortiropoulus, D.H. Huffman, *Nature* 347, 354 (1990).
[6] S. Iijima, *Nature* 354, 56 (1991).
[7] S. Iijima, T. Ichichashi, *Nature* 363, 603 (1993).
[8] D.S. Bethune, C.H. Klang, M.S. de Vries, G. Gorman, R. Savoy, J. Vasquez, R. Beyers, *Nature* 363, 605 (1993).
[9] C. Journet, W.K. Matser, P. Bernier, L. Laiseau, S. Lefrant, P. Deniard, R. Lee, J.E. Fischer, *Nature* 388, 756 (1997).
[10] Thess, R. Lee, P. Nikolaev, H. Dai, P. Petit, J. Robert, C. Xu, Y.H. Lee, R.E. Smaley, *Science* 276, 483 (1996) 483
[11] E. Flahaut, C. Laurent, A. Peigney, *Carbon* 43, 375 (2005).
[12] M. Meyyappan, *Carbon Nanotubes-Science and Applications*, Taylor and Francis, Boca Raton (2004).
[13] S. Wang, *Current Appl. Phys.* 9, 1146 (2009).
[14] Y.L. Zhao, J. Fraser Stoddart, *Acc. Chem. Res.* 42, 1161 (2009).
[15] C.Y. Hu, Y.J. Xu, S.W. Duo, R.F. Zhang, *J. Chin. Chem. Soc.* 56, 234 (2009).
[16] V. Valcárcel, S. Cárdenas, B.M. Simonet, *Anal. Chem.* 79, 4788 (2007).
[17] Y. Gogotsi, *Carbon Nanomaterials*, Taylor and Francis, Boca Raton (2006).
[18] M. Sato, *Procedia Engin.* 14, 2366 (2011).
[19] M. Trojanowicz, *Trends Anal. Chem.* 25, 480 (2006).
[20] C. Li, E.T. Thostenson, T.-S. Chou, *Compos. Sci. Technol.* 68, 1227 (2008).
[21] C.B. Jacobs, M.J. Peairs, B.J. Venton, *Anal. Chim. Acta* 662, 105 (2010).
[22] J. Li, J.E. Koehne, A.M. Cassell, H. Chen, H.T. Ng, Q. Ye, W. Fan, J. Han, M. Meyyppan, *Electroanalysis* 17, 15 (2005).
[23] L. Xiao, G.G. Wildgoose, R.G. Compton, *Anal. Chim. Acta* 620, 44 (2008).
[24] U. Injang, P. Noyrod, W. Siangproh, W. Dungchai, S. Motomizu, O. Chailapakul, *Anal. Chim. Acta* 668, 54 (2010).
[25] G. Chen, *Talanta* 74, 326 (2007).
[26] M. Pumera, A. Merkoci, S. Alegret, *Electrophoresis* 28, 1274 (2007).

[27] A.-H. Duan, S.-M. Xie, L-M- Yuan, *Trends Anal. Chem.* 30, 484 (2011).
[28] Safavi, N. Maleki, M.M. Doroodmand, *Anal. Chim. Acta* 675, 207 (2010).
[29] D. Merli, A. Speltini, D. Ravelli, E. Quartarone, L. Costa, A. Profumo, *J. Chromatogr. A* 1217, 7275 (2010).
[30] M. Valcárcel, S. Cárdenas, B.M. Simonet, Y. Moliner-Martínez, R. Lucena, *Trends Anal. Chem.* 27, 34 (2008).
[31] E. Mena, F. Della Negra, M. Prato, N. Tagmatarchis, A. Ciogli, F. Gasparrini, D. Misita, C. Villani, *Carbon* 44, 1609 (2006).
[32] S.D. Chambers, F. Svec, J.M.J. Frechet, *J. Chromatogr. A* 1218, 2546 (2011).
[33] B. Suárez, B.M. Simonet, S. Cárdenas, M. Valcárcel, *Electrophoresis* 28, 1714 (2007).
[34] Y. Xu, S.F.Y. Li, *Electrophoresis* 27, 4025 (2006).
[35] Y. Picó, M. Fernández, M.J. Ruiz, G. Font, *J. Biochem. Biophys. Methods* 70, 117 (2007).
[36] R.P. Belardi, J. Pawliszyn, *Water Pollut. Res. J. Canada* 24, 179 (1989).
[37] C.L. Arthur, J. Pawliszyn, *Anal. Chem.* 62, 2145 (1990).
[38] K. Ridgway, S.P.D. Lalljie, R.G. Smith, *J. Chromatogr. A* 1153, 36 (2007).
[39] M. Asensio-Ramos, A.V. Herrera-Herrera, M.A. Rodríguez-Delgado, J. Hernández-Borges, Sorbent-based techniques for the determination of pesticides in foods *Comprehensive sampling and sample preparation*, volume 4. Eds: J. Pawliszyn, L. Mondello, P. Dugo. Elsevier, Academic Press 7 Oxford, UK, pág. 263-312, 2012. ISBN: 9780123813732.
[40] L.M. Ravelo-Pérez, A.V. Herrera-Herrera, J. Hernández-Borges, M.A. Rodríguez-Delgado, *J. Chromatogr. A* 1217, 2618 (2010).
[41] C.M. Hussain, C. Saridara, S. Mitra, *Analyst* 134, 1928 (2009).
[42] K. Pyrzynska, *Trends Anal. Chem.* 29, 718 (2010).
[43] P. Liang, Q. Ding, F. Song, *J. Sep. Sci.* 28, 2339 (2005).
[44] C.R. Teixeira-Tarley, A.F. Barbosa, M. Gava-Segatelli, E. Costa-Figueiredo, P. Orival-Luccas, *J. Anal. At. Spectrom.* 21, 1305 (2006).
[45] A.F. Barbosa, M.G. Segatelli, A.C. Pereira, A. de Santana Santos, L. Tatsuo Kubota, P. Orival Luccas, C.R. Teixeira Tarley, *Talanta* 71, 1512 (2007).
[46] P. Liang, E. Zhao, Q. Ding, D. Du, *Spectrochim. Acta B* 63, 714 (2008).
[47] Stafiej, K. Pyrzynska, *Microchem. J.* 89, 29 (2008).
[48] S. Chen, C. Liu, M. Yang, D. Lu, L. Zhu, Z. Wang, *J. Hazard. Mater.* 170, 247 (2009).
[49] S. Ghaseminezhad, D. Afzali, M.A. Taher, *Talanta* 80, 168 (2009).
[50] H.-M. Yu, W. Sun, M.-L. Chen, Y. Tian, J.-H. Wang, *Chin. J. Anal. Chem.* 38, 1721 (2010).
[51] A.H. El-Sheikh, Y.S. Al-Degs, R.M. Al-As'ad, J.A. Sweileh, *Desalination* 270, 214 (2011).
[52] J.P. Xiao, Q.X. Zhou, H.H. Bai, *J. Environ. Sci.* 19, 1266 (2007).
[53] J.P. Xiao, Q.X. Zhou, H.H. Bai, *Chin. Chem. Lett.* 18, 714 (2007).
[54] B. Parodi, M. Savio, L.D. Martinez, R.A. Gil, P. Smichowski, *Microchem. J.* 98, 225 (2011).
[55] M. Savio, B. Parodi, L.D. Martinez, P. Smichowski, R.A. Gil, *Talanta* 85, 245 (2011).
[56] Z. Es'haghi, M. Khalili, A. Khazaeifar, G.H. Rounaghi, *Electrochim. Acta* 56, 3139 (2011).
[57] M. Tuzen, M. Soylak, *J. Hazard. Mater.* 147, 219 (2007).

[58] M. Tuzen, K.O. Saygi, M. Soylak, *J. Hazard. Mater.* 152, 632 (2008).
[59] Duran, M. Tuzen, M. Soylak, *J. Hazard. Mater.* 169, 466 (2009).
[60] H. Wu, X. Wang, B. Liu, Y. Liu, S. Li, J. Lu, J. Tian, W. Zhao, Z. Yang, *Spectrochim. Acta B* 66, 74 (2011).
[61] López-García, R.E. Rivas, M. Hernández-Córdoba, *Talanta* 86, 52 (2001).
[62] A.H. El-Sheikh, J.A. Sweileh, Y.S. Al-Degs, *Anal. Chim. Acta* 604, 119 (2007).
[63] H. Li, Y. Huang, Y. Wang, W. Wang, *Anal. Chim. Acta* 631, 182 (2009).
[64] Y. Liu, Y. Li, X.P. Yan, *Adv. Funct. Mater.* 18, 1536 (2008).
[65] P.H. Pacheco, P. Smichowski, G. Polla, L.D. Martinez, *Talanta* 79, 249 (2009).
[66] P.H. Pacheco, R.A. Gil, P. Smichowski, G. Polla, L.D. Martinez, *Anal. Chim. Acta* 656, 36 (2009).
[67] B. Parodi, M. Savio, L.D. Martinez, R.A. Gil, P. Smichowski, *Microchem. J.* 98, 225 (2011).
[68] M. Savio, B. Parodi, L.D. Martinez, P. Smichowski, R.A. Gil, *Talanta* 85, 245 (2011).
[69] Y. Liu, Y. Li, Z.Q. Wu, X.P. Yan, *Talanta* 79, 1464 (2009).
[70] M. Tuzen, K.O. Saygi, C. Usta, M. Soylak, *Bioresour. Technol.* 99, 1563 (2008).
[71] D. Afzali, A. Mostafavi, Anal. Sci. 24, 1135 (2008).
[72] S. Tajik, M.A. Taher, *Desalination* 278, 57 (2011).
[73] T. Shamspur, A. Mostafavi, *J. Hazard. Mater.* 168, 1548 (2009).
[74] Mostafavi, F. Etemadi, D. Afzali, H. Khabazzadeh, *Chem. Anal. (Warsaw)* 54, 459 (2009).
[75] D. Afzali, R. Jamshidi, S. Ghaseminezhad, Z. Afzali, *Arabian J. Chem.* (2001) doi: 10.1016/j.arabjc.2011.01.017.
[76] Z. Zang, Z. Hu, Z. Li, Q. He, X. Chang, *J. Hazard. Mater.* 172, 958 (2009).
[77] Y.-C. Chiang, W.-H. Lin, Y.-C. Chang, *Appl. Surf. Sci.* 631, 182 (2009).
[78] Z. Zhang, H. Zhang, Y. Hu, X. Yang, S. Yao, *Talanta* 82, 304 (2010).
[79] R.A. Gil, S.N. Goyanes, G. Polla, P. Smichowski, R.A. Olsina, L.D. Martinez, *J. Anal. At. Spectrom.* 22, 1290 (2007).
[80] R.A. Gil, P.H. Pacheco, P. Smichowski, R.A. Olsina, L.D. Martinez, *Microchim. Acta* 167, 187 (2009).
[81] J. Muñoz, M. Gallego, M. Valcárcel, *Anal. Chem.* 77, 5389 (2005).
[82] P. Liang, Y. Liu, L. Guo, *Spectrochimica Acta B* 60, 125 (2005).
[83] L. Li, Y. Huang. Y. Wang, W. Wang, *Anal. Chim. Acta* 631, 182 (2009).
[84] M. Soylak, O. Ercan, *J. Hazard. Mater.* 168, 1527 (2009).
[85] M. Soylak, Y.E. Unsal, *J. AOAC Int.* 92, 1219 (2009).
[86] B. Yang, Q. Gong, L. Zhao, H. Sun, N. Ren, J. Qin, J. Xu, H. Yang, *Desalination* 278, 65 (2011).
[87] J. Álvarez Méndez, J. Barciela García, R.M. Peña Crecente, S. García Martín, *Talanta* 85, 2361 (2011).
[88] C. Basheer, A.A. Alnedhary, B.S.M. Rao, S. Valliyaveettil, H.K. Lee, *Anal. Chem.* 78, 2853 (2006).
[89] D. Du, M. Wang, J. Zhang, J. Cai, H. Tu, A. Zhang, *Electrochem. Commun.* 10, 85 (2008).
[90] L.M. Ravelo-Pérez, J. Hernández-Borges, M.A. Rodríguez-Delgado, *J. Sep. Sci.* 31, 3612 (2008).

[91] L.M. Ravelo-Pérez, J. Hernández-Borges, M.A. Rodríguez-Delgado, *J. Chromatogr. A* 1211, 33 (2008).
[92] H. Katsumata, T. Matsumoto, S. Kaneco, T. Suzuki, K. Ohta, *Microchem. J.* 88, 82 (2008).
[93] M. Asensio-Ramos, J. Hernández-Borges, T.M. Borges-Miquel, M.A. Rodríguez-Delgado, *Anal. Chim. Acta* 647, 167 (2009).
[94] M.R. Hadjmohammadi, M. Peyrovi, P. Biparva, *J. Sep. Sci.* 33, 1044 (2010).
[95] Q. Zhou, J. Xiao, W. Wang, *J. Chromatogr. A* 1125, 152 (2006).
[96] Q. Zhou, J. Xiao, W. Wang, G. Liu, Q. Shi, J. Wang, *Talanta* 68, 1309 (2006).
[97] Q. Zhou, W. Wang, J. Xiao, J. Wang, G. Liu, Q. Shi, G. Guo, *Microchim. Acta* 152, 215 (2006).
[98] Q. Zhou, Y. Ding, J. Xiao, *Chromatographia* 65, 25 (2007).
[99] G. Min, S. Wang, H. Zhu, G. Fang, Y. Zhang, *Sci. Total Environ.* 396, 79 (2008).
[100] Y.S. Al-Degs, M.A. Al-Ghouti, A.H. El-Sheikh, *J. Hazard. Mater.* 169, 128 (2009).
[101] H. Katsumata, H. Kojima, S. Kaneco, T. Suzuki, K. Ohta, *Microchem. J.* 96, 348 (2010).
[102] Z.G. Yu, Z. Qin, H.R. Ji, X. Du, Y.H. Chen, P. Pan, H. Wang, Y.Y. Liu, *Chromatographia* 72, 1073 (2010).
[103] Q. Zhou, W. Wang, J. Xiao, *Anal. Chim. Acta* 559, 200 (2006).
[104] A.H. El-Sheikh, A.A. Insisi, J.A. Sweileh, *J. Chromatogr. A* 1164, 25 (2007).
[105] Q. Zhou, J. Xiao, W. Wang, *Anal. Sci.* 23, 189 (2007).
[106] Q. Zhou, J. Xiao, W. Wang, *Microchim. Acta* 157, 93 (2007).
[107] H. Niu, Y. Shi, Y. Cai, F. Wei, G. Jiang, *Microchim. Acta* 164, 431 (2009).
[108] Q. Wu, C.Wang, Z. Liu, C. Wu, X. Zeng, J. Wen, Z. Wang, *J. Chromatogr. A* 1216, 5504 (2009).
[109] V.H. Springer, A.G. Lista, *Talanta* 83, 126 (2010).
[110] M. Biesaga, K. Pyrzynska, *J. Sep. Sci.* 29, 2241 (2006).
[111] M.F. Dong, Y.Q. Ma, E.C. Zhao, C. Qian, L. Han, S. Jiang, *Microchim. Acta* 165, 123 (2009).
[112] R.S. Zhao, C.P. Diao, Q.F. Chen, X. Wang, *J. Sep. Sci.* 32, 1069 (2009).
[113] M. Dong, Y. Ma, F. Liu, C. Qian, L. Han, S. Jiang, *Chromatographia* 69, 73 (2009).
[114] Q. Zhou, Y. Ding, J. Xiao, *Anal. Bioanal. Chem.* 385, 1520 (2006).
[115] Q.X. Zhou, G.H. Xie, J.P. Xiao, W.D. Wang, Y.J. Ding, *Chin. Chem. Lett.* 19, 95 (2008).
[116] Q. Zhou, J. Xiao, G. Xie, W. Wang, Y. Ding, H. Bai, *Microchim. Acta* 164, 419 (2009).
[117] S. Wang, P. Zhao, G. Min, G. Fang, *J. Chromatogr. A* 1165, 166 (2007).
[118] Q. Zhou, J. Xiao, Y. Ding, *Anal. Chim. Acta* 602, 223 (2007).
[119] A.H. El-Sheikh, J.A. Sweileh, Y.S. Al-Degs, A.A. Insisi, N. Al-Rabady, *Talanta* 74, 1675 (2008).
[120] M. Asensio-Ramos, J. Hernández-Borges, L.M. Ravelo-Pérez, M.A. Rodríguez-Delgado, *Electrophoresis* 29, 4412 (2008).
[121] S. López-Feria, S. Cárdenas, M. Valcárcel, *J. Chromatogr. A* 1216, 7346 (2009).
[122] M. Asensio-Ramos, G. D'Orazio, J. Hernandez-Borges, A. Rocco, S. Fanali, *Anal. Bioanal. Chem.* 400, 1113 (2011).
[123] Y.S. Al-Degs, M.A. Al-Ghouti, *Int. J. Environ. Anal. Chem.* 88, 487 (2008).
[124] Q. Li, X. Wang, D. Yuan, *J. Environ. Monitor.* 11, 439 (2009).

[125] EU Council, Directive on the Quality of Water Intended for Human Consumption, 98/83/CE, European Union, Brussels 1998.
[126] J. Lü, J. Liu, Y. Wei, K. Jiang, S. Fan, J. Liu, G. Jiang, *J. Sep. Sci.* 30, 2138 (2007).
[127] Y. Ji, X. Liu, X. Jiang, H. Huang, Z. Xu, H. Zhang, C. Wang, *Chromatographia* 70, 753 (2009).
[128] F. Wu, W. Lu, J. Chen, W. Liu, L. Zhang, *Talanta* 82, 1038 (2010).
[129] Y. Sun, W.Y. Zhang, J. Xing, C.M. Wang, *Microchim. Acta* 173, 223 (2011).
[130] C. Pang, S. Xu, H. Zou, Z. Guo, Y. Zhang, B. Guo, *J. Am. Soc. Mass Spectrom.* 16, 263 (2005).
[131] L.Wang, H. Zhao, Y. Qiu, Z. Zhou, *J. Chromatogr. A* 1136, 99 (2006).
[132] G.Z. Fang, J.X. He, S.Wang, *J. Chromatogr. A* 1127, 12 (2006).
[133] B. Suárez, B.M. Simonet, S. Cárdenas, M. Valcárcel, *J. Chromatogr. A* 1159, 203 (2007).
[134] H. Zhao, L.Wang, Y. Qiu, Z. Zhou,W. Zhong, X. Li, *Anal. Chim. Acta* 586, 399 (2007).
[135] B. Suárez, B. Santos, B.M. Simonet, S. Cárdenas, M. Valcárcel, *J. Chromatogr. A* 1175, 127 (2007).
[136] H. Niu, Y. Cai, Y. Shi, F.Wei, J. Liu, S.Mou, G. Jiang, *Anal. Chim. Acta* 594, 81 (2007).
[137] M. Cruz-Vera, R. Lucena, S. Cárdenas, M. Valcárcel, *Anal. Bioanal. Chem.* 391, 1139 (2008).
[138] Y. Lu, Q. Shen, Z. Dai, H. Zhang, *Anal. Bioanal. Chem.* 398, 1819 (2010).
[139] G. Morales-Cid, A. Fakete, B.M. Simonet, R. Lehmann, S. Cárdenas, X. Zhang, M. Valcárcel, P. Schmitt-Kopplin, *Anal. Chem.* 82, 2743 (2010).
[140] A.V. Herrera-Herrera, L.M. Ravelo-Pérez, J. Hernández-Borges, M.M. Afonso, J.A. Palenzuela, M.A. Rodríguez-Delgado, *J. Chromatogr. A* 1218, 5352 (2011).
[141] Z. Zhang, X. Yang, H. Zhang, M. Zhang, L. Luo, Y. Hu, S. Yao, *J. Chromatogr. B* 879, 1617 (2011).
[142] L. Xi, Z. Zhao-Hui, Z. Hua-Bin, H. Yu-Fang, Y. Xiao, N. Li-Hua, *Chin. J. Anal. Chem.* 39, 839 (2011).
[143] Y. Cai, G. Jiang, J. Liu, Q. Zhou, *Anal. Chem.* 75, 2517 (2003).
[144] H.Y. Niu, Y.Q. Cai, Y.L. Shi, F.S. Wei, J.M. Liu, G.B. Jiang, *Anal. Bioanal. Chem.* 392, 927 (2008).
[145] Y. Cai, Y. Cai, S. Mou, Y. Lu, *J. Chromatogr. A* 1081, 245 (2005).
[146] M.A. Salam, R. Burk, *J. Sep. Sci.* 32, 1060 (2009).
[147] Z. Du, Y.L. Yu, X.W. Chen, J.H.Wang, *Chem. Eur. J.* 13, 9679 (2007).
[148] K.-J. Huang, C.-H. Han, C.-Q. Han, J. Li, Z.-W. Wu, Y.-M. Liu, *Microchim. Acta* 174, 421 (2011).
[149] J. Ding, Q. Gao, X.-S. Li, W. Huang, Z.-G. Shi, Y.-Q. Feng, *J. Sep. Sci.* 34, 2498 (2011).
[150] Z. Du, Y.L. Yu, J.H.Wang, *Macromol. Biosci.* 9, 55 (2009).
[151] Y.Q. Cai, G.B. Jiang, J.F. Liu, Q.X. Zhou, *Anal. Chim. Acta* 494, 149 (2003).
[152] Q.L. Li, D.X. Yuan, Q.M. Lin, *J. Chromatogr. A* 1026, 283 (2004).
[153] O. Sae-Khow, S. Mitra, *J. Chromatogr. A* 1216, 2270 (2009).
[154] W.D.Wang, Y.M. Huang,W.Q. Shu, J. Cao, *J. Chromatogr. A* 1173, 27 (2007).
[155] Z. Guan, Y. Huang,W.Wang, *Anal. Chim. Acta* 627, 225 (2008).
[156] K. Shrivas, H.F.Wu, *Anal. Chim. Acta* 628, 198 (2008).

[157] B. Jurado-Sánchez, E. Ballesteros, M. Gallego, *Talanta* 79, 613 (2009).
[158] Márquez-Sillero, E. Aguilera-Herrador, S. Cárdenas, M. Valcárcel, *J. Chromatogr. A* 1217, 1 (2010).
[159] D. Pardasani, P.K. Kanaujia, A.K. Purohit, A.R. Shrivastava, D.K. Dubey, *Talanta* 86, 248 (2011).
[160] G. Liu, J.Wang, Y. Zhu, X. Zhang, *Anal. Lett.* 37, 3085 (2004).
[161] X. Liu, Y. Ji, Y. Zhang, H. Zhang, M. Liu, *J. Chromatogr. A* 1165, 10 (2007).
[162] Q. Li, X.Wang, D. Yuan, *J. Chromatogr. A* 1216, 1305 (2009).
[163] W. Du, F. Zhao, B. Zeng, *J. Chromatogr. A* 1216, 3751 (2009).
[164] W. Chen, J. Zeng, J. Chen, X. Huang, Y. Jiang, Y.Wang, X. Chen, *J. Chromatogr. A* 1216, 9143 (2009).
[165] J.X.Wang, D.Q. Jiang, Z.Y. Gu, X.P. Yan, *J. Chromatogr. A* 1137, 8 (2006).
[166] X.Y. Liu, Y.S. Ji, H.X. Zhang, M.C. Liu, *J. Chomatogr. A* 1212, 10 (2008).
[167] W. Zhang, Y. Sun, C.Wu, J. Xing, J. Li, *Anal. Chem.* 81, 2912 (2009).

In: Applications of Carbon Nanotubes
Editor: Ajay Kumar Mishra
ISBN: 978-1-62081-918-0

Chapter 6

APPLICATIONS OF CARBON NANOTUBES IN NONTRADITIONAL MACHINING AND MICROSCOPY

***Y. H. Guu*[1], *C. C. Mai*[2] *and H. Hocheng*[3]**

[1]Department of Mechanical Engineering, National United University
Taiwan, Republic of China
[2]Department of Numerical Control Technology, Intelligent Machinery Technology
Division Mechanical and Systems Research Laboratories,
Industrial Technology Research Institute Taichung Industrial Area,
Taiwan, Republic of China
[3]Department of Power Mechanical Engineering,
National Tsing Hua University, Taiwan, Republic of China

ABSTRACT

Carbon nanotubes possess advantages over other materials due to their superior strength-to-weight ratios, tremendous stiffness, high conductivity, high flexibility, and low density. Many promising applications have been proposed for carbon nanotubes, including miniaturized electronic and mechanical devices. In this chapter, the applications on nontraditional machining and microscopy are introduced. Electrical discharge machining (EDM) is one of the most successful and widely accepted manufacturing processes for complicated shapes and tiny apertures with high accuracy including micro nozzle fabrication, drilling of composites and making of moulds and dies of hardened steels.

This method is considered suitable for machining of materials with extremely high hardness, strength, wear resistance and thermal resistance. Carbon nanotube powder is mixed in the dielectric of EDM, where the powders continuously float and uniformly disperse throughout the entire dielectric-filled cavity with little agglomeration during EDM.

Hence it reduces the insulating strength of the dielectric fluid and increases the spark gap between the tool and specimen. As a result, the EDM becomes more stable, thereby improving the material removal rate, the surface characteristics and the machining efficiency. The surface characteristics play an important role in micro/nanofabrication as well as understanding the nature of workpiece. In recent years, atomic force microscopy (AFM) has been widely used to measure the surface at the nanometer scale.

AFM is also a powerful technique to conduct lithography on the substrates acquiring the desired nanostructures. AFM can act as a nanomanipulator to pick up and release nanoscale structures on surfaces. Carbon nanotubes show great potential for AFM applications due to their smaller diameters than Si or metal tips result in a high resolution. The high aspect ratios of carbon nanotubes are very suitable to observe a specimen with the precipitous features and their physical and chemical robustness lead to a high abrasion resistance. This work will review the mechanical and electrical properties of carbon nanotubes and the applications in EDM and AFM.

INTRODUCTION

Carbon nanotubes (CNTs) are composed of graphene sheet that is rolled up spirally into a seamless cylinder with diameters ranging from 0.5 to 50 nm and lengths of many microns. The graphite sheet was constructed by the carbon atoms with honeycomb structure. There are two types of carbon nanotubes: single-walled carbon nanotubes (SWNTs) and multi-walled carbon nanotubes (MWNTs). SWNT is formed by rolling a grapheme sheet into a cylinder which have one atom thick and with diameters ranging from 0.6 to a few nm. MWNTs are made of a rolled-up stack of graphene sheets in concentric cylinders. MWNTs have diameters ranging from a few nm to 50 nm. Electron diffraction analysis showed that the graphene sheets had a helical arrangement relative to the tube axis [1-3]. In general, the SWNTs were put together to form bundles.

In a bundle, it is hexagonally arranged to form crystal-like structure. The carbon nanotubes (both MWCNTs and SWCNTs) possess extraordinary mechanical combined with low density, unique electrical properties, and excellent thermal properties. The carbon nanotubes are the candidate with great potential as an application in nanoscale devices and can be widely used in electronic, mechanical, electromechanical, chemical and scanning probe, powder mixed EDM (PMEDM), and materials for macroscopic composites [1-2]. A number of reviews on the applications of CNTs in electronic and optoelectronic [3], energy storage devices [4], nano electro mechanical system (NEMS), sensors [5], reinforced polymer [6], field emission and lighting [7], and biological [8]. Mechanical properties including excellent tensile strength and high Young's modulus and extreme flexibility with elastic buckling without occurring catastrophic failure under large loads hence it is a promising material for AFM tips applications.

In this review article, the authors will focus on the major research of the properties of carbon nanotube and to review the emerging applications of carbon nanotubes in EDM and AFM. This chapter is organized as follows. First, the properties of the carbon nanotubes are discussed. This includes the basic mechanical and electrical properties. A section is thus devoted to the applications in EDM after the description of the advantageous properties. Finally, a survey of the status of applications in AFM is provided. The report ends with a conclusion.

Mechanical Properties of Carbon Nanotubes

Carbon nanotube is composed of the discrete molecular structures linked by the carbon-to-carbon bonds. The carbon-carbon bond existed in graphite is one of the strongest structure in nature [9]. The carbon nanotubes possess the unique property: density lower than that of Al [10]. In the past, researchers used experimental method to measure mechanical properties of SWCNTs. The micro-Raman spectroscopy was used monitor the cooling-induced compressive deformation of carbon nanotubes embedded in an epoxy matrix. According to the measurement, the Young's modulus of SWCNT was 2.8–3.6 TPa, while for MWCNT was 1.7–2.4 TPa [11]. The values of the elastic moduli are consistent with the published experimental data from Treacy et al. [12]. The stiffness of SWCNTs is estimated by observation of their freestanding room-temperature vibrations in a transmission electron microscope. The average Young's modulus of SWCNTs, of 1.0-1.5 nm diameter, is 1.25 TPa [13]. Using an atomic force microscope to measure the elastic and shear moduli of SWNT ropes, the values of elastic and shear moduli of SWNTs are the order of 1 TPa and 1 GPa, respectively [14]. The Young's modulus of individual SWNTs under tensile load ranged from 0.32 to 1.47 TPa (mean 1.0 TPa). The breaking strength of SWNTs of 13 to 52 GPa (mean 30 GPa) was found on the perimeter of each rope. The mechanical properties of MWNTs surpassed those of any previously existing industry materials [15]. The correlation between the amplitude of the thermal vibrations of the free ends of anchored nanotubes as a function of temperature with the Young's modulus showed that the elastic modulus of MWNTs have an average value of 1.8 TPa while its strength is approximately 10-100 times greater than that of the strongest steel [12]. The elastic moduli of multi-walled nanotubes are sensitive to both tube diameter and structure. The elastic modulus of MWNTs has also been obtained from the bending force measured by using AFM. The bending force was measured versus displacement along the unpinned lengths. An AFM was applied to record the profile of a MWNT lying across an electrode array. The measured profile was found to be consistent with a Young's modulus of approximately 1 TPa [16]. The mechanical properties of MWCNTs were determined with a "nanostressing stage" which located within a scanning electron microscope. Experimental results indicate that the tensile strength was 11 to 63 GPa for the 19 MWCNTs and the Young's modulus varied from 0.27 to 0.95 TPa [17]. MWCNTs possess a tensile strength of 0.15 TP and a Young's modulus of 0.9 TPa when pulling and bending tests were conducted on MWCNTs in-situ in a transition electron microscope [18]. By tensile loading inside an SEM to study the mechanical properties of arc-grown MWCNT, the fracture strength of the outermost shell of 14 MWCNTs ranged from 10 to 66 GPa. The Young's moduli of the MWCNTs obtained from linear fit of the stress-strain curves were from 620 to 1,200 GPa with an average value of 940 GPa [19]. Furthermore, the mechanical properties of MWCNTs were performed in a SEM by nano-manipulator (Figure 1). The load response during the tensile test was obtained by the force sensor. The tensile strength of the MWCNTs was about 41.01 GPa and the elastic modulus was 0.98 TPa [20].

The theoretical approaches play an important role in helping us to understand mechanical properties of CNTs. The mechanical properties of CNTs have been addressed by means of computational simulation in a number of publications. An empirical lattice dynamics model was used to study the mechanical properties of single and mulyilayered nanotubes and found that Young's modulus and the shear modulus are about 1.0 and about 0.45 TPa, respectively [21]. A tight-binding (TB) total energy method was applied to calculate the elastic properties of SWNTs. The theoretical values for the Young's modulus was approximately 1.26 TPa [22], conforming with results reported by Krishnan and coworkers [13] for SWNTs (1.25 TPa).

The mechanical properties of SWNTs are strongly dependent on the tube thickness. The theoretical Young's modulus of SWNTs, of 0.25 nm thickness, is about 0.54-0.70 TPa. The shear modulus is 0.32-0.37 TPa [9]. MWCNTs produced by catalytic chemical vapour deposition (CCVD) weaken the Young's modulus (E<1 TPa) due to defects existing in the structure and it cannot be fully improve by high-temperature annealing [23]. A mathematic model was developed to characterize the mechanical properties of SWCNTs. The carbon-carbon bonds between two adjacent atoms are modeled as Euler beams. In terms of the conjunction of Tersoff-Brenner force field method and energy conservation law, the Young's modulus of graphite sheet is 1.04 TPa [24], which is close to the value of 1.033 TPa calculated by Li and Chou [25].

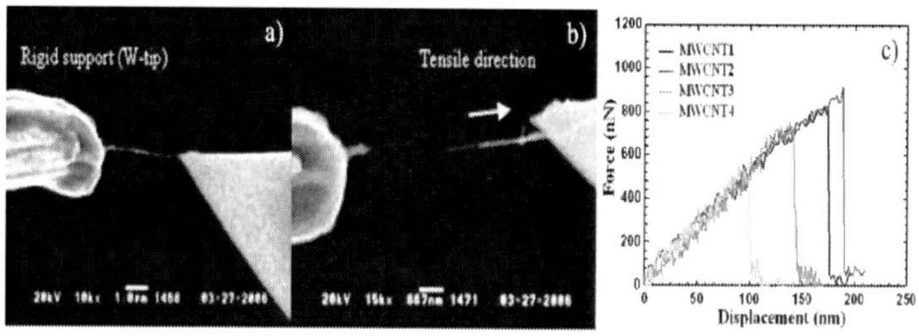

Figure 1 The SEM images show (a) before and (b) after tensile test of MWCNTs, and (c) the force-displacement curves of MWCNTs with different growth methods during tensile tests. The tensile loads of the MWCNT1 and MWCNT2 (produced by arc-discharge method) are 827 and 916 nN, respectively. The tensile loads of the MWCNT3 and MWCNT4 (grown with CVD) are 755 and 557 nN, respectively [20].

Carbon nanotubes have excellent elastic buckling property which can suffer large strains without evidence of collapse or failure under compressive loading. Compared with Si probe, the CNT probe shows better probe wearing resistance and lower sample damage. Such flexibility is important for their potential applications as nanoprobes and in electromechanical applications [26-27]. It has been reported that adding 1 wt% MWCNTs into polystyrene matrix by a simple solution-evaporation method led to increase of elastic modulus 36-42% [28]. As such, they are ideal candidates as reinforcement fibres in composite materials due to their high Young's modulus. From the literature, it is evident that the tensile strength, the elastic modulus, and buckling property of carbon nanotubes (both MWNTs and SWNTs) were far better than any other traditional material.

Electrical Properties of Carbon Nanotubes

The electrical resistence of CNTs is a central concern for CNT applications, particularly for those in nanoelectronics. The electrical properties of CNTs have been under both theoretical and experimental investigations. The early work reported on electrical resistance measurements of a single carbon nanotube bundle. The nanotubes exhibit a striking increase of the resistance followed by a broad maximum at very low temperatures [29]. SWCNT is fabricated by scalable purification process with microscopic density of 2 g/cm^3 and possesses electrical resistivity of (0.5-17) $\times 10^{-3}\Omega$cm [30]. Electrical resistivity of magnetically aligned single-walled carbon nanotubes exhibits moderate anisotropy with respect to the alignment axis. The value of electrical resistivity of magnetically aligned SWCNTs have identical temperature dependencies in the two orientations [31]. The electrical resistivity of SWCNTs prepared by the large-scale induction thermal plasma process was approximately $2\times 10^{-3}\Omega$cm. The resistivity varies across the sample, depending on the surface conditions such as roughness and thickness [32]. The bulk compact of carbon nanotubes with a density of 1.1 g/cm^3 by hot-pressing showed that the electrical resistivity increased with increasing temperature [33]. Four-probe resistance measurements showed that the resistivity of the sinter CNTs was (2-3) $\times 10^{-4}\Omega$cm . The change is about 20% from 285 to 5 K.

There have been several reports relating the electrical conductivity of MWNTs. Four-probe method was used to measure the electrical conductivity of bulk multi-walled carbon nanotube prepared by spark plasma sintering. The MWNTs exhibit the resistivity in the range of $(1.2\text{-}1.6)\times 10^{-2}\Omega$cm [34], which was much larger than that of the aligned SWCNTs reported by others [21]. The electrical conductivity perpendicular to the pressure direction is bigger than that parallel to the pressure direction are $8.3\times 10^{-3}\Omega^{-1}$m^{-1} and $6.2\times 10^{-3}\Omega^{-1}$m^{-1} at room temperature, respectively. The MWNTs have higher electrical conductivity along the long axis. The resistivity of the MWNTs under stress between two conducting plates was $(5\text{-}8)\times 10^{-1}\Omega$cm [35], which was three orders of magnitude greater than that of the hot-pressed sintered CNTs ($\times 10^{-4}\Omega$cm) [23]. High contact resistance would restrain the applications of CNTs in many areas including nanoelectronics. To clarify this issue, the electronic transport of individual SWNTs was studied and demonstrated SWNTs have unique conductivity properties [36]. SWNTs show either metallic or non-metallic behaviour depending on geometry and structural variations. At Cu–Cu interfaces using a MWCNT layer as an electrically conductive interfacial material obtained an 80% reduction in contact resistance. Resistance decreased with increasing contact force, thereby making it possible to use this arrangement as a small-scale force sensor [37]. Using focused ion beam (FIB) to make metallic contact between a tube end and an electrode (Figure 2). The contact resistance of MWNTs with gold was around 100 k$\Omega\mu$m. The contact resistance between MWNTs could be as low as 50 kΩ. The contact resistance of CNTs can be affected by many factors including the preparation of CNTs, the fabrication of electrodes, the assembly of CNTs, as well as the characterization of electrical properties of CNTs [38].

When the resistance of the nanotube junction is sufficiently low, it becomes superconducting and can carry high currents. Superconductivity is a phenomenon of zero electrical resistance below a characteristic temperature [39]. The ropes of SWNTs, contacted to nonsuperconducting metallic pads, exhibit superconductivity and strong non-linearity in the IV characteristics as temperature below 0.4K. The external magnetic field and the temperature could affect the electronic properties of SWNTs [40]. As temperature below 20K the ultra-small tubes exhibit superconducting behavior i.e. Meissner effect, with a superconducting gap and fluctuation super-current. Meissner effect is a characteristic of a conductor having superconductivity [41-42].

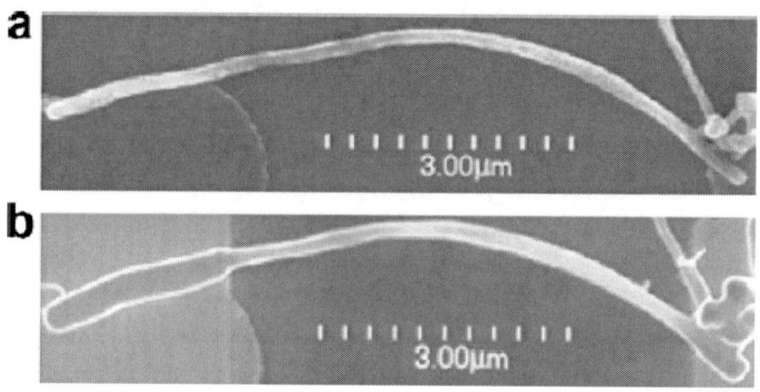

Figure 2. SEM images of a MWNT assembled across a pair of parallel electrodes (a) before and (b) after W-deposition by focused ion beam [38].

Carbon Nanotubes for Electrical Discharge Machining Applications

Electrical-discharge machining is extensively employed for machining of materials with extremely high hardness, strength, so-called difficult-to-cut materials, or for production of complicated shapes and tiny apertures with high accuracy [43-45]. The workpiece and the electrode do not directly contact, thus eliminating the force-induced chatter or vibration problems and allowing small or thin components to be machined without causing distortion [46-47]. During the EDM, the electrical breakdown of the dielectric occurs, an electrical arc is produced in the narrow gap between the workpiece and electrode [48]. The discharge energy produces a very high temperature on the point of the spark of the workpiece surface, and removes the material by melting and vaporization. A small amount of molten layer is expelled from the surface, while the residual melted layer that is not flushed out by the dielectric fluid is later solidified on the workpiece surface as recast layer. The electrical-discharge often will induce machining defects such as craters, microcracks, microvoids and pockmarks on the surface layer causing detrimental effect on the strength [49-50]. The microcracks are the consequence of the thermal stresses. The primary cause of the residual stress in the machined surface is that the machined surface experiences heating and cooling at a high rate and non-uniform temperature distribution. The microvoids observed on the surface resulted from gas bubbles expelled from the molten material during solidification. The discharge energy creates machining damages on the surface causing stress concentration eventually decreasing the service strength. Although an excellent machined finish can be obtained by setting the EDM parameters at a low pulsed current and a small pulse on-time to improve strength but this approach is more time consuming. In general, the conventional machining processes such as grinding, polishing, and lapping are adopted to remove the surface damage layer induced by EDM process. These operations will cause excessive tool wear and expense.

There have been numerous research efforts on the conventional powder-mixed EDM to produce fine finish and to minimize the damaged layer. Dielectric mixed with either aluminum (Al), chromium (Cr), copper (Cu), and silicon carbide (SiC) powder reduces the insulating strength of the dielectric fluid and increases the spark gap between the tool and workpiece, thus disperses the discharging energy to obtain superior surface roughness [51-

53]. Conductive powders added to kerosene can increase the microhardness, make the molten layer thinner and reduce cracks [54-56]. The suspended powder such as graphite (Gr) and silicon (Si) distributed in the dielectric can cause the effect of lowering the breakdown voltage so that the discharges can occur at a wider interelectrode gap, which facilitates flushing debris and reducing servo hunting so that machining is more stable with improved machining efficiency [57-58]. When the silicon powder is used, the surface roughness can be significantly reduced allowing the generation of mirror-like surfaces. The presence of silicon in the dielectric can eliminate the undesirable discharge conditions [59]. The particle size, the particle concentration, the particle density, the electrical resistivity and the thermal conductivity of powder are important characteristics that significantly affect the machining performance in EDM process [60].

While the smallest particles generate the best surface finish of the EDM work, literature survey also shows that powders of lower density, lower electrical resistivity and excellent thermal conductivity will benefit to improve the machining performance in the EDM process [61]. The surfactant molecules act as steric barrier to separate the agglomerated aluminum powder in the dielectric homogenously can improve the surface roughness of the EDM workpiece.

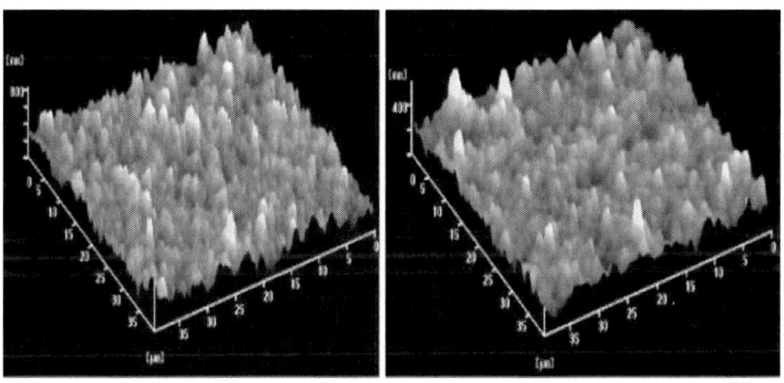

(a) die-sinking, without nano-powder (b) die-sinking, with nano-powder

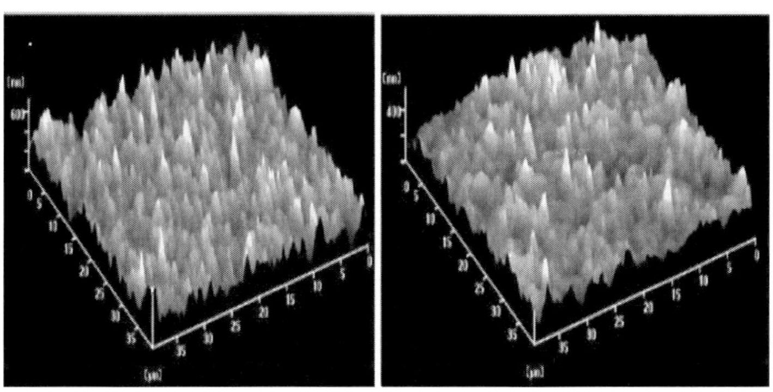

(c) milling, without nano-powder (d) milling, with nano-powder

Figure 3. AFM images of machined surface of the cemented tungsten carbide using (a) diesinking micro-EDM without nano-powder, (b) nano-powder-mixed sinking, (c) EDM milling without nano-powder, and (d) nano-powder-mixed micro-EDM milling at 60 V and stray capacitance. The crater heights for both sinking and milling have been reduced significantly by the addition of graphite nano-powders to the dielectric [68].

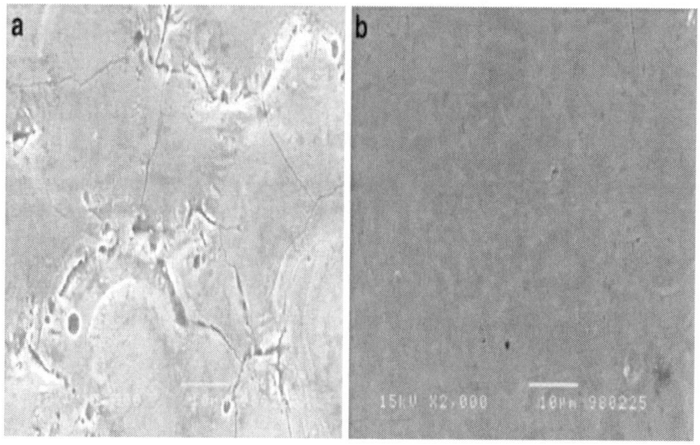

Figure.4 SEM micrographs of the machined surface in use of dielectrics of (a) kerosene and (b) kerosene+CNTs. It is clear that EDM process without the added CNTs in dielectric produces more damage such as pinholes, micro-voids and crater on the machined surface, while the surface machined under kerosene mixed with CNT powder was smoother with less micro-cracking [72].

The increase of viscosity and thermal conductivity of a powder suspended in dielectric alleviate the plasma heat flux into the electrodes leading to form a smaller-sized crater cavity [62]. The graphite, aluminium and alumina nanopowders mixed in dielectric on micro EDM of tool steel (SKH-51) and tungsten carbide (WC). The results show the presence of a conductive or semi-conductive nanopowder lowers the breakdown strength and increases the spark gap during powder-mixed micro-EDM. The surface finish and topography is improved significantly owing to the increased spark gap and uniform discharging. In addition, the lower breakdown strength of dielectric and increased spark gap contribute to higher MRR, lower tool wear, and better surface finish [63-64]. Mixing of boron carbide powder in deionized water dielectrics enhances machining performance in micro-EDM of Ti-6Al-4V alloy [65]. Silicon powder concentrations, peak current and pulse duration play important roles in improving the surface quality in EDM of metal matrix composites, while the supplied voltage is an insignificant variable [66]. The Al particle size and concentration significantly affect machining efficiency during EDM of Inconel 718 [67].

Recently, the nano powder-mixed EDM plays an important role in improving EDM performance. Before 2010's, the improvement of surface finish of EDM specimen was mainly by the use of micro powders (such as Al, Cr, Cu, SiC, Gr, MoS_2). However, there remains the concerns about flushing the controlled gap between the electrode and the workpiece owing to their heavy specific gravity and the associated non-uniform dispersion in the dielectric. The graphite nano-powder-mixed in the dielectric resulted in 20-30% reduction in surface roughness of WC-Co (Figure 3), lower the breakdown strength and facilitate the ignition process thus improving the material removal rate [68]. The reduction of 35% in machining time was obtained after using nano MoS_2 and nanographite powders suspension in dielectric fluid [69-70].

Carbon nanotubes (CNTs) are high aspect-ratio structures with a straight-pin shape and low density, coupled with very good electrical properties and excellent thermal conductivity, making them an ideal candidate for both electrode and powder-mixed EDM (PMEDM). MWCNTs as a miniscale electrodes for EDM can endure distortion providing precise surface

[71]. The technique of dielectric mixed with CNTs is a novel approach to attain a very fine finish (Figure 4). The average surface roughness of 0.09 μm was achieved within 1.2 h, and the material defects of the recast layer and the micro-cracks were considerably reduced by adding CNT powders to the dielectric at a concentration of 0.4 g/l. The surface finish and the machining time of the workpiece have been improved up to 70% and 60%, respectively [72]. Specimens sparked by MWCNTs have better surface finish, reduction of micro cracks and superior surface morphology as compared with specimens sparked without MWCNTs [73]. The electrical conductivity of hybrid Al_2O_3/CNTs composites has an increasing tendency as the CNTs content is increased. The conductivity of the materials and homogeneous distribution of CNTs in the matrix are important factors for micro-EDM of Al_2O_3/CNTs hybrid composites [74]. The AISI D2 tool steel sparked by SWCNTs has better surface finish and better surface morphology as compared with specimens sparked without SWCNTs. The SWCNTs can absorb the heat from electrical discharged material and minimize the white layer formed in the workpiece. A nanofinish can be obtained by setting the machining at low pulse energy [75-76].

Carbon Nanotubes for Atomic Force Microscopy Applications

Surface morphology is an important aspect in understanding the nature of sample surface. The surface textures of sample are usually examined with Alpha step, optical microscopy and scanning electron microscopy (SEM). The stylus of the Alpha step slides across the surface of the sample causing damage on the surface and have the limitations including the resolution in the lateral direction and the dynamic range in height, which may misrepresent the real surface topography. The optical microscopy provides a low resolution and is not able to obtain the information in depth. The sample needs to be placed in vacuum while measuring the surface features by SEM but it does not accurately measure the height of a sample surface with three dimensions [77]. With the progress of technology, the ultramicroscopic scale of surface measurement is highly desired. In 1986, atomic force microscopy (AFM) is firstly developed [78]. AFM operates by measuring the forces between a sharp scanning probe and a sample surface for analyzing structural surface characterization including topological, mechanical and chemical properties. It also can be used to fabricate nanoscale devices on various materials using the anodic oxidation lithography [79-80]. The advantages of AFM are that they can operate in a vacuum, in air or even in liquid environments to observe the nanostructure of the samples. AFM requires little sample preparation and it is a non-destructive technique [81-82]. In recent years, AFM has been widely used for biomedical research [83], chemistry [84-86] and nanometrology in semiconductor manufacturing processes [87-89] to obtain the surface image in the nanometer scale.

A typical AFM probe consists of a cantilever beam which is fabricated from a silicon or silicon nitride (Si_3N_4) and a pyramidal-shaped tip. Commercially available silicon probe tips have radii of curvature of 5-15 nm, silicon nitride of 20-60 nm and have cone angles of 20-30^0 [90]. The Si tips with large cone angle make it difficult for probing high aspect ratio features, thus limiting its use in applications. Hence, the AFM with Si tips needs to be improved to obtain an excellent metrology. To improve the resolution of AFM images, Walters et al. [91-92] tried to use focused ion beam to fabricate the super sharp silicon tips with small radii of curvature and small cone angles. However, these tips appear to be with limited stability and exhibit brittleness of the fine silicon tip, significant tip-to-tip variations as well as tip wear problem. As the tip becomes sharper and narrower, it tends to fracture much easier [77], and furthermore the silicon AFM probe tips are inherently brittle and easy to be abraded away which resulting in surface damage and blunting causing poor accuracy when interacting with a hard sample [93].

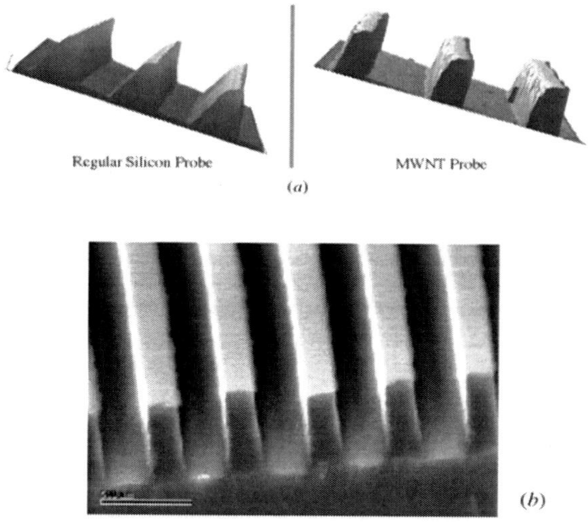

(b)

Figure 5. (a) AFM images of 280 nm line/space pattern of polymeric photoresist (300 nm thickness) on a silicon substrate measured by regular silicon probe (left image) and MWNT probe (right image), (b) SEM image of the pattern. The image acquired with the MWNT probe displays photoresist lines with steep side walls, which is the real pattern as confirmed by SEM [93].

In many imaging applications, especially molecular and biological sciences, materials science, and semiconductor metrology, sharper and higher aspect ratio tips are needed. Superior to the standard silicon AFM probes, carbon nanotubes (CNTs) have excellent features of smaller diameter, high strength, high aspect ratio, and good elastic buckling property, which make them ideal candidate as AFM probes [94-98]. It is shown that the average image resolution obtained by nanotube tips is about two times higher than that by the best Si tips [82, 99]. It is believed that the carbon nanotube tip can act as a compliant spring, thus minimizing the force exertion on samples. This is important for avoiding damage to soft samples, such as polymers and biomaterials [100-101].

MWNT consists of multiple layers of concentric nanotubes, which is much stiffer than SWNT, hence long MWNT tips are still able to produce high quality images. The image resolution acquired with MWNT probe is superior to that of conventional silicon probe (Figure 5). The lifetime of the MWNT tip is 5 times better than that of the silicon tip when continuously measuring the micro-roughness of a Czochralski (Cz) P-type (100) silicon wafer. The MWNT tip is able to trace a narrow space (width less than 1 μm) better than the conventional silicon tip because of its high aspect ratio [102]. The surface height maps by the MWNT probe have a better resolution than those by a conventional silicon tip due to the sharper tip with the larger aspect ratio of the MWNT. A MWNT probe is especially useful to observe surface height of the MEMS/NEMS structures that have larger aspect ratio [103]. Using MWCNTs with controlled and changed length to study the amplitude-distance responses in tapping mode AFM, experimental results have shown shorter MWCNTs with smaller tilting angle [104]. The room-temperature liquidphase dielectrophoresis provided an efficient method for fabrication of CNT-based AFM probes with controlled length and orientation. The CNTs have been shown to have a significantly longer lifetime and higher spatial resolution compared to commercial Si tips. The spatial resolution of AFM imaging is controlled by tip diameter and aspect ratio [105].

A SWNT is mechanically more flexible due to a smaller radius, thus having a smaller bending force constant. SWNT tip is promising for higher resolution because their radii (0.7-1.0 nm) is smaller than those of MWNTs (5-20 nm) [106]. A SWNT scanning probe tip with its small diameter affords high resolution images as compared to that obtained from a large diameter MWNT scanning probe tip [93]. As compared to MWNTs, SWNTs are better as AFM tips due to their high lateral resolution arising from the smaller diameters of SWNTs [107]. The CNT-AFM probes by CVD process have higher horizontal resolution than standard commercial AFM probes [108]. The Langmuir-Blodgett (LB) technique was developed to fabricate SWNT nanoprobes for AFM. The curvature diameter of the nanoprobes was less than 3 nm and a fine resolution was obtained than that from conventional AFM probes [109].

CONCLUSION

The CNTs have high mechanical strength and stiffness, high electrical conductivity, extremely small size, superior elasticity and durability. CNTs show remarkable potential in a wide variety of applications and novel technologies will emerge in the near future. The development of PMEDM is still in the sprouting stages. The effects of tool geometries, electrode materials, flushing conditions, electrode rotation, workpiece rotation, hybrid technique of ultrasonic vibration assisted EDM, multi electrode EDM on material removal rate, electrode wear and surface roughness require further study. The great potential for the AFM tips have been demonstrated. The CNT-AFM tips provide better resolution for imaging, high aspect ratio features and less wear rate than conventional AFM probes. The main challenge now is to develop a rapid and cost-effective technique to fabricate CNT-AFM tips with high quality and wide variety of shapes and material properties for further expansion of the scanning probe technology.

REFERENCES

[1] A. Bachtold, P. Hadley, T. Nakanishi and C. Dekker, *Science* 294, 1317 (2001).

[2] H. Dai, *Surf. Sci.* 500, 218 (2002).

[3] J. Chen, "Carbon nanotubes for potential electronic and optoelectronic applications", 2006 *International Conference on Computer-Aided Design, ICCAD*, November 5, 2006 - November 9, 2006, (San Jose, CA, United states) pp649-650.

[4] A. C. Dillon, *Chem. Rev.* 110, 6856 (2010).

[5] O. Kanoun and Z. Fahem, "Potential of carbon nanotubes for sensor applications", 2008 *5th International Multi-Conference on Systems, Signals and Devices, SSD'08*, July 20, 2008 - July 23, 2008, (Amman, Jordan).

[6] X. Li, H. Gao, W. A. Scrivens, D. Fei, X. Xu, M. A. Sutton, A. P. Reynolds and M. L. Myrick, *J. Nanosci. Nanotechno.* 7, 2309 (2007).

[7] D. S. Chung, W. B. Choi, J. H. Kang, H. Y. Kim, I. T. Han, Y. S. Park, Y. H. Lee, N. S. Lee, J. E. Jung and J. M. Kim, *J. Vac. Sci. Technol. B: Microelectronics and Nanometer Structures* 18, 1054 (2000).

[8] D. Qian, G. J. Wagner, W. K. Liu, M.-F. Yu and R. S. Ruoff, *Appl. Mech. Rev.* 55, 495 (2002).

[9] T. Natsuki, K. Tantrakarn and M. Endo, *Carbon* 42, 39 (2004).

[10] S. Iijima, "Carbon nanotubes: Past, present, and future", *Proceedings of the Tsukuba Symposium on Carbon Nanotube in Com (CNT10)*, October 3, 2001 - October 5, 2001, (Tsukuba, Japan) pp1-5.

[11] O. Lourie and H. D. Wagner, *J. Mater. Res.* 13, 2418 (1998).

[12] M. M. J. Treacy, T. W. Ebbesen and J. M. Gibson, *Nature* 381, 678 (1996).

[13] A. Krishnan, E. Dujardin, T. W. Ebbesen, P. N. Yianilos and M. M. J. Treacy, *Phys. Rev. B* 58, 14013 (1998).

[14] J.-P. Salvetat, G. A. D. Briggs, J.-M. Bonard, R. R. Bacsa, A. J. Kulik, T. Stockli, N. A. Burnham and L. Forro, *Phys. Rev. Lett.* 82, 944 (1999).

[15] M.-F. Yu, B. S. Files, S. Arepalli and R. S. Ruoff, *Phys. Rev. Lett.* 84, 5552 (2000).

[16] J. Muster, M. Burghard, S. Roth, G. S. Duesberg, E. Hernandez and A. Rubio, *J. Vac. Sci. Technol. B: Microelectronics Processing and Phenomena* 16, 2796 (1998).

[17] M.-F. Yu, O. Lourie, M. J. Dyer, K. Moloni, T. F. Kelly and R. S. Ruoff, *Science* 287, 637 (2000).

[18] B. G. Demczyk, Y. M. Wang, J. Cumings, M. Hetman, W. Han, A. Zettl and R. O. Ritchie, *Mat. Sci. Eng. A* 334, 173 (2002).

[19] W. Ding, L. Calabri, K. M. Kohlhaas, X. Chen, D. A. Dikin and R. S. Ruoff, *Exp. Mech.* 47, 25 (2007).

[20] H.-S. Jang, S. K. Jeon, H. J. Lee and S. H. Nahm, " Mechanical and electrical properties of multi-walled carbon nanotubes by nano-manipulator", *11th International Conference on the Mechanical Behavior of Materials, ICM11*, June 5, 2011 - June 9, 2011, (Como, Italy) pp2917-2922.

[21] J. P. Lu, *J. Phys. Chem. Solids* 58, 1649 (1997).

[22] E. Hernandez, C. Goze, P. Bernier and A. Rubio, *Appl. Phys. A-Mater.* 68, 287 (1999).

[23] B. Lukic, J. W. Seo, E. Couteau, K. Lee, S. Gradecak, R. Berkecz, K. Hernadi, S. Delpeux, T. Cacciaguerra, F. Beguin, A. Fonseca, J. B. Nagy, G. Csanyi, A. Kis, A. J. Kulik and L. Forro, *Appl. Phys. A-Mater.* 80, 695 (2005).

[24] X. Lei, T. Natsuki, J. Shi and Q.-Q. Ni, *J. Nanomater.* 2011, (2011).

[25] C. Li and T.-W. Chou, *Int. J. Solids Struct.* 40, 2487 (2003).

[26] T. Hertel, R. Martel and P. Avouris, *J. Phys. Chem. B* 102, 910 (1998).

[27] Z. Xu, S. Zhang, F. Fang, S. Dong and Y. Chen, *Chinese J. Sci. Instrum.* 30, 157 (2009).

[28] D. Qian, E. C. Dickey, R. Andrews and T. Rantell, *Appl. Phys. Lett.* 76, 2868 (2000).

[29] L. Langer, L. Stockman, J. P. Heremans, V. Bayot, C. H. Olk, C. Van Haesendonck, Y. Bruynseraede and J. P. Issi, *J. Mater. Res.* 9, 927 (1994).

[30] A. G. Rinzler, J. Liu, H. Dai, P. Nikolaev, C. B. Huffman, F. J. Rodriguez-Macias, P. J. Boul, A. H. Lu, D. Heymann, D. T. Colbert, R. S. Lee, J. E. Fischer, A. M. Rao, P. C. Eklund and R. E. Smalley, *Appl. Phys. A-Mater.* 67, 29 (1998).
[31] J. Hone, M. C. Llaguno, N. M. Nemes, A. T. Johnson, J. E. Fischer, D. A. Walters, M. J. Casavant, J. Schmidt and R. E. Smalley, *Appl. Phys. Lett.* 77, 666 (2000).
[32] K. S. Kim, M. Imris, A. Shahverdi, Y. Alinejad and G. Soucy, *J. Phys. Chem. C* 113, 4340 (2009).
[33] R. Z. Ma, C. L. Xu, B. Q. Wei, J. Liang, D. H. Wu and D. J. Li, *Mater. Res. Bull.* 34, 741 (1999).
[34] C. Qin, X. Shi, S. Q. Bai, L. D. Chen and L. J. Wang, *Mat. Sci. Eng. A* 420, 208 (2006).
[35] P. Singjai, S. Changsarn and S. Thongtem, *Mat. Sci. Eng. A* 443, 42 (2007).
[36] T. W. Ebbesen, H. J. Lezec, H. Hiura, J. W. Bennett, H. F. Ghaemi and T. Thio, *Nature* 382, 54 (1996).
[37] M. Park, B. A. Cola, T. Siegmund, J. Xu, M. R. Maschmann, T. S. Fisher and H. Kim, *Nanotechnology* 17, 2294 (2006).
[38] L. An and C. R. Friedrich, (2011).
[39] A. Y. Kasumov, R. Deblock, M. Kociak, B. Reulet, H. Bouchiat, I. I. Khodos, Y. B. Gorbatov, V. T. Volkov, C. Journet and M. Burghard, *Science* 284, 1508 (1999).
[40] A. Y. Kasumov, M. Kociak, M. Ferrier, Y. A. Kasumov, S. Gueron, B. Reulet, I. I. Khodos, Y. B. Gorbatov, V. T. Volkov, L. Vaccarini and H. Bouchiat, "Superconductivity in ropes of single-walled carbon nanotubes", pp1321-1322.
[41] Z. K. Tang, L. Y. Zhang, N. Wang, X. X. Zhang, J. N. Wang, G. D. Li, Z. M. Li, G. H. Wen, C. T. Chan and P. Sheng, "Ultra-small single-walled carbon nanotubes and their superconductivity properties", *ISCOM 2001*, September 10, 2001 - September 14, 2001, (Rusutsu, Hokkaido, Japan) pp689-693.
[42] Z. K. Tang, L. Zhang, N. Wang, X. X. Zhang, G. H. Wen, G. D. Li, J. N. Wang, C. T. Chan and P. Sheng, *Science* 292, 2462 (2001).
[43] J. Y. Kao, C. C. Tsao, S. S. Wang and C. Y. Hsu, *Int. J. Adv. Manuf. Tech.* 47, 395 (2010).
[44] S. Singh, S. Maheshwari, A. Dey and P. C. Pandey, *Int. J. Manuf. Tech. Manage.* 21, 67 (2010).
[45] K. M. Patel, P. M. Pandey and P. V. Rao, *J. Eng. Mater. Technol, T. ASME* 133, (2011).
[46] T. Y. Tai and S. J. Lu, *Int. J. Fatigue* 31, 433 (2009).
[47] J. Marafona and C. Wykes, *Int. J. Mach. Tool. Manu.* 40, 153 (2000).
[48] H. Yoshida and M. Kunieda, *J. Jpn. Soc. Precis. Eng.* 62, 1464 (1996).
[49] Y. H. Guu and H. Hocheng, *Int. J. Mater. Prod. Tec.* 16, 642 (2001).
[50] Y. H. Guu, H. Hocheng, C. Y. Chou and C. S. Deng, *Mat. Sci. Eng. A* 358, 37 (2003).
[51] B.-H. Yan and S.-L. Chen, *J. Chin. Soc. Mech. Eng.* 14, 307 (1993).
[52] H.-M. Chow, B.-H. Yan, F.-Y. Huang and J.-C. Hung, *J. Mater. Process. Tech.* 101, 95 (2000).
[53] B. H. Yan, Y. C. Lin, F. Y. Huang and C. H. Wang, *Mater. Trans.* 42, 2597 (2001).
[54] Q. Y. Ming and L. Y. He, *J. Mater. Process. Tech.* 52, 44 (1995).
[55] H. K. Kansal, S. Singh and P. Kumar, *J. Mater. Process. Tech.* 184, 32 (2007).

[56] F. Hu, B. Y. Song, Y. F. Guo, X. D. Yang, J. C. Bai and D. Li, " Study on powder mixed EDM of an aluminium matrix composite", *2011 International Conference on Environmental Biotechnology and Materials Engineering, EBME 2011*, March 26, 2011 - March 28, 2011, (Harbin, China) pp1947-1951.

[57] Y. S. Wong, L. C. Lim, I. Rahuman and W. M. Tee, *J. Mater. Process. Tech.* 79, 30 (1998).

[58] W. S. Zhao, Q. G. Meng and Z. L. Wang, " The application of research on powder mixed EDM in rough machining", *10th International Manufacturing Conference in China (IMCC 2002)*, October 11, 2002 - October 11, 2002, (Fujian, China) pp30-33.

[59] P. Pecas and E. Henriques, *Int. J. Mach. Tool. Manu.* 43, 1465 (2003).

[60] Y. F. Tzeng and C. Y. Lee, *Int. J. Adv. Manuf. Tech.* 17, 586 (2001).

[61] T. Yih-Fong and C. Fu-Chen, *J. Mater. Process. Tech.* 170, 385 (2005).

[62] S. H. Yeo, P. C. Tan and W. Kurnia, *J. Micromech. Microeng.* 17, N91 (2007).

[63] M. P. Jahan, M. M. Anwar, Y. S. Wong and M. Rahman, *Proceedings of the Institution of Mechanical Engineers, Part B: Journal of Engineering Manufacture* 223, 1127 (2009).

[64] M. P. Jahan, M. Rahman and Y. S. Wong, *Proceedings of the Institution of Mechanical Engineers, Part B: Journal of Engineering Manufacture* 224, 1725 (2010).

[65] G. Kibria, B. R. Sarkar, B. B. Pradhan and B. Bhattacharyya, *Int. J. Adv. Manuf. Tech.* 48, 557 (2010).

[66] H. Kumar and J. P. Davim, *J. Compos. Mater.* 45, 133 (2011).

[67] A. Kumar, S. Maheshwari, C. Sharma and N. Beri, *Mater. Manuf. Process.* 26, 1011 (2011).

[68] M. P. Jahan, M. Rahman and Y. S. Wong, *Int. J. Adv. Manuf. Tech.* 53, 167 (2011).

[69] G. S. Prihandana, T. Sriani, K. Prihandana, Y. Prihandana, M. Mahardika, M. Hamdi and K. Mitsui, " Study on the effect of nano and micro MoS2 powder in micro-electrical discharge machining",*International Conference on Advances in Materials and Processing Technologies, AMPT 2009*, October 26, 2009 - October 29, 2009, (Malaysia) pp1450-1455.

[70] G. S. Prihandana, M. Mahardika, M. Hamdi, Y. S. Wong and K. Mitsui, *Int. J. Adv. Manuf. Tech.* 56, 143 (2011).

[71] H. Y. Miao, J. T. Lue and M. S. Ouyang, "Surface modification by electric discharge implemented with electrodes composed of carbon nanotubes", pp1375-1380.

[72] C. Mai, H. Hocheng and S. Huang, *Int. J. Adv. Manuf. Tech.*, 1 (2011).

[73] S. Prabhu and B. K. Vinayagam, *Arch. Civil Mech. Eng.* 11, 149 (2011).

[74] H.-S. Tak, C.-S. Ha, H.-J. Lee, H.-W. Lee, Y.-K. Jeong and M.-C. Kang, *T. Nonferr. Metal. Soc. China* 21, s28 (2011).

[75] S. Prabhu and B. K. Vinayagam, *Int. J. Mach. Machina. Mater.* 10, 99 (2011).

[76] S. Prabhu and B. K. Vinayagam, *Int. J. Comput. Mater. Sci. Surf. Eng.* 4, 23 (2011).

[77] R. M. Stevens, *Mater. Today* 12, 42 (2009).

[78] Q. C. F. Binnig G., *Phys. Rev. Lett.* 56, 930 (1986).

[79] R. Held, T. Heinzel, P. Studerus and K. Ensslin, *Physica E: Low-Dimensional Systems and Nanostructures* 2, 748 (1998).

[80] K. Kumar, O. Sul, S. Strauf, D. S. Choi, F. Fisher, M. G. Prasad and E.-H. Yang, *IEEE T. Nanotechnol.* 10, 849 (2011).

[81] R. W. Phillips, *Surf. Coat. Tech.* 68-69, 770 (1994).
[82] S. Akita, H. Nishijima, Y. Nakayama, F. Tokumasu and K. Takeyasu, *J. Appl. Phys. D: Appl. Phys.* 32, 1044 (1999).
[83] J. H. Choi, F. T. Nguyen, P. W. Barone, D. A. Heller, A. E. Moll, D. Patel, S. A. Boppart and M. S. Strano, *Nano Lett.* 7, 861 (2007).
[84] M. X. Pulikkathara, O. V. Kuznetsov, I. R. G. Peralta, X. Wei and V. N. Khabashesku, *Nanotechnology* 20, (2009).
[85] R. Pokrop, I. Kulszewicz-Bajer, I. Wielgus, M. Zagorska, D. Albertini, S. Lefrant, G. Louarn and A. Pron, *Syn. Met.* 159, 919 (2009).
[86] J. Zhang, X. Li, W. Guo, T. Hreid, J. Hou, H. Su and Z. Yuan, *Electrochim. Acta* 56, 3147 (2011).
[87] H.-C. Liu, D. Fong, G. A. Dahlen, M. Osborn, S. Hand and J. R. Osborne, " Carbon nanotube AFM probes for microlithography process control", *Metrology, Inspection, and Process Control for Microlithography* XX, January 20, 2006 - January 23, 2006, (San Jose, CA, United states) ppSPIE.
[88] M. Edamura, Y. Kunitomo, T. Morimoto, S. Sekino, T. Kurenuma, Y. Kembo, M. Watanabe, S. Baba and K. Hidaka, " New inline AFM metrology tool suited for LSI manufacturing at the 45-nm node and beyond", *Metrology, Inspection, and Process Control for Microlithography* XXI, Febrary 26, 2007 - March 1, 2007, (San Jose, CA, United states) ppSPIE.
[89] T. Bao and D. Fong, " Recent advances in atomic force microscope metrology for fab-wide dimension control", *6th International Conference on Semiconductor Technology, ISTC2007*, March 18, 2007 - March 20, 2007, (Shanghai, China) pp513-537.
[90] I. R. S. Lawrence A. Wade, Ziyang Ma, Stephen R. Quake, C. Patrick Collier, *Nano Lett.* 4, 725 (2004).
[91] D. A. Walters, D. Hampton, B. Drake, H. G. Hansma and P. K. Hansma, *Appl. Phys. Lett.* 65, 787 (1994).
[92] H. G. Hansma, Laney, D.E., Bezanilla, M., Sinsheimer, R.L. and Hansma, P.K., *Biophys. J.* 68, 1672 (1995).
[93] C. V. Nguyen, K.-J. Chao, R. M. D. Stevens, L. Delzeit, A. Cassell, J. Han and M. Meyyappan, " Carbon nanotube tip probes: Stability and lateral resolution in scanning probe microscopy and application to surface science in semiconductors", 8th *Foresight Conference on Molecular Nanotechnology*, November 2, 2000 - November 5, 2000, (Bethesda, MR, United states) pp363-367.
[94] S. Yamanaka, T. Okawa, S. Akita and Y. Nakayama, " Carbon nanotube atomic force microscopy probes", Metrology, Inspection, and Process Control for Microlithography XIX, Febrary 28, 2005 - March 3, 2005, (San Jose, CA, United states) pp1450-1456.
[95] H. U. Danzebrink, L. Koenders, G. Wilkening, A. Yacoot and H. Kunzmann, *CIRP Ann.-Manuf. Techn.* 55, 841 (2006).
[96] M. Ishikawa, M. Yoshimura and K. Ueda, " Simultaneous measurement of topography and contact current by contact mode atomic force microscopy with carbon nanotube probe", Microoptics Scanning Probe Microscopy, pp4908-4910.
[97] Y. Nakayama, *Ultramicroscopy* 91, 49 (2002).
[98] C. T. Gibson, S. Carnally and C. J. Roberts, *Ultramicroscopy* 107, 1118 (2007).
[99] F. Z. Fang, Z. W. Xu, G. X. Zhang and X. T. Hu, *CIRP Ann.-Manuf. Techn.* 58, 455 (2009).

[100] S. S. Wong, A. T. Woolley, T. Wang Odom, J.-L. Huang, P. Kim, D. V. Vezenov and C. M. Lieber, *Appl. Phys. Lett.* 73, 3465 (1998).

[101] J. S. Bunch, T. N. Rhodin and P. L. McEuen, *Nanotechnology* 15, S76 (2004).

[102] M. Yasutake, Y. Shirakawabe, T. Okawa, S. Mizooka and Y. Nakayama, *Ultramicroscopy* 91, 57 (2002).

[103] B. Bhushan, T. Kasai, C. V. Nguyen and M. Meyyappan, *Microsyst. Technol.* 10, 633 (2004).

[104] A. N. Jiang, S. Gao, X. L. Wei, X. L. Liang and Q. Chen, *J. Phys. Chem. C* 112, 15631 (2008).

[105] J. Tang, G. Yang, Q. Zhang, A. Parhat, B. Maynor, D. Liu, L.-C. Din and O. Zhou, *Nano Lett.* 5, 11 (2005).

[106] N. Choi, T. Uchihashi, H. Nishijima, T. Ishida, W. Mizutani, S. Akita, Y. Nakayama, M. Ishikawa and H. Tokumoto, *Japanese J. Appl. Phys., Part 1: Regular Papers and Short Notes and Review Papers* 39, 3707 (2000).

[107] C. V. Nguyen, Q. Ye and M. Meyyappan, *Meas. Sci. Technol.* 16, 2138 (2005).

[108] K. Takagahara, Y. Takei, E. Iwase, K. Matsumoto and I. Shimoyama, " Batch fabrication of carbon nanotubes at AFM probe tips and AFM imaging", *21st IEEE International Conference on Micro Electro Mechanical Systems, MEMS 2008* Tucson, January 13, 2008 - January 17, 2008, (Tucson, AZ, United states) pp713-716.

[109] J.-H. Lee, W.-S. Kang, B.-S. Choi, S.-W. Choi and J.-H. Kim, *Ultramicroscopy* 108, 1163 (2008).

In: Applications of Carbon Nanotubes
Editor: Ajay Kumar Mishra

ISBN: 978-1-62081-918-0

Chapter 7

CNT Nanocomposites and CNT-Hybrid Materials for Membrane-Based Separation Processes

Roman Konietzny and Claudia Staudt*

Institute for Organic and Macromolecular Chemistry,
Department Functional Materials,
Heinrich-Heine-University of Duesseldorf,
Duesseldorf, Germany

Abstract

The nowadays challenging task for researchers as well as for industry is to develop membrane materials that offer high fluxes combined with high selectivities. Nanocomposites or hybrid materials containing carbon nanotubes (CNTs) appear to be a promising attempt in this field. CNTs can delegate their unique properties to membrane materials and significantly improve their separation characteristics. In the last few years nanocomposites and hybrid membranes showed improved properties and very promising results in different fields of membrane-based research. In comparison to polymeric and inorganic membranes they show a higher chemical and thermal stability, an excellent mechanical strength and they are robust to handle. This paper reviews various developments of CNT membranes for the separation of liquid and gaseous mixtures. Thereby applications for the separation of aqueous/organic mixtures and also the separation of organic/organic and gaseous mixtures are mentioned. Furthermore the preparation of the CNTs as additive for the preparation of nanocomposites or hybrid membranes and their influence on the transport properties of different components are discussed briefly.

* E-mail: claudia.staudt@basf.com.

1. Introduction

Since the discovery in the early 1990's carbon nanotubes (CNTs) gain more and more relevance in various fields of science and it is nowadays one of the fastest growing area of research. Formally CNTs can be described as a graphene sheet rolled into a tube or cylinder, whereby different ways of rolling are known, like achiral, zigzag and armchair tubes. CNTs are classified into two major groups. Single-walled carbon nanotubes (SWNTs) consist of a single sheet of graphene rolled seamlessly to form a tube with a diameter ranging from 1-2 nm to centimeters in length. Multi-walled carbon nanotubes (MWNTs) on the other hand consist of an array of such graphene tubes, which are concentric arranged with interlayer distances of approximately 0.35 nm. The diameter of MWNTs range between 2-100 nm and a total length of tens of microns is possible [1]. The structures of SWNTs and MWNTs are shown in figure 1. CNTs can be mainly prepared by three different techniques, like arc discharge, laser vaporization and chemical vapor deposition (CVD). The first two techniques are mostly used for the preparation of SWNTs, where in case of the arc discharge method electricity is used as energy source which implies arc evaporation of pure or metal-doped carbon electrodes [2]. High intensity light is used at laser vaporization of metal-doped carbon targets [3]. The CVD method implies the decomposition of carbon-containing gases at catalytic nanoparticles and offer more control over the nanotube structures on an industrial scale. This method is often used for the preparation of MWNTs [4].

CNTs obtain excellent chemical and thermal stability, as well as extraordinary mechanical and electrical properties. Some of the unique properties are thermal stability higher than 1000 °C as well as its thermal conductivity of 3000 W/m·K, which is two times higher than the conductivity of diamonds. Beside that CNTs have a 100 times higher tensile strength compared to steel and a ten times higher one than Kevlar. CNTs show an unique elastic Young's modulus which is 50-500 GPa for SWNTs and 10-60 GPa for MWNTs [1, 5].

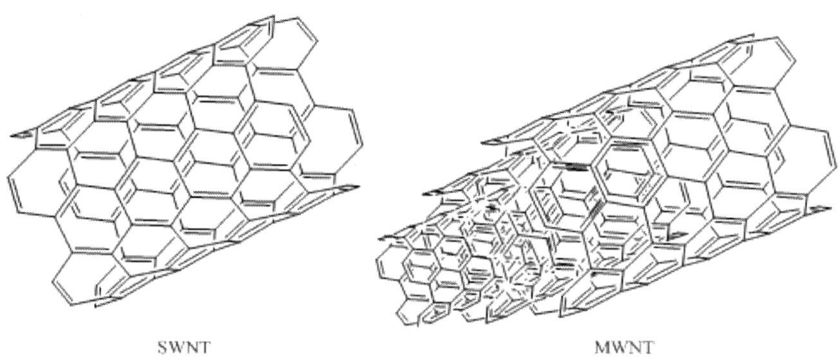

Figure 1. Different carbon nanotube structures.

Because of their outstanding mechanical properties CNTs are very interesting as additives for the preparation of high performance membrane materials which can be used in industrial applications. If CNTs are incorporated into a polymer matrix, the resulting composite offers a combination of different properties which can lead to a significant improvement of the separation characteristics. Functionalized CNTs, e. g. can act as a cross-linking agent between polymer chains and depending on the functional groups covalent bonds or hydrogen bond interactions are formed between functional groups of the CNTs and the polymer leading to a higher chemical and thermal resistance of the material [6]. Furthermore the incorporation of CNTs into a polymer matrix leads to a higher mechanical stability of the hybrid material and due to the bulky nanoscale objects embedded in the polymer matrix a higher free fractional volume is created, leading to higher permeabilities. From an economic point of view this modification is very desirable, since for a specific separation process less membrane area and a lower number of modules is necessary. Therefore hybrid membranes consisting of CNTs and a polymer matrix are important precursors for developing a new type of membranes that offer higher fluxes compared to commercially available materials. Furthermore improvement of selectivity can be simultaneously reached if functionalized CNTs are used as cross-linkers for polymers with suitable side groups. Beneath that the resulting materials provide good physical properties in terms of chemical and swelling resistance against feed components. It is expected that hybrid materials will contribute to the market entrance for several membrane based separation processes, which have not been commercialized so far because of a lack of suitable materials.

In general three different classes of polymer/CNT materials are known; CNT nanocomposites, mixed matrix membranes and CNT hybrid membranes. In the first two fields the pure CNTs are dispersed in the polymer, while in hybrid materials the CNTs are for example covalently bonded to the polymeric chains or polymerized into the polymer chains and due to that approve a specific arrangement between functional groups or atoms. Compared to mixed matrix membranes in nanocomposites at least one sidewall of the particles range in the nanometer scale.

2. Membrane-Based Separations

2.1. Theoretical Background

Membranes are present in our daily life. Cell membranes enable the transport of specific substances from the outside to the inside of cells and vice versa.

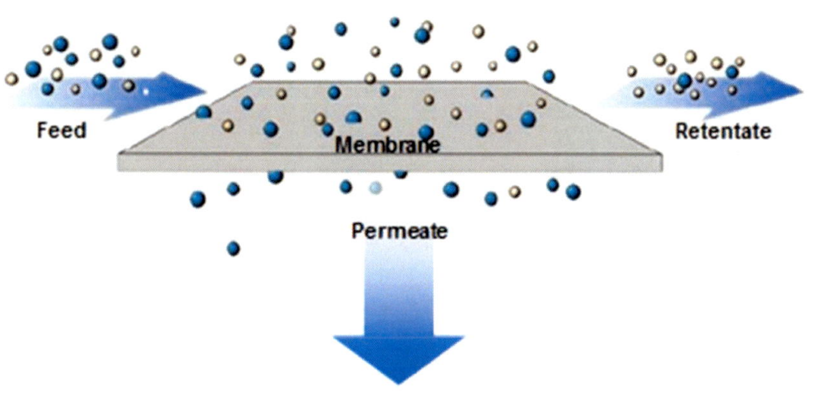

Figure 2. Principle of a membrane-based separation process.

For industrial applications a similar concept can be applied in order to separate gaseous, vaporous or liquid mixtures. Membranes are in general thin layers, which have specific structures and compositions and act as a semipermeable barrier, where certain components are transported through the membrane (permeate) whereas the others are held back (retentate) as shown in Figure 2.

Generally there are two types of membranes, porous and solution diffusion membranes. As shown in Figure 3 there is a floating crossover from processes with porous to those using solution-diffusion membranes.

In filtration processes where particles, bacterias, viruses or small molecules with a diameter between 10 and 1000 nanometers are separated mostly from an aqueous phase, porous membranes are well established. The separation is based on a sieve mechanism.

Components with diameters of less than 1 nm can be separated using solution diffusion membranes, which are in most cases made out of polymers [7]. The mechanism of this type of membrane can be explained by the so called solution-diffusion-model [8], where the components are permeating through a polymeric membrane by a three step process.

In the first step one of the components is preferentially sorbed on the feed side of the membrane surface depending on the interactions between the component and membrane material. In the second step the component which is preferably sorbed is diffusing through the membrane.

The diffusion rate is depending on the size of the component as well as on the mobility of the polymer backbone. In the last step the component is desorbed on the permeate-side of the membrane. In figure 4 the principle of the transport mechanism for a solution diffusion membrane is shown schematically.

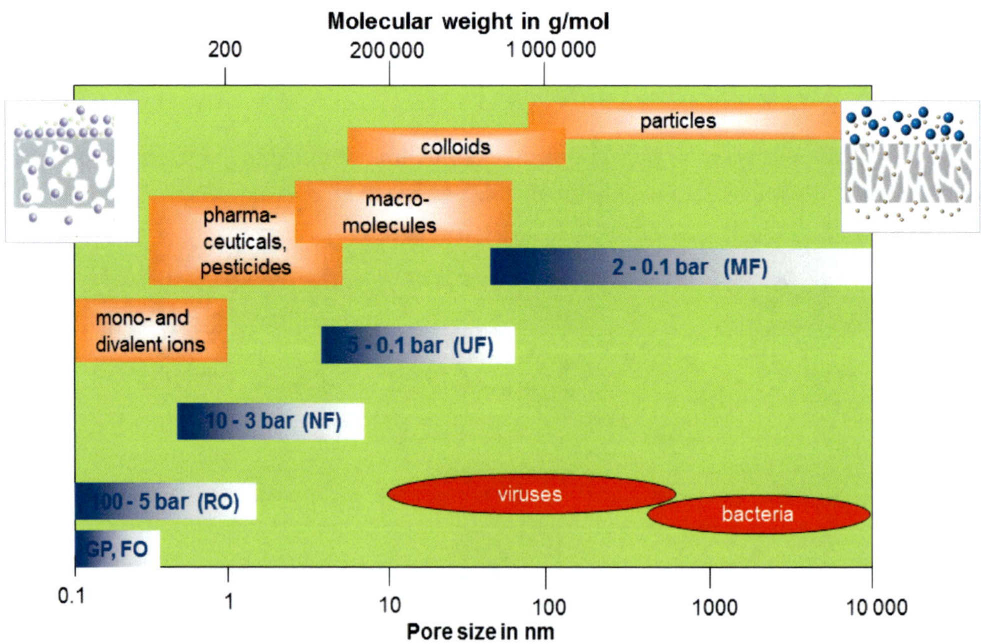

Figure 3. Different membrane-based separation processes.

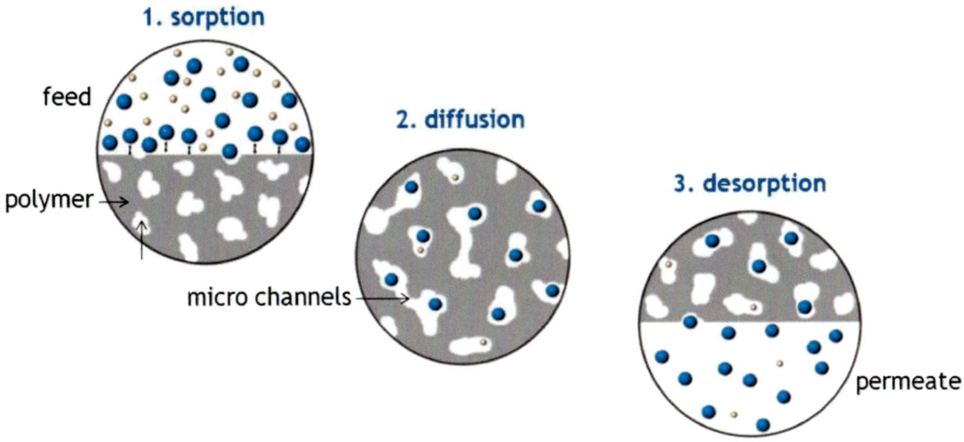

Figure 4. Separation mechanism based on a solution diffusion membrane.

In separation experiments, the membrane is placed in a module which can be a flat sheet, spiral wound or hollow fiber module. As feed, a binary mixture or multi-component mixtures can be used. However in most cases for the development of new membrane materials exclusively binary mixtures are used since in multicomponent mixtures the single components are influencing each other making correlations between membrane structure and separation properties difficult. The retentate which is depleted in the components permeating is leaving the module on the same side of the membrane as the feed. The components permeating preferentially through the membrane are collected on the back side of the membrane in a stream which is called permeate. The driving force for the mass transport through a membrane is the difference in chemical potential between the components on the feed side and permeate side, respectively. The driving force is depending on parameters like temperature, pressure, and mole fractions of all components present in the mixture.

In chemical industry over 50% of the energy costs are due to the separation of gaseous or liquid mixtures [9]. Most of the separation processes are highly energy intensive, i.e. distillation of olefin/paraffin mixtures or the extractive distillation for the benzene production. If these separation processes would be more effective the costs of basic chemicals could be reduced. However since the demand for base chemicals is continuously increasing, energy issues are also increasing and new energy saving separation processes need to be developed.

Due to this fact, membrane-based separations are growing rapidly and are gaining more and more importance in industry. Membrane devices can be used for the separation of gaseous, vaporous and liquid separations. They are much smaller and can be processed at lower temperatures compared to conventional distillation units, which are highly energy intensive. Membrane technologies offer high efficiency with low energy requirement at low operating costs and investment. Therefore membrane-based separations are often mentioned to be eco-friendly and are known as clean-technologies since they play an important role in reducing environmental impacts [10, 11].

For example it has been estimated for the separation of olefin/paraffin mixtures that membrane based hybrid processes could save approximately 40-50% of the production costs [12, 13].

Also so called hybrid processes have been established in industry, which are a combination of two separation technologies like a distillation unit and a membrane device in order to obtain a high purity with low energy consumption and cost savings [14, 15].

2.2. Pervaporation

Pervaporation is a membrane-based process for the separation of liquid mixtures, where solution diffusion membranes are used. Thus the separation is based on the solution-diffusion mechanism as described in paragraph 2.1 [8]. Pervaporation is the only membrane-based process with a phase transition between the feed and the permeate side as shown in figure 5. The feed as well as the retentate is liquid, whereas the permeate stream is vaporous and has to be condensed by a heat exchanger. For the phase transition evaporation energy is required, which can be fed by additional heating. In case of a pervaporation processes as described in this work the driving force or the difference in chemical potential is mainly achieved by keeping the pressure and temperature on the feed side much higher than on the permeate side.

In pervaporation processes the key parameters for the efficiency of the membrane material is the flux and selectivity. The flux J quantifies the mass fraction of permeate m_p which is transported through a certain membrane area A during a time period t.

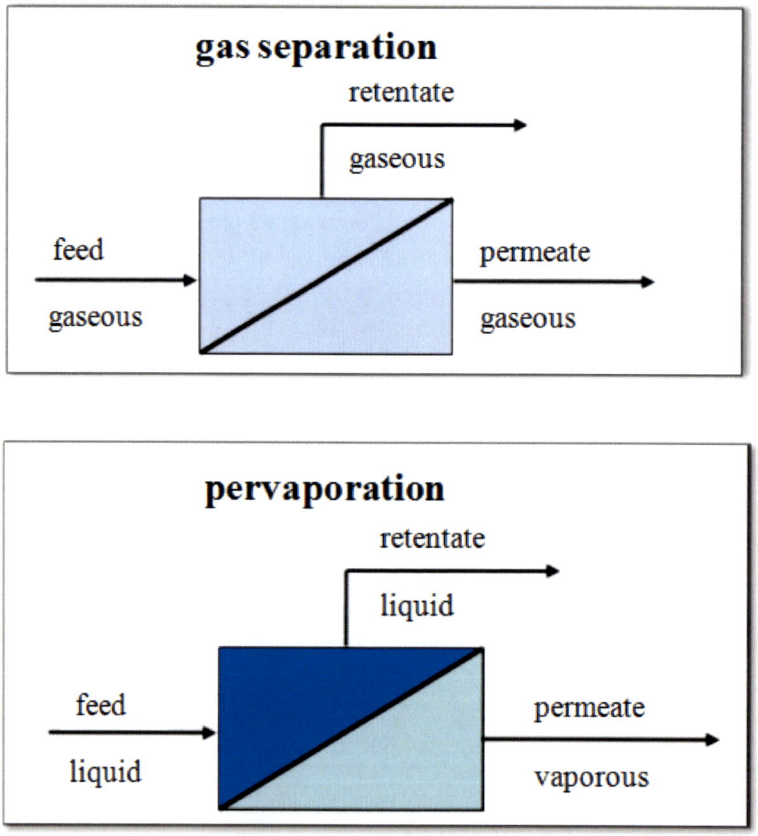

Figure 5. Comparison of membrane-based gas separation and pervaporation processes.

In order to evaluate different membrane materials, the flux is normalized to the membrane thickness (based on a 1 micrometer thick membrane). J_{norm} is defined according to equation (1) and given in kg·µm·m^{-2}·h^{-1}.

$$J_{norm} = \frac{m_p \cdot \delta_M}{t \cdot A} \qquad \text{(eq. 1)}$$

The separation efficiency of a membrane is described by the selectivity, whereby the concentrations of the components in the feed and permeate have to be determined. The weight fraction w_i/w_j characterizes the considered components in the permeate and in the feed, whereby w_i is the preferentially permeating component. In equation (2) the selectivity is shown for a binary mixture.

$$\alpha = \frac{w_i^{permeate} / w_j^{permeate}}{w_i^{feed} / w_j^{feed}} \qquad \text{(eq. 2)}$$

Generally pervaporation can be used for three different approaches depending on the type of membrane - hydrophilic, hydrophobic or organophilic, respectively and target-organophilic applications.

Hydrophilic pervaporation is used for the removal of water from aqueous/organic mixtures, e.g. in equilibrium reactions or for drying of solvents. Thereby a hydrophilic membrane type as polyvinylalcohol (PVA) or a composite type of membrane consisting of PVA on a polyacrylnitril (PAN) support layer is applied, where very high selectivities for water can be reached.

Organophilic membranes are used for the removal of organic traces from waste or process water or for the purification of drinking water. In this application membranes with high selectivities for organic compounds are necessary. Organophilic membranes are for example polydimethylsiloxane (PDMS) or polybutadiene (PB). With the target-organophilic pervaporation organic mixtures are separated, where one compound of the mixture is preferentially permeating through the membrane. Membranes made out of PDMS, composites of PVA/PAN and also polyimides (section 4.2.4) can be used for the separation of organic mixtures, e.g. aromatic/aliphatic separation [16]. Some typical membrane materials used for pervaporation processes are shown in figure 6.

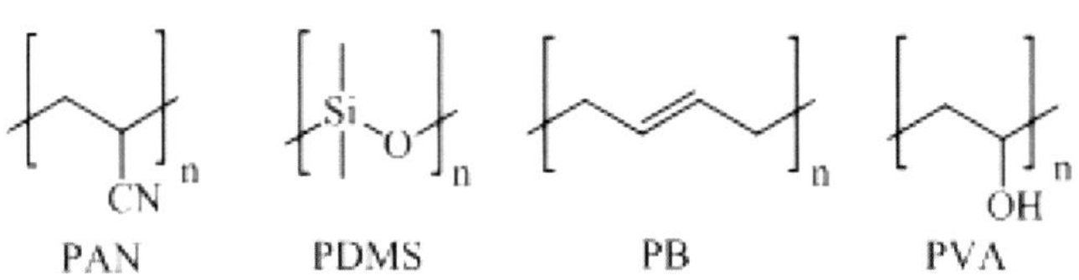

Figure 6. Typical membrane materials for hydrophilic, hydrophobic and target-organophilic membranes.

For the evaluation of new membrane materials, pervaporation experiments can be performed using small modules where membrane samples with an area of 33.2 cm^2 are used. Figure 7 shows a membrane module which is commonly used for determining the separation properties.

Figure 7. Membrane module for pervaporation experiments (upper part) as well as sinter plate (lower part left side) and CNT-membrane (lower part right side).

In figure 8 a schematic diagram of the complete pervaporation equipment is shown. The liquid feed mixture is stored in a feed tank and can be heated. The feed is transported into the heated membrane module with a pump and recirculated back to the feed tank.

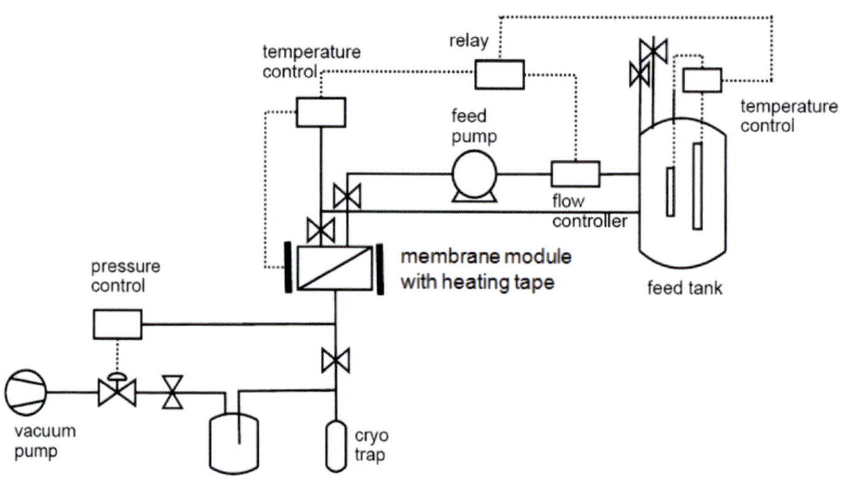

Figure 8. Schematic diagram of a pervaporation set-up.

On the permeate-side a vacuum pump is used for keeping the pressure low and the vaporous permeate is condensed via a cryo trap.

3. CNT-Membranes for Membrane-Based Separations

3.1. Pervaporation

The challenging task for researchers as well as for industry nowadays is to develop membrane materials that offer high fluxes combined with high selectivities. Furthermore these membranes should provide superior physical properties and a long lifetime. Nanocomposites or hybrid membranes containing CNTs appear to be a promising attempt in this field. CNTs can delegate their unique properties to membrane material and influence their separation characteristics. In the last few years nanocomposites and hybrid membranes showed improved properties and very promising results in different fields of membrane-based research.

In comparison to polymeric membranes they show a higher chemical and thermal stability and in comparison to inorganic membranes an excellent mechanical strength and they are robust to handle.

The price for the preparation of nanocomposite membranes is assumed to be moderate in contrast to the production of pure inorganic membranes, where very high fabrication costs have to be accepted [17].

Since CNTs as nanoscale objects tend to agglomerate due to van der Waals forces, membrane preparation is sometimes very difficult as shown in figure 9, where a membrane with strong agglomeration of CNTs is shown on the left side, as well as a proper membrane sample with homogeneous distribution of the right side. In figure 10 a scanning electron microscope (SEM) image of a homogeneous CNT/hybrid membrane is shown.

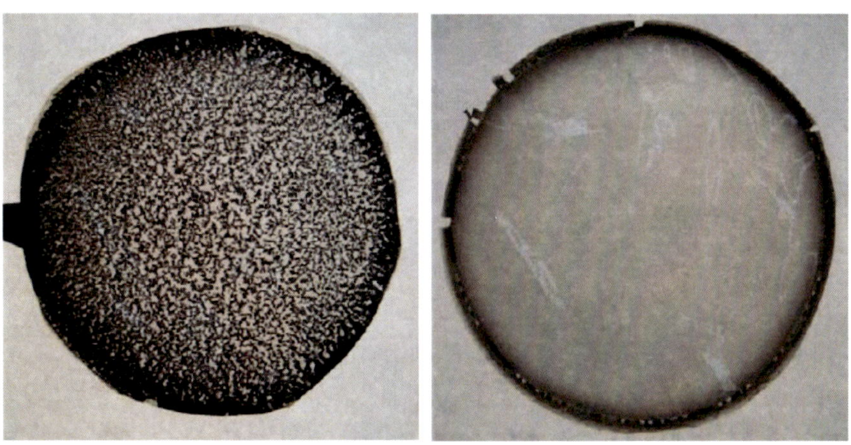

Figure 9. Nanocomposite membranes made from CNTs and a copolyimide with strong agglomeration of CNTs (left) and homogeneous distribution (right).

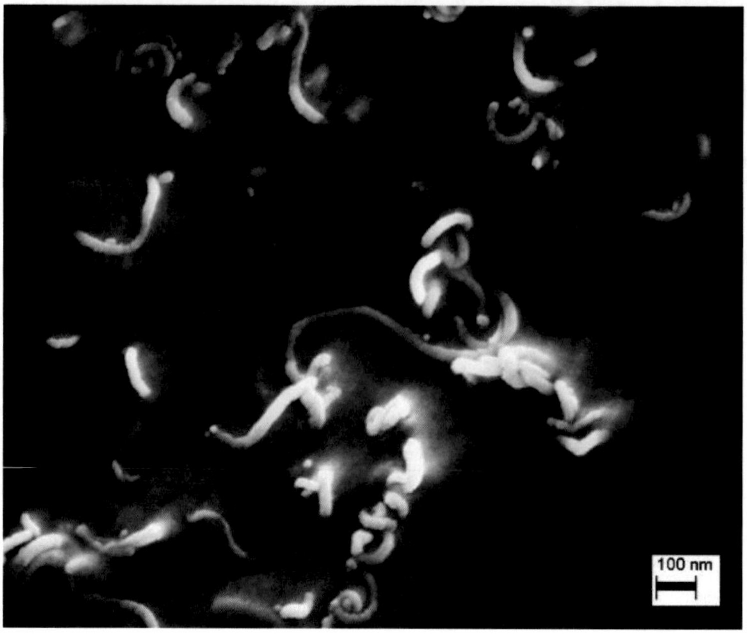

Figure 10. SEM image of the homogenous CNT/hybrid membrane (enlargement of small area from sample shown in Figure 9, right side).

The functionalization of CNTs is a promising way to prevent agglomeration and to obtain homogeneous and well dispersed membranes. Generally, functionalized CNTs can be obtained by chemical modification e.g. with strong acids at the end of the tubes or at defect sites respectively [18]. Beside the sidewall treatment of CNTs a non-covalent functionalization is possible by functional polymers or organic molecules, e.g. cyclodextrin (CD) or chitosan. The main advantage of this method is to keep the original properties of the CNTs.

So far in most cases CNT nanocomposites with dispersed or non-covalently functionalized CNTs have been investigated in the literature. Some possibilities of CNT functionalization are shown in figure 11.

In recent publications it is claimed that CNTs are toxic towards bacteria present in natural water streams. This is generally of very high interest since the production of clean potable water will be one of the major challenges in the next decade. Since microorganisms and their degradation products in water treatment processes lead to biofouling layers on the membrane surface and as a result to reduced water fluxes or increasing pressure, developing of new powerful anti-fouling strategies is urgently necessary. If CNTs as additives are able to reduce this undesirable fouling effects, maintenance costs can be reduced and the lifetime of the modules can be extended [19, 20].

For the preparation of high flux CNT membranes with vertically aligned CNTs as shown in figure 12 which are fixed in a non-permeable polymeric matrix, new methods are developing fast. This approach is very attractive since due to their inherent smoothness of the channels and the extreme distinct channel sizes of around 2-8 nm, these materials are highly interesting for the preparation of new membrane materials for water treatment processes.

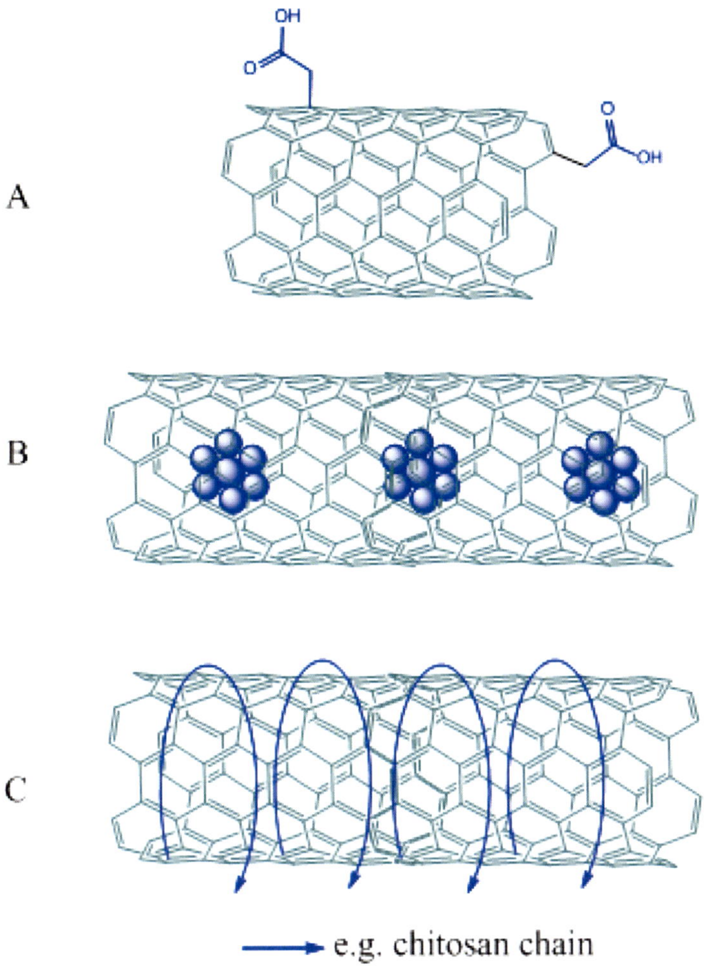

Figure 11. Functionalization possibilities for CNTs, A covalent sidewall/end tube functionalization, B endohedral functionalization with, for example fullerenes, and C non-covalent exohedral functionalization with polymers, e.g. wrapping with chitosan.

Model calculations have shown that water as well as gas molecules will be able to move through nanotube pores with orders of magnitude faster than through other pores of comparable size.

The proposed water transport mechanism show similar transport mechanism as biological ion channels [21, 22]. Membrane filtration experiments indicated that CNT membranes showed indeed very high fluxes of over 155 $L \cdot m^{-2} \cdot h^{-1}$ which is 10-100 times higher than that of commercial nanofiltration membranes with attractive rejection rates e.g. for methylene blue.

Furthermore regeneration was successfully applied so that a high potential in water purification technology will be expected from these materials [23]. However scale-up and defect free preparation of membranes using this approach will be a great challenge.

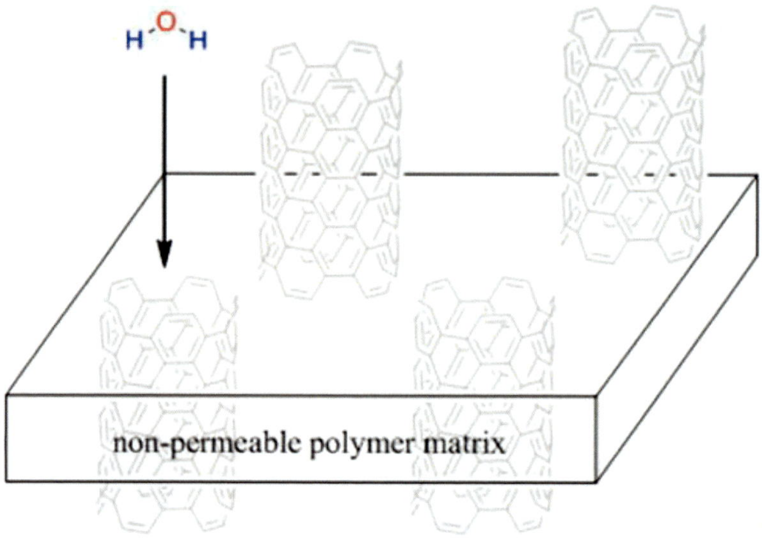

Figure 12. Vertically aligned CNTs in a non-permeable polymer matrix.

Furthermore latest research on modified CNTs with very small pore openings of 2 nm showed that if the pore openings are e.g. decorated with negative charged groups by plasma treatment, significant ion exclusion of up to 98% can be reached [24].

3.2. Gas Separation

Most of the commercially applied membrane materials for gas separation processes are polymers, e. g. polyimides, polysulfones and cellulose acetate [25]. However inorganic membranes like zeolites offer certain advantages, e.g. high selectivity combined with high permeability and in many cases better chemical resistance compared to polymers can be reached. However pure inorganic membranes are difficult to process in large scale and modules are still much more expensive compared to polymers. Therefore considerable research has been performed during the last years in order to produce mixed matrix or nanocomposite membranes consisting of an inorganic material, e, g, zeolites, MOFs (metal organic frameworks) or CNTs as additives as well as a bulk polymer. In many cases improvement of the separation properties were possible showing permeability/selectivity data beyond the so-called Robeson upper bound which was initially set up in 1991 and updated in 2008 [26].

As inorganic additives, CNTs are extremely promising for preparation of new membrane materials since computer simulations proposed that if alignment of the carbon nanotubes will be successful and the gaps between the CNTs can be filled by a polymeric phase gas as well as water fluxes can be increased tremendously [27, 28]. For example it has been predicted by Sholl and co-workers that gas diffusivities are a magnitude faster for SWNTs than in zeolites having similar pore sizes [28]. The reason for this is that the carbon nanotubes obtain extremely narrow pore size distribution of several nanometers and the inner surface of the tubes is inherently smooth so that the diffusion rate of molecules is very fast.

Recent experimental studies indeed showed that nanocomposites made from CNTs and poly (imidesiloxane) showed a facilitate hydrogen transport compared to the pure polymeric material [29]. Similar approaches showed also an increase of gas permeation for other gases like carbon dioxide nitrogen and oxygen of 10-20% without a loss in selectivity [30]. Other studies by Ge et al. showed that CNT nanocomposites made from polyethersulfone and modified CNTs showed 66% higher carbon dioxide permeation compared to the pure polymer materials whereas the selectivity for carbon dioxide over nitrogen did not change significantly [26].

In principle the increase of gas permeability can be reached due to the fast transport through the tubes but in order to reach that, an alignment of the nanotubes is necessary. Although several approaches have been investigated realizing a controlled growing of the tubes [29, 31, 32] it is still a challenge to upscale this process to a continuously working large scale membrane production.

However even if the nanotubes are disordered in the polymer phase, they are also useful additives leading to a strong improvement of permeability because of the reduction of the packing density. Thereby it is essential that good adhesion between the nanotubes and the bulk polymer material is reached to avoid micro gaps between additive and polymer which can lead to a lower selectivity. Proper adhesion can be reached e.g. if functionalized CNTs and polymers with functionalities are used. Fonseca et al. for example tested amino functionalized CNTs which were mixed with a polyetherblockamide to prepare nanocomposite membranes [33]. Thereby the CNTs were shortened by using a ball mill so that the average length did not exceed 300 nm [34]. This procedure was necessary to prepare thin film samples which were tested in gas separation. It was shown that by introduction of amino groups onto the CNT, carbon dioxide permeability is strongly increasing (40% if 0.5 wt. % of modified CNTs are used in the membrane preparation). For the selectivity of carbon dioxide over methane no significant change was observed, thus it is assumed that proper adhesion between polymer and CNT was reached. Comparison between single and multi-walled carbon nanotobes as additives in nanocomposite materials showed that MWNTs were more effective as additive for the improvement of the gas permeability than SWNTs, whereas the diameter and length of MWNTs had little effect on the separation properties of carbon dioxide and nitrogen. A maximum increase of permeability by a factor of around 2 could be reached with 5 wt. % of MWNTs [29].

4. Pervaporation Applications

4.1. Water/Organic Separations

For water/organic separations membrane materials with a strong hydrophilic character are necessary in order to remove the water from the mixture. As a result dried organic solvents without any impurities can be reached or water can be removed out of equilibrium reactions.

4.1.1. Chitosan-based CNT Nanocomposite Membranes

Chitosan (CS) is a biodegradable polymer, which can be produced from natural renewable resources (figure 13). It has good film-forming properties, a significant hydrophilicity and a high chemical resistance. It can be derived via deacetylation of chitin under alkaline conditions. Membranes made of pure CS offer a high flux which is very desirable. However since the high flux is caused due to high swelling a comparably low selectivity is obtained [35]. In the literature several approaches have been made to reduce the swelling, e. g. cross-linking or blending with additives.

Beneath that, the incorporation of nano-fillers as MWNTs was investigated in order to form so-called “green” nanocomposite membranes, which is a promising way to obtain a better membrane efficiency and at the same time ecofriendly membrane materials since chitosan is belonging to the class of renewable raw materials.

Tan et al. synthesized a poly (3-hydroxybutyrate)-functionalized (PHB) MWNTs /chitosan for the preparation of nanocomposite membranes and investigated these materials for the dehydration of 1,4-dioxane by pervaporation. 1,4-dioxane is an organic solvent and widely used in chemical and pharmaceutical industries. Conventional separation techniques like distillation can be used for the separation of 1,4-dioxane and water but additives like entrainers are needed because an azeotrop is formed at a water composition of 18 wt. % [36]. In this case a pervaporation process would be very advantageous because separation temperatures can be kept far beyond the boiling points of both components and no additives are necessary [37, 38].

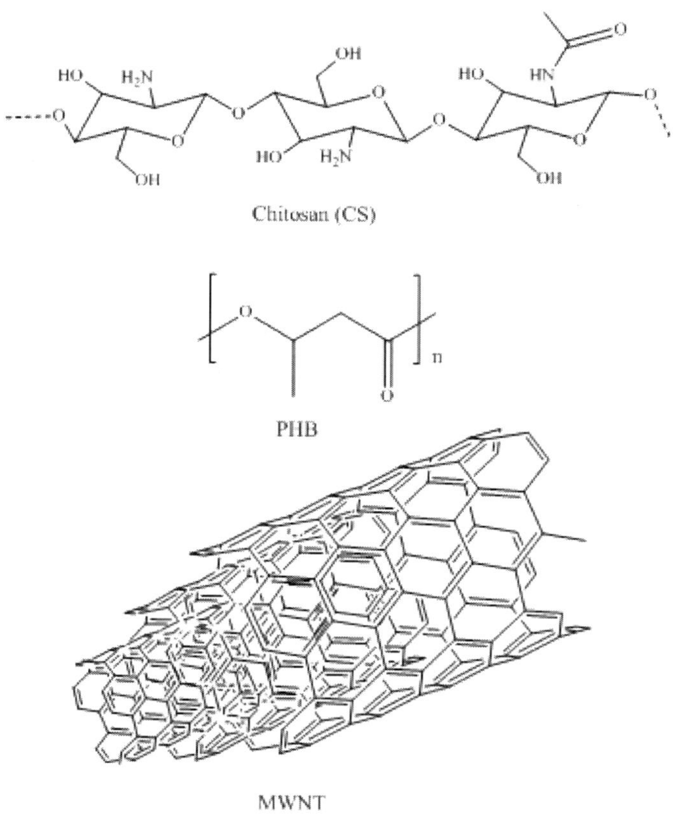

Figure 13. Chemical structures of chitosan (CS), poly(3-hydroxybutyrate) (PHB) and MWNT.

In order to reduce the above mentioned aggregation of the MWNTs Tan et al. have oxidized the MWNTs using a solution of sulfuric acid and hydrogen peroxide and in the second step functionalized them by adding it to the PHB solution. The oxidized MWNTs contain hydroxyl and carboxyl groups and the PHB solution contain hydroxyl groups in order to improve interaction between both components and the chitosan polymer. The PHB-MWNTs solutions were then aligned before chitosan was incorporated into the matrix. A detailed description of the membrane preparation can be found elsewhere [39].

In figure 14 for this type of membrane temperature depending pervaporation experiments for water/1,4-dioxane mixtures (5/95 wt. %) with a chitosan and PHB-MWNTs/chitosan nanocomposite membranes are shown. At higher temperatures an increase of the flux for both types of membranes is observed, whereas the nanocomposite membrane obtained higher fluxes than the native chitosan membrane. With increasing temperature a strong decrease of the selectivity for both types of membranes was found. However the selectivity of the nanocomposite membrane is lower compared to the native chitosan membrane. It is assumed that at higher temperatures the thermal mobility of the molecules inside the membrane increases generating extra free volume which is causing a higher flux but also a lower selectivity.

Another approach was investigated using chitosan nanocomposite membranes for the separation of ethanol and water. Zhang et al. incorporated functionalized MWNTs into the chitosan matrix and prepared nanocomposite membranes by the solution blending method [40].

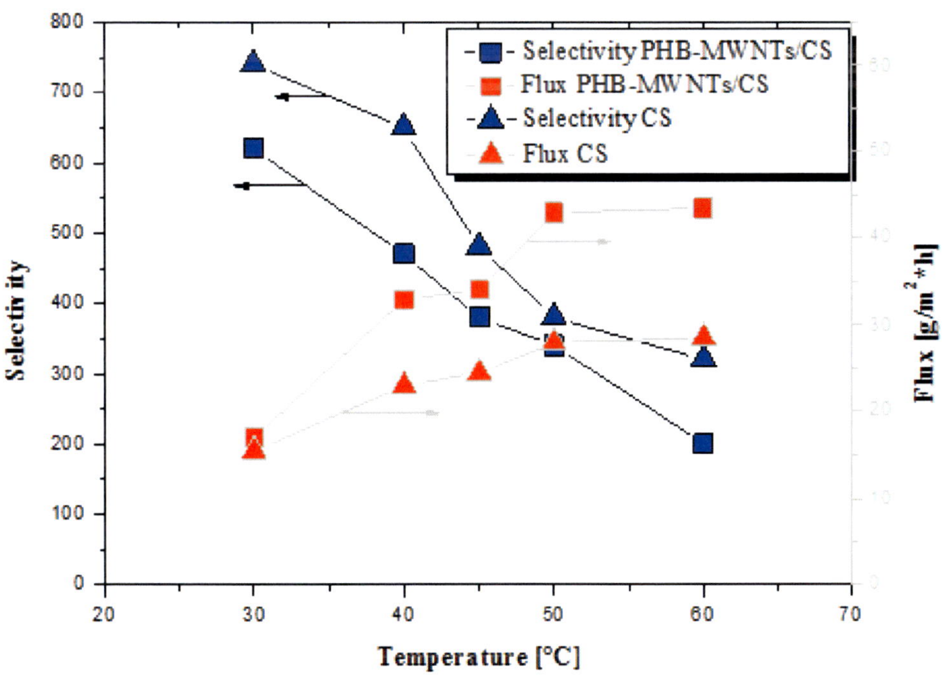

Figure 14. Temperature depending pervaporation experiments with a chitosan (CS) and a PHB-MWNTs/CS membrane using water/1,4-dioxane mixtures (5/95 wt. %) as feed, permeate pressure 6-33 mbar (modified after [39].

The functionalized MWNTs contained carboxyl groups and were prepared by treating pure MWNTs in a solution of sulfuric acid and nitric acid (3:1) for a defined period of time at 80 °C. After purification the oxidized MWNTs were dispersed in DMF and reacted with diisobutyryl peroxide afterwards to form MWNTs with carboxylic groups (MWNT-COOH). The diisobutyryl peroxide is heated to decay into radicals. After heating and elimination of carbon dioxide a CH_2 radical is formed, which than reacts with the MWNTs. The MWNT-COOH was then added with different ratios to an acetic acid solution of chitosan and casted on a PAN substructure to obtain the nanocomposite membranes. In this process the carboxyl groups of the MWNTs can react with free hydroxyl groups and amino groups of the chitosan structure and build up covalent bonds as shown in figure 15. It has to be mentioned that Zhang et al. describe their membranes as nanocomposites and not hybrid membranes, although a covalent bond between MWNTs and chitosan is presumably formed.

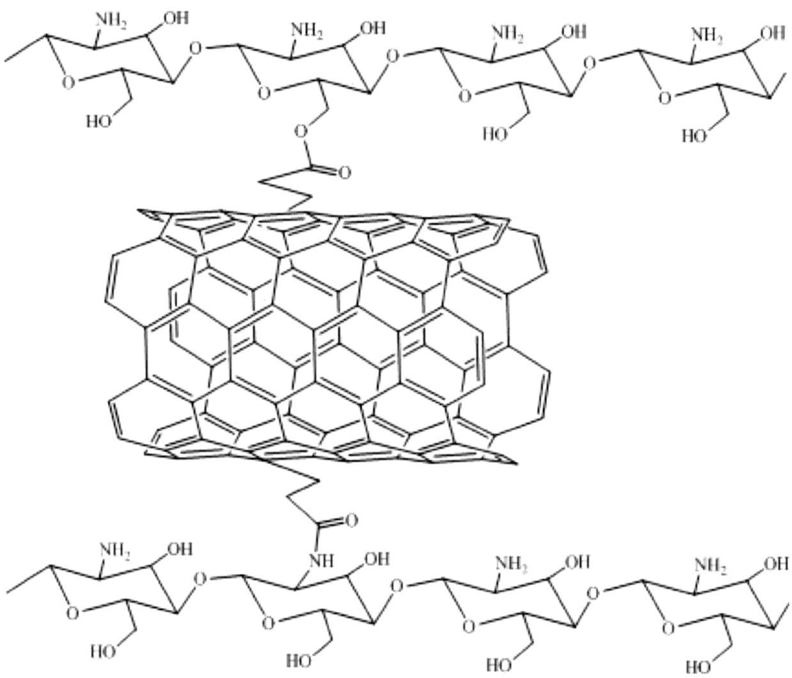

Figure 15. Schematic illustration of covalently bonded MWNT-COOH to chitosan.(modified after [40])

Table 1. Pervaporation results for a pure chitosan (0 wt. %) and CS/MWNTs-COOH nanocomposite membranes using water/ethanol mixtures (10/90 wt. %)

Nanotubes [wt. %] in chitosan	Flux [$g/m^2 \cdot h$]	Selectivity
0	112	580
0.5	233	576
1	293	574
1.5	340	573
2	337	570

Pervaporation experiments were carried out for the separation of water/ethanol mixtures (10/90 wt. %) with a pure chitosan membrane and chitosan/MWNTs-COOH nanocomposite membranes with different amounts of MWNTs-COOH. Unfortunately no further information such as operating temperature and permeate pressure are mentioned in the publication [40]. Selectivity and flux for nanocomposites with a different amount of nanotubes are shown in table 1.

The results show a higher flux for the nanocomposite membranes in comparison to the pure chitosan membrane. At higher contents of MWNTs-COOH in the membranes an increase of the flux is observed. In case of the selectivity nearly no decrease in selectivity has been found. Zhang et al. suggested the possibility that both components of the feed, water and ethanol, can permeate through the inner tubes of the MWNTs. The kinetic diameters of a water molecule (0.3 nm) and an ethanol molecule (0.43 nm) are much lower than the inner diameter of the MWNTs (around 5 nm). Both molecules can pass through the MWNTs which results in an increase of the flux. However no decrease in selectivity was observed due to the fact that the ethanol molecule is much larger compared to the water.

4.1.2. Nanocomposites Made of PVA and CNTs

In pervaporation experiments for the dehydration of isopropanol from aqueous mixtures Mohammadi et al. investigated PVA-CNT nanocomposite membranes. PVA is a commercially available polymer and one of the most studied membrane materials. Its hydrophilic character makes PVA very attractive as membrane material which can be used for dehydration applications. Furthermore it has a very good film-forming ability and a high chemical stability. As mentioned in paragraph 4.1.1 swelling behavior in aqueous solutions can be repressed by cross-linking of the PVA.

In the work published, the CNTs were synthesized by CVD method from cyclohexanol as carbon source and treated after purification with nitric acid to remove encapsulated catalyst particles and to prevent agglomeration of the CNTs in the PVA solution. The nanocomposite membranes consisting of PVA and CNTs were prepared by solution casting and solvent evaporation technique [41]. To achieve an in situ cross-linking, glutaraldehyde as the cross-linking agent and hydrochloric acid as catalyst were added to the aqueous solution (Figure 16).

PVA-CNT nanocomposites have been produced with various amounts of CNTs in the nanocomposites. The prepared membranes showed thicknesses of approximately 25 μm. The pervaporation experiments have been performed with a water/isopropanol mixture with a concentration of 10 wt. % water. The operating temperature was kept at 30 °C and the permeate pressure was 1 mbar.

The results of the comparison between a pure PVA membrane (0 wt. % CNTs content) and PVA-CNT nanocomposite membranes with a different amount of CNTs can be found in figure 17.

For pervaporation experiments PVA-CNT nanocomposite membranes with a CNT content of 0.5, 1, 2 and 4 wt. % have been analyzed. The results show a higher selectivity for the PVA-CNT nanocomposite membrane in comparison to a pure PVA membrane, although lower fluxes have been found. This is presumably due to the fact that the PVA is cross-linked via acetal groups and the CNTs are fixed within that network.

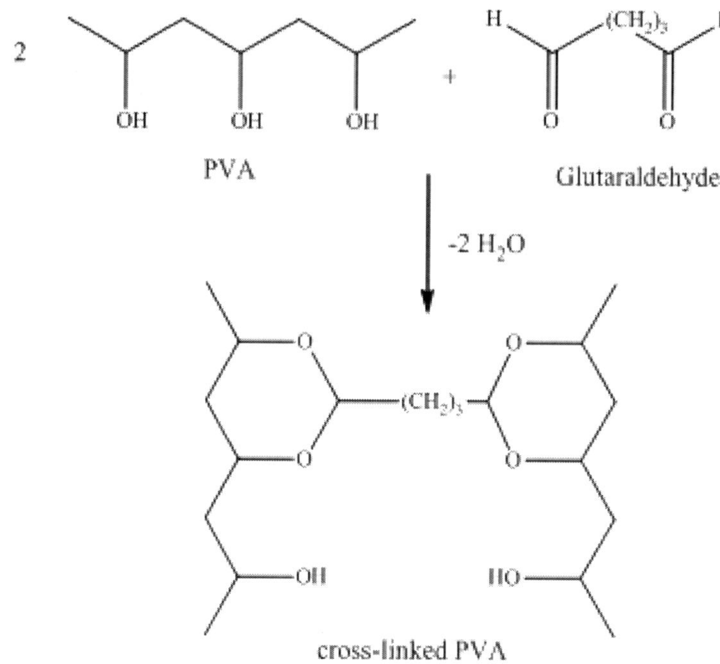

Figure 16. Scheme of cross-linked PVA.

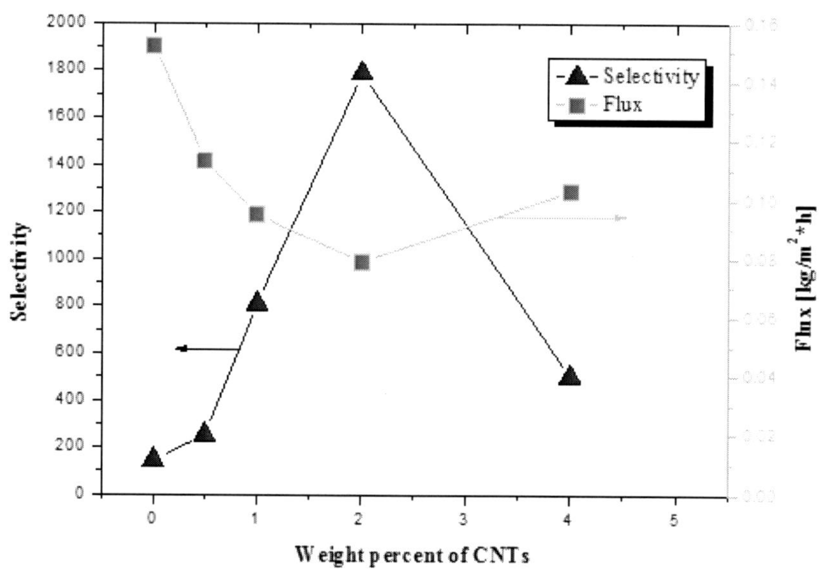

Figure 17. Pervaporation results of a pure PVA membrane and a PVA-CNT nanocomposite membrane with various amounts of CNTs for a water/isopropanol mixtures (10/90 wt. %), feed temperature 30 °C, permeate pressure 1 mbar (modified after 41]).

The most promising results were obtained with a PVA-CNT nanocomposite membrane with an amount of 2 wt. % CNTs, where a very high selectivity (α~1800) with an adequate flux of around 0.08 kg/m^2·h (which correlates to 2 kg·µm/m^2·h normalized to the membrane thickness) was found. At higher weight percent of CNTs in the nanocomposite membrane, an enormous decrease in selectivity is observed with a 20 % increase in flux. Beside the pervaporation experiments it was found that with incorporation of CNTs into the PVA polymer the degree of swelling can be significantly reduced.

Measurements at 30 °C were performed with all type of membranes, where the PVA-CNT 2 wt. % nanocomposite membrane showed an almost three times lower swelling degree than the pure PVA membrane [41]. This might be due to the fact, that the CNTs during the treatment with nitric acid are functionalized at the end of the tubes and on defect sites so that carboxyl groups are present. These carboxyl groups are able to react with the PVA and lead to an additional cross-linking.

A similar approach was used by Jegal et al. where MWNTs were incorporated into PVA membranes and investigated with regard to the dehydration of ethanol [42]. The incorporation of CNTs into polymer matrix leads to a higher free volume resulting in a higher flux and lower selectivity. In case of cross-linking generally a lower flux and an increase in selectivity is expected.

For membrane preparation, the MWNTs were firstly treated with sulfuric acid and nitric acid to obtain hydroxyl and carboxyl groups at the end of the pipes or at defect structures. The pretreated MWNTs were then blended with a PVA solution in water and casted onto a glass plate to receive MWNTs/PVA membranes by evaporation. Pervaporation experiments with MWNTs/PVA membranes with different contents of MWNTs were carried out for water/ethanol mixtures (10/90 wt. %). The operating temperature was kept at 40 °C and the permeate pressure was 0.1 mbar. The results are shown in figure 18.

From the results of the pervaporation measurements general trends in membrane performance can be observed. The pure PVA membrane shows a flux of ~32 $g/m^2 \cdot h$ and a selectivity of around 800.

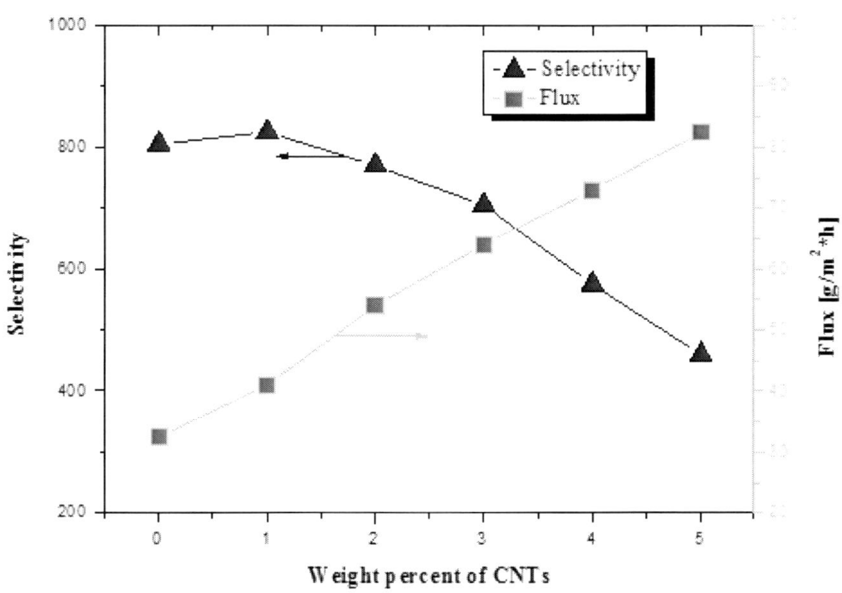

Figure 18. Pervaporation results for a pure PVA and MWNTs/PVA membranes for the separation of water/ethanol mixtures (10/90 wt. %) at 40 °C, permeate pressure 0.1 mbar (modified after [42]).

With increasing amount of MWNTs in the PVA membranes, an increase of flux is observed (1 wt. % ~41 $g/m^2 \cdot h$ in comparison to 4 wt. % ~73 $g/m^2 \cdot h$), while a decrease in selectivity is obtained (1 wt. % ~825, 4 wt. % ~575). An incorporation of the MWNTs into the polymeric matrix lead to an increase of the free volume. The transport of molecules through the membrane is relieved and is even possible through the CNTs, which explains the higher flux and lower selectivity.

In temperature depending pervaporation experiments with related MWNTs/PVA membranes Jegal et al. found out that with increasing temperature and additionally with higher content of MWNTs a higher flux and a lower selectivity was obtained [43] which is most likely due to the increasing chain mobility at higher temperatures.

4.2. Organic/Organic Separations

The separation of organic mixtures is very important in chemical industry to obtain solvents with a high purity. For example in the petrochemical industry membranes might be used in the future to reduce the sulfur amount in gasoline. Membranes with a high selectivity for organic compounds are necessary for the separation of organic mixtures, e.g. the separation of aromatics/aliphatics.

4.2.1. CNT Nanocomposites for Benzene/Cyclohexane Separation

Peng et al. used novel nanocomposite membranes containing MWNTs for the separation of benzene/cyclohexane mixtures by pervaporation. On the one hand they used membranes composed of PVA and chitosan-wrapped MWNTs (PVA-MWNT(CS)) (already mentioned in paragraph 4.1.1) and on the other hand PVA membranes filled with CNTs dispersed by β-cyclodextrin (β-CD). Both types of membranes were investigated in order to improve the separation properties of pure PVA membranes.

Beneath the dehydration of organic solvents, PVA as membrane material can be used for the separation of organic/organic mixtures, like benzene and cyclohexane. This is possible because of the polar and hydrophilic character of the PVA which offers a significant higher solubility for benzene compared to cyclohexane [44, 45]. Since aggregation often occurs, it is difficult to achieve a homogenous incorporation of CNTs into polymeric matrices. A strong hydrophobicity of the membrane material leads to a decrease in selectivity [35, 46]. Therefore as mentioned in paragraph 4.1.1, a functionalization of the CNTs with a hydrophilic agent seems to be a promising way to prevent a loss in selectivity. Due to amino and hydroxyl groups in the structure chitosan has a high hydrophilicity and good solubility properties. The dispersion and solubility behavior of CNTs in a PVA matrix can be improved by adding chitosan [47]. In figure 19 an illustration of a chitosan-wrapped MWNT structure is shown.

The composite membrane consisting of PVA and chitosan-wrapped MWNT in this approach is prepared of a chitosan solution in an acetic acid aqueous media, where a certain amount of MWNTs was added to the solution. This solution was then added to an aqueous PVA solution with glutaraldehyde as cross-linking agent to form a well dispersed PVA-MWNT(CS) membrane with a thickness of around 80 μm for the use in pervaporation (4.1.2, Figure 16). A detailed description of the membrane preparation can be found elsewhere [48].

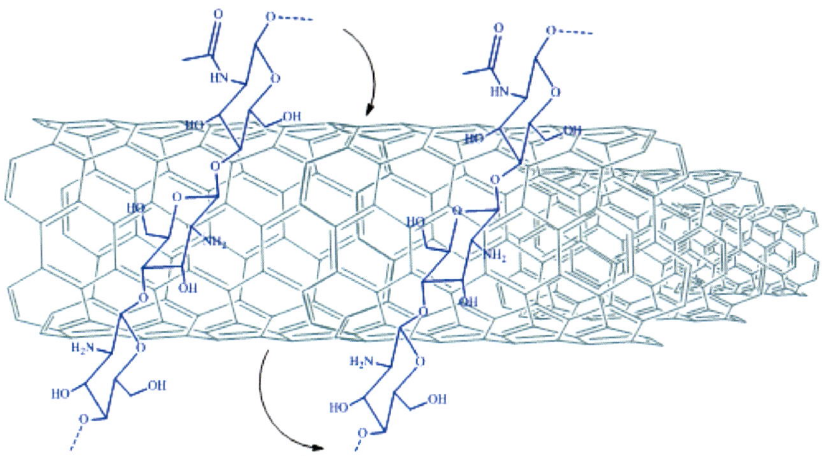

Figure 19. Illustration of a chitosan-wrapped MWNT (modified after [48]).

In the described approach Peng et al. investigated PVA-MWNT chitosan membranes with different amounts of MWNTs (0.5 to 2.5 wt. %) for the separation of a binary mixture of benzene/cyclohexane (50/50 wt. %) at 40 °C. The permeate pressure in the experiments was kept at 1 mbar. As shown in figure 20 the pervaporation experiments showed that with incorporation of MWNTs into the PVA matrix the separation characteristics changed and could be improved significantly. With increasing amount of chitosan-wrapped CNT a higher flux is observed and simultaneously a higher selectivity was found.

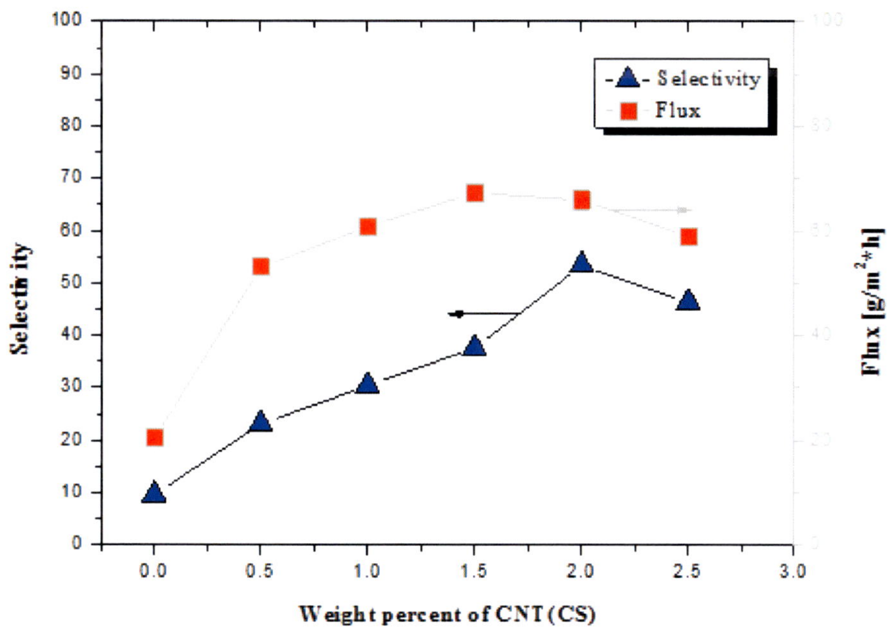

Figure 20. Pervaporation experiments of a pure PVA membrane and PVA membranes containing different amounts of chitosan-wrapped MWNTs for the separation of benzene/cyclohexane mixture (50/50 wt. %) at 40 °C, permeate pressure 1 mbar.

In comparison to a pure PVA a separation factor which is more than 5 times higher (53.4 compared to 9.6) and a flux more than 3 times higher (65.9 compared to 20.3) was found for a 2 wt. % PVA-CNT(CS) membrane. It seems as if a higher amount of chitosan-wrapped CNT is added, for example 2.5 wt. %, selectivity and flux are decreasing. These results can be explained due to a stronger affinity of CNTs towards benzene, because of π-π interactions or stackings causing an increased free volume [49-51].

4.2.2. CNT Hybrid Membranes for Benzene/Cyclohexane Separation

In another work Peng et al. synthesized hybrid membranes made of PVA and CNTs dispersed in β-cyclodextrin (β-CD) [52]. The β-cyclodextrine molecules are cyclic oligosaccharides consisting of seven glucose units, which offer hydrophobic cavities. Host-guest interactions with several components, e.g. benzene, toluene, but also polymers are well known.

However it is not expected that CNTs can form such host-guest complexes due to their average outer diameter of 2-100 nm, since the inner cavity of the β-cyclodextrin is only 0.6 nm. However in previous studies it was found that with incorporation of β-CD the dispersion and solubility of CNTs can be remarkably improved [53-55]. This is presumably due to the fact that the β-cyclodextrin can form micelles with a hydrophilic outer layer. In figure 21 the structure of β-CD is shown.

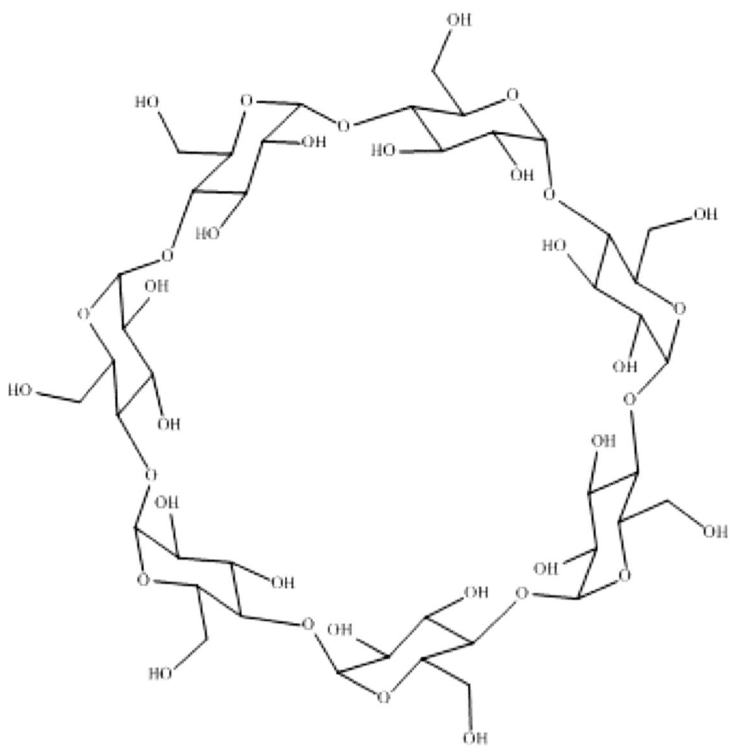

Figure 21. Chemical structure of β-cyclodextrin (β-CD).

β-CD-CNT/PVA hybrid membranes were made by the grinding method, where a powder of β-CD and CNTs is added to an aqueous PVA solution which was mentioned before [52]. The ratio between β-CD and CNTs was 30:1 and the obtained membranes showed a thickness of around 80 µm. The feed mixture was a binary mixture of benzene/cyclohexane (50/50 wt. %), while the permeate pressure was kept at 1 mbar and the operating temperature was 40 °C. An excerpt of the pervaporation results showing the effect of β-CD-CNT content in β-CD-CNT/PVA hybrid membranes is shown in figure 22.

The results show an increase of flux and selectivity for nearly all membranes. A higher flux and even a higher selectivity were observed with an increasing content of β-CD-CNT in the membrane material. In comparison to the pure PVA membrane (0 wt. % β-CD-CNT) a flux of around 2 times higher was obtained and a 2 up to 3-times higher selectivity was achieved. From these results it can be concluded that β-CD-CNT can enhance the sorption selectivity of benzene in the membrane due to π-π interactions between CNTs and benzene molecules. Thus the distribution of β-CD-CNT interfere the polymer chain packing in the membrane and increase the transport of the feed components.

In other experiments Peng et al. showed that with an increasing operating temperature and at higher feed flow rates higher selectivities and fluxes have been found using β-CD-CNT/PVA membranes [52].

However in this approach it should be also mentioned that two factors might lead to an increase in flux combined with an increase in selectivity.

Not only the CNT is affecting the separation properties, but also the cyclodextrin is enhancing the transport of benzene through the membrane materials, due to the formation of host-guest interactions which are well known for many aromatic components [55, 56].

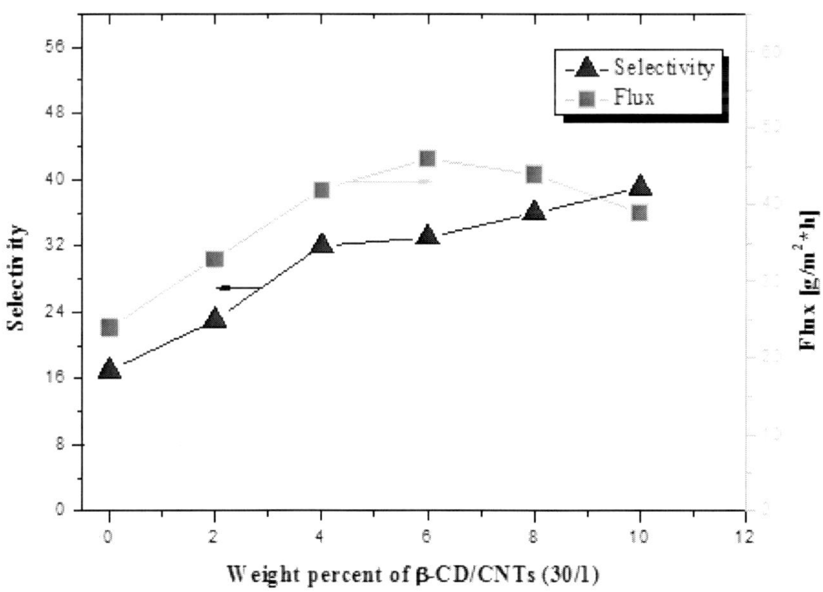

Figure 22. Pervaporation experiments of β-CD-CNT/PVA with different β-CD-CNT content for a benzene/cyclohexane mixture (50/50 wt. %) at 40 °C, permeate pressure 1 mbar.(modified after [52].

4.2.3. Polyamide/CNT Nanocomposites

Penkova et al. developed nanocomposites consisting of poly (phenylene isophtlamide) (PA) and CNTs for the separation of methanol/methyl *tert.*-butyl ether (MeOH/MTBE) mixtures [57]. MTBE gains more and more interest in chemical and petrochemical industry, e.g. as reagent and additive in gasoline due to its unique property increasing the octane number of gasolines. As polymeric material PA is used in this work, which is known as commercial membrane material with high chemical resistance and good mechanical properties (Figure 23).

The PA/CNTs nanocomposites were obtained by mixing of PA powder and MWNTs with different ratios.

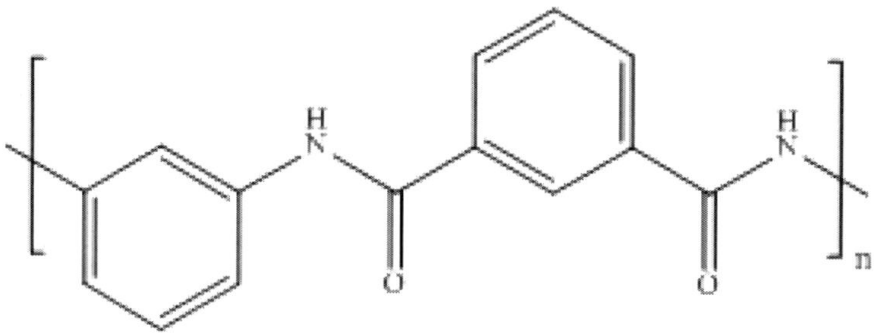

Figure 23. Structure of poly(phenylene isophtlamide) (PA).

After dissolving in N,N-dimethylaccetamide a small amount of lithiumchloride has been added in order to stabilize the solution and to avoid gel formation which can occur between carboxyl groups present on the CNTs depending on the production process and the amine end groups on the polymer backbone. After the membrane preparation the PA/CNTs nanocomposites obtained thicknesses of around 20 to 40 μm and were used in pervaporation experiments for comparison with native PA membranes. Pervaporation experiments were carried out using a varying feed concentration of MeOH (14-70 wt. %) at 50 °C with a permeate pressure of 1 mbar.

In figure 24 the results of pervaporation experiments depending on the concentration of MeOH in the feed mixture are shown for the separation of MeOH/MTBE with a native PA membrane and two PA/CNT nanocomposite membranes (2 wt. % and 5 wt. % of CNTs). It is obvious that with a higher concentration of MeOH in the feed mixture, a decrease in selectivity for all types of membranes occured, whereby the 2 wt. % PA/CNT nanocomposite membrane indicates a higher selectivity than the native PA membrane. Concerning the fluxes a higher amount of CNTs in the membrane led to a higher flux in comparison to the native PA membrane. Penkova showed that with the investigation of PA/CNT nanocomposite membranes incorporation of CNT into the polymer matrix is a useful tool for increasing the flux and improve the selectivity for the removal of methanol from a MeOH/MTBE mixture.

4.2.4. Copolyimdes/CNT Hybrid Materials

In previous work it has been shown that incorporation of CNTs and MWNTs into a polyimide matrix often lead to strong agglomeration effects as shown in figure 9. Therefore in a recent work hybrid materials consisting of functionalized MWNTs and copolyimides were investigated. The copolyimides were made from an anhydride 6FDA (4,4'-hexafluoroisopropylidene diphthalic acid anhydride) with two diamines 4MPD (2,3,5,6-tetramethylphenylene-1,4-diamine) and DABA (3,5-diaminobenzoic acid). The DABA monomer contains a carboxyl group, where further modification is possible, e.g. cross-linking. The copolyimide is synthesized in a two-step reaction. A detailed mechanism of the reaction can be found elsewhere [58]. The structure of the polyimide is shown in figure 25.

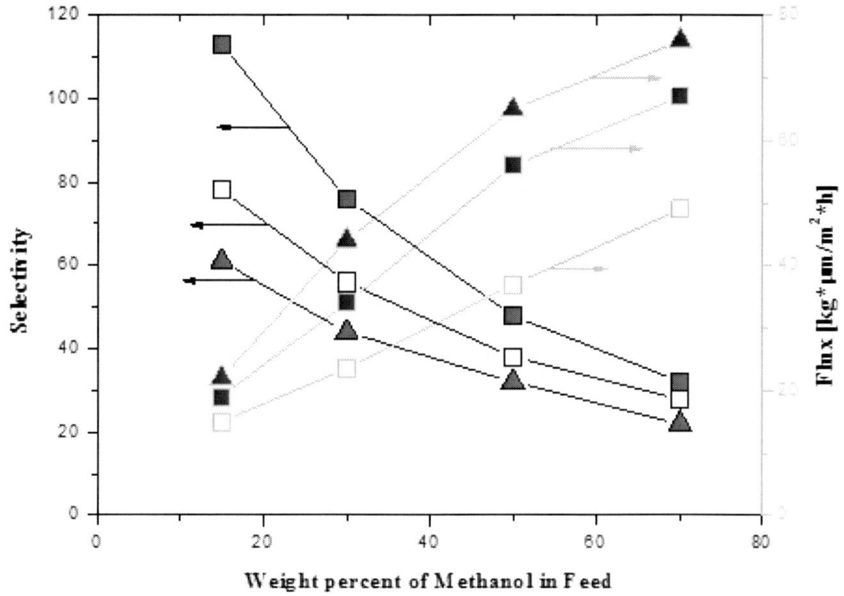

Figure 24. Concentration depending pervaporation measurements for the separation of MeOH/MTBE with a native PA (□) and PA/CNT nanocomposite membranes with different amounts of CNTs (■/■ 2 wt.%, ▲/▲ 5 wt.%) at 50 °C, permeate pressure 1 mbar.(modified after [57]

Copolyimides are highly interesting for membrane applications due to their outstanding properties. They offer a high chemical resistance, superior mechanical properties over a wide temperature range and good separation performance. Membranes made of copolyimides have been studied in many different fields of separation processes. They showed very promising results for the separation of aromatic/aliphatic mixtures [59-63] and for desulfurization of sulfur aromatic compounds from organic mixtures [64-66], but also for gas separations [67-69].

In the approach described, MWNTs were used without further purification. Membranes prepared from non-modified MWNTs did not form well dispersed membranes since agglomeration occured. The MWNTs were functionalized by a treatment with concentrated nitric acid (65%) for 150 min at 120 °C under reflux. After purification with water and acetone the functionalized MWNTs were dried at 150 °C and analyzed by elementary analysis. The results showed a 6.6 times higher amount of oxygen for the carboxyl containing MWNTs-COOH (~2 wt. %) in comparison to the pure MWNTs (0.3 wt. %). After functionalization the MWNTs contain carboxyl groups, for example at the end of the pipes and if defect structures are present at the tubes. Hydrogen bond interactions can be formed between the carboxyl groups of the MWNTs and the carboxyl groups of the polymer chains (DABA group). The amount of MWNTs-COOH in the copolyimide used in this work was one weight percent (1 wt. %) based on the solid polymer. In figure 25 a structure of 6FDA-4MPD/6FDA-DABA copolyimides with a ratio of 49:1 is shown with functionalized MWNTs-COOH.

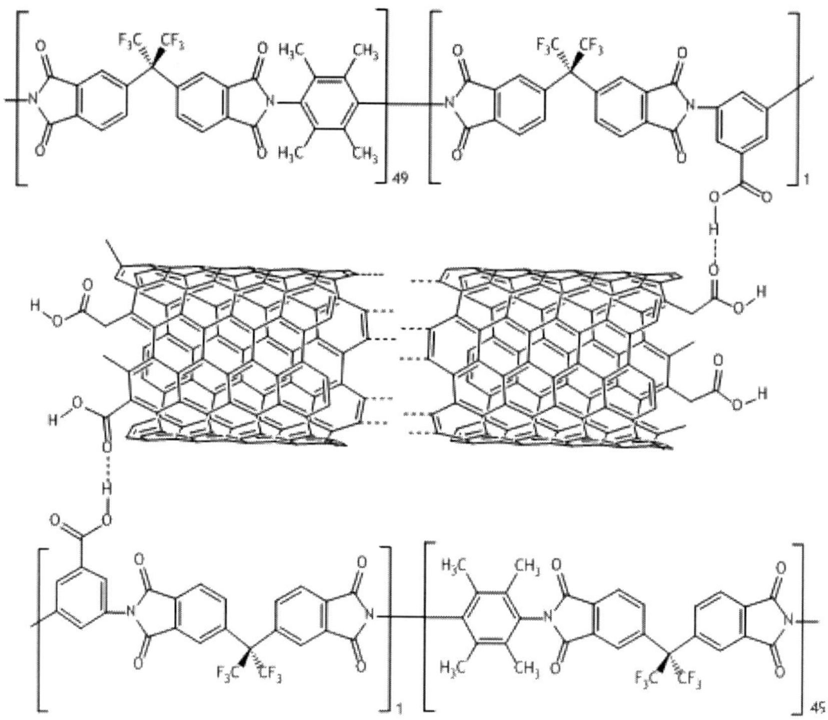

Figure 25. Schematic structure of MWNT-COOH/copolyimide hybrid material.

Temperature depending pervaporation experiments have been performed with a binary mixture of benzothiophene/*n*-dodecane (0.25/99.75 wt. %) using 6FDA-4MPD/6FDA-DABA 49:1 copolyimide membranes (Figure 26). Thereby a native copolyimide was compared to a hybrid membrane consisting of the copolyimide and functionalized MWNTs containing carboxyl groups. The hybrid membranes showed well dispersed MWNTs-COOH in the polymer matrix. All membranes have been prepared with a thickness of approximately 30 μm and have been pretreated in a benzothiophene/*n*-dodecane mixture (10/90 wt. %) with a higher content of sulfur compound. Therefore the membrane was kept for 24 h at room temperature in the mixture. The pretreatment of the membrane leads to a certain degree of swelling, thus the microvoids in the polymer matrix are enlarged, which results in a higher free volume involving higher permeabilities.

It has been found that at elevated temperature the MWNT-COOH/copolyimide hybrid membrane obtained as desired, much higher fluxes compared to the native membrane, while no significant change in selectivity was observed.

The hybrid membrane shows a stronger increase of the flux at higher temperatures in comparison to the native copolyimide membrane. It is assumed that the MWNTs act as spacer between the polymer chains which leads to a higher free volume and because of that a higher flux resulted.

Additionally temperature dependent pervaporation experiments were performed investigating the effects of possible hysteresis on hybrid membranes. The operating conditions were the same as described above. A MWNT-COOH/copolyimide hybrid membrane has been studied from 80 to 100 °C and back from 100 to 80 °C in 2.5 °C steps. It has been found that no significant change in flux as well as in selectivity occurs. The membrane showed a high stability during an operating period of nearly 3 months.

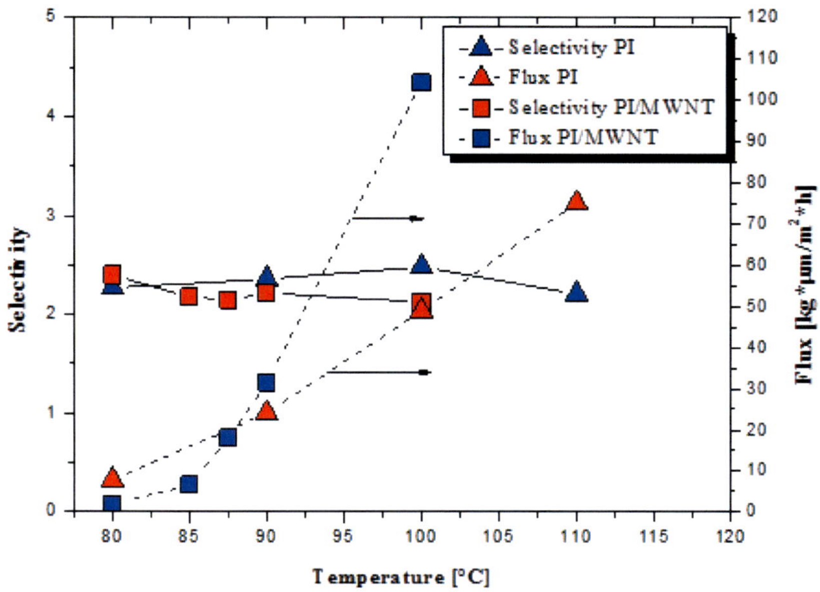

Figure 26. Temperature dependent pervaporation experiments using a benzothiophene/*n*-dodecane mixture (0.25/99.75 wt. %) and pretreated copolyimide native (▲/▲) and MWNT-COOH (■/■) hybrid membranes, permeate pressure 19-30 mbar.

Conclusion

The development of suitable and high efficient membranes for membrane-based separation processes in industry is of great interest. The improvement of existing membranes to overcome limitations for example of conventional polymeric membranes is the main target of researchers. The fabrication of polymeric membranes with incorporated CNTs is a promising way to obtain a new class of membranes with improved separation characteristics. Since their unique properties and great potential CNTs became an attractive candidate for the preparation of nanomaterials in membrane technology. Due to their structure CNTs are able to enhance significantly the transport through membranes. Simulation models have shown that molecules can pass through the inner channels of CNTs and increase the flux dramatically. Functionalization of CNTs opens up the opportunity to prevent agglomeration and to obtain well dispersed membranes. Covalent bonds and non-covalent interactions can be formed between CNTs and polymer to prevent a drastical loss in selectivity due to defect sites and to reduce the degree of swelling.

In the current work hybrid material and nanocomposite membranes made of CNTs and polymers for different applications in pervaporation have been discussed. Thereby polymer/CNT membranes showed impressive results and performance improvement for the separation of water/organic as well as organic/organic mixtures. In all discussed studies the flux could be increased through CNT nanocomposites or CNT hybrid membranes and an adequate selectivity has been reached. Furthermore the potential of CNTs in gas separation has been discussed.

CNT nanocomposites and CNT hybrid membranes have been successfully investigated in several applications in pervaporation and gas separation, but there are still more experimental studies necessary, especially concerning the long term stability and mechanical properties during separation processes running under harsh conditions to implement these materials in large scale membrane plants.

REFERENCES

[1] Herranz A., Martin N., Polymer Based on Carbon Nanotubes, in Fullerene Polymers: Synthesis, Properties and Applications (eds N. Martín and F. Giacalone), Wiley-VCH Verlag GmbH and Co. KGaA, Weinheim, Germany, (2009).

[2] Journet C., Maser W.K., Bernier P., Lee R., Fischer J.E., Large scale production of single-walled carbon nanotubes by electric-arc technique, *Nature* 388, 756-758, (1997).

[3] Smalley R.E., Guo T., Colbert D.T., Catalytic growth of single-walled nanotubes by laser vaporization, *Chem. Phys. Lett.* 243, 49-54, (1995).

[4] Smalley R.E., Hafner J.H., Smith K.A., Bronikowski M.J., Catalytic Growth of Single-Wall Carbon Nanotubes From Metal Particles, *Chem. Phys. Lett.* 296, 195-202, (1998).

[5] Saeed K., Review on Properties, Dispersion and Toxicology of Carbon Nanotubes, J. *Chem. Soc. Pak.* 32, 4, 449, (2010).

[6] Hirsch A., Functionalization of single-walled carbon nanotubes, *Angew. Chem. Intern. Ed.* 41, 1853-1859, (2002).

[7] Maier G, Gasseparation with polymer membranes, *Angew. Chem.* 37, (21), 2961, (1998).

[8] Graham T., On the Absorption and Dialytic Separation of Gases by Colloid Septa, *Philos. Mag.* 32, 401-420, (1866).

[9] Ullman´s Encyclopedia of Industrial Chemistry, C. Judson King University of California, Berkeley, Wiley-VCH Verlag GmbHand Co. KgaA, (2002).

[10] Smitha B., Suhanya D., Sridhar S., Ramakrishna M., Separation of organic-organic mixtures by pervaporation - a review, *Journal of Membrane Science* 241, 1-21, (2004).

[11] Jonquieres A., Clement R., Lochon P., Neel J., Dresch M., Chretien B., Industrial state-of-the-art of pervaporation and vapour permeation in the western countries, *J. Membr. Sci.* 206, 87, (2002).

[12] J.C. Davis, R.J. Valus, R. Eshragi and A.E. Velikoff, Facilitated Transport Membrane Hybrid Systems for Olefin Purification, *Sep. Sci. Techn.* 28, 463, (1993).

[13] H. Strathmann, Membrane separation processes: current relevance and future opportunities, *AIChE Journal* 47 (5), 1077, (2001).

[14] F. Lipinzki, R. W. Field, P. K. Ten, Pervaporation-based hybrid process: a review of process design, application and economics, *Journal of Membrane Science* 153, 183-210, (1999).

[15] R. Rautenbach, R. Albrecht, The Separation Potential of Pervaporation, 2. Progress Design and Economics, *J. Membr. Sci.* 25, 25, (1985).

[16] Lipinzki F., Hausmanns S., Ten P.K., Field R.W., Laufenberg G., Organophilic pervaporation: prospects and performance, *Chemical Engineering Journal* 73, 113-129, (1998).
[17] Ismail A.F., Goh P.S., Sanip S.M., Aziz M., Transport and separation properties of carbon nanotube-mixed matrix membrane, *Separation and Purification Techn.* 70, 12-26, (2009).
[18] Sieffert D., Staudt C., Preparation of hybrid materials containing copolyimides covalently linked with carbon nanotubes, *Sep. and Purif. Tech.* **77**, 99-103, (2011).
[19] Narayan R.J., Berry C.J., Brigmon R.L., Structural and biological properties of carbon nanotube composite films, *Mater. Sci. Eng. B* 123, 123-129, (2005).
[20] O´Cornell M.J., Boul P., Ericson L.M., Huffman C., Wang Y., Kuper C., Ausman K.D., Smalley R.E., Reversible water-solubilization of single-walled carbon nanotubes by polymer wrapping, *Chem. Phys. Letter* 342, 265-271, (2001).
[21] Noy A., Park H.G., Fornasiero F., Holt J.K., Grigoropoulos C.P., Bakajin O., Nanofluidics in carbon nanotubes, *Nanotoday* 2, (6) 22-29, (2007).
[22] Wang Y., Artyukhin A.B., Noy A., Park H.G., Fornasiero F., Holt J.K., Grigoropoulos C.P., Bakajin O., Fast Mass Transport Through Sub-2-Nanometer Carbon Nanotubes, *Science* 312, 1034-1037, (2006).
[23] Liang H.W., Cao X., Zhang W.J., Lin H.T., Zhou F., Chen L.F., Yu S.H., Robust and highly efficient free-standing carbonaceous nanofiber membranes for water purification, *Advanced Functional Materials* 21, (20) 3851-3858, (2011).
[24] Fornasiero F., Park H.G., Holt J.K., Stadermann M., Grigoropoulos C.P., Noy A., Bakajin O., Ion exclusion by sub-2-nm carbon nanotube pores, *PNAS* 105, (45) 17250-17255, (2008).
[25] Baker R.W., Future Directions of Membrane Gas Separation Technology, *Ind. Eng. Chem. Res.* 41, 1393, (2002).
[26] Robeson L.M., The upper bound revisited, *Journal of Membrane Sci. 320*, 2008, 390.
[27] Henk V., Schillo C.M., Li J., Fast mass transport through carbon nanotube membranes, *Small* 3, (12) 1996-2004, (2007).
[28] Ackerman D.M., Skolidas A.I., Sholl D.S., Johnson J.K., Diffusivities of Ar and Ne in carbon nanotubes, *Mol. Sim.* 29, 677, (2003).
[29] Sharma A., Kumar S., Tripathi B., Singh M., Vijay Y.K., Aligned CNT/Polymer nanocomposite membranes for hydrogen separation, *Intern. Journal of Hydrogen Energy* 34, 3977, (2009).
[30] Kim S., Pechar T.W., Marand E., Poly(imide siloxane) and carbon nanotube mixed matrix membranes for gas separation, *Desalination* 192, 330, (2006).
[31] Hinds B.J., Chopra N., Rantell T., Andrews R., Gavalas V., Bachas L.G., Aligned Multiwalled Carbon Nanotube Membranes, *Science* 303, (2004).
[32] Yu M., Funke H.H., Falconer J.L., Noble R.D., High Density Vertically-Aligned Carbon Nanotube Membranes, *Nano Lett.* 9, (1), (2009).
[33] Fonseca A., Reijerkerk S., Potreck J., Nijmeier K., Mekhalif Z., Delhalle J., Very short functionalized carbon nanotubes for membrane applications, *Desalination* 259, 1150, (2010).
[34] Ge L., Zhu Z., Li F., Liu S., Wang L., Tang X., Rudolph V., Investigation of Gas Permeability in Carbon Nanotube (CNT) Polymer Matrix Membranes via Modifying

CNTs with Functional Groups/Metals and Controlling Modification Location, *J. Phys. Chem.* 115, 6661, (2011).

[35] Wang S.F., Shen L., Zhang W.D., Preparation mechanical properties of chitosan/carbon nanotubes composites, *Biomacromolecules* 6, 3067-3072, (2005).

[36] Baker R.W., Koros W.J., Membrane Separation Systems: Recent Developments and Future Directions, Noyes Data Corp., Park Ridge, *NJ.* (1991).

[37] Shao P., Huang R.Y.M., Polymeric membrane pervaporation, *J. Membr. Sci.* 287, 162-179, (2007).

[38] Dubey V., Pandey L.K., Saxena C., Pervaporative separation of ethanol/water azetrope using a novel chitosan-impregnated bacterial cellulose membrane and chitosan-PVA blend, *J. of Membr. Sci.* 251, 131-136, (2005).

[39] Ong Y.T., Ahmad A.L., Zein S.H.S., Tan S.H., Poly(3-hydroxybutyrate)-fuctionalised multi-walled carbon nanotubes/chitosan green nanocomposite membranes and their application in pervaporation, *Separation and Purification Tech.* 76, 419-427, (2011).

[40] Qiu S., Wu L., Shi G., Zhang L., Chen H., Gao C., Preparation and Pervaporation Property of Chitosan Membrane with Functionalized Multiwalled Carbon Nanotubes, *Ind. Eng. Chem. Res.* 49, 11667-11675, (2010).

[41] Shirazi Y., Tofighy M., Mohammadi T., Synthesis and characterization of carbon nanotubes/PVA nanocomposite membranes for dehydration of isopropanol, *J. Membr. Sci.* 378, 551-561, (2011).

[42] Choi J., Jegal J., Kim W., Choi H., Incorporation of multiwalled carbon nanotubes into poly(vinyl alcohol) membranes for use in the pervaporation of water/ethanol mixtures, *Journal of Applied Polymer. Sci.* 111, 2186-2193, (2009).

[43] Choi J., Jegal J., Kim W.N., Modification of Performances of Various Membranes Using MWNTs as a Modifier, *Macromol. Symp.* 249-250, 610-617, (2007).

[44] Robeson L.M., Correlation of separation factor versus permeability for polymeric membranes, *J. Membr. Sci.* 62, 165, (1991).

[45] Peng F.B., Lu L.Y., Sun H.L., Jiang Z.Y., Organic-inorganic hybrid membranes with simultaneously enhanced flux and selectivity, *Ind. Eng. Chem. Res.* 46, 2544-2549, (2007).

[46] Delozier D.M., Watson K.A., Smith J.G., Clancy T.C., Connell J.W., Optically active photochromic polymers with three-arm star structure by transfer radical polymerziation, *Macrom.* 39, 1721-1739, (2006).

[47] Peng F.B., Lu L.Y., Wang J., Jiang Z.Y., Pervaporation of benzene/cyclohexane mixtures through PVA/chitosan blend membranes, *J. Appl. Polym. Sci.* 101, 167-173, (2006).

[48] Peng F.B., Lu L.Y., Sun H.L., Pan F., Novel nanocomposite pervaporation membranes composed of PVA and chitosan-wrapped carbon nanotube, *J. Membr. Sci.* 300, 13-19, (2007).

[49] Sanders J., Hunter C.A., The nature of pi-pi interactions, *J. Am. Chem. Soc.* 112, 5525-5534, (1990).

[50] Zhao J., Lu J.P., Han J., Yang C.K., Noncovalent functionalization of carbon nanotubes by organic molecules, *Appl. Phys. Let.* 82, 3746, (2003).

[51] Winey K.I., Moniruzzaman M., Polymer nanocomposites containing carbon nanotubes, *Macrom.* 39, 5194-5205, (2006).

[52] Peng F., Hu C., Jiang Z., Novel poly(vinyl alcohol)/carbon nanotube hybrid membranes for pervaporation separation of benzene/cyclohexane mixtures, *J. Membr. Sci.* 297, 236-242, (2007).
[53] Chen J., Dyer M.J., Yu M.F., Cyclodextrin-mediated soft cutting of single-walled carbon nanotubes, *J. Am. Chem. Soc.* 123, 6201, (2011).
[54] Farrell G.F., Carroll C., Chamber G., Characterization of the interaction of gamma cyclodextrin with single-walled carbon nanotubes, *Nano Lett.* 3, 843, (2003).
[55] Peng F.B., Lu L.Y., Wang J., Jiang Z.Y., Wu H., Pervaporation of benzene/cyclohexane mixtures through PVA membranes without and with β-cyclodextrin, *Desalination* 193, 182, (2006).
[56] Rölling P., Lamers M., Staudt C., Cross-linked membranes based on acrylated cyclodextrins and polyethylene glycol dimethacrylates for aromatic/aliphatic separations, *Journal of Membrane Sci.* 362, 154, (2010).
[57] Penkova A., Polotskaya G.A., Gavrilova V.A., Polyamide Membranes Modified by Carbon Nanotubes: Application for Pervaporation, *Sep. Sci. and Tech.* 45, 35-41, (2010).
[58] Jizhong R., Staudt-Bickel C., Lichtenthaler R.N., Separation of aromatics/aliphatics with cross-linked 6FDA-based copolyimides, *Sep. and Purif. Techn.* 22-23, 31, (2001).
[59] Katarzynski D., Staudt C., Temperature dependent separation of naphthalene/*n*-decane mixtures using 6FDA-DABA-copolyimide membranes, *Journal of Membrane Science,* 348, 84–90, (2010).
[60] Pithan F., Staudt-Bickel C., Hess S., Lichtenthaler R.N., Polymeric membranes for aromatic/aliphatic separation processes, *Chem.Phys.Chem.* 3, 856-862, (2002).
[61] Katarzynski D, Pithan F., Staudt C., Pervaporation of Multi Component Aromatic/Aliphatic Mixtures through Copolyimide Membranes, *Separation Science and Technology* 43, 59-70, (2008).
[62] Schmeling N., Konietzny R., Sieffert D., Rölling P., Staudt C., Functionalized Copolyimide Membranes for the Separation of Gaseous and Liquid Mixtures, *Beilstein Journal of Organic Chemistry 2010*, 6. doi: 10.3762/bjoc.6.86.
[63] Suding M., Staudt C., Sulfur containing polyimide membranes in aromatic/aliphatic separation, *Macromolecules Rapid Communications* 31, (2) F27-F28, (2010).
[64] Bettermann I., Staudt C., Desulphurization of kerosene: pervaporation of benzothiophene/*n*-dodecane mixtures, *Journal of Membrane Science* 343, 119-127, (2009).
[65] Bettermann I., Staudt C., Permeation of binuclear aromatic and sulfur containing aromatic compounds, *Desalination* 250, 1144–1146, (2010).
[66] Konietzny R., Bettermann I., Staudt C. ,Membrane based removal of sulfur aromatic compounds from jet fuel, Bulletin of the Australian Institute of High Energetic Materials, USBN: 978-0-9806811-3-0, 61-66, (2010).
[67] Koros W.J., Paul D.R., Qiu W., Cui L., Physical aging of 6FDA-based polyimide membranes monitored by gas permeability, *Polymer* 52, (15) 3374, (2011).
[68] Esekhile O., Qiu W., Koros W.J., Permeation of butane isomers through 6FDA-DAM dense films, *Journal of Polymer Sci.* 49, (22) 1605, (2011).
[69] Bighane N., Koros W.J., Novel silica membranes for high temperature gas separations, *Journal of Membrane Sci.* 371, 254, (2011).

In: Applications of Carbon Nanotubes
Editor: Ajay Kumar Mishra
ISBN: 978-1-62081-918-0

Chapter 8

SINGLE-WALLED CARBON NANOTUBES: DIFFERENT DIAMETERS, DIFFERENT PROPERTIES AND DIFFERENT APPLICATIONS

Jie Ma[1]*, Fei Yu[2] and Buyong Ma[3]
[1]State Key Laboratory of Pollution Control and Resource Reuse, College of Environmental Science and Engineering, Tongji University, Shanghai, P. R. of China
[2]School of Materials Environmental Science and Engineering, Shanghai Jiao Tong University, Shanghai, P. R. China
[3]Center for Cancer Research Nanobiology Program, SAIC-Frederick, Inc.NCI-Frederick, NIH, Frederick, US

ABSTRACT

The synthesis of carbon nanotube recently made progresses to efficiently control the diameter and the size of single-walled carbon nanotubes (SWCNTs). The diameters of SWCNTs are closely related to the electronic structure and surface chemistry of SWCNTs, and subsequently affect the interaction of SWCNTs with membrane, protein, and DNA which underlay complex biological responses. Large-diameter SWCNTs can be used as the catalyst support for fuel cells; with much better performance than the original SWCNTs, multi-walled carbon nanotubes and the commercial catalyst from Johnson Matthey Company as well. Different-diameter SWCNTs were synthesized controlablly effectively from ~1.0 to ~5.8 nm by the floating chemical vapor deposition method; the growth process and mechanism of SWCNTs were studied to understand why tubes with relatively larger diameters can be produced. Furthermore, it would also help us to improve the growth efficiency and control electrical structure of SWCNTs. SWCNTs in their as-grown state contain a large quantity of impurities such as catalytic particles and encapsulating graphitic shells. The new efficient purification methods are discussed; experimental results show that catalytic particles and encapsulating graphitic shells can be removed. More importantly, the efficient purification induces little damage to the graphitic structure of SWCNTs.

* E-mail: jma@tongji.edu.cn.

The present method may be applied for highly efficient and nondestructive purification of CNT materials of any origin to be used in wide areas. The water-dispersible properties, electric properties, thermal-oxidative stability of purified SWCNTs have been also discussed deeply.

1. Introduction

Since the discovery of carbon nanotubes (CNTs)[1,2], numerous studies have shown that single-wall carbon nanotubes (SWCNTs) have many unique chemical, electrical, and mechanical properties[3-6]· The current and potential commercial applications of SWCNTs include hydrogen gas storage in hydrogen-powered fuel cells, catalyst supports, tips for scanning electron microscopes, electron field emitters, capacitors, and membranes for filtration devices [7-9]. The physico-chemical and mechanical properties of SWCNTs strongly depend on the diameter and chirality of SWCNTs. The diameter of SWCNTs is closely related to the energy gap of the semiconducting nanotubes and modulate the interaction between the nanotubes and adsorbed molecules[10]. CNTs tend to form bundles due to the strong van der Waals interactions between the surfaces of small-diameter (SD) CNTs[11]. Previous studies on the properties and applications of SWCNTs were mostly limited to SWCNTs with diameters less than 3 nm. SWCNTs with a large diameter (LD) appear to have special properties[12,13] and wide application potentials. The LD-SWCNTs will be promising in gas-storage application, such as a hydrogen-storage material for fuel-cell operated electric vehicles[3]. Thus, the diameter-controllable technique is very important for the investigation of physical properties and future applications of SWCNTs. Yet, limited progress has been realized in controlling the diameters of SWCNTs. In this chapter, we describe a method to use sulfur-containing additive (thiophene) to enhance the growth of SWCNTs, and different-diameter SWCNTs can be synthesized from ~1.0 to ~5.8 nm by the floating chemical vapor deposition method. The SWCNTs products can be brought out of the reaction zone continuously by Ar gas, rather than deposited in the reaction zone. We also developed a new efficient purification method. We also discuss the electric properties, thermal-oxidative stability of different-diameter purified SWCNTs. The chapter ends with a novel application in fuel cells and biological responses with membrane, protein and DNA to illustrate the potential applications of different-diameter SWCNTs.

2. Synthesis Methods of Different-Diameter SWCNTs

Many efforts have been made toward controlling the diameter distribution of SWCNTs by adjusting parameters such as carbon sources [14], carbon feeding rate [15], pressure [16] etc. For example, Barreiro reported [17] that the total gas flow can control the carbon nanotubes diameter, with increasing gas flow leading to decreasing of the mean diameter of SWCNTs. However, the adjusting range of diameter distribution is limited to only 1.25 and 1.55 nm. In reference [16] the mean diameter distributions of SWCNTs can be increased from 0.93 to 1.78 nm. Matthieu et al. [18] used temperature and partial pressure to control SWCNTs growth and diameter distribution, with the cost of reducing the yield of SWCNTs; and the largest diameter of SWCNTs is less than 3 nm. In this chapter, different-diameter

SWCNTs were prepared by spray pyrolysis. Ethanol was used as carbon feedstock, ferrocene as catalyst, and thiophene as growth promoter. Ferrocene and thiophene were dissolved in ethanol to prepare solutions with different concentrations of ferrocene and thiophene. The concentration of ferrocene ranges from 6.7 to 20 mg mL^{-1}, and that of thiophene from 0 to 5% (vol.-%, the same hereafter). The solution was sonicated for 10 min and then transferred to a reservoir for spray pyrolysis. A quartz tube of 3 cm in diameter and 1 m in length was heated up to temperatures between 1000-1150 °C using a tubular furnace with a heating length of 70 cm. In raising the temperature, Ar flow was initiated in the quartz tube in order to eliminate oxygen from the reaction chamber. After 15 min holding at a desired temperature, the ethanol solution with ferrocene and thiophene was supplied by an electronic micro pump at the rate of 0.4 mL min^{-1} and sprayed through a nozzle with an Ar flow of 160 L h^{-1}. After several hours of pyrolysis, the supply of ethanol was terminated, and the sample was collected from a glass bottle connected to the quartz tube. The experimental set-up is presented in Figure 1. At high temperature (1150°C), diameters of SWCNT increase with the concentration of thiophene. The TEM images of SD-SWCNs and LD-SWCNTs were shown in Figure 2. We found that the average diameter of SWCNTs is ~1.2 nm for SD-SWCNTs with a narrow distribution, ~2.3 nm for middle-diamter(MD)-SWCNTs, ~5.8 nm for LD-SWCNTs. The average diameter of SWCNTs is increased from ~1.0 to ~5.8 nm with increasing the thiophene from 0.1% to 1%.

Note that without thiophene, singe wall carbon nanotubes could be collected only from the internal wall of quartz tube in the low temperature region in the form of thin film. But when thiophene was included, carbon nanotubes in the form of powder could be collected from the sample collector connected with the quartz tube.

Micro-Raman spectra, especially the radial breathing mode (RBM), can be effectively employed to study the properties of SWCNTs. The low-frequency micro-Raman spectrum for SD-SWCNTs, MD-SWCNTs and LD-SWCNT samples are shown in Figure 3a.

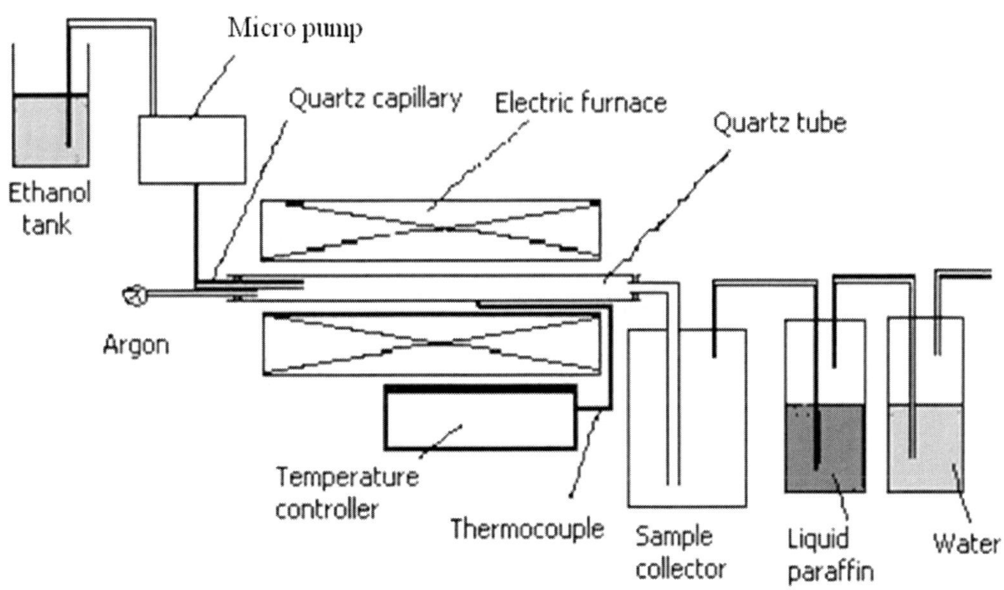

Figure 1. Experimental setup used for the present synthesis of carbon nanotubes.

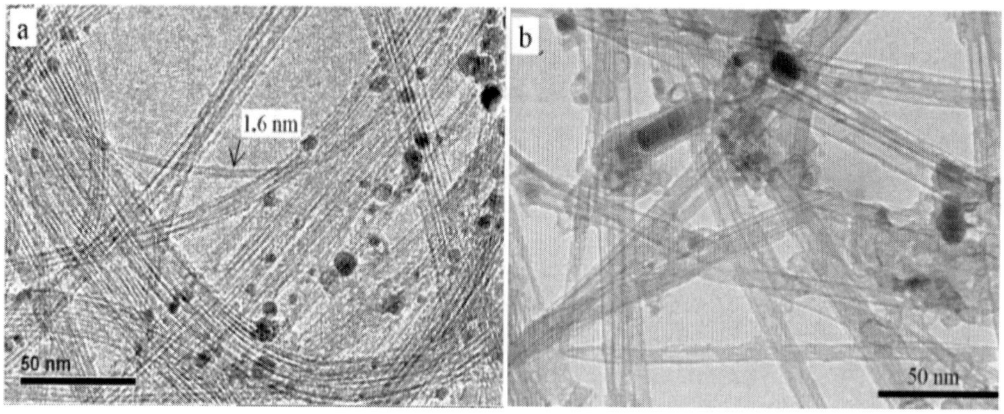

Figure 2. HRTEM images of the samples synthesized at 1150°C containing SD-SWCNTs (a) and LD-SWCNTs (b).

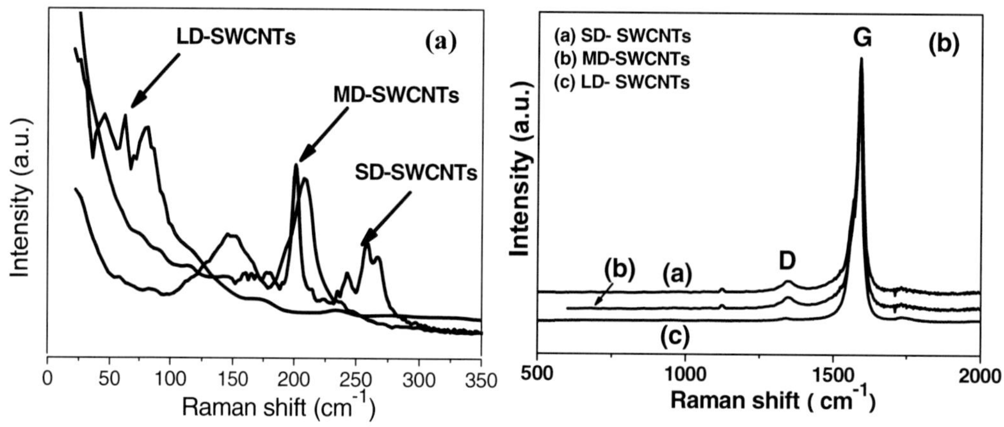

Figure 3. Raman spectra of purified SWCNTs at different thiophene concentrations.

The micro-Raman spectrum was recorded with a laser excitation wavelength of 514.5 nm. The peak positions for the SD-SWCNTs are at 200-270 cm^{-1}, and those for the LD-SWCNTs are at 45-85 cm^{-1}. With increasing thiophene addition amount, the Raman peak position decreased to the low frequency, indicating that the diameters of SWCNTs are significantly enlarged with the addition of more thiophene. The observed peaks correspond to the RBMs, which frequencies ω_{RBM} depend on the diameter of SWCNTs d by

$$\omega_{RBM} = A/d + B \qquad (1)$$

where A and B are two constants. For typical SWCNTs, A = 234 $cm^{-1}\bullet$ nm and B = 10 cm^{-1} have been found [19,20]. Based on eqn.(1) and the observed peaks at low frequencies, the diameters are calculated to be in the range of 0.91–1.56 nm for the SD-SWCNTs, but in the range of 3.2-7.0 nm for LD-SWCNTs, in consistent with HRTEM images analysis.

The second group of Raman peaks appears at high frequencies (Figure 4b). The G peak at 1585 cm^{-1} is related to E_{2g} graphite mode [21-23].

The strong intensity of this peak indicates good graphitization of CNTs. The D-line at around 1345 cm^{-1} is induced by defective structures. These could include the minor amorphous carbon and some hollow graphite particles seen in the sample. The intensity ratio of the G and D peaks (I_G/I_D) is an indicator for estimating the purity and structure quality of SWCNT sample. As shown (Figure 4b), therefore, the higher I_G/I_D ratio means a higher structure quality and purity of SWCNTs after the purification process. Therefore, we can see that LD-SWCNTs are less defective than SD-SWCNTs.

Comparing SD-SWCNTs sample with LD-SWCNTs sample shows that thiophene is a key element in enhancing the productivity of CNTs by the present spray-pyrolysis method. Thiophene may control the decomposition of ethanol in the presence of the catalyst formed from ferrocene. Without thiophene, the flow rate of the gaseous phase might be too fast for ethanol to be decomposed even in the presence of catalyst. The ferrocene would decompose in the floating process with Ar gas. Fe metal particles would deposit on the inner surface of the quartz tube. [24] At such deposited particles, high-temperature ethanol would be catalytically decomposed, leading to growth of CNTs. As a result, the produced CNTs could be collected only from the inner wall of the quartz tube in a film-like form, as in previous studies[25,26]. This is actually similar to previous catalyst-supported methods where catalysts were either impregnated on ceramic particles or deposited on a quartz plate first and then transferred into the reactor for catalytic growth of CNTs[27]. In addition to the beneficial effect of thiophene on ethanol decomposition and floating-catalyst growth of CNTs, thiophene addition was found also to have an effect on nanotube diameter as demonstrated in Figure 2. The general trend is that a higher addition of thiophene leads to a larger tube diameter. Previous studies suggested that when CNTs were synthesized at different conditions with the introduction of pre-formed metal catalyst particles, the formation and diameter of CNTs were strongly dependent on the size of the catalyst particles[28-30]. Therefore, it may be inferred that the effect of thiophene addition on tube diameter observed in this study took place through its effect on the size of catalyst particles, especially, the size of catalytic Fe-S regions[28,29,31]. It may be possible thay a higher addition resulted in larger catalytic Fe-S regions and thus larger tube diameters. However, alternative mechanism could exist.

The effect of thiophene on growth of MWCNTs or SWCNTs is not independent, but coupled with the effect of temperature. A higher temperature is an important factor in initiating the growth of SWCNTs and suppressing the growth of MWCNTs from floating catalysts. The reason may be that the surface of the catalyst particle could reconstruct with the added sulfur at a sufficiently high temperature. This would be coupled with the blocking action of the sulfur atoms so that carbon atoms were depleted and the formation of the graphitic layers from the particle was suppressed[32].

The beneficial effect of a high temperature may also be related to the short growth time, another key factor determining the growth of LD-SWCNTs. The following expression describes the residence time t of a gas phase in a high-temperature zone[33]:

$$t = (ToV) / (TF) \qquad (2)$$

where V is the volume of reactor space, $T_o = 273$ K, T is the reaction temperature (K), and F is Ar flow rate. According to this expression, the growth time of the present SWCNTs from floating catalysts at 1423 K is estimated to be shorter than 0.12 s. If the evaporation of ethanol was also considered, the growth time would be much shorter than this value. The reason for the growth of the present LD-SWCNTs may be that the growth process took place in a short time. This interpretation is consistent with the previous observation that LD-SWCNTs formed in the high-temperature zone only if the CVD process was terminated in a very short time, otherwise MWCNTs formed[25].

3. Growth Mechanism of SWCNTs

Unlike the progress in the synthesis of CNTs, the knowledge about the details of kinetics and microscopic growth mechanisms is still seriously missing [34]. Computational modeling [35] and experimental in situ observations[34,36] are adding more clues to the nucleation and growth of the formation of CNTs.

In a study of the beginning of the process to generate atomic carbons by exposing Ni tip to acetylene in high temperature, mass spectrum revealed that the most abundant species are C_1, C_2, and C_3, with occasionally C_4 as well[34]. The atomic C_1 forms first, and then followed by C_2 and C_3. Combined with Monte Carlo simulation, it was found that adsorption of carbon is preferentially at the step edge leading to linear and connected atomic structure, finally to defective graphene like sheet on the relatively large Ni surface (Figure 4(a), according to Ref.[34]).

When using Ni_{50} cluster for simulation, the similar process leads to the formation of carbon cap (Figure 4(b), according to Ref.[37]), which can be viewed as formation of CNT. In these models catalyzed by Ni, there is no chemical bond formed between Ni and C.

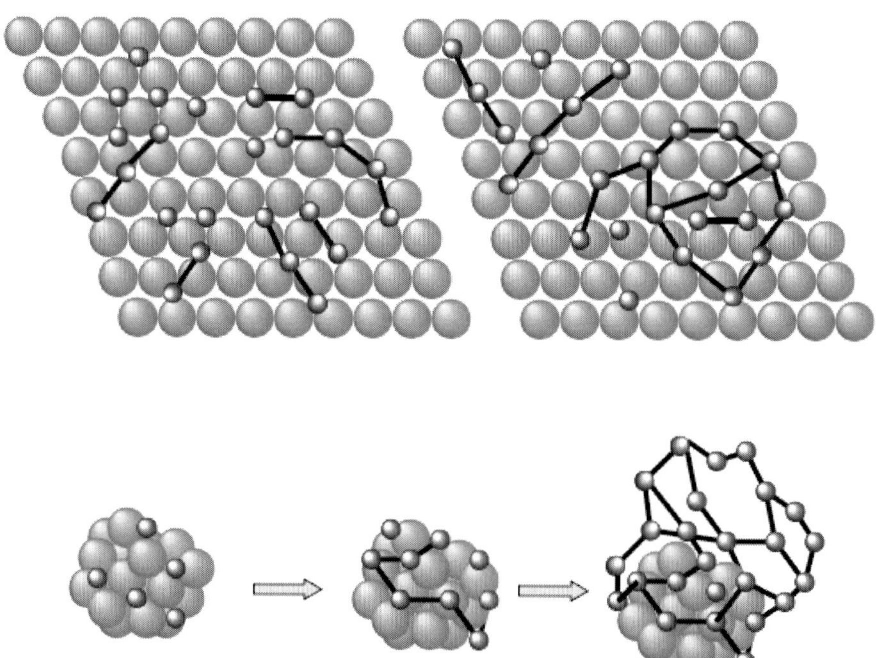

Figure 4. Growth mechanism of SWCNT (a) on flat Ni surface and (b) around Ni cluster.

But in a work using density-functional tight-binding molecular dynamics (DFTB/MD) approach [35], direct Fe-C bond can form on a Fe_{38} cluster. Similar to Ni clusters, the formation of carbon sheet also was found wrapping the iron cluster. Thus, CNT seed can be directly formed on the iron cluster surface or containing all iron cluster. Three types of rings (five-ring, six-ring, and seven ring) can coexist from the DFTB/MD simulation.

In the simulations of the Ni and Fe clusters, CNTs grow through adding carbons on the open edge of the nanotube seeds [35,37]. The simulation of the iron cluster demonstrated that long polyyne chains with more than 4 carbons are not necessarily required for nanotube growth. Instead, short C_1 and C_2 extensions are more frequent. In a fascinating in situ monitoring of CNT growth, it was observed that closed cap is able to grow or shrink[36]. Since the round-shaped cap has no contact with catalyst, one may infer that the catalyst is only needed for the CNT seed formation or for generating atomic carbons. Jin et al. proposed two schemes of the possible pathways to incorporate a C_2 into a hexagon and a heptagon. Following the Jin-Suenaga-Iijima schemes, we tested to add C_2 into a closed cap to simulate a detailed procedure.

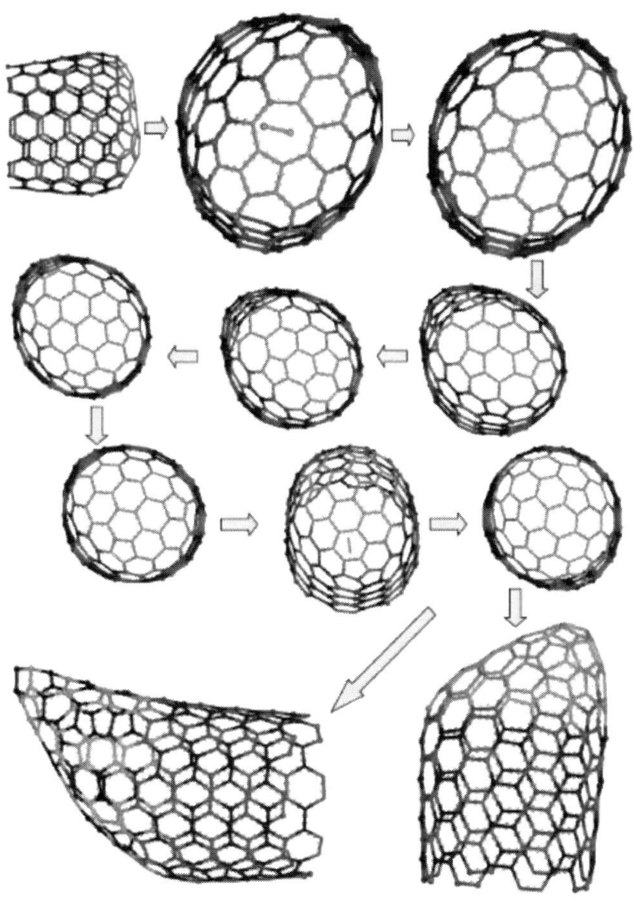

Figure 5. Growth mechanism of SWCNT through addition of C2 into existing nanotube. The red and blue carbons are walls of SWCNT, and the ray carbons are caps. The newly inserted carbons are shown in green.

As shown in Figure 5, the repeated adding of C_2 gradually extends the CNT wall. If adding of C_2 is evenly distributed, a round cap may be preferred. However, if C_2 breaks existing symmetry of the nanotube wall, a nanohorn-like cap may be formed. SWCNTs can be "cloned" via open-end growth mechanism[38]. Active open-end SWCNT can be prepared by cutting long SWCNT using electron beam lithography or other methods, and the duplicate SWCNT can be made by putting the open-end seeds in CVD furnace.

There are two key characteristics of the "clone" experiment: (i) there is no other catalyst involved; and (ii) the "cloned" SWCNTs have the same (n, m) chirality as the initial seeded short SWCNTs. Thus, Yao's experiments [38] again support the idea that the catalyst is only important for the initial seed formation stage. After seed is formed with the original (n, m) chirality parameter fixed for an SWCNT, the after-growth mostly will follow the pattern in the seed.

4. An Efficient and Nondestructive Purification Approach of SWCNTs

In order to obtain the optimal performance of SWCNTs in various applications, high purity is a criterion. However, initial SWCNTs synthesized by electric arc-discharge [39], laser vaporization, [40] and vapor deposition methods[41,42] often contain a large quantity of impurities, including small catalytic metal particles and carbonaceous by-products such as fullerenes, amorphous or graphitic carbon particles, multi-walled carbon nanotubes (MWCNTs). While these carbonaceous by-products can be minimized by optimizing experimental conditions, the catalytic metal particles are always present as they are essential to growth of SWCNTs. The current methods for purification can be categorized into three major classes: physical separation, gas-phase oxidation and liquid-phase oxidation. The physical separation method includes procedures of dispersion of SWCNTs in a surfactant solution, and then separation of SWCNTs by filtration[43], centrifugation[44], or a chromate-graphic technique[45]. Highly pure and length-selected SWCNTs in aqueous solutions can be obtained by column chromatography. However, this method is only useful for the preparation of a very small amount of purified samples but not for large-scale applications. The gas-phase oxidation under air, oxygen, ozone, H_2S and CO_2 atmospheres at high temperature have been used to selectively eliminate the unwanted amorphous carbon materials[46]. However, the oxidation temperature and time could not be easily controlled, owing to the close similarity of the chemical properties of the CNTs with the remaining carbonaceous materials. Alternatively, liquid-phase oxidation was reported, with strong oxidative agents such as HNO_3, $HClO_4$ and $KMnO_4/H_2SO_4$. But, it had been shown that such acid treatments led to partial destruction of the nanotube sample with the walls of the SWCNTs being damaged, distorted, or even segmented[47]. Furthermore, the "purified" SWCNT bundles could be covered by debris from oxidation[48].

The main problem with previous methods is that the metal particles which played a catalytic role on CNT growth are encapsulated by graphitic carbon cages or shells. The graphitic structure presents high resistance to oxidation or acid treatment, and thus carbon cages and the enclosed metal particles usually remain after purification. In order to achieve a sufficient purification, severe oxidation had to be imposed. But in this case, there was an inevitable big loss and damage to the CNTs. Therefore, to date, it is still a great challenge to seek a method that can purify SWCNTs efficiently but with the graphitic structure of the tubes remaining intact.

Here we describe a simple and efficient method for purifying SWCNT raw materials. Our work focuses on the removal of the graphitic carbon cages and enclosed metal particles with little destruction to the SWCNT structure. This was done by transformation of the metallic particles at a low temperature to iron oxide and then removal of the graphitic cages by reaction with the enclosed metal oxides at a high temperature. With a final treatment in the mild acid of HCl, the original catalytic particles with graphitic cages could be completely eliminated.

In order to remove the impurities contained in the pristine sample (SWCNTs-0), such sample was treated by four different processes for the purpose of comparison (Table 1). All processes started with an air-oxidation step.

To do this, the pristine sample in a quartz boat was put in a quartz tube and heated in air at 400 °C for 45 min to oxidize Fe nanoparticles. Air was introduced into the quartz tube at a slow rate to provide oxygen continuously.

After air oxidation, the sample was treated by different processes. In process 1, the sample was directly refluxed in 12 M HCl at 100 °C for 4 hrs (SWCNTs-1). In process 2, the sample was directly refluxed in 14 M HNO3 at 100 °C for 4 hrs (SWCNTs-2). In process 3, however, the sample was further heat-treated at 800 °C or 900 °C for 60 mins under the protection of Ar gas, and then refluxed in 12 M HCl at 100 °C for 4 hrs (SWCNTs-3). In process 4, the SWCNTs-3 sample was further heat-treated in air at 400 °C for 45 mins (SWCNTs-4). The purpose of this additional heat treatment is to test if air oxidation at 400 °C can induce damage to the graphitic structure of SWCNTs.

Figure 6 shows the TEM images of the pristine sample. As can be seen in Figs. 6a and 6b, entangled SWCNT bundles were mixed with a high density of metallic particles, as indicated by the numerous black dots. Except for these particles, multi-walled CNTs and carbon nanofibers were not observed. HRTEM (Figure 6c) revealed that the metallic nano-particles were covered by carbon cages or shells with diameters ranging from 2 to 10 nm, and these cages consisted of several graphitic layers. These nano-particles were presumably the result of the decomposition of ferrocene and ethanol. They were widely observed in the synthesis of SWCNTs by all previous methods [39-42]. The purpose of the present purification is to eliminate these particles.

Table 1. Different purification processes

Process	Air oxidation (400^0C,45min)	High-temperature treatment (800^0C/900^0C, 60min)	HCI reflux (100^0C,4hr)	HNO_3 reflux (100^0C,4hr)	Air oxidation (400^0C,45min)
Process-1	yes		yes		
Process-2	yes			yes	
Process-3	yes	yes	yes		
Process-4	yes	yes	yes		yes

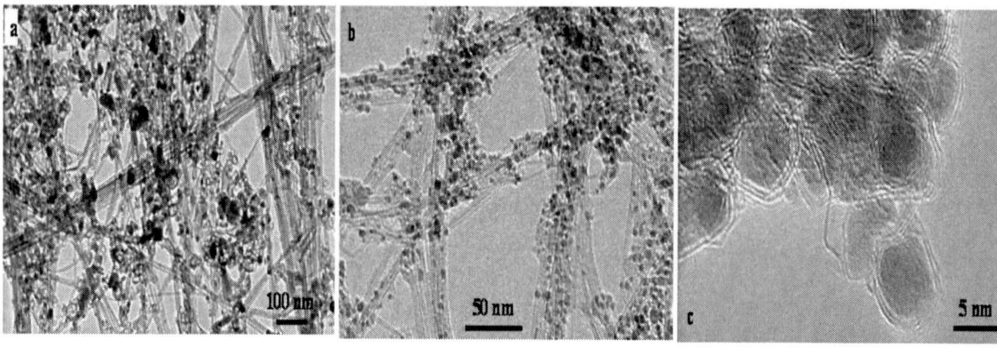

Figure 6. TEM images of the pristine sample (a, b) and iron particles embedded in graphitic shells (c).

Table 2. Analyses of the pristine and differently purified SWCNTs

Sample	Yield (wt.%)[a]	Fe impurity (wt.%)[b]	P_{NIR} (wt.%)[c]	$P_{OVERALL}$ (wt.%)[d]	C impurity (wt.%)[e]	I_G/I_D
SWCNTs-0		27	46	33.6	39.4	7.3
SWCNTs-1	65	12	62	54.6	33.4	11
SWCNTs-2	35	2.5	28	27.3	70.2	3
SWCNTs-3	55	3	96	93.1	3.9	19
SWCNTs-4						18

a The yield of material by weighing after purification.
b The Fe impurity contained in the pristine or purified sample, estimated from TGA results (Figure 5a).
c The purity of SWCNTs estimated by using NIR spectroscopy.
d The purity of SWCNTs in the overall sample when the residual Fe is also taken into account.
e The impurity of non-CNT carbons in the overall sample.

The TEM images of the pristine samples after the three different purification processes are shown in Figure 7. It can be seen that after the purification by process 1 (air oxidation and HCl reflux), many metallic particles and hollow carbon cages remained (Figure 7a). After the purification by process 2 (air oxidation and HNO_3 reflux), even though the metallic particles were removed efficiently, empty carbon cages still persisted.

Only after the purification by process 3 (high-temperature treatment and HCl reflux) could both the metallic particles and graphitic carbon cages be completely removed (Figs. 7c and 7d). These observations suggest that high-temperature treatment assisted the purification and was critical to complete removal of impurities. The yield of material appears different after different purification processes. The material loss is the largest (65 wt.%) for the case of purification by HNO_3 and 55 wt.% for the case involving high-temperature treatment (Table 2).

XRD analysis was used to understand the phase changes taking place during the processes of air oxidation and high-temperature treatment. Figure 8a shows the XRD pattern of the sample after air oxidation. There was no diffraction for graphite, indicating that there were very few or no MWCNTs in the sample. Only peaks which can be assigned to α-Fe_2O_3 were observed. Figure 8b shows the pattern for the sample after additional heat treatment in

Ar at 800 °C. It can be seen that clear FeO peaks could be distinguished in addition to those for γ-Fe_2O_3.

Considering the presence of only α-Fe_2O_3 in the air-oxidized sample (Figure 8a), the appearance of FeO suggests partial reduction of α- Fe_2O_3 at 800 °C. Figure 8c illustrates the result of the sample after heat treatment in Ar at 900 °C. In this case, clear Fe peaks could be identified although γ-Fe_2O_3 peaks were also found. Particularly, the FeO peaks seen for the sample heat treated at 800 °C were no longer observed.

Such observation suggests reduction of some α-Fe_2O_3 particles to Fe and transformation of the rest α-Fe_2O_3 particles to γ-Fe_2O_3 during heat treatment at 900 °C. More detailed features of the present SWCNTs, such as purity, structural integrity and diameter distribution, were shown in table 2.

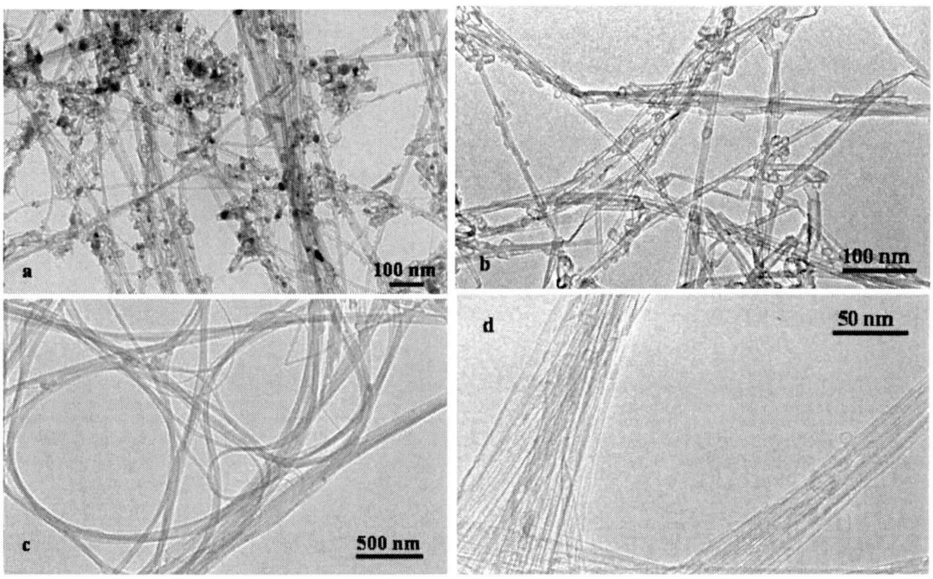

Figure 7. Typical TEM images of purified SWCNTs. (a) SWCNTs-1 (air oxidation and HCl reflux). (b) SWCNTs-2 (air oxidation and HNO3 reflux). (c, d) SWCNTs-3 (high-temperature treatment and HCl reflux).

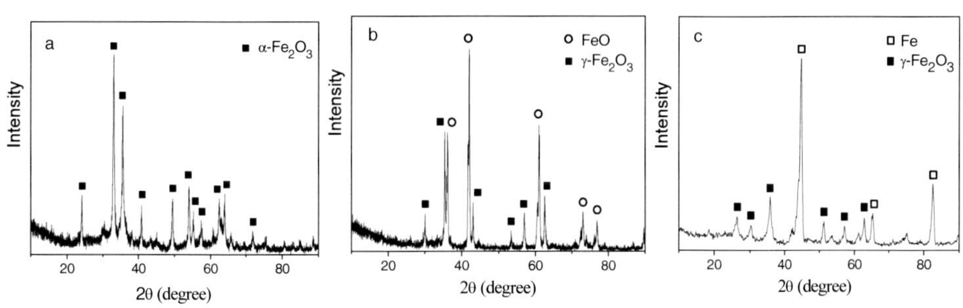

Figure 8. XRD patterns of the samples after heat treatments at different temperatures. (a) Air oxidation at 400 °C. (b) Air oxidation at 400 °C and heat treatment at 800 °C. (c) Air oxidation at 400 °C and heat treatment at 900 °C.

Complete removal of impurities, in particular, catalytic particles, encapsulating carbon cages, and amorphous carbon, is the first goal for purifying SWCNTs.

The present study shows that only after the purification by process 3 (high-temperature treatment) could both the metallic particles and carbon cages be almost completely removed (Figure 7c). In contrast, the purification by either process 1 (HCl reflux) or process 2 (HNO_3 reflux) couldn't eliminate the metallic particles and carbon cages completely (Figs. 7a and 7b). These observations suggest that the high-temperature treatment assisted the purification. Complete removal of impurities with minimal destruction of the graphitic structure of SWCNTs is the second goal. Previous gas-phase or liquid-phase chemical oxidation is not able to easily remove the graphite particles and shells since these particles are resistant to oxidation. In order to remove them and obtain high-purity SWCNTs, severe oxidation conditions, such as a high concentration of HNO_3 acid or a long reaction time, must be imposed as in recent studies[49,50]. However, the severe conditions invariably introduced defects into SWCNTs. Complete removal of impurities with minimal loss of SWCNTs is the third goal. In order to remove the catalytic and graphite particles and obtain high-purity SWCNTs, severe oxidation conditions must be imposed in previous gas-phase or liquid-phase chemical methods. However, the severe conditions inevitably introduced not only defects but loss of SWCNTs as well.

The present purification approach mainly relies on the chemical reaction occurring at high temperature between Fe_2O_3 nanoparticles and the encapsulating graphitic carbon: $Fe_2O_3 + C \rightarrow Fe + CO_X$.

It is the reduction of Fe_2O_3 by C that led to the consumption of C and thus the purification of SWCNTs, as shown by TEM in Figure 7c and XRD in Figs. 8b and 8c. The basic concept of the present purification is illustrated in Figure 9. First, as the sample is heated in air, Fe particles encapsulated in graphitic shells are oxidized to Fe_2O_3 (Figure 8a). When the sample is heated at 800 °C or 900 °C under Ar gas protection, the ferric oxide is reduced by the encapsulating carbon to FeO (800 °C, Figure 8b) or Fe (900 °C, Figure 8c).

Since the previous encapsulating carbon (graphitic shells) has been completely oxidized, the remaining and exposed Fe or Fe oxide particles can be removed easily even by HCl. This present method may be applied for high-efficient and non-destructive purification of CNT materials of any origin to be used in wide areas.

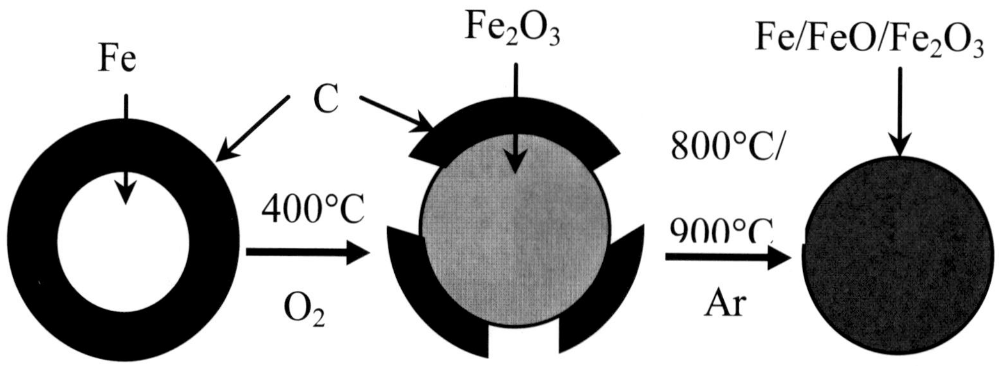

Figure 9. Schematic illustration of the removal of graphitic shells by air oxidation and high-temperature reaction.

5. Diameter-Dependent Properties of SWCNTs

5.1. Dispersing Properties of Different-Diameter SWCNTs

SWCNTs are almost always present in the form of bundles, because of the huge Van der Waals binding energy of ~500 eV μm^{-1} between tubes[51]. These bundles can be quite large with diameters of 10-200 nm, which makes them difficult to disperse in either aqueous solvents or non-aqueous matrixes. As a result, water-dispersible SWCNTs have been the subject of many investigations, and several approaches have been reported to obtain such a material. The reported methods could be separated into two strategies: direct solubilization, achieved by shortening the length of SWCNTs[52]; and dispersion of integral SWCNTs, provided by surface functionalization[53], surfactants and polymeric additives[54].

A number of methods have been reported to cut SWCNTs to short lengths to improve their dispersion. Mechanical cutting is induced by ball-milling [55,56] or sonication with the use of diamond particles as an abrasive [57,58]. Chemical oxidation is carried out under ultra-sonication or reflux condition using a mixture of strong acids[59-61]. In addition, SWCNTs can be chemically cut by partially surface functionalizing with fluorine[62]. However, these methods have several disadvantages, such as introduction of contamination, loss of materials, increase of defects and decrease of quality. Although the shortened length is effective in improving the dispersion of SWCNTs to some extent, the bundles are not easily separated into small bundles because of the strong Van der Waals binding energy.

Other methods which involve non-cutting schemes have also been employed to disperse SWCNTs. One method involves the attachment of various functional groups to the side walls and ends of the SWCNTs. These groups act both as handles for further reaction as well as separation between the tube surfaces [63-65]. But the surface has to be modified significantly, which may alter electrical properties of the SWCNTs. The other method is non-covalent modifications on the surface of SWCNTs and appears to be the best way without adversely affecting the properties of the SWCNTs. But surfactants are used with a high concentration (20 mg ml^{-1}), which induces bad effects on the next application[66]. For instance, when water-dispersible SWCNTs with surfactants are used as catalyst support, the existence of surfactants reduces the adhesion force between Pt particles and side walls of SWCNTs. Furthermore, aggregation of Pt particles happens to decrease the catalytic properties [67,68], in the extra process of removing those surfactants by annealing.

Bundling of SWCNTs modifies the electronic structure of carbon nanotubes[69], which has been a major obstacle for potential applications of SWCNTs in nanodevices. To overcome this problem, many researchers have explored selective chemistry to obtain mono-dispersed SWCNTs, including electrophoretic separation[70] or chromatographic methods [71]. However, these chemical routes based on solution-phase involve not only polymer or surfactants, but also multi-step chemical and physical processes. In the meantime, these chemical approaches have doping effects on properties of SWCNTs[72].

5.1.1. Water-Dispersible LD-SWCNTs

Dispersion of SWCNTs in a solution is important for many applications. The present purified LD-SWCNTs were dispersed in water with a concentration of 0.7 mg mL^{-1} by ultrasonic vibration.

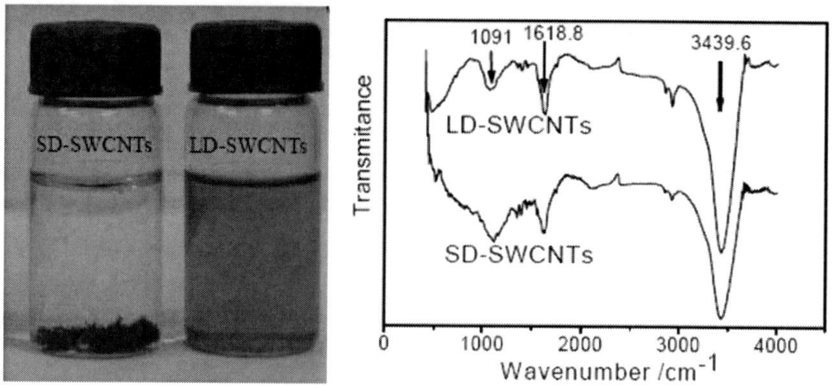

Figure 10. The dispersion of the present SD- and LD-SWCNTs in water with a concentration of 0.7 mg mL^{-1}) (left) and infrared spectra of SD-SWCNTs and LD-SWCNTs (right).

The purified LD-SWCNTs could be dispersed in water uniformly after ultrasonic vibration for 15 min, and agglomeration and settling were not observed (Figure 10) after several days. In contrast, the purified SD-SWCNTs could not be dispersed even after hours of ultrasonic vibration. Instead, they tended to deposit on the bottom of glass bottle. The reason for this may be that such SWCNTs appear isolated (Figure 1b), and they do not form entangled bundles.

FT-IR and XPS spectra of the SD-SWCNTs and LD-SWCNTs are shown in Figure 10 and Figure 11, to further study the possible modification of the surface chemistry during the purification process. In Figure 10, both samples show characteristic peaks at 1091, 1618.8 and 3439.6 cm^{-1}, which could be attributed to –C–O, –C=O and O–H stretching, respectively.

This observation suggests that the SD- and LD-SWCNTs have the same oxygenous groups on their surfaces. Typical XPS survey scans of the SWCNTs are shown in Figure 11a. The asymmetric nature and intensity of the C1s peak remained almost unchanged between SD- and LD-SWCNTs (Figure 11b). The observed asymmetric tailing was a result of the presence of some functionalization of the surfaces of both kinds of SWCNTs. The SD-SWCNTs showed a small O1s signal with an atomic content of 5.27 atom %, and that of the LD-SWCNs is 5.97 atom % (Figure 11c). It can be revealed that the same oxygen content on the surface of LD-SWCNTs and SD-SWCNTs.

Figure 11d is the principal deconvoluted component of the C1s region recorded for the LD-SWCNTs (that for the SD-SWCNTs looked similar). We can see the strongest peak at 284.6 eV assigned to double bonding carbons for SWCNTs[73], which resulted from non-functionalized carbon. The peak with a binding energy around 286.2 eV can be attributed to oxygen atoms bound to carbon by –C–O, and the peak at the binding energy of about 288.1 eV is a consequence of functional groups wherein the oxygen is attached to carbon by a double bond (–C=O) [74]. The curve fitting of O1s for the LD-SWCNTs revealed two separated peaks as shown in Figure 11e. All O1s spectra consist of two peaks which are assigned to oxygen single bonding(–C–O; 533.3 eV) and double bonding(–C=O; 531.7 eV) to carbon[75], This result is consistent with that obtained from FT-IR spectroscopy, indicating minimal introduction of new functional groups on the surfaces of SWCNTs. Considering the minimal difference of functional groups on the surface of the purified SD- and LD- SWCNTs (Figure 10, Figure 11), the good dispersion of purified LD-SWCNTs in water could be attributed to their large diameters.

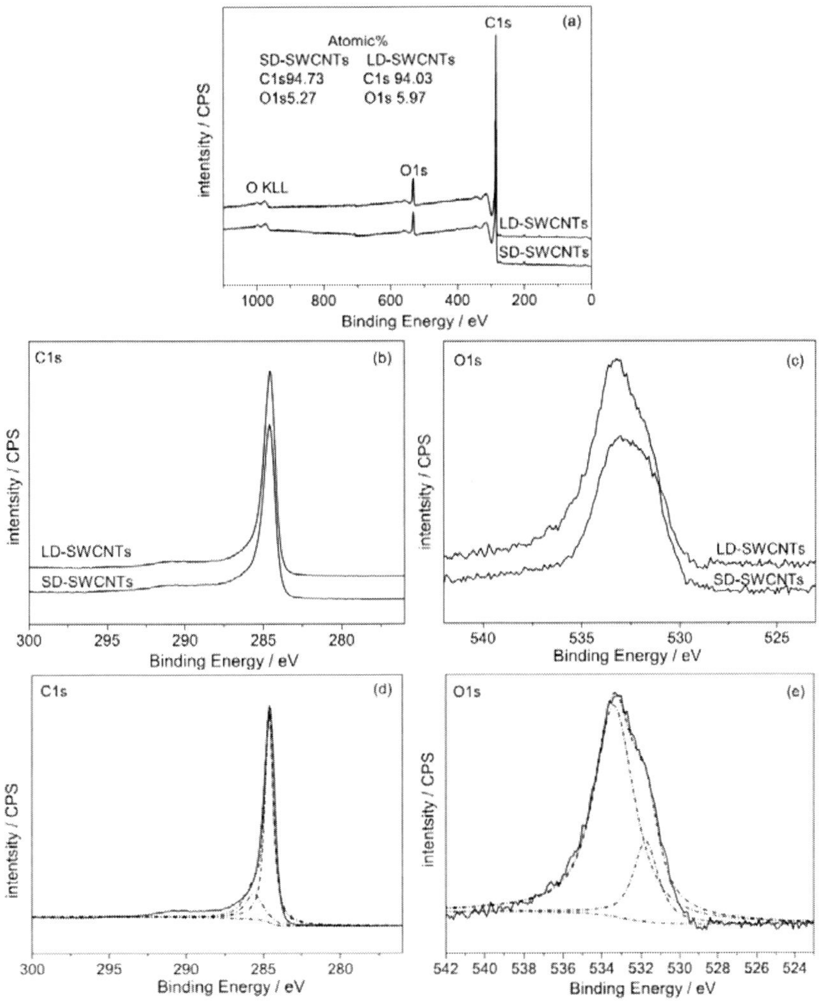

Figure 11. XPS survey scans of the SD- and LD-SWCNTs (a), XPS for C1s region of SD- and LD-SWCNTs (b), for O1s region of SD- and LD-SWCNTs (c), the C1s deconvolution of the LD-SWCNTs (d), and the O1s deconvolution of the LD-SWCNTs (e).

5.1.2. Water-Dispersible SD-SWCNTs by Freeze-Smashing Method

Those SWCNTs with small diameters (<2 nm) are usually bundled together to form ropes, and such ropes are further entangled, as observed in many previous studies and also in the present study. Thus, it becomes impossible or at least very difficult to disperse such SWCNTs in any solutions without imposition of sophisticated treatment (e.g., functionalization). In this chapter, we have reported a new and simple freeze-smashing way for preparing water-dispersible SWCNTs with little damage of the graphitic structure and no addition of any dispersant. The freeze-smashing method involves two steps. Firstly, the purified SWCNTs were ultrasonicated in water of 100 mL, and then frozen by liquid nitrogen. Secondly, the ice block with SWCNTs frozen in it was smashed using a crusher with sharp blades operated at a speed of 25000 rpm for 2 minutes. This process was repeated for 3-5 times. After the ice was melted, the product was filtered, washed and dried. Such a material is designated as "SWCNTs+F", whereas that without the above-mentioned treatment is designated as "SWCNTs+N". The whole process of freeze-smashing approach can be sketched in Figure 12. In the smashing process, the force applied on the ice block can be resolved to a vertical force (Figure 12a) and a parallel force with respect to the tube axis (Figure 12b). Under the action of the parallel force, the large SWCNT bundles could be separated to some small bundles or even individual SWCNT in the smashing process. Comparing SWCNTs+N with SWCNTs+F suggests that the bundle diameter of SWCNTs decreased from ~200 to ~20 nm after the freeze-smashing process (comparing Figure 13b with Figure 13d). More importantly, some ends of SWCNT bundles can be seen in SWCNTs+F samples (Figure 13c). The reason for this may be that the ice block was smashed and the SWCNTs were cut by the vertical force as shown in Figure 12a.

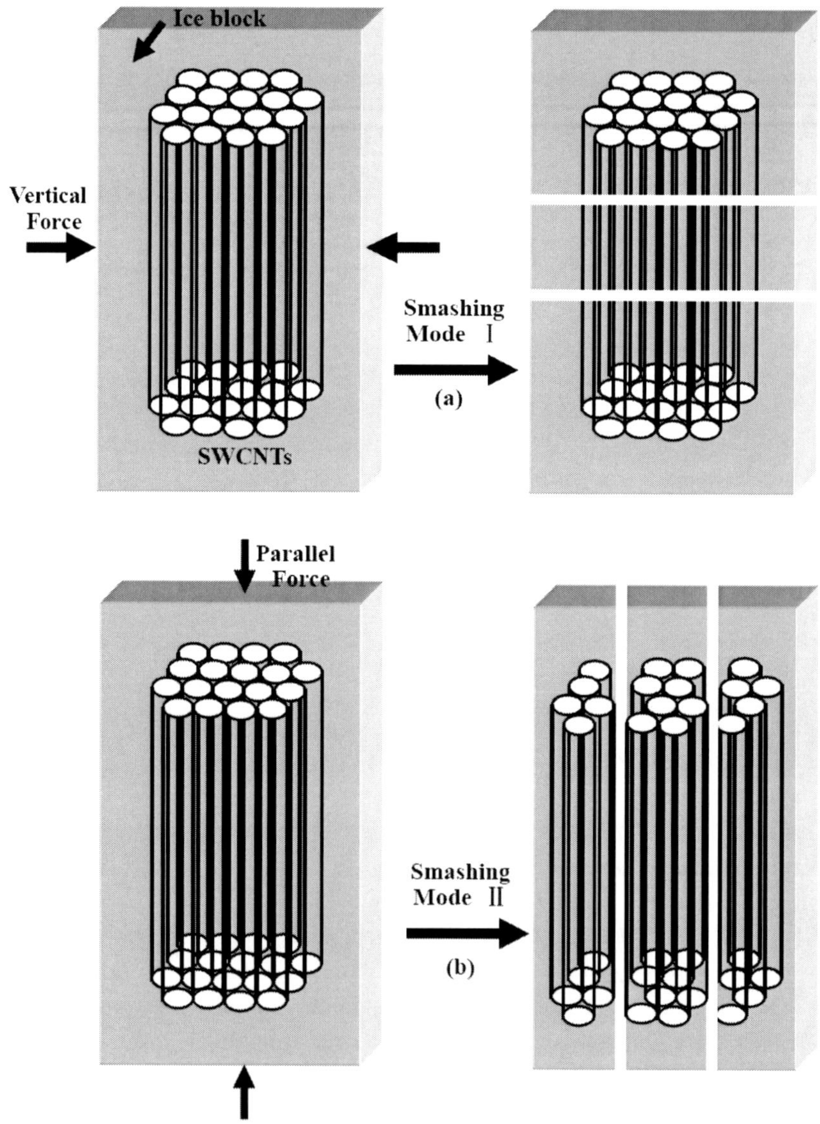

Figure 12. Schematic illustration of smashing process for SWCNTs.

Figure 13a, b shows the image of the purified single-walled carbon nanotubes (SWCNTs+N) before the freeze-smashing process. It can be seen that these SWCNTs are curved and tangled, aggregated to large bundles with diameters of 20–200 nm and lengths of a few micrometers.

This is a common feature of SWCNTs synthesized by chemical vapor deposition. Figure 13c, d shows the images of the single-walled carbon naotubes after the freeze-smashing process (SWCNTs+F). From these images, some distinct characteristics could be found. First, the diameter of SWCNT bundles decreased to ~20 nm from 20-200nm. Second, some ends of SWCNT bundles can be seen in Figure 13c as circled. To further study the possible modification of the graphitic structure during the freeze-smashing process, Figure 14a presents Raman spectra of SWCNTs+N and SWCNTs+F at high frequencies.

The *G* peak at 1585 cm^{-1} is related to E_{2g} graphite mode [76-78]. The strong intensity of this peak indicates good graphitization of CNTs. The D-line at around 1345 cm^{-1} is induced by defective structures. The intensity ratio of the G and D peaks (I_G/I_D) is an indicator for estimating the graphitic structure of a SWCNT sample.

The original sample of SWCNTs+N had a G/D ratio of 14. After the freeze-smashing process, the G/D ratio remained high (13).

To investigate the dispersion of the SWCNTs+N and SWCNTs+F, water solutions with different concentrations were prepared by ultrasonic vibration. Figure 10 shows that the SWCNTs+N could not be dispersed even after hours of ultrasonic vibration. Instead, they tended to deposit on the bottom of a glass bottle after several minutes. In contrast, it was found that thermodynamically stable solutions of SWCNTs+F formed after 5 min sonication.

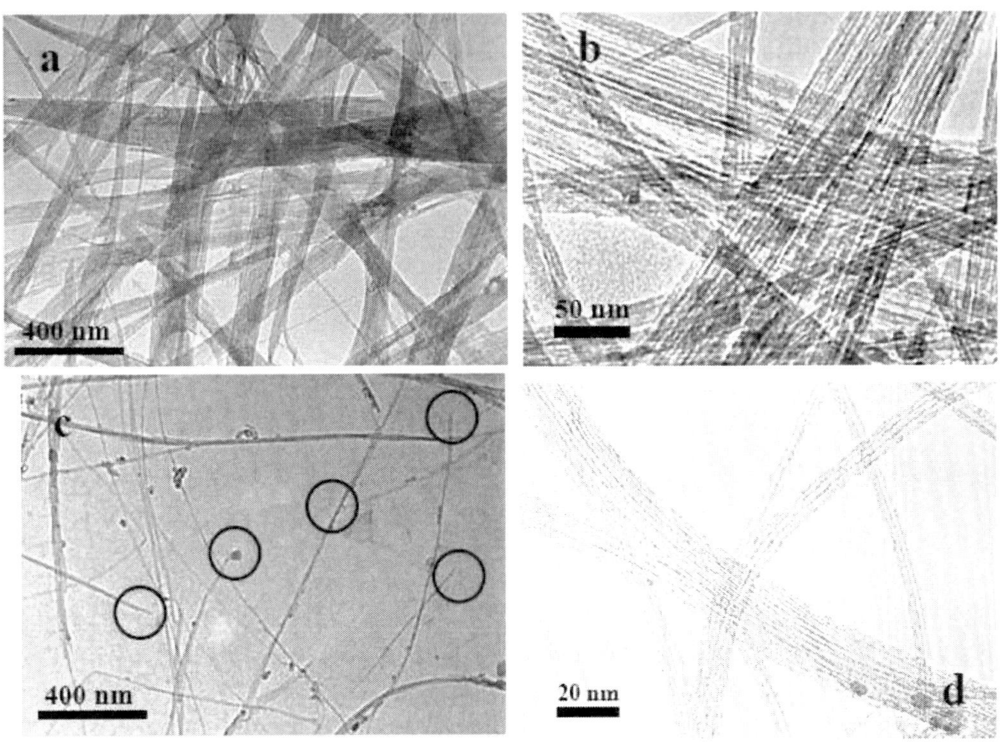

Figure 13. TEM images of SWCNTs+N (a, b) and SWCNTs+F (c, d).

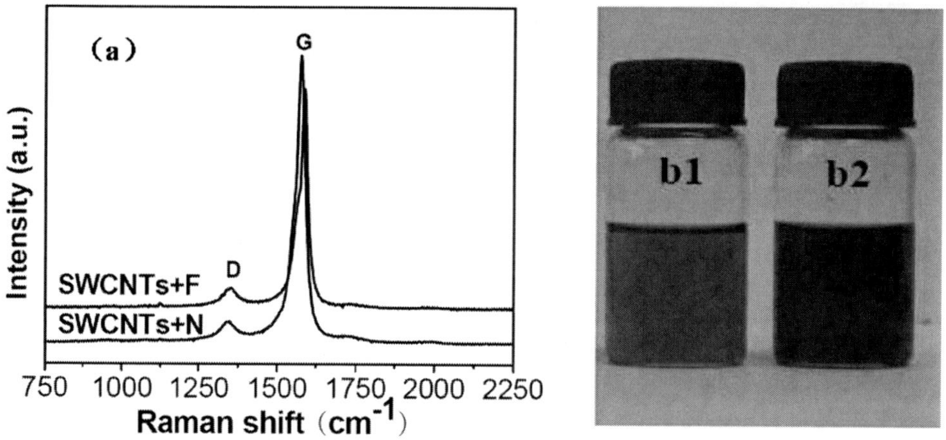

Figure 14. Raman spectra of SWCNTs+N and SWCNTs +F after purification at high frequencies (a), and dispersion situation of SWCNTs in water (b1, b2) SWCNTs+F (from left to right: 0.5, 1.6 mg mL^{-1}).

Visual inspection of these solutions revealed no indication of aggregation and settling after many hours of standing in Figure 14 (b1-b2). Considering that no other chemical materials were added except for water in the smashing process, the improvement in dispersion could be attributed to the freeze-smashing process.

This chapter reports a new and simple way for obtaining dispersible SWCNTs, which were obtained successfully by the method of smashing the freezing ice block including SWCNTs. No chemical materials were used in the freeze-smashing process. In the freeze-smashing process, weighing the pristine and final powder samples showed that the material loss was less than 5 wt. %. Raman spectroscopic measurement showed that the original sample of SWCNTs+N had a G/D ratio of 14. After the freeze-smashing process, the G/D ratio remained high (13), suggesting the maintenance of the good graphitic structure. This result indicates that almost no destruction was introduced to the graphitic structure of SWCNTs through the freeze-smashing process. The present freeze-smashing treatment might be a new method that can separate nanotubes by a simple mechanical means, which could be performed directly on the frozen SWCNTs and avoiding chemical modifications.

5.2. Electronic Properties of Different-Diameter SWCNTs

The structure of CNTs can be visualized as the cylindrical roll-up of one or more flat graphene sheets containing carbon atoms in a honeycomb arrangement (Figure 15). For SWCNTs and MWCNTs, the diameter ranges from 0.4 to 6 nm and 2 to 50 nm, respectively. The diameter and helicity are the key parameters that control the unusual physical, mechanical, and chemical properties of CNTs. The structure of an ideal straight, infinitely long SWCNTs can be uniquely described by two integers (n, m), which refer to the number of unit vectors in the carbon wall lattice. The tube diameter d is related to the (n, m) by $d = a (n_2 + m_2 + nm)^{1/2}/\pi$, where a = 0.249 nm. In Figure 15, we illustrated four types of SWCNT, with both atomic structures (left panels) and distributions of electron density (right panels). Figure 15(a) is an SWCNT with lattice of (6, 6) and diameter of 0.8 nm, and Figure 15(b) with lattice of (8, 4) and also diameter of 0.8 nm. Figures 15(c) and (d) are the SWCNTs with diameters of 4 nm, one with lattice of (30, 30) (Figure 15(c)) and the other of (40, 20) (Figure 15(d)).

The electronic structure of SWCNTs is related to the tube diameter d and depends on the pair (n, m) values. SWCNTs exhibit either zero bandgap (i.e., metallic for n = m, like Figure 15(a) and (c)), small bandgap (ca. 10 meV, i.e., semimetallic for n –m = 3k, where k is an integer), or large bandgap (0.6 eV and above, i.e., semiconducting for n –m≠3k, as shown in Figure 15(b) and (d))[79].

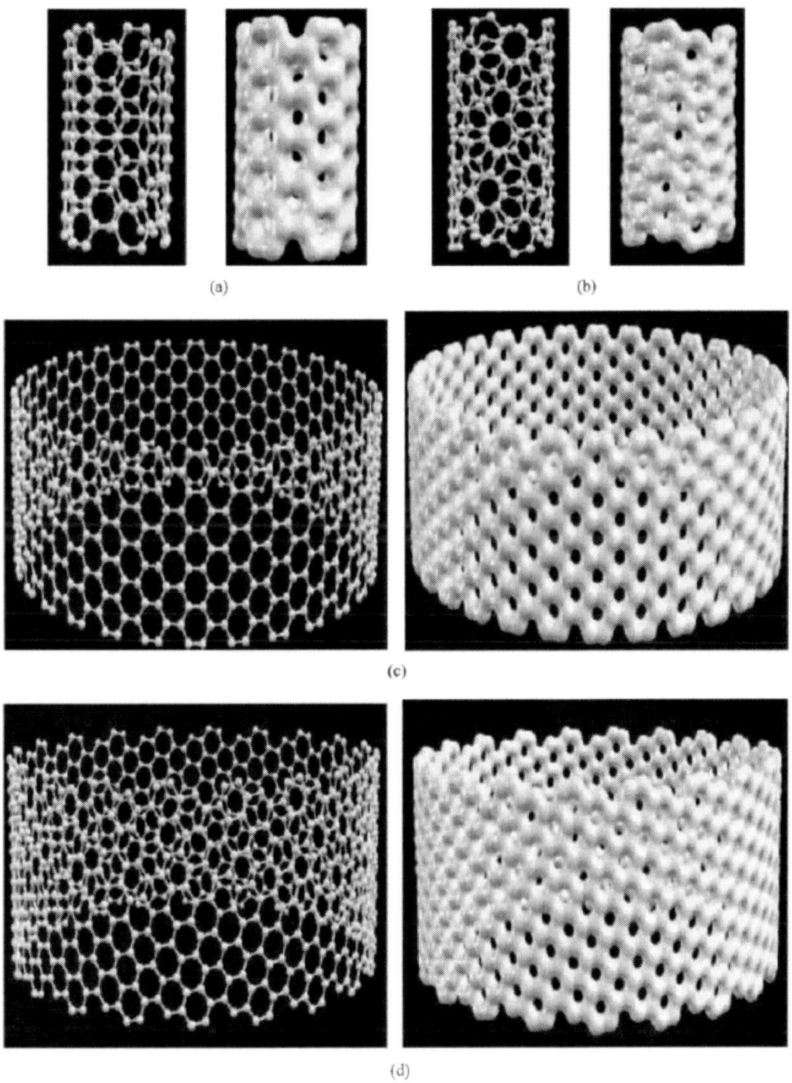

Figure 15. CNTs (left panels) and their electron density distribution on surface (right panels); (a) SWCNTwith lattice of (6, 6) and diameter of 0.8 nm; (b) SWCNT with lattice of (8, 4) and also diameter of 0.8 nm; (c)(d) SWCNTs with diameters of 4 nm, one with lattice of (30,30) (c) and another of (40, 20) (d).

Typically, semimetallic tubes are treated as metallic owing to their small bandgap at room temperature. The electronic band structure also relates to the optical absorptions of SWCNTs. These electronic transitions produce prominent features in the NIR (near infra-red) spectral range, which can be used to analyze the type, purity, and diameter distribution of SWCNTs.

The unique feature of optical excitation of SWCNT at NIR range makes it possible to be used as thermal therapy to target cancer cell with NIR irradiation to heat CNT (reviewed in Ref.[80]). For example, NIR light can kill targeted cell without damaging receptor-free cells[81], and the SWCNT functionalized with specific monoclonal antibodies can target breast cancer cells and kill the cancer cell upon the near-infrared phototherapy (reviewed in Ref.[80]). The CNT surface contains sp_2-hybridized carbon atoms. For an infinite, flat grapheme sheet, these p-orbital electrons organize in broad valence (π) and conduction (π_*) bands, providing a semimetallic character owing to their theoretically zero bandgap [82]. On the other hand, when the graphene sheet is rolled to the cylindrical structure of SWCNT, the π and π_* electron clouds experience significant curvature, which causes partial σ-π hybridization [83]. As a result, the electron density around SWCNT with smaller diameter is more diffuse, and that for LD-SWCNT is more confined to the atomic core region. In terms of biologically relevant surface properties, we can see that the CNTs with smaller diameters (Figure 15(a) and (b)) are more polar, and the surfaces of the CNT with large diameters are more hydrophobic (Figure 15(c) and (d)). SWCNTs are usually synthesized as a mixture of metallic and semiconducting CNTs. Several methods were reported to separate metallic and semiconducting CNTs. For example, the semiconducting CNTs can be etched away by a gas reaction, while metallic ones can be retained [84]. On the other hand, the metallic SWCNT can be suppressed and converted to semiconducting SWCNTs by cycloaddition reactions of fluorinated olefins[85] The metallic or semiconducting properties of the SWCNTs can also be modulated by adsorption of anionic surfactants on nanotube surface in ionic liquids. It was demonstrated that imidazolium-based ionic liquid can be used to effectively disperse of SWCNTs and modify their band structures [86]. The band structure changes depend on SWCNT type and diameter, and also depend on the anionic surfactant concentration. Further studies have shown that only surfactant with small cations (H^+, Li^+, and Na^+) can selectively modify the band structure, indicating a possible mechanism of pi-cation interactions. The significance of this study also illustrates that upon interacting with protein or DNA (reviewed in next sections), the electronic structure of SWCNTs can be changed as well. The phenomena not only underlie that mechanism to use SWCNT as biosensors, but also imply possible interplay of electronic properties of SWCNTs and the biomolecules adsorbed on SWCNT surface.

5.3. Thermal Stability Properties of Different-Diameter SWCNTs

CNTs so far have been demonstrated for a wide range of applications due to a favorable combination of various unique properties such as the relatively high heat conduction performance and thermal stability [87]. Pang et al.[88] and Ajayan et al.[89] reported firstly the use of thermo gravimetric methods for CNTs in the presence of high contents of carbon nanoparticles. Pang et al. [88] reported that CNTs and nanoparticles were more resistant to oxidation than other forms of carbon, such as diamond, soot, graphite and C_{60} studied, under the same condition. However, CNTs exhibit very poor oxidation resistance in air at high temperatures. One influence is the metal and metallic salts which are active catalysts for graphite oxidation. The relative activities of these catalysts probably depend on the chemical nature of the catalyst particle and the localized interaction between catalyst particle and CNTs [90]. Several studies on thermal-oxidative stability of CNTs due to the presence of residual metals have been reported [91-94] Wu et al. [95] reported the thermal-oxidative stability of SWCNTs affected by various metal particles, such as naked metal particles, metal nanoparticles embedded inside multi-shelled graphite particles or in amorphous carbon. The results demonstrated that the presence of residual metal particles could reduce the thermal stability of SWCNTs in air by catalyzing their low-temperature oxidation.

The thermal-oxidative stability of purified SWCNTs is an important issue, because these CNTs are expected to be used in composites, electronics or as sorbents that require chemical thermal robustness[96]. The thermal oxidation of SWCNTs leads to the formation of patches and circular pits on the exposed walls [97,98]. Most prior studies focused on the thermal stability of CNTs affected by purification conditions and chemical modification [99]; however, diameter-dependent thermal-oxidative stability of purified SWCNTs has been rarely reported so far. The main reason is that the present state-of-the-art technology for CNT synthesis always produces CNT samples with different diameters. More importantly, the larger diameter SWCNTs (>2 nm) can not be readily grown with control and a narrow diameter distribution.

In our previous work, different-diameter SWCNT samples have been synthesized successfully using a floating catalytic chemical vapor deposition method [100], and a highly efficient and nondestructive purification approach has been developed[101-103]. Based on our previous research results, in this chapter, purified SWCNTs with three different diameters were obtained, and then the diameter-dependent thermal-oxidative stability of purified SWCNTs were studied.

These three kinds of different-diameter representative samples (SD-, MD-, LD-SWCNTs) have been selected to study the thermal-oxidative stability detailed in the following content. TG and DTA have been widely utilized to make qualitative and quantitative studies on the oxidative behaviour of various materials. Here, main thermal event temperature (T_m) for chemical oxidative reaction, can be defined and measured precisely and is associated with burning of SWCNTs.

Firstly, the thermal analytic curves of as-prepared samples are presented in Figure 16a. The residual weights (%) after the heating process show the presence of metallic impurities in pristine samples.

Assuming that all carbonaceous materials have been oxidized and Fe particles were oxidized to Fe_2O_3, the amount of Fe in each sample can be estimated. Figure 16a shows that SD-SWCNTs contain the nearly 21% Fe in weight, attributed to catalyst Fe nanoparticles. The LD-SWCNTs sample contains 39% Fe in weight, which is more than that of MD-SWCNTs (around 30%). The weight-loss behaviour of purified samples is shown in Figure 16c. The purified samples contain a residual weight of Fe about 2%. This indicates that there were far fewer catalytic particles left in the purified samples as confirmed by HRTEM, suggesting efficient removal of catalyst particles.

The temperatures at which the main thermal events took place during the whole oxidation process may be identified from the DTA curves (Figure 16b, 16d).

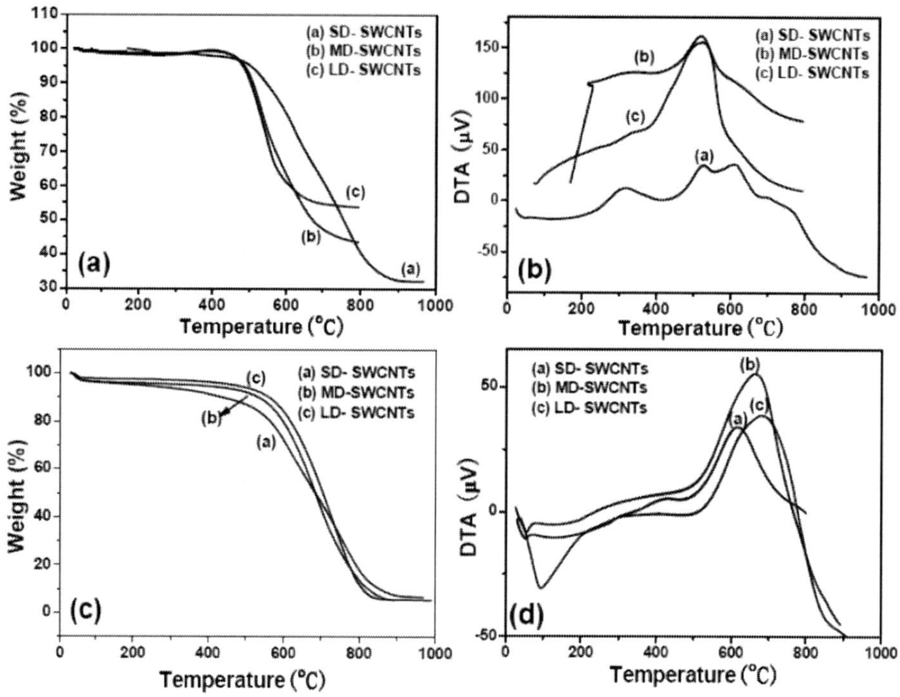

Figure 16. TG curves of different-diameter SWCNTs (a) pristine SWCNTs, (c) purified SWCNTs, DTA curves of different-diameter SWCNTs (b) pristine SWCNTs, (d) purified SWCNTs.

For the purified SD-SWCNT sample, the events might occur at temperatures from 550 °C to 650 °C with the main thermal oxidation taking place at ~610 °C. Main thermal events occured at ~685 °C for the LD-SWCNTs, which has undergone higher temperature treatment.

Comparing as-prepared sample (Figure 16b) with purified sample (Figure 16d), it is clearly seen that T_m goes up with decreasing residual metal weight content. The trend indicates that the metal content strongly affects the thermal-oxidative stability of SWCNTs. The higher metal content is, the lower the thermal-oxidative stability of SWCNTs is, which is in good agreement with previous reports [104,105]. Comparing the line (a), (b) and (c) of Figure 16d, it is also clearly seen that the T_m goes up from 610 °C to 685 °C with the increasing tube diameter. The results indicate that SD-SWCNTs (~1 nm) are less stable and can burn at a lower temperature; however, the lLD-SWCNTs (~5 nm) can survive after burning at a higher temperature. The peak shape of figure 16d show that the oxidation rate varies inversely with the tube diameter of SWCNTs. This is an indication that the LD-SWCNTs were more oxidation-resistant than MD-SWCNTs and SD-SWCNTs.

Sawada and Hamada [106] calculated the value of the bond-bending energy (E_{bend}) of carbon naonotube for optimized atomic configuration of each structure using Tersoff's potential. They found the E_{bend} almost depends only on a single parameter, the diameter of the tubule d: for d larger than 10Å nm, which can be expressed in a relation derivable from a continuum elastic theory:

$$E_{bend}=C_1(1/d)^2 \tag{3}$$

For d smaller than 10Å, the deviation from the continuum elastic theory appears, and the relation is modified as:

$$E_{bend} = C_1(1/d)^2 + C_2(1/d)^4 \quad (4)$$

where C_1 and C_2 are two constants. C_1 = 5.64eV•Å^2, and C_2=25.1eV•Å^2. The coefficient C_1 does not depend on the helix pitch due to the six-fold symmetry of a graphite sheet. Based on eqns. (3) and (4), the E_{bend} of different-diameter SWCNTs are calculated to be 0.0392 eV for the SD-SWCNTs, 0.0107 eV for the MD-SWCNTs and 0.0017 eV for LD-SWCNTs, respectively. The calculation results show that the larger-diameter SWCNTs have lower curvature-induced strain by rolling the planar graphene sheet. Our research results of the oxidation rate varying inversely with the tube diameter of SWCNTs are also in agreement with the calculation results.

The thermodynamic stability of carbon nanotubes is determined by a competition between the increase in energy due to the bond-bending strain introduced by rolling the planar graphene sheet into a tube and the decrease in energy due to the dangling bonds on the edges of the graphene sheet bonding together upon the formation of the tube. Since the strain energy within continuum elasticity theory is expected to vary inversely with the square of the tube diameter[12], small tubes will become thermodynamically unstable, a simple Tersoff-like atomic potential [107] predicting a critical tube diameter of 0.4 nm [106]. Some or all of tube diameter effects may be responsible for conflicting statements in the literature. One reference [108] reported that small tubes burn first due to curvature-induced strain, smaller tubes are less stable and burn at lower temperature, and only the largest diameter tubes survive after burning at high temperatures, which is in agreement with our results. However, another study reported that large diameter tubes are less stable due to a higher defect density [109]. In this chapter, a non-destructive purified approach has been adopted, so the excellent graphite structure has been retained. The higher oxidation-resistant temperature of LD-SWCNTS may be attributed to the high graphite quality and lower curvature-induced strain at a larger diameter. Therefore, our results can reconciliate these two conflicting studies.

5. The Size of CNT and Related Biological Issues

5.1. Solubility and Biodistributions Related to the Size of SWCNTs

The two types of nanotube (MWCNTs and SWCNTs) differed significantly with respect to their size, shape, length, and chemical surface state, and therefore, they have different biocompatibility. Fraczek et al.[110] implanted two types of CNTs (SWCNTs and MWCNTs) into the skeletal rat muscle and found that the reactions of tissue carbon materials are dependent on the tendency of CNTs to group together. SWCNTs created smaller particles, which were easily dispersed in the tissue environment and were subsequently transported to the local lymph nodes by macrophages. In contrast, MWCNTs created aggregates of relatively large size in the vicinity of the implantation site, and such aggregates increased with time. Undesirable cytotoxicity may be caused by the presence of abundant multinucleated cells attached to the MWCNTs agglomerations and the transport of SWCNTs from the implant sites to the lymph nodes[110].

The large-diameter nanotubes may form less aggregation, however, with less mobility. After effectively dispersed, CNT with smaller size may have more mobility, especially helped by lipid molecules. In a study to test diameter-selective solubility of SWCNTs by lipid micelles, it was found that solubilization occurs through wrapping of lipid molecules around the CNTs, yielding lipid monolayers on the graphitic sidewalls. Raman spectroscopy showed that the dispersion and centrifugation process leads to an effective enrichment of the stable aqueous suspension in carbon nanostructures with smaller diameters[111].

The high solubility and mobility of smaller SWCNTs may imply higher membrane penetration. It is good for nanodelivering system. Furthermore, the membrane damage ability could be a double-sided knife. It may imply higher toxicity for SWCNT with smaller diameters; however, it can also be used to kill cells in purpose. By using pristine SWCNTs with a narrow diameter distribution, it was demonstrated that cell membrane damage resulting from direct contact with SWCNTs aggregates could damage bacterial cell membrane and kill the bacterial cell. Direct comparison of the size effects confirmed that the diameter of CNTs is a key factor governing their antibacterial effects. This finding may be useful in the application of SWCNTs as building blocks for antimicrobial materials[112]. Experiments with wellcharacterized SWCNTs and MWCNTs demonstrate that SWCNTs are much more toxic to bacteria than MWCNTs. Gene expression data show that in the presence of both MWCNTs and SWCNTs, Escherichia coli expresses high levels of stress-related gene products, with the quantity and magnitude of expression being much higher in the presence of SWCNTs[113].

The large attraction area of MWCNTs and resulted aggregates may be carcinogenic. If the MWCNTs form fibrous or rod-shaped particles of length around 10 to 20 μm with an aspect ratio of more than 3, it could have similar carcinogenic effects as asbestos. This has been demonstrated in a study of p53 heterozygous mice which are sensitive to asbestos, indicating the possibility that carbonmade fibrous or rod-shaped micrometer particles may share the carcinogenic mechanisms postulated for asbestos[114]. Similar effects were also found for SWCNTs. In a cell behavior study of SWCNT raw material and purified SWCNT (SWCNT bundles), it was found that the cell proliferation, cell activity, cytoskeleton organization, apoptosis, and cell adhesion were dependent on cell type, SWCNT quality (purified or not), and SWCNT concentration [115].

5.2. CNT-Protein Interactions

Both functionalized CNT and nonfunctionalized CNT can interact with protein, for example, hemoglobin[116]. CNTs are often functionalized to increase either dispersion in water or biocompatibility. Introduction of polar or charged functional group increases the interaction of CNT with charged polypeptide like polylysine[117,118]. In the case of nonfunctionalized CNT, the protein-CNT interactions can be dominated by hydrophobic and aromatic π-π interactions for peptides with aromatic side chains. Such effect cannot be overestimated. It seems that the aromatic interactions of peptides with CNT surface are even stronger than charged interactions of polylysine with functionalized CNT, as evidenced by stronger interaction of polytryptophan with both MWCNT and SWCNT[117].

The general preference of tryptophan to interact with the CNT can be illustrated by selection of SWCNT-binding peptides. Phage-displayed libraries have been used to selected peptides with high-affinity binding to CNTs, including SWCNTs and MWCNTs. One unique feature of these peptides is that their amino acid sequences are rich in tryptophan and histidine residues. Analysis of experimental and computational data suggests that the peptide-SWCNT interaction can be defined by the electronic nature of tryptophan and the peptide conformation. It was found that the highest occupied molecular orbital of the tryptophan residue in the peptide interacts with the lowest unoccupied molecular orbital (LUMO) from the SWCNT[119]. It was also revealed[120] that the high affinity binding of the motif to SWCNTs required constrained peptide conformations, which can be achieved through either the extension of the amino acid sequence or the addition of a constrained disulfide bond.

The involvements of molecular orbital of aromatic residues with the LUMO of SWCNT [119] suggest that SWCNT with smaller diameters may have stronger LUMO interactions and thus stronger aromatic preference. By this argument, MWCNTs may have less orbital interaction due to overall larger outer wall diameter.

The SWCNT with large diameter may also have less orbital interaction and have more hydrophobic interactions in nature. If this is the case, the protein-CNT interaction may not be necessarily dominated by aromatic interaction. Indeed, in a genetic analysis of CNT-binding proteins[121] enrichment by cell-surface display, a different trend was found. Trp residue was found not to be absolutely required for a peptide to bind CNT, and no histidine was presented in the optimized sequence. Meantime, we noticed that in the surface-selected CNT-binding peptides, Met, Phe, and Trp often exist[121]. As found in our study, Trp, Phe, and Met are hotspot residues enriched in normal proteinprotein interactions[122]. If the hotspot residues can be selectively targeted by CNT, CNT might interfere with normal protein-protein interaction network. For example, the selective interaction of SWCNT with the hotspot residues may also change protein aggregation pattern to either promote [123] protein-protein aggregation or the other way around. Generally, nanoparticles have large surface areas and could enhance the rate of protein fibrillation by decreasing the lag time for nucleation [123]. On the other hand, the CNT hotspot residue interaction may decrease protein aggregation by shielding the hydrophobic residues, which are often critical to protein aggregation [124,125].

The interactions of CNTand protein have been shown to be selective to protein types. Nonwoven SWCNTs clearly displayed greater adsorption preference for fibrinogen than for albumin. As a result, the function of adsorbed fibrinogen to mediate platelet recognition, adhesion, activation, and aggregation was significantly suppressed, which induced extremely low levels of platelet adhesion and activation[126]. Therefore, the binding behavior of CNT with proteins needs to be carefully evaluated before the full potential of CNT in biological studies can be realized.

5.3. Easy Disturbance of DNAs by CNT Interactions

The selective binding of hydrophobic and aromatics groups by CNT may also interfere with DNA conformation and function. Since DNA molecules are genetic in nature, the effect of the CNT-DNA interaction goes beyond the molecular interaction and could trigger large-scale biological response related to the gene network.

At the molecular level, it was found that SWCNTs bind to human telomeric i-motif DNA under molecular crowding conditions. The results indicate that SWCNTs may have the potential to modulate the structure of human telomeric DNA in vivo, like DNA B-A transitions and B-Z changes on SWCNTs in live cells[127]. Indeed, molecular dynamics simulations have shown that the hydrophobic end groups of DNA are attracted to the hydrophobic SWCNT surface, while the hydrophilic backbone of DNA does not bind to the uncharged SWCNT. The adsorption process appears to have negligible effect on the internal stacking structure of the DNA molecule but significantly affects the A to B form conversion of A-DNA. The adsorption of A-DNA onto an uncharged SWCNT inhibits the complete relaxation of A-DNA to B-DNA within the time scale of the simulations. In contrast, binding of the ADNA onto a positively charged SWCNT may promote slightly the A to B conversion[128]. In another simulation, it was found that SWCNT induces ssDNA (single-stranded DNA) to undergo a spontaneous conformational change that enables the hybrid to self-assemble via the π-π stacking interaction between ssDNA bases and SWCNT sidewall. ssDNA is observed to spontaneously wrap about SWCNT into compact right- or left-handed helices within a few nanoseconds[129]. DNA destabilization and conformational transition induced by SWCNTs are sequence dependent[117,130]. It was found that for GC homopolymer, DNA melting temperature was decreased by 40°C by SWCNTs; but no change for AT-DNA was observed. Conformational transition caused the melting temperature change. SWCNTs can induce B-A transition for GC-DNA, but AT-DNA resisted the transition, possibly because SWCNTs bind to the DNA major groove with GC preference [117,130].

Such DNA interaction can change the cell cycle, indicating that the interaction of the SWCNT with its host biological system goes beyond DNA molecule level. In the case of the SWCNT binding to human telomeric imotif DNA, such interaction leads to significant acceleration of S1 nuclease cleavage rate[131]. Even though it was hoped that SWCNTs might have the intriguing potential to modulate human telomeric DNA structures in vivo, its biological outcome must be carefully balanced. In a study to evaluate the genotoxic potential of SWCNT, cytotoxicity tests showed loss of viability in a concentration- and time-dependent manner after exposure of cells to SWCNT. The induction of DNA damage was found after only 3 h of incubation with 96 μg/cm2 of SWCNT [132]. Other types of stresses also have been observed after applying SWCNT in a cell, for example, inducing oxidative stress in rat lung epithelial cells and also activating specific signaling pathway[133]. DNA damage induced by MWCNTs was also found in mouse embryonic stem cells. These results suggest that careful scrutiny of the genotoxicity of nanomaterials is needed even for those materials that have been previously demonstrated to have limited or no toxicity at the cellular level [134].

6. Application of Different-Diameter SWCNTs in Fuel Cells

To investigate the performance of the present CNTs as a catalyst support for the catalysts in proton exchange membrane fuel cells, an electrochemical study of a Pt catalyst (45 wt.-%) supported on different CNTs was performed. Three kinds of catalyst were prepared with three different supports. They were the present LD- and SD-SWCNTs and commercial MWCNTs. The MWCNTs were purchased from Chengdu Organic Chemicals Co., Ltd., (Chengdu, China). They had well-defined multi-walls and an external diameter of 20~30 nm. In a typical experiment, the CNTs were sonicated in ethylene glycol to form slurry, which was heated up to a desired temperature around 140 °C. Then, chloroplatinic acid (H_2PtCl_6) dissolved in ethylene glycol was added slowly into the CNT slurry. The ethylene glycol here acted as both a reducing medium as well as a surfactant avoiding the conglomeration of Pt nanoparticles. Such prepared mixture was refluxed at 140 °C for 3 h. After having been cooled to the room temperature, the catalyst was then filtered, washed with excess deionized water and dried for the following test. The Pt loading was controlled to be 45 wt.- % (weight percentage). Platinum particles aggregated on the surface of SD-SWCNT bundles (Figure 17a). However, platinum particles were generally uniformly distributed on LD-SWCNTs (Figure 17b). The mean diameters of Pt particles counted from more than one hundred particles were 5.02 nm for Pt on LD-SWCNTs and 6.2 nm on SD-SWCNTs, respectively. These values are in close agreement with those estimated by the Scherrer formula based on the XRD data shown in Figure 17c According to the average diameter of Pt particles, the surface area of Pt particles was calculated and listed in Table 3. It can be seen that the surface area of Pt on LD-SWCNTs is higher than the surface area of Pt on SD-SWCNTs.

Figure 18 shows the cyclic voltamograms of the present catalysts. As can be seen, the catalyst supported on LD-SWCNTs demonstrates stronger hydrogen desorptions and adsorptions and thus a higher activity than the catalyst supported on the MWCNTs and SD-SWCNTs. When compared with the commercial catalyst from Johnson Matthey Co., the present catalyst (Pt/LD-SWCNTs) also shows an outstanding electroactivity.

The electroactive surface area α (cm2 mg^{-1}) for a catalyst can be estimated from the following equation: [135,136]

$$\alpha = Q/(m\tau), \tag{5}$$

where Q is the charge for hydrogen desorption (mC cm^{-2}), m the quantity of Pt used (= 0.4 mg cm^{-2} in the present study), and the charge required to oxidize a monolayer of H_2 on bright Pt (= 0.21 mCcm^{-2}). The Q value for each catalyst can be calculated from its cyclic voltamogram. In doing this, the contribution of the charge from the electric double layer to the overall amount of charge exchanged during the electro-adsorption/desorption of H_2 on Pt sites was deducted. Such calculated effective Q and values are included in Table 3.

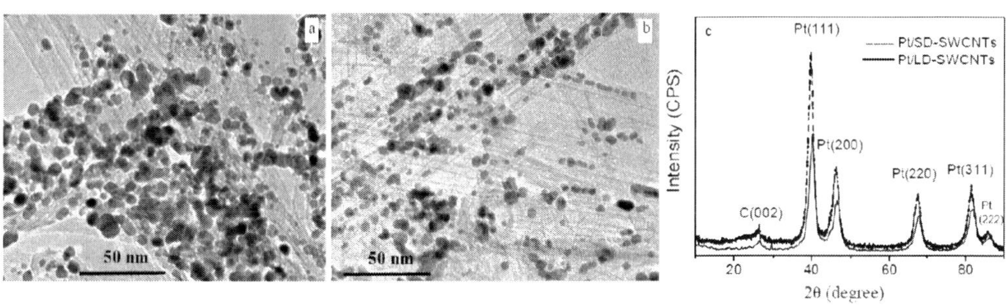

Figure 17. TEM images of Pt particles supported on SD-SWCNTs (a) and LD-SWCNTs (b), and XRD patterns (c).

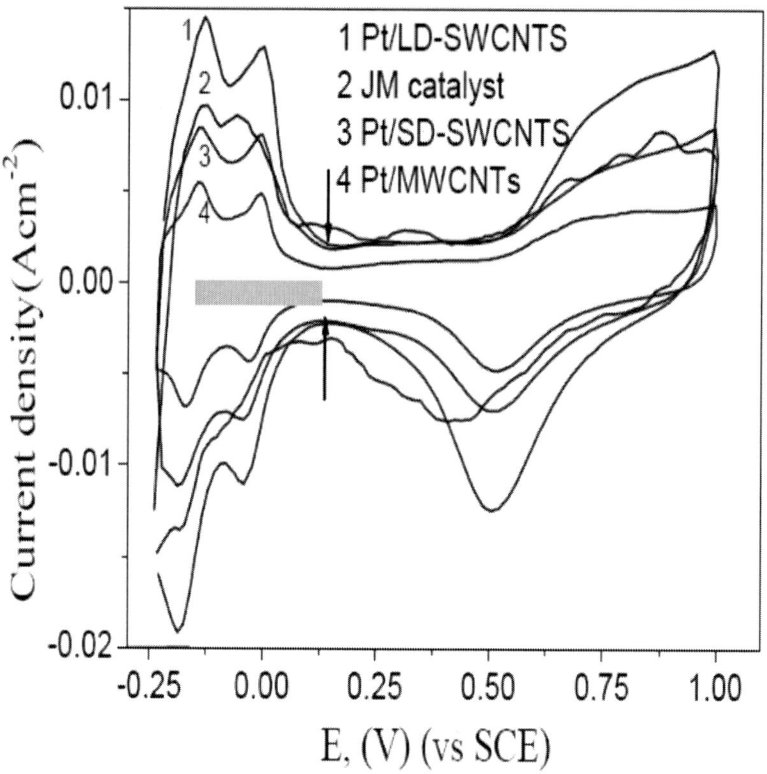

Figure 18. Cyclic voltamograms of the four catalysts. The catalyst based on SD-SWCNTs, LD-SWCNTs, MWCNTs and that from Johnson Matthey Co. (JM) are included for comparison. The Pt loading on each electrode was 0.4 mg cm^{-2}, and the potential sweep rate was 100 mV s^{-1}. The electric double layer of the catalyst of Pt/MWCNTs is arrowed. Q represents the amount of charge exchanged during the electro-desorption of H_2 on Pt sites, and the shaded area represents the contribution of the double layer charge to the overall amount of charge exchanged during the electro-adsorption/desorption of H_2 on Pt sites. Similar markings can be applied for all other catalysts.

Table 3. List of the characteristics of different CNT based catalysts

Catalyst	LC(mn)[a]	S(m^2g^{-1})[b]	Q(mC cm^{-2})[c]	a(m^2g^{-1})[d]	U(%)[e]
JM			15.89	18.92	
SD-SWCNTs	6.2	45.2	12.70	15.12	33.4
LD-SWCNTs	5.02	56.1	23.70	28.21	50.3
MWCNTs			12.10	14.40	

[a] The average size of Pt particles Lc ,which is from the histogram of diameter distribution of Pt particles.

[b] The surface area S = 6000/(ρLc), where ρ represents Pt density (21.4 g cm^{-3}).

[c] The charge Q for hydrogen desorption without the contribution from the electric double layer marked with shadow in Figure 4c.

[d] The electroactive surface area $\alpha = Q/(m\tau)$, where m is the quantity of Pt used (= 0.4×10^{-3} g cm^{-2} in this study), and τ the charge required to oxidize a monolayer of H2 on bright Pt (= 0.21 mC cm^{-2}).

[e] The utilization ratio of platinum $u = \alpha/S$.

Calculations based on the cyclic voltamograms (Figure 18) show that the electroactive surface area of the Pt catalyst based on LD-SWCNTs is the highest among all the catalysts studied and about 2 times of that based on SD-SWCNTs (Table 3).

As a support for Pt nanoparticles, the present LD-SWCNTs showed a better performance than MWCNTs, SD-SWCNTs and the commercial catalyst from Johnson Matthey Co. This is evidenced by the stronger hydrogen desorption/adsorption peaks (Figure 18) and a higher electroactive surface area (Table 3). The reason for this may be that LD-SWCNTs with larger diameter prepared here can be well dispersed in solution. This is an important requirement for uniform deposition of small Pt nanoparticles on the surface of SWCNTs and for uniform deposition of the catalyst particles on the electrode. Severe entangling of SD-SWCNTs usually leads to the opposite results. Furthermore, LD-SWCNTs maybe provide more channels for electrolyte than SD-SWCNTs, and facilitate the diffusion of H^+ in the catalyst layer to contact with Pt nanoparticles. As a result, the utilization ratio of Pt nanoparticles is sharply enhanced (Table 3), and the electrochemical activity is improved.

It should be noted that the present study demonstrated only one example as a catalyst support for the application of the present LD-SWCNTs. Actually, they could have many other applications in wide areas as adsorbents, electrode materials, and energy storage media, [3-6,12,13] because they have a unique dimension, structure and good dispersion. In terms of the good dispersion in water, the LD-SWCNTs can be widely used in biology, medicine and pharmaceutics.

Summary and Outlook

Recent advances in the catalytic CVD methods make it possible to systematically control the diameter and the size of CNT, mainly by controlling the thiophene added. Such control enables us to synthesize uniform different diameter SWCNTs at high yield, including those with a large-diameter (more than 5 nm). These progresses promote us to examine the relationship of the diameter of CNT, the electronic structure and surface chemistry of SWCNT, and the interactions of SWCNT with biomolecules. In this chapter, we examined several aspects of the SWCNTs prepared by CVD processes using ethanol with the additions of ferrocene and thiophene, focusing on the potential mechanisms leading to the differences in the diameters, structures and conductivities of the SWCNTs formed. Different diameter SWCNTs appear to have many special properties. The larger-diameter SWCNTs (>5 nm) have a higher thermal stability than smaller-diameter SWCNTs. With increasing diameter, the electrical properties change from a majority of semi-conducting SWCNTs to a majority of metallic SWCNTs. This chapter introduced two simple and effective purified and dispersive method of SWCNTs: One new purified method is demonstrated, which involves transformation of catalytic iron particles to iron oxide at low temperature, subsequent removal of the graphitic shells by reaction with the encapsulated iron oxide at high temperature, and final elimination of iron oxide particles by HCl. Experimental results show that catalytic particles and encapsulating graphitic shells can be efficiently removed. More importantly, the high-efficient purification induces little damage to the graphitic structure of SWCNTs. The present method may be applied for high-efficient and non-destructive purification of CNT materials of any origin to be used in wide areas. The other dispersive mthod was introduced to prepare water-dispersible SD-SWCNTs with little damage of the graphitic structure and no addition of any dispersant. After the freeze-smashing treatment, the diameter of SWCNT bundles decreased to ~20 nm from ~200 nm. This process might be very useful to separate long, entangled SWCNT bundles to small ones without destruction of the graphitic structure. Due to its unique features, the present approach might be suitable for large-scale production of water-dispersible SWCNTs. In the biology, we discussed the size of CNT and related biological issues. SWCNTs have been observed to interact directly with membrane, protein, and DNA, the three important classes of biomolecules in cell. Such interactions depend on the size and diameter of SWCNT. The SWCNT with smaller size and diameter is easier to interact with lipids and has larger mobility in cell. The interactions of SWCNT with protein are mostly hydrophobic, with Trp, Phe, and Met being preferred to bind on the SWCNT surface. Since the Trp/Phe/Met are hotspot residues controlling protein-protein interaction interface, we suspect that SWCNT may disrupt cell protein-protein interaction network. Another significant consequence of SWCNT interaction is selectively interfering GC-rich DNA conformation, resulting in cell responses similar to DNA damage (for example, by irradiations). Such interactions may underlie the toxic mechanism of SWCNT, but also provide the methods to kill cancer cell when SWCNT can be delivered to cancer cell. The possible ways include NIR thermal therapy or taking advantage of DNA damage effect of SWCNT.

As an application example, the SWCNTs+F and LD-SWCNTs show much better performance as a catalyst support than original SWCNTs+N, MWCNTs and the commercial catalyst from Johnson Matthey Company. It is suggested that SWCNTs+F and LD-SWCNTs might be applied in potential applications in information technology, biomedicine, environmental and energy industries.

Clearly, more systematic studies are needed to evaluate the CNT at both molecular and cellular levels. The application of CNT in biological system must be carefully planned to minimize the possible long-term and genetic damages. The selection of CNT in biological application should consider purity, size distribution, and surface properties. This may need close collaborations among material scientists, biochemists, and biologists.

ACKNOWLEDGMENTS

This work was financially supported by the National Natural Science Foundation of China (51072135) and Program for Young Excellent Talents at Tongji University (2010KJ026). B.M has been funded in whole or in part with Federal funds from the National Cancer Institute, National Institutes of Health, under contract number HHSN261200800001E.

REFERENCES

[1] S. Iijima, *Nature* 354, 56 (1991).
[2] S. Iijima and T. Ichihashi, (1993).
[3] Dillon, K. Jones, T. Bekkedahl, C. Kiang, D. Bethune, and M. Heben, *Nature* 386, 377 (1997).

[4] M. Menon and D. Srivastava, *Physical Review Letters* 79, 4453 (1997).
[5] T. Hertel, R. Martel, and P. Avouris, *The Journal of Physical Chemistry B* 102, 910 (1998).
[6] S. J. Tans, A. R. M. Verschueren, and C. Dekker, *Nature* 393, 49 (1998).
[7] H. Lee, N. Johnson, J. Drelich, and Y. K. Yap, *Carbon* 49, 669 (2011).
[8] I.Mukhopadhyay, Y. Suzuki, T. Kawashita, Y. Yoshida, and S. Kawasaki, *Journal of Nanoscience and Nanotechnology* 10, 4089 (2010).
[9] H. Suga, Y. Naitoh, M. Tanaka, M. Horikawa, H. Kobori, and T. Shimizu, *Applied Physics Express* 2 (2009).
[10] J. Wang and Y. Li, *Journal of the American Chemical Society* 131, 5364 (2009).
[11] T. Hertel, R. E. Walkup, and P. Avouris, *Physical Review B* 58, 13870 (1998).
[12] D. Robertson, D. Brenner, and J. Mintmire, *Physical Review B* 45, 12592 (1992).
[13] C. H. Kiang, J. S. Choi, T. T. Tran, and A. D. Bacher, *The Journal of Physical Chemistry B* 103, 7449 (1999).
[14] Y. Chen, B. Wang, L. J. Li, Y. H. Yang, D. Ciuparu, S. Y. Lim, G. L. Haller, and L. D. Pfefferle, *Carbon* 45, 2217 (2007).
[15] C. Lu and J. Liu, *The Journal of Physical Chemistry B* 110, 20254 (2006).
[16] Z. Y. Wang, Z. B. Zhao, and J. S. Qiu, *Carbon* 44, 1845 (2006).
[17] A. Barreiro, C. Kramberger, M. Rummeli, A. Gruneis, D. Grimm, S. Hampel, T. Gemming, B. Buchner, A. Bachtold, and T. Pichler, *Carbon* 45, 55 (2007).
[18] M. Picher, E. Anglaret, R. Arenal, and V. Jourdain, *Acs. Nano.* 5, 2118 (2011).
[19] M. Milnera, J. Kürti, M. Hulman, and H. Kuzmany, *Physical Review Letters* 84, 1324 (2000).
[20] A. Jorio, R. Saito, J. H. Hafner, C. M. Lieber, M. Hunter, T. McClure, G. Dresselhaus, and M. S. Dresselhaus, *Physical Review Letters* 86, 1118 (2001).
[21] Z. H. Yu and L. E. Brus, *Journal of Physical Chemistry B* 105, 6831 (2001).
[22] A. G. Souza, A. Jorio, G. G. Samsonidze, G. Dresselhaus, M. A. Pimenta, M. S. Dresselhaus, A. K. Swan, M. S. Unlu, B. B. Goldberg, and R. Saito, *Physical Review B* 67 (2003).
[23] A. Gruneis, R. Saito, T. Kimura, L. G. Cancado, M. A. Pimenta, A. Jorio, A. G. Souza, G. Dresselhaus, and M. S. Dresselhaus, *Physical Review B* 65, 155405 (2002).
[24] L. Su, J. Wang, F. Yu, Z. Sheng, H. Chang, and C. Pak, *Chemical Physics Letters* 420, 421 (2006).
[25] Q. H. Yang, S. Bai, J. L. Sauvajol, and J. B. Bai, *Advanced Materials* 15, 792 (2003).
[26] F. Lupo, J. Rodriguez-Manzo, A. Zamudio, A. Elias, Y. Kim, T. Hayashi, M. Muramatsu, R. Kamalakaran, H. Terrones, and M. Endo, *Chemical Physics Letters* 410, 384 (2005).
[27] G. G. Tibbetts, *Carbon* 27, 745 (1989).
[28] H. Ago, T. Komatsu, S. Ohshima, Y. Kuriki, and M. Yumura, *Applied Physics Letters* 77, 79 (2000).
[29] T. Saito, S. Ohshima, W. C. Xu, H. Ago, M. Yumura, and S. Iijima, *The Journal of Physical Chemistry B* 109, 10647 (2005).
[30] A. G. Nasibulin, P. V. Pikhitsa, H. Jiang, and E. I. Kauppinen, *Carbon* 43, 2251 (2005).
[31] A. Jorio, M. Pimenta, A. Souza Filho, G. G. Samsonidze, A. Swan, M. Ünlü, B. Goldberg, R. Saito, G. Dresselhaus, and M. Dresselhaus, *Physical Review Letters* 90, 107403 (2003).

[32] H. Kanzow and A. Ding, *Physical Review B* 60, 11180 (1999).

[33] G. G. Tibbetts, C. A. Bernardo, D. W. Gorkiewicz, and R. L. Alig, *Carbon* 32, 569 (1994).

[34] M. Moors, H. Amara, T. V. de Bocarme, C. Bichara, F. Ducastelle, N. Kruse, and J. C. Charlier, *Acs. Nano.* 3, 511 (2009).

[35] Y. Ohta, Y. Okamoto, S. Irle, and K. Morokuma, *Acs. Nano.* 2, 1437 (2008).

[36] C. Jin, K. Suenaga, and S. Iijima, *Acs. Nano.* 2, 1275 (2008).

[37] H. Amara, C. Bichara, and F. Ducastelle, *Physical Review Letters* 100 (2008).

[38] Y. G. Yao, C. Q. Feng, J. Zhang, and Z. F. Liu, *Nano Letters* 9, 1673 (2009).

[39] M. Yao, B. Liu, Y. Zou, L. Wang, D. Li, T. Cui, G. Zou, and B. Sundqvist, *Carbon* 43, 2894 (2005).

[40] J. L. Blackburn, Y. Yan, C. Engtrakul, P. A. Parilla, K. Jones, T. Gennett, A. C. Dillon, and M. J. Heben, *Chemistry of Materials* 18, 2558 (2006).

[41] L. Song, L. J. Ci, L. F. Sun, C. H. Jin, L. F. Liu, W. J. Ma, D. F. Liu, X. W. Zhao, S. D. Luo, Z. X. Zhang, Y. J. Xiang, J. J. Zhou, W. Y. Zhou, Y. Ding, Z. L. Wang, and S. S. Xie, *Adv. Mater.* 18, 1817 (2006).

[42] K. Hata, D. N. Futaba, K. Mizuno, T. Namai, M. Yumura, and S. Iijima, *Science* 306, 1362 (2004).

[43] L. A. Montoro, C. A. Luengo, J. M. Rosolen, E. Cazzanelli, and G. Mariotto, *Diam. Relat. Mater.* 12, 846 (2003).

[44] G. Duesberg, J. Muster, H. Byrne, S. Roth, and M. Burghard, *Applied Physics A: Materials Science and Processing* 69, 269 (1999).

[45] M. Holzinger, A. Hirsch, P. Bernier, G. Duesberg, and M. Burghard, *Applied Physics A: Materials Science and Processing* 70, 599 (2000).

[46] J. L. Zimmerman, R. K. Bradley, C. B. Huffman, R. H. Hauge, and J. L. Margrave, *Chemistry of Materials* 12, 1361 (2000).

[47] M. Monthioux, B. W. Smith, B. Burteaux, A. Claye, J. E. Fischer, and D. E. Luzzi, *Carbon* 39, 1251 (2001).

[48] E. Dujardin, T. W. Ebbesen, A. Krishnan, and M. M. J. Treacy, *Adv. Mater.* 10, 1472 (1998).

[49] H. Hu, B. Zhao, M. E. Itkis, and R. C. Haddon, *J. Phys. Chem. B* 107, 13838 (2003).

[50] T. J. Park, S. Banerjee, T. Hemraj-Benny, and S. S. Wong, *J. Mater. Chem.* 16, 141 (2006).

[51] A. Thess, R. Lee, P. Nikolaev, H. Dai, P. Petit, J. Robert, C. Xu, Y. H. Lee, S. G. Kim, and A. G. Rinzler, *Science* 273, 483 (1996).

[52] L. Cai, J. L. Bahr, Y. Yao, and M. James, *Chemistry of Materials* 14, 4235 (2002).

[53] W. Huang, S. Fernando, Y. Lin, B. Zhou, L. F. Allard, and Y. P. Sun, *Langmuir.* 19, 7084 (2003).

[54] Y. Lin, S. Taylor, H. Li, K. A. S. Fernando, L. Qu, W. Wang, L. Gu, B. Zhou, and Y. P. Sun, *Journal of Materials Chemistry* 14, 527 (2004).

[55] N. Pierard, A. Fonseca, J. F. Colomer, C. Bossuot, J. M. Benoit, G. Van Tendeloo, J. P. Pirard, and J. Nagy, *Carbon* 42, 1691 (2004).

[56] A. Kukovecz, T. Kanyó, Z. Kónya, and I. Kiricsi, *Carbon* 43, 994 (2005).

[57] G. Maurin, I. Stepanek, P. Bernier, J. F. Colomer, J. Nagy, and F. Henn, *Carbon* 39, 1273 (2001).

[58] L. Chen, M. Qu, G. Zhou, B. Zhang, and Z. Yu, *Materials Letters* 58, 3737 (2004).

[59] J. Liu, A. G. Rinzler, H. Dai, J. H. Hafner, R. K. Bradley, P. J. Boul, A. Lu, T. Iverson, K. Shelimov, and C. B. Huffman, *Science* 280, 1253 (1998).
[60] K. J. Ziegler, Z. Gu, H. Peng, E. L. Flor, R. H. Hauge, and R. E. Smalley, *Journal of the American Chemical Society* 127, 1541 (2005).
[61] K. J. Ziegler, Z. Gu, J. Shaver, Z. Chen, E. L. Flor, D. J. Schmidt, C. Chan, R. H. Hauge, and R. E. Smalley, *Nanotechnology* 16, S539 (2005).
[62] Z. Gu, H. Peng, R. Hauge, R. Smalley, and J. Margrave, *Nano Letters* 2, 1009 (2002).
[63] M. Aizawa and M. S. P. Shaffer, *Chemical Physics Letters* 368, 121 (2003).
[64] C. A. Dyke and M. James, *Journal of the American Chemical Society* 125, 1156 (2003).
[65] M. Holzinger, J. Abraham, P. Whelan, R. Graupner, L. Ley, F. Hennrich, M. Kappes, and A. Hirsch, *Journal of the American Chemical Society* 125, 8566 (2003).
[66] M. Islam, E. Rojas, D. Bergey, A. Johnson, and A. Yodh, *Nano Letters* 3, 269 (2003).
[67] Y. Wang, X. Xu, Z. Tian, Y. Zong, H. Cheng, and C. Lin, *Chemistry–A European Journal* 12, 2542 (2006).
[68] C. Hui, X. Li, and I. M. Hsing, *Electrochimica acta* 51, 711 (2005).
[69] A. Rochefort, D. R. Salahub, and P. Avouris, *Chemical Physics Letters* 297, 45 (1998).
[70] D. A. Heller, R. M. Mayrhofer, S. Baik, Y. V. Grinkova, M. L. Usrey, and M. S. Strano, *Journal of the American Chemical Society* 126, 14567 (2004).
[71] D. Chattopadhyay, S. Lastella, S. Kim, and F. Papadimitrakopoulos, *Journal of the American Chemical Society* 124, 728 (2002).
[72] H. J. Shin, S. M. Kim, S. M. Yoon, A. Benayad, K. K. Kim, S. J. Kim, H. K. Park, J. Y. Choi, and Y. H. Lee, *Journal of the American Chemical Society* 130, 2062 (2008).
[73] J. H. W. D.M. Poirier, *Surf.Sci.Spectra* 2 (1993).
[74] Z.-M. Wang, H. Kanoh, K. Kaneko, G. Q. Lu, and D. Do, *Carbon* 40, 1231 (2002).
[75] S. Biniak, G. Szymański, J. Siedlewski, and A. Świa□tkowski, *Carbon* 35, 1799 (1997).
[76] A. Grüneis, R. Saito, T. Kimura, L. Cançado, M. Pimenta, A. Jorio, A. Souza Filho, G. Dresselhaus, and M. Dresselhaus, *Physical Review B* 65, 155405 (2002).
[77] A. Souza Filho, A. Jorio, G. G. Samsonidze, G. Dresselhaus, M. Pimenta, M. Dresselhaus, A. K. Swan, M. Ünlü, B. Goldberg, and R. Saito, *Physical Review B* 67, 035427 (2003).
[78] M. Dresselhaus, G. Dresselhaus, A. Jorio, A. Souza Filho, and R. Saito, *Carbon* 40, 2043 (2002).
[79] N. Hamada, S. Sawada, and A. Oshiyama, *Physical Review Letters* 68, 1579 (1992).
[80] A. Bianco, K. Kostarelos, and M. Prato, *Expert Opinion on Drug Delivery* 5, 331 (2008).
[81] N. W. S. Kam, M. O'Connell, J. A. Wisdom, and H. J. Dai, *Proceedings of the National Academy of Sciences of the United States of America* 102, 11600 (2005).
[82] G. Painter and D. Ellis, *Physical Review B* 1, 4747 (1970).
[83] X. Blase, L. X. Benedict, E. L. Shirley, and S. G. Louie, *Physical Review Letters* 72, 1878 (1994).
[84] H. L. Zhang, Y. Q. Liu, L. C. Cao, D. C. Wei, Y. Wang, H. Kajiura, Y. M. Li, K. Noda, G. F. Luo, L. Wang, J. Zhou, J. Lu, and Z. X. Gao, *Advanced Materials* 21, 813 (2009).
[85] M. Kanungo, H. Lu, G. G. Malliaras, and G. B. Blanchet, *Science* 323, 234 (2009).
[86] J. Y. Wang and Y. Li, *Journal of the American Chemical Society* 131, 5364 (2009).
[87] Q. Zhu, X. Qiu, and C. Ma, *Carbon* 37, 1475 (1999).

[88] L. S. K. Pang, J. D. Saxby, and S. P. Chatfield, *The Journal of Physical Chemistry* 97, 6941 (1993).

[89] P. Ajayan, T. Ebbesen, T. Ichihashi, S. Iijima, K. Tanigaki, and H. Hiura, *Nature* 362, 522 (1993).

[90] Z. Klusek, P. Datta, and W. Kozlowski, *Corrosion Science* 45, 1383 (2003).

[91] A. M. F. Lima, A. W. Musumeci, H. W. Liu, E. R. Waclawik, and G. G. Silva, *J. Therm. Anal. Calorim.* 97, 257 (2009).

[92] A. R. Adhikari, M. B. Huang, H. Bakhru, R. Vajtai, C. Y. Ryu, and P. M. Ajayan, *J. Appl. Phys.* 100 (2006).

[93] R. Brukh and S. Mitra, *J. Mater. Chem.* 17, 619 (2007).

[94] R. Brukh, O. Sae-Khow, and S. Mitra, *Chemical Physics Letters* 459, 149 (2008).

[95] C. X. Wu, J. X. Xu, J. X. Li, G. F. Dong, and L. H. Guan, *Physica E.* 41, 1591 (2009).

[96] C. Saridara, R. Brukh, Z. Iqbal, and S. Mitra, *Analytical chemistry* 77, 1183 (2005).

[97] X. Chu and L. Schmidt, *Surface Science* 268, 325 (1992).

[98] X. Chu and L. D. Schmidt, *Ind. Eng. Chem. Res.* 32, 1359 (1993).

[99] S. Kundu, Y. M. Wang, W. Xia, and M. Muhler, *J. Phys. Chem. C* 112, 16869 (2008).

[100] J. Ma, J. N. Wang, and X. X. Wang, *J. Mater. Chem.* 19, 3033 (2009).

[101] J. Ma and J. N. Wang, *Chem. Mater.* 20, 2895 (2008).

[102] J. Ma, F. Yu, and J. N. Wang, *Journal of Materials Chemistry* 20, 5742 (2010).

[103] J. Ma, J. N. Wang, C. J. Tsai, R. Nussinov, and B. Ma, *Frontiers of Materials Science in China* 4, 17 (2010).

[104] E. Illeková and K. Csomorová, *J. Therm. Anal. Calorim.* 80, 103 (2005).

[105] B. J. Landi, C. D. Cress, C. M. Evans, and R. P. Raffaelle, *Chem. Mater.* 17, 6819 (2005).

[106] S. Sawada and N. Hamada, *Solid State Communications* 83, 917 (1992).

[107] J. Tersoff, *Physical Review Letters* 61, 2879 (1988).

[108] W. Zhou, Y. H. Ooi, R. Russo, P. Papanek, D. E. Luzzi, J. E. Fischer, M. J. Bronikowski, P. A. Willis, and R. E. Smalley, *Chemical Physics Letters* 350, 6 (2001).

[109] Y. Zhang, Z. Shi, Z. Gu, and S. Iijima, *Carbon* 38, 2055 (2000).

[110] A. Fraczek, E. Menaszek, C. Paluszkiewicz, and M. Blazewicz, *Acta Biomaterialia* 4, 1593 (2008).

[111] D. Tasis, K. Papagelis, D. Douroumis, J. R. Smith, N. Bouropoulos, and D. G. Fatouros, *Journal of Nanoscience and Nanotechnology* 8, 420 (2008).

[112] S. Kang, M. Pinault, L. D. Pfefferle, and M. Elimelech, *Langmuir.* 23, 8670 (2007).

[113] S. Kang, M. Herzberg, D. F. Rodrigues, and M. Elimelech, *Langmuir.* 24, 6409 (2008).

[114] A. Takagi, A. Hirose, T. Nishimura, N. Fukumori, A. Ogata, N. Ohashi, S. Kitajima, and J. Kanno, *Journal of Toxicological Sciences* 33, 105 (2008).

[115] J. P. Kaiser, P. Wick, P. Manser, P. Spohn, and A. Bruinink, *Journal of Materials Science-Materials in Medicine* 19, 1523 (2008).

[116] X. C. Wu, W. J. Zhang, R. Sammynaiken, Q. H. Meng, Q. Q. Yang, E. Zhan, Q. Liu, W. Yang, and R. Wang, *Colloids and Surfaces B-Biointerfaces* 65, 146 (2008).

[117] X. J. Li, W. Chen, Q. W. Zhan, L. M. Dai, L. Sowards, M. Pender, and R. R. Naik, *Journal of Physical Chemistry B* 110, 12621 (2006).

[118] V. Z. Poenitzsch, D. C. Winters, H. Xie, G. R. Dieckmann, A. B. Dalton, and I. H. Musselman, *Journal of the American Chemical Society* 129, 14724 (2007).

[119] Z. Su, K. Mui, E. Daub, T. Leung, and J. Honek, *The Journal of Physical Chemistry B* 111, 14411 (2007).
[120] Z. D. Su, T. Leung, and J. F. Honek, *Journal of Physical Chemistry B* 110, 23623 (2006).
[121] S. Brown, T. S. Jespersen, and J. Nygård, *Small* 4, 416 (2008).
[122] B. Y. Ma, T. Elkayam, H. Wolfson, and R. Nussinov, *Proceedings of the National Academy of Sciences of the United States of America* 100, 5772 (2003).
[123] S. Linse, C. Cabaleiro-Lago, W. F. Xue, I. Lynch, S. Lindman, E. Thulin, S. E. Radford, and K. A. Dawson, *Proceedings of the National Academy of Sciences of the United States of America* 104, 8691 (2007).
[124] B. Y. Ma and R. Nussinov, *Current Opinion in Chemical Biology* 10, 445 (2006).
[125] B. Ma and R. Nussinov, *Current Topics in Medicinal Chemistry* 7, 999 (2007).
[126] J. Meng, L. Song, H. Xu, H. Kong, C. Wang, X. Guo, and S. Xie, *Nanomedicine: Nanotechnology, Biology and Medicine* 1, 136 (2005).
[127] C. Zhao, J. S. Ren, and X. G. Qu, *Chemistry-a European Journal* 14, 5435 (2008).
[128] X. Zhao and J. K. Johnson, *Journal of the American Chemical Society* 129, 10438 (2007).
[129] R. R. Johnson, A. T. C. Johnson, and M. L. Klein, *Nano Letters* 8, 69 (2008).
[130] X. Li, Y. H. Peng, and X. G. Qu, *Nucleic Acids Research* 34, 3670 (2006).
[131] Y. H. Peng, X. Li, J. S. Ren, and X. G. Qu, *Chemical Communications*, 5176 (2007).
[132] E. R. Kisin, A. R. Murray, M. J. Keane, X. C. Shi, D. Schwegler-Berry, O. Gorelik, S. Arepalli, V. Castranova, W. E. Wallace, V. E. Kagan, and A. A. Shvedova, *Journal of Toxicology and Environmental Health-Part a-Current Issues* 70, 2071 (2007).
[133] C. S. Sharma, S. Sarkar, A. Periyakaruppan, J. Barr, K. Wise, R. Thomas, B. L. Wilson, and G. T. Ramesh, *Journal of Nanoscience and Nanotechnology* 7, 2466 (2007).
[134] L. Zhu, D. W. Chang, L. M. Dai, and Y. L. Hong, *Nano Letters* 7, 3592 (2007).
[135] S. Lee, S. Mukerjee, J. McBreen, Y. Rho, Y. Kho, and T. Lee, *Electrochimica Acta* 43, 3693 (1998).
[136] E. Antolini, L. Giorgi, A. Pozio, and E. Passalacqua, *Journal of Power Sources* 77, 136 (1999).

In: Applications of Carbon Nanotubes
Editor: Ajay Kumar Mishra
ISBN: 978-1-62081-918-0

Chapter 9

SYNTHESIS, PROPERTIES AND APPLICATIONS OF HETEROATOMS-DOPED CARBON NANOTUBES

Yougui Chen and Xueliang Sun*
Department of Mechanical and Materials Engineering,
University of Western Ontario, London, Ontario, Canada

ABSTRACT

Carbon nanotubes (CNTs) with unique structure exhibit outstanding properties, which promise a wide range of potential applications. However, the difficulty of controllable synthesis of semiconducting and conducting single-wall CNTs, and the inert surface make some applications formidable. A new class of CNTs, the heteroatom-doped CNTs, has been developed recently. The heteroatoms in CNTs include nitrogen, boron and phosphorus. Due to the doping of the heteroatoms, the electronic properties of the CNTs are significantly tuned. The surface of CNTs is activated for various applications such as fuel cells and batteries. In this chapter, the synthesis of the new kind of CNTs will be introduced first in details. The doping effects of heteroatoms into CNTs are discussed. Emphasis will be put on the practical applications of the heteroatom-doped CNTs, which show superior advantages over general CNTs, especially in fuel cells and lithium batteries.

1. INTRODUCTION

Since the discovery by Iijima in 1991, [1] carbon nanotubes (CNTs) have been extensively studied both fundamentally and for wide applications. However, the controllable synthesis and inert surface of CNTs limit the real applications. Functionalization of CNTs for specific applications are widely studied at present. Alternatively, the properties of CNTs can be tuned by the doping of heteroatoms into the hexagonal carbon matrix. Doped heteroatoms as electron donors or acceptors result in a shift of the Fermi level and change the intrinsic

* E-mail: xsun@eng.uwo.ca.

electronic properties. There have been numerous reports about the general CNTs including single wall carbon nanotubes (SWCNTs) and multi-wall carbon nanotubes (MWCNTs). [2, 3]

In this chapter, we will review the recent work on heteroatoms doped CNTs, including the synthesis, doping effects and applications. Some of them are covered in recent reviews. [4, 5] The doping here is strictly defined as foreign atoms formed in the matrix of CNTs, rather than the absorption of foreign atoms on the surface. Nitrogen or boron doping is the major topic based on the large volume of literature and our research interests.

2. Synthesis of Heteroatoms Doped CNTs

Many studies have attempted to synthesize doping heteroatoms in CNTs. Although N_2 is commonly applied as a carrier gas in the synthesis of CNTs, the ordinary temperature (700-900°C) for CNTs is not high enough to break the N≡N bond to incorporate the N atoms into CNTs. Therefore, extra precursors containing the heteroaotms or other experimental conditions such as high temperature or plasma are needed. The various synthesis methods of heteroatoms doped CNTs can be categorized into two methods: ex-situ doping with post treatment of pre-synthesized CNTs and in-situ doping during the growth of the CNTs. Between them, the in-situ doping of heteroatoms into CNTs is more widely reported.

2.1. Ex-Situ Doping of Heteroatoms

B or N doped CNTs have been synthesized with the post treatment of CNTs with a B or N precursor at high temperature. N-CNTs were synthesized with CNTs exposed under nitrogen plasma of 3.1 kW microwave input. [6] B-CNTs were obtained by the partial substitution of C by B atoms by thermal treatment of a mixture of boron trioxide and CNTs. [7] The B concentration in the B-CNTs can be as high as 10 at%. With nitrogen as the flow gas, B and N can be co-doped into CNTs at very high temperature (1503 – 1773 K). [8, 9]

This post treatment doping technique maintains the morphology and diameter of the precursor CNTs. However, the nitrogen content is usually low. The highest nitrogen content Golberg got is only 2 at%. [9]

2.2. In-Situ Synthesis of Heteroatoms-Doped CNTs

The arc discharge method is applied to synthesize the heteroatoms doped CNTs. Droppa *et al* produced N-CNTs by the evaporation of a mixture of graphite and Ni, Fe, Co catalysts in a nitrogen-helium atmosphere in an arc discharge experiment. [10] With an anode made of homogeneous BC4N, B and N can be co-doped into CNTs. [11]

The chemical vapor deposition (CVD) method is the most widely studied method to grow heteroatoms-doped CNTs due to its large scale production and simple setup. A typical CVD process involves the chemical reaction of the precursors of carbon and heteroatoms on the catalysts to form heteroatoms-doped CNTs at a temperature. Based on different experimental setups, floating catalyst CVD, aerosol-assisted CVD, spray pyrolysis CVD, injection CVD, and plasma enhanced CVD are developed.

Our group carried out some work on N-CNTs. We have synthesized N-CNTs by floating catalyst CVD method with melamine, ethylene and ferrocene as the nitrogen, carbon and the catalyst precursor respectively. [12] The nitrogen contents in N-CNTs varied from 0 to 8.4 at% depending on the melamine amount in the precursors. The tube morphology and chemical configuration of the nitrogen atoms differ among the N-CNTs with different nitrogen contents.

Table 1. A summary of the precursors and contents of the heteroatoms in some typical references

Precursors	Dopants	Ref.
Ethylene, Melamine	0-8.4 at% N	[12]
Melamine	1.7 and 10.4 at% N	[13]
Acetonitrile, imidazole	3.2, 5.2 % N	[14]
Imidazole	12.1-25.7 at% N	[16]
Imidazole, triphenylphosphine	1.9 % P, 7.1 % N	[15]
Acetonitrile	0.9-6 at% N	[18-23]
Methane, acetonitrile	2.6, 5.4 at% N	[24]
CoPc, FePc, NiPc	0.018-8 at% N	[25, 26]
FePc, NH_3	9 % N	[27]
Benzene, Acetonitrile, BCl_3	N/C: 0.025 at. B/C: 0.040 at.	[28]
Pyridine	2.0-4.6 at% N	[29]
Pyridine, FePc	1.59 -3.22 at% N	[30]
Pyridine, pyrimidine	1-3.2 % N	[31]
Xylene, pyridine	--	[32]
Pyridine, melamine	3.5% N	
Turpentine oil, 4-tert-butylpyridine	1.6, 2 at% N	[33]
C_2H_2, NH_3	0.4-6.6 at%	[34, 35]
CH_4, H_2, trimethylborate	--	[36]
Acetonitrile, tetrahydrofuran	~ 20 at% N	[37]
Benzylamine in toluene, triethylborane in hexane	2.2 at% N, 0.5 at% B	[38, 39]
Pyridine, NH_3, 1,2-ethylenediamine, n-propylamine	4.32, 4.52, 5.34 at% N	[40]
Ethylenediamine	5.4-8.8 at%	[41, 42]
Ethanol, ethylenediamine	1.2 at% N	[43]
Monoethanolamine	4.8-7.8 at% N	[44, 45]
CH_4, H_2, NH_3	0.6, 3.2 at%	[46]
Polystyrene-block-poly(methyl methacylate), NH_3	0-8 at% N	[47]
4-ferrocenylaniline, aniline	1 at% N	[48]
Dimethylformamide	2-16 at% N	[49]
Ethanol, benzylamine	--	[50]
BF_3, methanol, acethylene	<0.5% B	[51]
CH_4, B_2H_6, H_2	B/C of 1/24, 1/16 and 1/10	[17]
CH_4, pyridine or NH_3, B_2H_6	1 at% N and B	[52]
Acetonitrile and borane-pyridine complex	17 at% N and 9 at% B	[53]
Triphenyl-phosphine, benzylamine	0.5 at%P, 1.5 at% N	[54]

An even higher nitrogen concentration (10.4 at%) was obtained with only melamine as the precursor of both nitrogen and carbon, and without other hydrocarbons. [13] It demonstrated that a lower synthesis temperature resulted in a higher nitrogen concentration. We also reported N-CNTs synthesized with a mixture of imidazole and acetonitrile as the precursors with ultrasonic spray pyrolysis CVD. [14] The structures and nitrogen contents of the N-CNTs can be tuned by the imidazole amounts. With the triphenylphosphine (TTP, $P(C_6H_5)_3$) as the precursor of P and mixed with imidazole, N and P were co-doped into CNTs. [15] The addition of TTP changed the shape of the catalyst particles and resulted in different tube diameters and wall thicknesses.

A lot of experimental work has be attempted to prepare heteroatoms doped CNTs, both in SWCNTs and MWCNTs. The precursors and the contents of the heteroatoms from some typical references are summarized Table 1. A great deal of effort is devoted to the high concentration of heteroatoms and large scale production. X-ray photoelectron spectroscopy and Electron energy loss spectroscopy are usually applied to characterize the concentrations of the dopants. Up to now, the highest nitrogen content in N-CNTs achieved is 25.7 at% with imidazole as the nitrogen and carbon precursor, and ferrocene as the catalyst precursor, by CVD at 850°C. [16] The highest B content is about 10%. [17]

There is not a universal rule to control the synthesis of heteroatoms doped CNTs. It can be learned from Table 1 that the precursors play an important role in the heteroatoms doping. Various chemicals containing the heteroatoms have been tried and different contents of the heteroatoms are obtained. The catalyst effect has also been investigated. N-CNTs were synthesized by pyrolysis of acetonitrile over Co-Ni bimetallic catalysts with different atomic ratios. [23] The highest nitrogen content was achieved on the catalysts ratio of 1:1. This was ascribed to the highest nitrogen solubility on this catalyst. We demonstrated the temperature effect in the synthesis of N-CNTs. [13] The CVD of melamine results in a higher nitrogen concentration at 800°C than at higher temperatures. But the tube obtained at higher temperatures is longer. Ghosh et al. obtained the highest nitrogen concentration in CVD of imidazole at 850°C. Cui found that the addition of sulfur as an additive in the precursors can promote the incorporation of nitrogen atoms in N-CNTs. [19] However, the addition of hydrogen in the precursors resulted in lower nitrogen contents in N-CNTs. [21, 22]

3. Morphology, Structure and Composition of Heteroatoms-Doped CNTs

3.1. Morphological Changes

Upon doping with heteroatoms, the morphology is modified compared with pristine CNTs. As shown in Figure 1, N-CNTs exhibit a bamboo-like structure with well-defined compartments separated by graphene layers, different from the hollow tubular structure of pristine CNTs. The density of the compartments increases with the nitrogen concentrations. [12] Simply, the density of the compartments is a direct indicator to compare the nitrogen concentrations in similar synthesis conditions. At high nitrogen concentrations, the interlinks become irregular and connected with each other.

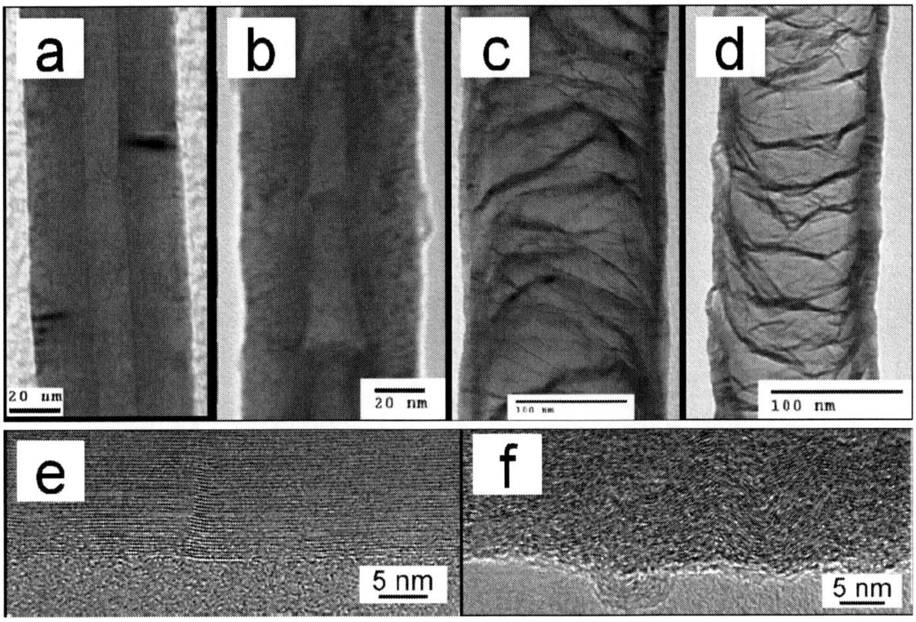

Figure 1. The TEM images of synthesized (a) CNTs and N-CNTs with the nitrogen concentration of (b) 1.5, (c) 5.4 and (d) 8.4 at%, and high resolution TEM images of the surface of (e) CNTs and (f) N-CNTs with a nitrogen content of 8.4 at% in our groups [55].

High resolution TEM images also reveal the differences of surfaces between pristine CNTs and N-CNTs. Pristine CNTs show a smooth surface with fine graphite layers. However, the surface of N-CNTs becomes rough and disordered. As mentioned above, the doping of nitrogen atoms form pyridinic, pyrrolic, and oxidized nitrogen sites, which disrupts the planar hexagonal arrangement of carbon atoms in CNTs and results in the rough surface of N-CNTs. We observed an increase of the tube diameter after nitrogen doping. [12] The tube diameter increases from 60-80 nm to 100-120 nm with the increase of the nitrogen contents. However, a decrease of the tube diameter was also obtained by other researchers. [50]

3.2. Structural Changes

Raman spectroscopy is a powerful tool to analyze structural change between the pristine CNTs and heteroatoms doped CNTs.

The analysis of the peak positions and intensities gives information about defectiveness, resulting from the doping of heteroatoms. Raman spectra of CNTs and N-CNTs are compared in Figure 2. The G-band (~1579 cm^{-1}) indicates the presence of crystalline graphite carbon, whereas the D-band (~1348 cm^{-1}) is attributed to disorders in the graphite structure. The D'-band, a shoulder of the G-band at a higher frequency, corresponds to second-order Raman scattering from the variation of the D-band. The ratio of the intensities between the two bands (I_D/I_G) is an indicator of the degree of disorder in carbon materials. In Figure 2, the I_D/I_G ratio of CNTs and N-CNTs is 0.886 and 1.553, respectively. Furthermore, the I_D/I_G ratio increases with the nitrogen contents in the N-CNTs. [12]

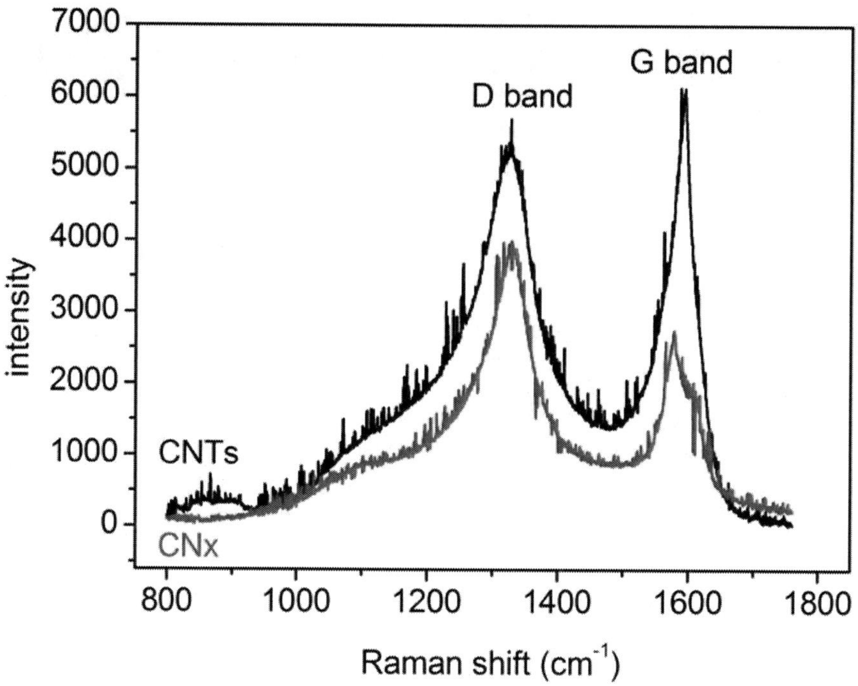

Figure 2. Raman spectroscopy of CNTs and N-CNTs (8.4 at% N).

The much higher I_D/I_G ratio and D' band intensity of N-CNTs suggest that the doping of nitrogen atoms brings more defects in N-CNTs than the pristine CNTs, which can be observed by the high resolution TEM images above. And the defectiveness increases with the contents of nitrogen in N-CNTs. Furthermore, there is a downshift of the G band center of N-CNTs compared to CNTs. This can be explained by the donor charge transfer from nitrogen dopants to graphite. [56]

3.3. Chemical Configurations of Doped Nitrogen Atoms

The chemical bonding configurations of the dopants have been investigated with XPS. Figure 3a summarizes the structures of nitrogen atoms doped in N-CNTs. [57] The two main kinds are quaternary (or graphitic, or substitutional, or graphite-like) and pyridinic (or pyridine-like) nitrogen atoms. Quaternary nitrogen is a nitrogen atom, which replaces a carbon atom within a graphite plane and is bonded to three carbon atoms. Nitrogen doped at the quaternary sites keeps the structure of graphite structure in CNTs. Pyridinic nitrogen is bonded to two carbon atoms and donates one p electron to the aromatic π system. Oxidized nitrogen is oxidized pyridinic nitrogen, which is bonded to two carbon atoms and one oxygen atom. Pyrrolic (or pyrrole-like) nitrogen is bonded with one hydrogen atom and two carbon atoms and forms a five-atom aromatic ring. [58] The pyrrolic nitrogen is not stable over 600°C and transforms to pyridinic or quaternary nitrogen. The pyridinic, oxidized and pyrrolic nitrogens break the hexagonal arrangement of the carbon atoms in graphite structures and results in surface defects and bamboo-like structures. Besides, tetrahedral and molecular nitrogen atoms are also detected.

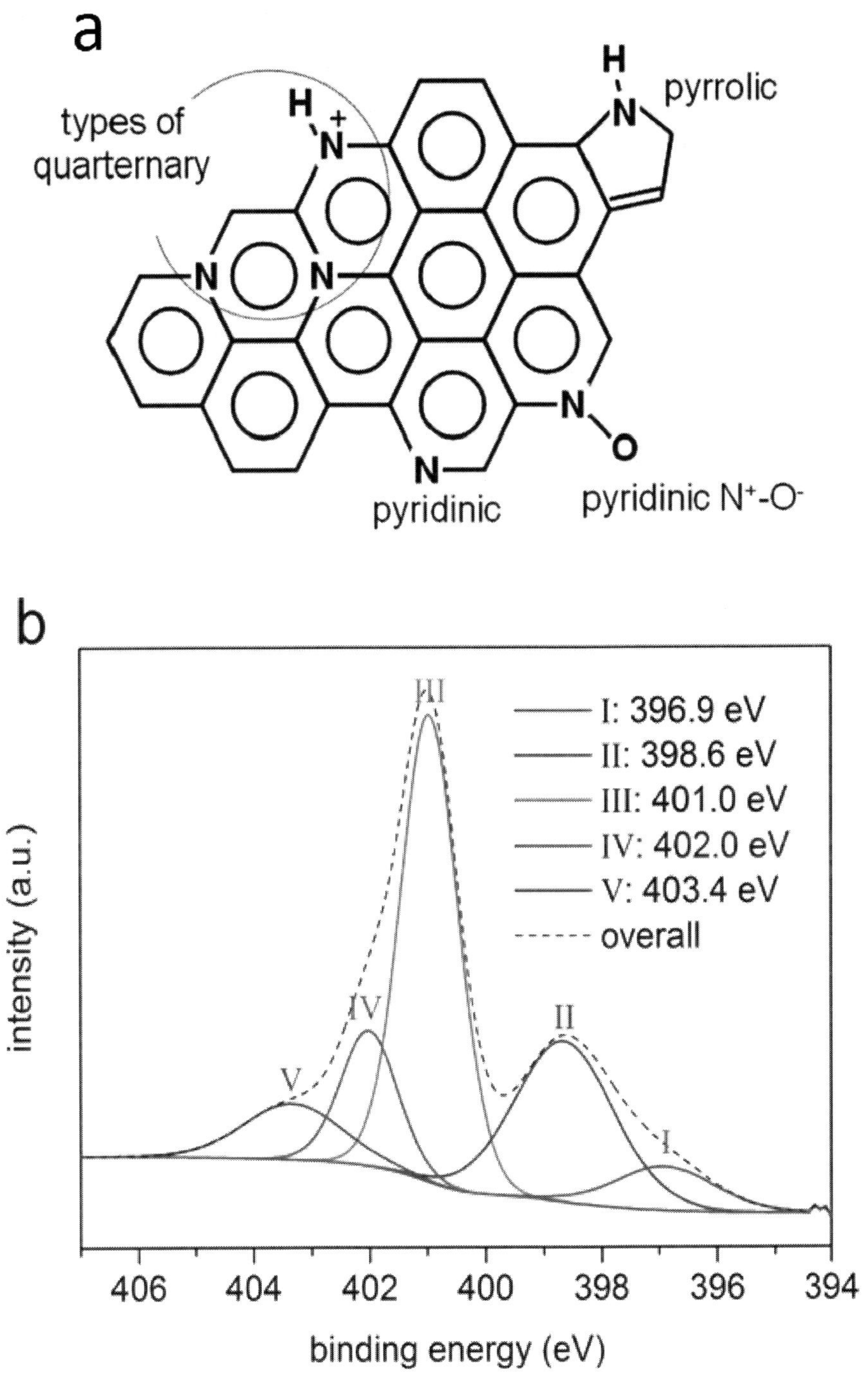

Figure 3. (a) Chemical natures of nitrogen atoms in N-CNTs [57] and (b) high resolution XPS spectroscopy of N 1s in N-CNTs [55].

Tetrahedral nitrogen is bonded to sp^3 C due to an undecomposed N-H bond in the precursors such as melamine. [59] Molecular N_2 are nitrogen molecules adsorbed/intercalated between the carbon walls or trapped in the compartments, which has been confirmed by the X-ray absorption near-edge structure analysis [60] and our direct scanning transmission X-ray microscopy observations [61]. Figure 3b shows the high resolution XPS spectroscopy of N 1s of N-CNTs synthesized in our lab. [55] The asymmetric N 1s peak centered at 401 eV can be deconvoluted into five peaks at 396.9, 398.6, 401.0, 402.0, and 403.4 eV from low to high binding energy. They are ascribed as tetrahedral, pyridinic, graphitic, oxidized, and molecular nitrogen, respectively.

4. Applications of Heteroatoms Doped CNTs

Heteroatoms doped CNTs have demonstrated a wide range of applications due to their superior properties over pristine CNTs, including energy storage, sensors and field emissions.

4.1. Catalyst Supports in Fuel Cells

CNTs provide a large surface area, high electrical conductivity and high corrosion resistance as good catalyst supports. However, the inert surface makes the deposition of metal catalyst particles a formidable challenge. Usually, a pretreatment by acid is conducted to functionalize the surface of CNTs but this will reduce their mechanical properties. The doping of foreign atoms in CNTs offers a feasible way to activate the surface for the deposition of metal nanoparticles and meanwhile, enhances interactions between the surface and deposited metal particles. We compared CNTs and N-CNTs (8.4 at% N) as the supports of Pt nanoparticles for the cathode of proton-exchange membrane fuel cells. [55, 62] As shown in Figure 4a and b, the diameters of the Pt nanoparticles on CNTs range from 4 to 15 nm with a mean size of 9.1 nm. While on N-CNTs, Pt nanoparticles show a uniform distribution from 2 to 6 nm with a mean size of 4.2 nm. The homogeneously dispersed nitrogen sites on the surface provide nucleation sites and result in higher dispersion of Pt nanoparticles. Our study on the time-dependent growth of Pt nanowires on N-CNTs confirmed that Pt prefers to nucleate on the nitrogen sites. [63] In addition, the hydrophilicity is improved by nitrogen doping as can be observed during experiments. The enhanced access of Pt ions to N-CNTs surface leads to higher dispersion of Pt nanoparticles. The smaller size of Pt nanoparticles results in a much larger electrochemical surface area (ECSA) of Pt/N-CNT than Pt/CNTs (Figure 4c). The ECSA of Pt/CNTs and Pt/N-CNTs is 26.1 and 41.8 m^2/g_{Pt}, respectively. As the cathode catalyst, Pt/N-CNTs exhibit higher activity toward oxygen reduction reaction (ORR) (Figure 4d) and better fuel cell performance than Pt/CNTs. The half-wave potential of ORR on Pt/N-CNTs is 32 mV more positive than that on Pt/CNTs. Thus, N-CNTs are better catalyst supports than CNTs, featuring higher dispersion of deposited Pt nanoparticles and higher electrocatalytic activity toward ORR. The electrochemical stabilities of Pt/N-CNTs and Pt/CNTs were also compared with accelerated durability tests. [64]

The electrochemical stability of Pt catalysts is improved by doping nitrogen into the CNTs support.

The stability increases with the nitrogen contents in N-CNTs. After 4000 cycles of potential cycling, Pt/CNTs only remains 11.2% of its initial ECSA. While, Pt/N-CNTs with a nitrogen content of 8.4 at% keeps 42.5% of its initial ECSA.

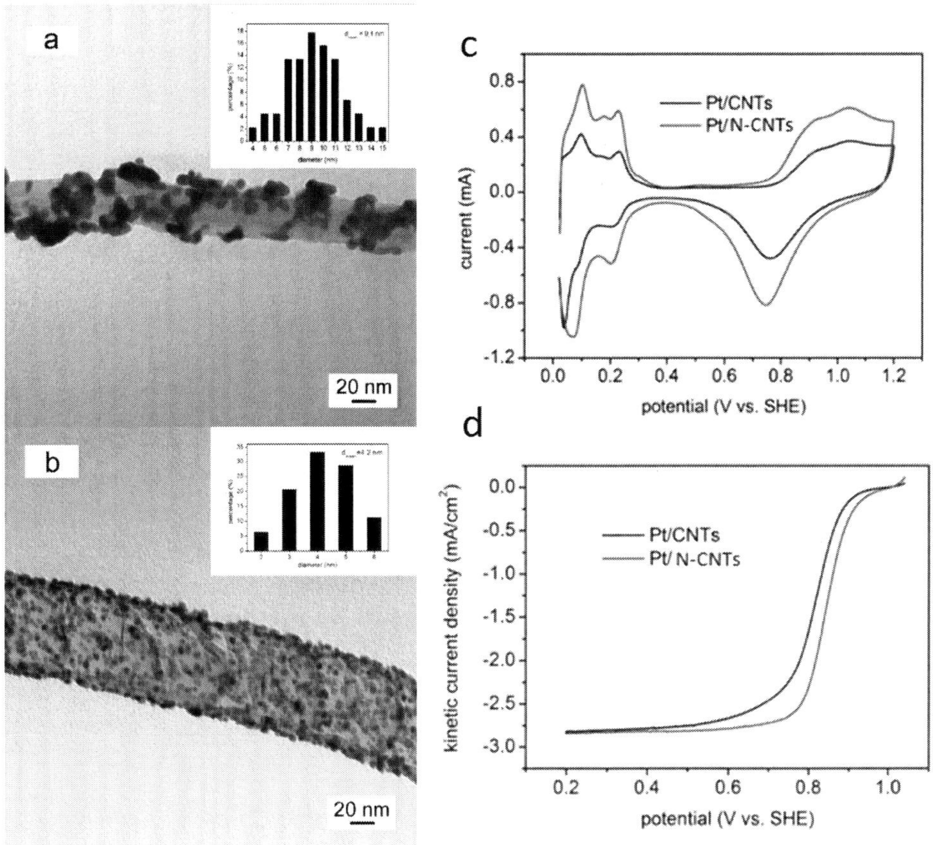

Figure 4. TEM images and size distribution histograms (inserts) of Pt nanoparticles deposited on (a) CNTs and (b) N-CNTs; (c) cyclic voltammograms of Pt/CNTs and Pt/N-CNTs in Ar saturated 0.5 M H_2SO_4 at 50 mV/s; (d) ORR activities on Pt/CNTs and Pt/N-CNTs in the cathodic sweep at 1600 rpm in O_2 saturated 0.5M H_2SO_4 at 5 mV/s and 25°C.

The improved electrochemical stability is credited to the protection of Pt nanoparticles by the lone electron pair of nitrogen atoms in CNTs and the stronger bonding between Pt and the N-CNTs supports. With N-CNTs as the supports, we obtained ultrathin single crystalline Pt nanowires with a diameter of 2.5 nm, which provides a wide range of applications including fuel cell catalysts. [63] Other researchers also investigate the application of N-CNTs synthesized by various methods as the catalyst support of Pt [65-69], PtRu [70-72], PtSn [73] and PtRuNi [74] nanoparticles for ORR [65, 67, 69], methanol electrooxidation [66, 70-72, 74] and ethanol electrooxidation [68, 73] in fuel cells.

4.2. Non-Noble Metal ORR Catalyst in Fuel Cells

The doping of heteroatoms breaks the inertness of the pristine CNTs and results in an active surface. The heteroatoms doped CNTs have been intensively studied for ORR as non-noble metal catalysts. We have demonstrated that the ORR activity of N-CNTs synthesized from ethylene, melamine and ferrocene increases with their nitrogen contents in both alkaline and acidic media. [75] As presented in Figure 5, the ORR onset potential of CNTs in alkaline solution is −0.013 V. The ORR polarization curves shift positively with the doping of nitrogen. For N-CNTs with a nitrogen content of 5.1 or 7.7 at%, the ORR onset potential is about 0.170 V. This indicates that nitrogen doping in CNTs can effectively increase the electrocatalytic activity over ORR. Although the activity in acidic solution is still not comparable with Pt based catalysts, the N-CNTs with a nitrogen concentration of 7.7 at% showed comparable ORR activity with a commercial Pt/C catalyst in alkaline solution. We also demonstrated that N-CNTs have high stability based on potential cycling tests in alkaline solution. [76] N-CNTs catalyst shows almost identical voltammetric curves before and after 3500 cycles of continuous potential cycling.

N-CNTs synthesized by various precursors [77-86] and B-CNTs [87] have also been widely studied as the non-noble metal ORR catalysts by other researchers.

Besides the superior electrocatalytic activity over pristine CNTs, heteroatoms doped CNTs also exhibit higher CO and methanol tolerance compared with Pt based catalysts. [85, 87] This is of great importance in real fuel cell applications because Pt is easily poisoned by CO and the activity is extensively sacrificed due to the methanol cross-over from the anode in direct methanol fuel cells. The underlying reason for the high ORR activities of heteroatoms doped CNTs is unclear. Although Fe, Ni and Co is widely used as the catalysts for the synthesis of heteroatoms doped CNTs, they are mainly encapsulated in the tube and their amount on the surface can be negligible. Moreover, N-CNTs synthesized without metal catalysts also present high ORR activity. [78] Therefore, the conventional FeN_4 or FeN_2 active sites [88] are not applicable in this case. Rao *et al* pointed out that the ORR activity of N-CNTs has a direct relationship with the content of pyridinic nitrogen sites. [84]

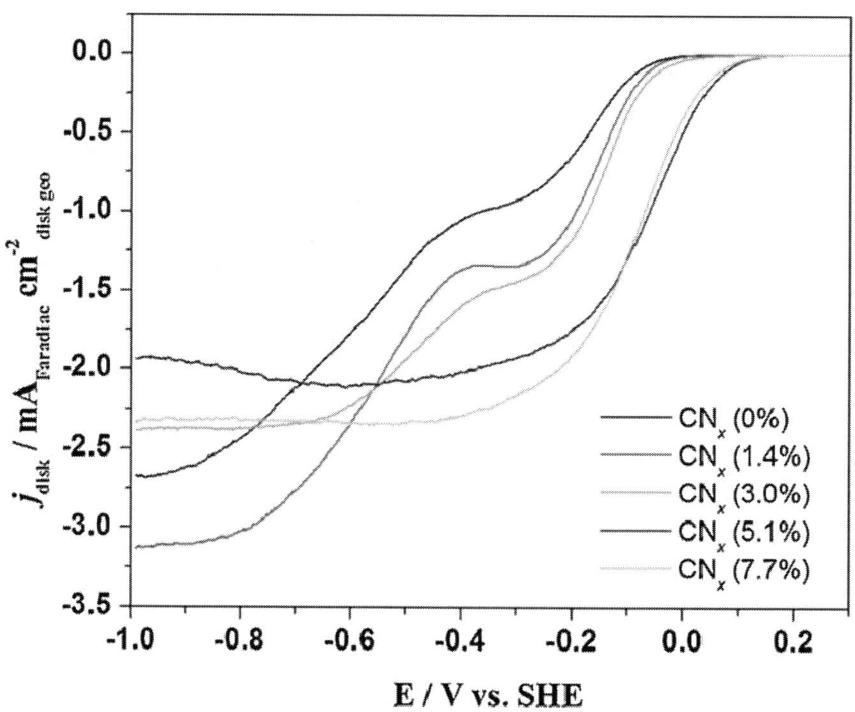

Figure 5. The polarization curves of oxygen reduction on N-CNTs catalysts with different nitrogen content. Electrolyte: 0.1M KOH, scan rate: 5 mVs^{-1}, rotation speed: 1600 rpm, and catalyst loading: 160 μgcm^{-2}.

However, it is still unknown whether the high ORR activity results from the exact pyridinic nitrogen sites or more edge planes brought by these sites. The oxygen adsorption and activation process on the surface of N-CNTs has been investigated by theoretical calculations to elucidate the enhanced ORR activity. [77, 87, 89-91] Due to the differences between elctronegativities, the nitrogen dopant is negatively charged and the neighboring carbon atoms are positively charged. To the contrary, the boron dopant is positively charged and the neighboring carbon atoms are negatively charged. The absorption of O_2 is energetically favored on the B atoms in B-CNTs [87] and on nitrogen neighboring carbon atoms in N-CNTs [89]. The absorption of O_2 cannot be achieved spontaneously on pristine CNTs without charged sites. The chemisorption mode of O_2 is changed from the end-on adsorption on CNTs to side-on adsorption on N-CNTs. The side-on adsorption of O_2 on N-CNTs could effectively reduce the O-O dissociation barrier and thus enhance ORR on N-CNTs. [77]

4.3. Metal-Air Battery

In addition to non-noble metal catalysts in fuel cells, heteroatoms doped CNTs have been studied as the cathode in metal-air batteries. The nonaqueous lithium-oxygen/air battery is one of the most promising energy storage systems for hybrid electric vehicles (HEVs) and electric vehicles (EVs) because of its extremely high theoretical energy density. The porosity of the air electrode plays an important role in the lithium-air battery performance because the insoluble products are deposited in the electrode, which block O_2 from diffusing to the reaction sites. Recent work also showed that the ORR in the carbon electrode significantly affected its performance. Therefore, it is important to develop new carbon electrodes to improve the kinetics and enhance the energy capacity.

We compared CNTs and N-CNTs as the cathode in Li-air batteries, as shown in Figure 6. [92] N-CNTs exhibit an initial discharge capacity of 866 $mAhg^{-1}$, while 590 $mAhg^{-1}$ for CNTs. That is, the initial discharge capacity of N-CNTs is about 1.5 times of CNTs as the cathode in Li-air battery. The discharge average voltage plateau is about 2.52 and 2.41 V for N-CNTs and CNTs respectively. The 0.1 V higher voltage indicates a higher ORR activity on N-CNTs than CNTs. The reversibility is also improved by the nitrogen doping in CNTs.

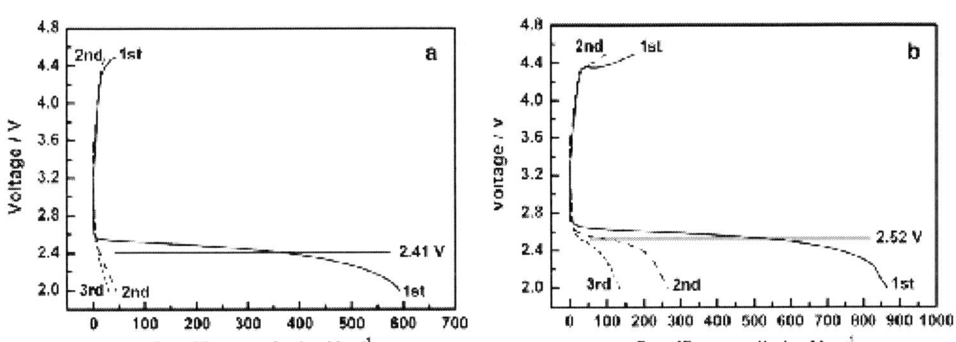

Figure 6. The voltage profiles of (a) CNTs and (b) N-CNTs electrodes cycled in a voltage range of 2.0–4.5 V at a current density of 75mA g^{-1} in the first three cycles.

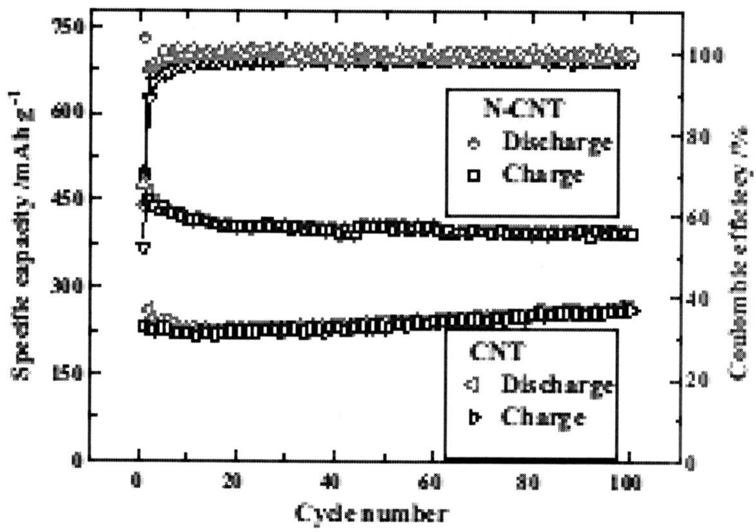

Figure 7. reversible charge/discharge capacity verse cycle numbers of CNTs and N-CNTs at a current density of 100 mA g^{-1}.

Zhu exemplified Zin-air batteries with N-CNTs as the cathode in aqueous alkaline electrolyte and showed a maximum cell power density of 69.5 mW/cm^2. [93]

4.4. Li-Ion Battery

As a kind of carbon material, the heteroatoms doped CNTs have been investigated as the anode material in the Li ion batteries. We demonstrated N-CNTs as the anode material in Li ion battery, as compared with pristine CNTs, as presented in Figure 7. [94] CNTs and N-CNTs exhibit a capacity of 440 and 730 mAh g^{-1}, respectively, during first discharge; and 231 and 494 mAh g^{-1} reversible capacity respectively during the first charge. N-CNTs display higher capacity than CNTs over 100 cycles. The charge transfer resistance decreases with N-CNTs as the anode material. Moreover, N-CNTs display higher rate capability than CNTs. Zhong *et al* also showed a much higher reversible Li storage of N-CNTs than commercial carbon materials. [95] Electrochemical Li insertion into B-CNTs was investigated in a nonaqueous medium. [96] The reversible capacity is 156 mAh/g for pristine and 180 for mAh/g for B-CNTs, respectively. EIS measurements showed that N-CNTs had smaller charge-transfer resistance and larger exchange current density. [97] The significant higher performance of N-CNTs than CNTs is contributed to the higher electrical conductivity and more surface defects.

CONCLUSION AND OUTLOOK

Various methods have been studied for the heteroatoms doped CNTs, particularly the doping of nitrogen and boron. Great efforts have been devoted to the control over the concentration and chemical configuration of the heteroatoms. The doping of heteroatoms results in morphological and structural changes of CNTs. Heteroatoms doped CNTs exhibit superior advantages over pristine CNTs in various applications.

The N-CNTs alkaline medium showed high ORR activity performance and durability for fuel cells although it is still much lower than Pt in acidic medium. Further, it is very promising to use N-CNTs as Pt support for PEM fuel cells, and also as electrodes for Li-Air batteries and Li-ion batteries. Research is ongoing on the increase of ORR activity by controllable synthesis including increase of the nitrogen content. The elucidation of the active sites for ORR should be emphasized, which can guide the synthesis of CNTs of high ORR activity. We believe that this is still a new area with many potential applications to be explored.

REFERENCES

[1] S. Iijima, Nature 354 (1991) 56.
[2] F. Beguin, P. Ehrburger, Carbon 40 (2002) 1619.
[3] R.C. Haddon, Acc. Chem. Res. 35 (2002) 997.
[4] L.S. Panchakarla, A. Govindaraj, C.N.R. Rao, *Inorg. Chim. Acta* 363 (2010) 4163.
[5] P. Ayala, R. Arenal, M. Rummeli, A. Rubio, T. Pichler, *Carbon* 48 (2010) 575.
[6] L.H. Chan, K.H. Hong, D.Q. Xiao, T.C. Lin, S.H. Lai, W.J. Hsieh, H.C. Shih, *Phys. Rev.* B 70 (2004) 125408.
[7] W. Han, Y. Bando, K. Kurashima, T. Sato, *Chem. Phys. Lett.* 299 (1999) 368.
[8] D. Golberg, Y. Bando, W. Han, K. Kurashima, T. Sato, *Chem. Phys. Lett.* 308 (1999) 337.
[9] D. Golberg, Y. Bando, L. Bourgeois, K. Kurashima, T. Sato, *Carbon* 38 (2000) 2017.
[10] R.D. Jr., P. Hammer, A.C.M. Carvalho, M.C.d. Santos, F. Alvarez, *J. Non-Cryst. Solids* 302 (2002) 874.
[11] P. Redlich, J. Loeffler, P.M. Ajayan, J. Bill, F. Aldinger, M. Riihle, *Chem. Phys. Lett.* 260 (1996) 465.
[12] H. Liu, Y. Zhang, R. Li, X. Sun, S. De´silets, H. Abou-Rachid, M. Jaidann, L.-S. Lussier, *Carbon* 48 (2010) 1498.
[13] Y. Zhong, M. Jaidann, Y. Zhang, G. Zhang, H. Liu, M.I. Ionescu, R. Li, X. Sun, H. Abou-Rachid, L.-S. Lussier, *J. Phys. Chem. Solids* 71 (2010) 134.
[14] J. Liu, Y. Zhang, M.I. Ionescu, R. Li, X. Sun, *Appl. Surf. Sci.* 257 (2011) 7837.
[15] J. Liu, H. Liu, Y. Zhang, R. Li, G. Liang, M. Gauthier, X. Sun, *Carbon* 49 (2011) 5014.
[16] K. Ghosh, M. Kumar, T. Maruyama, Y. Ando, *J. Mater. Chem.* 20 (2010) 4128.
[17] Z. Wang, C.H. Yu, D.C. Ba, J. Liang, *Vacuum* 81 (2007) 579.
[18] T. Cui, R. Lv, F. Kang, Q. Hu, J. Gu, K. Wang, D. Wu, *Nanoscale Res. Lett.* 5 (2010) 941.
[19] T. Cui, R. Lv, Z.-h. Huang, F. Kang, K. Wang, D. Wu, *Nanoscale Res. Lett.* 6 (2011) 77.
[20] P. Ayala, A. Grüneis, T. Gemming, B. Büchner, M.H. Rümmeli, D. Grimm, J. Schumann, R. Kaltofen, F.L.F. Jr., H.D.F. Filho, T. Pichler, *Chem. Mater.* 19 (2007) 6131.
[21] M. He, S. Zhou, J. Zhang, Z. Liu, C. Robinson, *J. Phys. Chem.* B 109 (2005) 9275.

[22] Y. Hao, L. Qingwen, Z. Jin, L. Zhongfan, *Chem. Phys. Lett.* 380 (2003) 347.
[23] A.G. Kudashov, A.V. Okotrub, L.G. Bulusheva, I.P. Asanov, Y.V. Shubin, N.F. Yudanov, L.I. Yudanova, V.S. Danilovich, O.G. Abrosimov, *J. Phys. Chem.* B 108 (2004) 9048.
[24] R. Xue, Z. Sun, L. Su, X. Zhang, *Catal. Lett.* 135 (2010) 312.
[25] H.C. Choi, J. Park, B. Kim, *J. Phys. Chem.* B 109 (2005) 4333.
[26] L. Zhi, T. Gorelik, R. Friedlein, J. Wu, U. Kolb, W.R. Salaneck, K. Mullen, *Small* 1 (2005) 798.
[27] X. Wang, Y. Liu, D. Zhu, L. Zhang, H. Ma, N. Yao, B. Zhang, *J. Phys. Chem.* B 106 (2002) 2186.
[28] Q. Yang, W. Xu, A. Tomita, T. Kyotani, *J. Am. Chem. Soc.* 127 (2005) 8956.
[29] Y. Zhang, L. Pan, B. Wen, X. Song, C. Liu, T. Li, *J. Mater. Res.* 26 (2011) 443.
[30] W.-X. Lv, R. Zhang, T.-L. Xia, H.-M. Bi, K.-Y. Shi, *J. Nanopart. Res.* 13 (2011).
[31] J. Liu, R. Czerw, D.L. Carroll, *J. Mater. Res.* 20 (2005) 538.
[32] E. Xu, J. Wei, K. Wang, Z. Li, X. Gui, Y. Jia, H. Zhu, D. Wu, *Carbon* 48 (2010) 3097.
[33] P. Ghosh, T. Soga, K. Ghosh, R.A. Afre, T. Jimbo, Y. Ando, *J. Non-Cryst. Solids* 354 (2008) 4101.
[34] J.W. Jang, C.E. Lee, S.C. Lyu, T.J. Lee, C.J. Lee, *Appl. Phys. Lett.* 84 (2004) 2877.
[35] C.J. Lee, S.C. Lyu, H.-W. Kim, J.H. Lee, K.I. Cho, *Chem. Phys. Lett.* 359 (2002) 115.
[36] C.F. Chen, C.L. Tsai, C.L. Lin, *Diamond Relat. Mater.* 12 (2003) 1500.
[37] M. Glerup, M. Castignolles, M. Holzinger, G. Hug, A. Loiseau, P. Bernier, *Chem. Commun.* (2003) 2542.
[38] A.A. Koo´s, F. Dillon, E.A. Obraztsova, A. Crossley, N. Grobert, *Carbon* 48 (2010) 3033.
[39] A.A. Koo´s, M. Dowling, K. Jurkschat, A. Crossley, N. Grobert, *Carbon* 47 (2009) 30.
[40] R. Zhang, L. Li, L. Chen, G. Zhang, K. Shi, *J. Alloys Compd.* 509 (2011) 8620.
[41] Y. Cao, Y. Zhao, Q. Jiao, *Mater. Chem. Phys.* 122 (2010) 612.
[42] Y. CAO, Y. ZHAO, Q. LI, Q. JIAO, *J. Chem. Sci.* 121 (2009) 225.
[43] Y.-L. Li, F. Hou, Z.-T. Yang, J.-M. Feng, X.-H. Zhong, J.-Y. Li, *Mater. Sci. Eng.* B 158 (2009) 69.
[44] P. Ghos, M. Subramanian, R.A. Afre, M. Zamri, T. Soga, T. Jimbo, V. Filip, M. Tanemura, *Appl. Surf. Sci.* 255 (2009) 4611.
[45] P. Ghosh, M. Zamri, M. Subramanian, T. Soga, T. Jimbo, R. Katoh, M. Tanemura, *J. Phys. D: Appl. Phys.* 41 (2008) 155405.
[46] X.Y. Tao, X.B. Zhang, F.Y. Sun, J.P. Cheng, F. Liu, Z.Q. Luo, *Diamond Relat. Mater.* 16 (2007) 425.
[47] D.H. Lee, W.J. Lee, S.O. Kim, *Nano Lett.* 9 (2009) 1427.
[48] E.N. Nxumalo, V.o. Nyamori, N.J. Coville, *J. Organomet. Chem.* 693 (2008) 2942.
[49] C. Tang, Y. Bando, D. Golberg, F. Xu, *Carbon* 42 (2004) 2625.
[50] B.G. Sumpter, V. Meunier, J.M. Romo-Herrera, E. Cruz-Silva, D.A. Cullen, H. Terrones, D.J. Smith, M. Terrones, *ACS Nano* 1 (2007) 369.
[51] K.C. Mondal, N.J. Coville, M.J. Witcomb, G. Tejral, *J. Havel, Chem. Phys. Lett.* 437 (2007) 87.
[52] L.S. Panchakarla, A. Govindaraj, C.N.R. Rao, *ACS Nano* 1 (2007) 494.
[53] S.m. Enouz-Ve´drenne, O. Ste´phan, M. Glerup, J.-L. Cochon, C. Colliex, A. Loiseau, *J. Phys. Chem.* C 112 (2008) 16422.

[54] E. Cruz-Silva, D.A. Cullen, L. Gu, J.M. Romo-Herrera, E. Muñoz-Sandoval, F. López-Urías, B.G. Sumpter, V. Meunier, J.-C. Charlier, D.J. Smith, H. Terrones, M. Terrones, *ACS Nano* 2 (2008) 441.
[55] Y. Chen, J. Wang, H. Liu, M.N. Banis, R. Li, X. Sun, T.-K. Sham, S. Ye, S. Knights, *J. Phys. Chem.* C 115 (2011) 3769.
[56] M.S. Dresselhaus, P.C. Eklund, *Adv. Phys.* 49 (2000) 705.
[57] P.H. Matter, L. Zhang, U.S. Ozkan, *J. Catal.* 239 (2006) 83.
[58] J.R. Pels, F. Kapteijin, J.A. Moulijn, Q. Zhu, K.M. Thomas, *Carbon* 33 (1995) 1641.
[59] K. Ghosh, M. Kumar, T. Maruyama, Y. Ando, Carbon 47 (2009) 1565.
[60] H.C. Choi, S.Y. Bae, W.-S. Jang, J. Park, H.J. Song, H.-J. Shin, S. Jung, J.-P. Ahn, *J. Phys. Chem.* B 109 (2005) 1683.
[61] J. Zhou, J. Wang, H. Liu, M.N. Banis, X. Sun, T.-K. Sham, *J. Phys. Chem. Lett.* 1 (2010) 1709.
[62] M.S. Saha, R. Li, X. Sun, S. Ye, *Electrochem. Commun.* 11 (2009) 438.
[63] S. Sun, G. Zhang, Y. Zhong, H. Liu, R. Li, X. Zhou, X. Sun, *Chem. Commun.* (2009) 7048.
[64] Y. Chen, J. Wang, H. Liu, R. Li, X. Sun, S. Ye, S. Knights, *Electrochem. Commun.* 11 (2009) 2071.
[65] R.I. Jafri, N. Rajalakshmi, S. Ramaprabhu, *J. Power Sources* 195 (2010) 8080.
[66] W.-C. Fang, *Nanoscale Res. Lett.* 5 (2010) 68.
[67] D.C. Higgins, D. Meza, Z. Chen, *J. Phys. Chem.* C 114 (2010) 21982.
[68] Z. Zhu, J. Wang, A. Munir, H.S. Zhou, *Electrochim. Acta* 55 (2010) 8517.
[69] G. Vijayaraghavan, K.J. Stevenson, *Languir.* 23 (2007) 5279.
[70] R. Chetty, S. Kundu, W. Xia, M. Bron, W. Schuhmann, V. Chirila, W. Brandl, T. Reinecke, M. Muhler, *Electrochim. Acta* 54 (2009) 4208.
[71] S. Jiang, L. Zhu, Y. Ma, X. Wang, J. Liu, J. Zhu, Y. Fan, Z. Zou, Z. Hu, *J. Power Sources* 195 (2010) 7578.
[72] R. Lv, T. Cui, M.-S. Jun, Q. Zhang, A. Cao, D.S. Su, Z. Zhang, S.-H. Yoon, J. Miyawaki, I.M. Kang, *Adv. Funct. Mater.* 21 (2011) 999.
[73] X. Wang, H. Xue, L. Yang, H. Wang, P. Zang, X. Qin, Y. Wang, Y. Ma, Q. Wu, Z. Hu, *Nanotechnology* 22 (2011) 395401.
[74] C.-L. Sun, Y.-K. Hsu, Y.-G. Lin, K.-H. Chen, C. Bock, B. MacDougall, X. Wu, L.-C. Chene, *J. Electrochem. Soc.* 156 (2009) A1249.
[75] D. Geng, H. Liu, Y. Chen, R. Li, X. Sun, S. Ye, S. Knights, *J. Power Sources* 196 (2011) 1795.
[76] H. Li, H. Liu, Z. Jong, W. Qu, D. Geng, X. Sun, H. Wang, *Int. J. Hydrogen Energy* 36 (2011) 2258.
[77] K. Gong, F. Du, Z. Xia, M. Durstock, L. Dai, *Science* 323 (2009) 760.
[78] D. Yu, Q. Zhang, L. Dai, *J. Am. Chem. Soc.* 132 (2010) 15127.
[79] W. Xiong, F. Du, Y. Liu, A.P. Jr., M. Supp, T.S. Ramakrishnan, L. Dai, L. Jiang, *J. Am. Chem. Soc.* 132 (2010) 15839.
[80] Z. Chen, D. Higgins, Z. Chen, *Carbon* 48 (2010) 3057.
[81] Z. Chen, D. Higgins, Z. Chen, *Electrochim. Acta* 55 (2010) 4799.
[82] D. Higgins, Z. Chen, Z. Chen, *Electrochim. Acta* 56 (2011) 1570.
[83] T.C. Nagaiah, S. Kundu, M. Bron, M. Muhler, W. Schuhmann, *Electrochem. Commun.* 12 (2010) 338.

[84] C.V. Rao, C.R. Cabrera, Y. Ishikawa, *J. Phys. Chem. Lett.* 1 (2010) 2622.
[85] Z. Wang, R. Jia, J. Zheng, J. Zhao, L. Li, J. Song, Z. Zhu, *ACS Nano* 5 (2011) 1677.
[86] N. Alexeyeva, E. Shulga, V. Kisand, I. Kink, K. Tammeveski, *J. Electroanal. Chem.* 648 (2010) 169.
[87] L. Yang, S. Jiang, Y. Zhao, L. Zhu, S. Chen, X. Wang, Q. Wu, J. Ma, Y. Ma, Z. Hu, Angew. *Chem. Int. Ed.* 50 (2011) 7132.
[88] M. Lefevre, J.P. Dodelet, P. Bertrand, *J. Phys. Chem.* B 106 (2002) 8705.
[89] X. Hu, Y. Wu, H. Li, Z. Zhang, *J. Phys. Chem.* C 114 (2010) 9603.
[90] S. Yang, G.-L. Zhao, E. Khosravi, *J. Phys. Chem.* C 114 (2010) 3371.
[91] B. Shan, K. Cho, *Chem. Phys. Lett*. 492 (2010) 131.
[92] Y. Li, J. Wang, X. Li, J. Liu, D. Geng, J. Yang, R. Li, X. Sun, *Electrochem. Commun.* 13 (2011) 668.
[93] S. Zhu, Z. Chen, B. Li, D. Higgins, H. Wang, H. Li, Z. Chen, *Electrochim. Acta* 56 (2011) 5080.
[94] X. Li, J. Liu, Y. Zhang, Y. Li, H. Liu, X. Meng, J. Yang, D. Geng, D. Wang, R. Li, X. Sun, *J. Power Sources* 197 (2012) 238.
[95] D.Y. Zhong, G.Y. Zhang, S. Liu, E.G. Wang, Q. Wang, H. Li, X.J. Huang, *Appl. Phys. Lett*. 79 (2001) 3500.
[96] I. Mukhopadhyay, N. Hoshino, S. Kawasaki, F. Okino, W.K. Hsu, H. Touhara, *J. Electrochem. Soc*. 149 (2002) A39.
[97] L.G. Bulusheva, A.V. Okotrub, A.G. Kurenya, H. Zhang, H. Zhang, X. Chen, H. Song, *Carbon* 49 (2011) 4013.

In: Applications of Carbon Nanotubes
Editor: Ajay Kumar Mishra
ISBN: 978-1-62081-918-0

Chapter 10

ZINC-RICH PAINT COATINGS CONTAINING EITHER IONIC SURFACTANT-MODIFIED OR FUNCTIONALIZED MULTI-WALLED CARBON NANOTUBE-SUPPORTED POLYPYRROLE UTILIZED TO PROTECT COLD-ROLLED STEEL AGAINST CORROSION

András Gergely[1,*], Tamás Török[2], Zoltán Pászti[3], Imre Bertóti[4], Judith Mihály[5] and Erika Kálmán[1]

[1]Department of Surface Modification and Nanostructures, Institute of Nanochemistry and Catalysis, Chemical Research Center of the Hungarian Academy of Sciences, Budapest, Hungary
[2]Department of Metallurgical and Foundry Engineering, Faculty of Materials Science and Engineering, University of Miskolc, Hungary
[3]Laboratory of Hydrogen Energy, Institute of Nanochemistry and Catalysis, Chemical Research Center of the Hungarian Academy of Sciences, Budapest, Hungary
[4]Department of Plasma Chemistry, Institute of Materials and Environmental Chemistry, Chemical Research Center of the Hungarian Academy of Sciences, Budapest, Hungary
[5]Department of Biological Nanochemistry, Institute of Nanochemistry and Catalysis, Chemical Research Center of the Hungarian Academy of Sciences, Budapest, Hungary

ABSTRACT

The intense anodic action of sacrificial zinc pigments ensured viable galvanic function of the highly porous liquid zinc-rich paints (ZRPs) result in deteriorated long-term corrosion resistance often accompanied by cathodic delamination phenomena. In

[*] E-mail: doohan11@chemres.hu.

our approach, such a efficacy problem related to the corrosion preventive function of ZRPs is addressed by the application of intimately structured anodic inhibitor particles composed of nano-size alumina and either polyelectrolyte-modified or chemically functionalized multi-walled carbon nanotubes (MWCNT) supported polypyrrole (PPy) in one specific zinc-rich hybrid paint formulation providing balanced active–passive protective functionality.

High dispersity of the nanotube-free PPy-deposited inhibitor particles (PDIPs) with uneven polymer distribution on the alumina carrier was confirmed by transmission electron microscopy (TEM) observations. Furthermore, the MWCNT-embedded PDIPs indicated almost complete surface coverage of the alumina-nanotube carriers by PPy with decreased microstructure dispersity which is attributed to the effect of double-flocculants type co-deposition of the oppositely charged polymers causing coalescence of the modified particles. Depending on the amount of the nanotubes and their proportion to the quantities of the deposited PPy and polyelectrolyte as well as the concentration of the surfactant, varied micron-scale association of the PDIPs in the suspensions of dissolved alkyd matrix was disclosed by rheology characterization carried out at particular solid contents similar to hybrid paint formulation. The evenly distributed but less densely packed nano-structure of PPy was evidenced on the polyelectrolyte-modified nanotubes by Fourier-transform infrared (FTIR) spectroscopy whereas more compact polymer film formation was confirmed on the surface of functionalized nanotubes. According to the greater electrical conductivity, enhanced electroactivity and reversibility of the nanotube-embedded PDIPs were indicated over the nanotube-free particles by cyclic voltammetry, depending on the type and the amount of the nanotubes and their modification.

Protection function of the hybrid paint coatings (formulated with spherical zinc pigment at 70 wt.%) was investigated by immersion and salt-spray chamber tests over 254 and 142 day periods, respectively. Firm barrier nature of the nanotube-embedded PDIP contained hybrids was proved by electrochemical impedance spectroscopy (EIS) and radio-frequency glow-discharge optical-emission-spectroscopy (RF-GD-OES). Furthermore, due to the increased conductivity of the nanotube-embedded PDIPs cemented in epoxy primers optimally at 0.4 and 0.6 wt.%, altered corrosion preventive behaviour of the hybrid coatings was indicated by the positively polarized open-circuit potentials (OCPs) and the X-ray photoelectron spectroscopy (XPS) detected lower relative quantities of the interfacially accumulated zinc corrosion products, moderate oxidative degradation of the epoxy vehicle. Decreasing oxidative conversion of iron at the surface was indicated by XPS found to correlate with the increasing intensity of zinc corrosion and decreasing oxidative degradation of the epoxy binder, according to the higher nanotube contents of hybrid coatings. In addition, inhibited zinc corrosion caused low rate of oxidative degradation of epoxy, allowing increased durability of coating adhesion and cohesion thereby ensuring reliable protection by zinc-rich compositions.

As a conclusion, modified or functionalized MWCNTs acting as unexchangeable doping agents promote enhanced reversibility and increased conductivity of PPy, forming nano-size inhibitor particles with advanced features. Thus, such inhibitor nano-particles in zinc-rich hybrid compositions afford improved barrier and high efficiency galvanic–cathodic corrosion preventive function, exceeding long-term protection capability of the conventional ZRPs.

INTRODUCTION

The number of applications related to ZRPs is due to the exceptional protection behaviour and preventive performance afforded to steel alloys against corrosion. Traditional liquid ZRPs comprise around 90 wt.% of spherical anodic zinc pigments to ensure viable cathodic protection of metallic substrates. The diminishing galvanic function over long test periods is associated with the decreased surface area ratio of zinc and the substrate as a consequence of the intense zinc corrosion and the loss of electric contacts between anodic pigments [1]. However, passive protection function is sustained based on the enhanced barrier nature of the coating reinforced by the accumulated zinc corrosion products. This is due to the fact that zinc oxide passivates the surface of zinc pigments and inhibits the chloride-induced corrosion by sealing coating pores [2]. Thus, several zinc-rich compositions were investigated in detail examining the effect of the binder [**3**], focusing on the size and shape of the anodic pigments [4-7], as well as paint formulations with various pigment compositions [8]. Lots of work has been devoted to improve protection function of the ZRPs by delaying the fading strong cathodic polarization – galvanic immunization of metallic substrates [9,10], achieving progress by alternative formulations [11,12] using organic [13,14] or inorganic binders [15,16]. Amid strict environmental regulations, any development in the field of metal-rich coatings is urged to take place to lessen the amount of zinc in primers, owing to toxicity reasons [17]. On the other hand, this is required to undergo to meet industry demand for wide selection of reliable choices whilst targeting growing consumer interests for high performance solutions.

Despite the reliable corrosion resistive behaviour, there are number of factors limiting the protection durability of ZRPs [8,11] such as electrical conductivity of the primers. Furthermore, ionic conductivity of the coatings alters the accessibility of anodic pigments since their corrosion is varied, whereas additives can mitigate the effects of organic vehicle degradation processes. Having a strong influence on active–passive function of the ZRPs [9], percolation and porosity of the primers are connected to the pigment volume concentration (PVC) and its ratio to the critical CPVC which depends on the morphology of anodic pigments having enormous effect on the protection performance [15]. Although, lamellar pigments are better than spherical ones but their combination lead to coatings performing even better because of the resultant feature of low electrical percolation threshold and the increased surface energy based intense anodic action of pigments and the related enhanced galvanic function of the coatings [12]. Primers containing zinc above 80 wt.% provides fine active protection but contents between 60 and 70 wt.% showed no valuable barrier nature or galvanic action [11]. Furthermore, the efficacy of long-term galvanic function may utterly be impaired because of the early deadhesion phenomenon (cohesive failure of binder at the interface [18]) recognized as cathodic delamination. The corrosion-induced cathodic deadhesion of ZRPs is a result of interfacial electron transfer reactions, evolution of oxidative intermediates and the formation of nucleophile agents (in the course of cathodic reactions) along with the accumulation of corrosion products. Reactive agents provoke oxidative degradation of the organic vehicles in the forms of growing yield of oxidized spices, mainly carboxyl moieties. Nonetheless, cathodic delamination can be controlled by additives mitigating the effect of alkaline pH on the organic binders [19].

Aside from addressing the porosity related issues of the ZRPs by introducing intrinsically conducting polymers (ICPs) as auxiliary inhibitors helping to substitute a part of the high zinc content [20], further improvements have been made using free-standing ICPs [21] in zinc-rich compositions obtaining organic–inorganic hybrid coatings with improved protection functionalities [20,22]. Enhanced barrier nature of the hybrids and the inhibited sacrificial and self-corrosion of anodic pigments [23,24] are attributed to the effect of anodic ICPs applied either as carrier supported additives in inorganic matrix [25] or stand-alone alternative in zinc-rich powder coatings [26]. The notion to combine ICPs with anodic pigments in hybrids arises from retaining galvanic and enhancing barrier nature of the ZRPs [22]. As a consequence, PPy is a worthy candidate to be utilized in hybrid coatings due to their ability to inhibit zinc corrosion [27] and enhance coating barrier functionality especially in the form of highly dispersed, nano-sized particles [26,28]. Inhibition function of PPy usually requires its partially oxidized form doped with unexchangeable counter ions [29,30]. On the other hand, undoped ICPs are more beneficial to be used in coatings [31,32] due to the advantage of optimal compatibility with the organic vehicles. However, long-term protection by PPy faces tough challenges in terms of rapidly fading conductivity and redox-activity. In addition, one of the toughest challenges related to coating preparations is the dispersity of ICPs in various matrices [33] which can be resolved by using carrier-supported polymers [34-36] to obtain highly dispersed particles with increased electroactivity, conductivity and processibility [37] with improved corrosion protection abilities [38].

In these aspects, nano-size alumina and MWCNT-supported PPy is applied in epoxy paint coatings to obviate any wetting or dispersability issues. Polyelectrolyte modification and functionalisation of the nanotubes serve as an unexchangeable polymer doping which in turn promotes matrix compatibilization, high dispersity and even distribution of the inhibitor particles.

In this chapter, the application of nano-size anodic inhibitor particles is investigated in zinc-rich paint compositions analyzing the contribution of the various particles to the protection mechanism of zinc-rich primers. Thus, experimental results are presented focusing on the micro and nanostructure, spectroscopic characteristics and electroactivity of the polypyrrole-modified inhibitor particles. Corrosion resistive function of the hybrid coatings is discussed to gain adequate interpretation about the synergistic effects of the MWCNT-embedded PDIPs contained paint compositions.

EXPERIMENTAL

Preparation of Polypyrrole-Deposited Alumina/MWCNT Inhibitor Particles

The 1-amino-2-naphthol-4-sulfonic acid was used as precursor to functionalise NC7000 nanotube (Nanocyl s.a., Belgium) with naphthol-sulfonic acid through diazotation activation and coupling procedure. Decomposition of the diazonium salt was initiated by quantitative amount of ammonium iron(II) sulphate. The reaction was carried out in aqueous acetic acid at 0 °C for 12 h. Then the solid was filtrated, washed thoroughly with essential acetic acid and dried. Chemically grafted nanotubes were dispersed in oleum (20 wt.%) by 10 min ultrasonication. Then the mixture was vigorously stirred and refluxed for either 2 or 5 h at 130 °C to alter the extent of nanotube functionalisation. Finally, the suspension was poured onto ice, neutralized with sodium carbonate solution and the solid was dialyzed for three times. Thus, low and highly sulfonated nanotube samples were obtained which are referred to as LS and HS NC7000, respectively. These samples contained sulphur atom at 0.4 and 1 atomic % detected by XPS in the form of carbon bound sulfonate and sulphate groups.

Modification of the NC7000 and MWNT-30 (Shenzhen Nanotech Port Co. Ltd., China) nanotubes was performed by 30 min ultrasonication in aqueous ethanol and acetic acid dispersion (volume mixture of 10:1:3) containing polyelectrolyte like poly(4-ammonium styrene-sulfonic acid, Aldrich) and ionic surfactants such as sodium dodecyl sulphate (Aldrich) referred to as PSS and SDS, respectively. For the preparation of inhibitor particles, pyrrole (Aldrich) was dissolved in the water dispersed sol of fumed alumina (AluC, Evonik Industries AG, Germany) and the colloid was stirred intensively for 2 h. In the case of sample *p2*, the procedure of polymerisation and deposition was carried out in aqueous ethanol (at a content of 13% by volume) dispersed alumina sol. As an oxidative agent of pyrrole monomer, the solution of iron(III) nitrate (Fluka) was added to the vigorously stirred sol in one step. When the mixture became homogeneous, 1M nitric acid solution (~1 ml) was added immediately to set the pH to 3 (for the sake of optimally balanced conditions to realize highly dispersed and stable alumina sol, moderate rate of pyrrole polymerisation with compact PPy structure and high deposition yield). Then the mixture was stirred slowly for 6 h then let stagnant for 16 h. Suspensions were filtrated and the solid was washed 12 times (to remove SDS) then particles were dried and ground. Composition of the inhibitor samples is summarised in Table 1.

Preparation of Paint Coatings

Paint coatings were layered on standard low-carbon, cold-rolled steel panels (RS type CRS, roughness; 25–65 micro-inches, complying with ASTM A1008.1010, A-109 and QQS-698 standards) were used as received from Q-Lab Ltd. PDIPs were dispersed by 20 min mortar grinding and milling in dissolved epoxy (at a polymer concentration of 50 wt.%).

Table 1. Elemental analysis determined component contents of PPy-deposited particles (wt.%) prepared at an initial molar ratio of iron(III)/pyrrole. Coulombic efficiency of PPy is derived from the voltammetry charging currents (on the average of five measurements) normalized according to the elemental analysis determined polymer contents of the PDIP samples based on nitrogen element

Samples	PSS and SDS in the preparation mixtures (in wt.% and mol dm^{-3})	Types of MWCNTs in preparation mixtures and PDIPs (wt.%)	PPy in preparation mixtures (mol dm^{-3}) and PDIPs (wt.%), deposition yield of PPy (%)	Charging efficiency of PPy (%)
p1	-	-	0.10, (4.1) 83	3.7×10^{-2} ($\pm4\times10^{-3}$)
p2	-	-	0.10, (3.2) 64	3.3×10^{-2} ($\pm3\times10^{-3}$)
p3	2.34×10^{-2}, 3.5×10^{-3}	MWCNT-30, 1.28 (17.5)	0.10, (3.3) 29	3.7 ($\pm4\times10^{-1}$)
p4	1.35×10^{-2}, 1.7×10^{-3}	LS NC7000, 0.84 (12.4)	0.12, (5.4) 42	3.9 ($\pm4\times10^{-1}$)
p5	2.21×10^{-2}, 2.1×10^{-3}	LS NC7000, 0.10 (1.6)	0.10, (7.1) 61	1.8 ($\pm2\times10^{-1}$)
p6	2.64×10^{-2}, 2.1×10^{-3}	HS NC7000, 0.20 (3.2)	0.10, (5.2) 40	1.7 ($\pm2\times10^{-1}$)

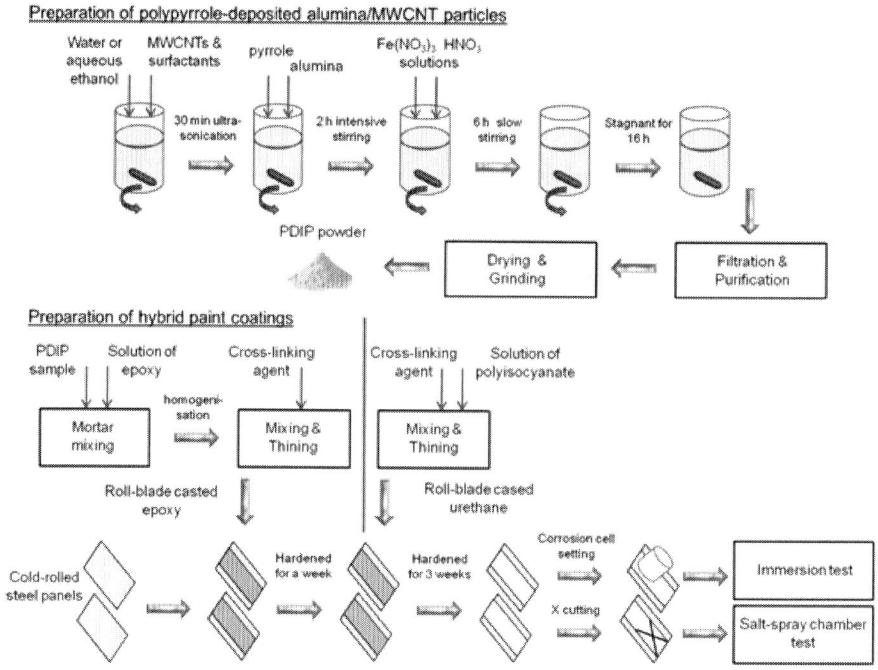

Figure 1. Representation of preparation procedures of the polypyrrole deposited alumina/multi-walled carbon nanotube particles and paint coatings.

Table 2. Composition of cured epoxy paint coatings in weight (wt.%) and volume fraction (φ), where hybrid C1–C6 contained PPy-deposited alumina/MWCNT particles at 3.21 wt.%

Coatings	PDIPs	PPy (φ) in the primer	PPy (φ) in epoxy resin	MWCNT (φ) in the primer	MWCNT (φ) in epoxy resin	Zinc
E	-	-	-	-	-	-
Z	-	-	-	-	-	90
ZA	-	-	-	-	-	70
C1	*p1*	0.133 (3.5×10^{-3})	0.456 (4.8×10^{-3})	-	-	70
C2	*p2*	0.103 (2.7×10^{-3})	0.354 (3.8×10^{-3})	-	-	70
C3	*p3*	0.114 (3.0×10^{-3})	0.392 (4.2×10^{-3})	0.417 (7.1×10^{-3})	1.428 (9.8×10^{-3})	70
C4	*p4*	0.186 (4.9×10^{-3})	0.637 (6.8×10^{-3})	0.429 (5.0×10^{-3})	1.472 (7.0×10^{-3})	70
C5	*p5*	0.242 (6.4×10^{-3})	0.830 (8.8×10^{-3})	0.056 (7.0×10^{-4})	0.191 (9.0×10^{-4})	70
C6	*p6*	0.175 (4.6×10^{-3})	0.600 (6.4×10^{-3})	0.111 (1.3×10^{-3})	0.381 (1.8×10^{-3})	70

The organic solvent was composed of 80:10:10 volume% of xylene (Fluka), 1-methoxy-2-propanol (Fluka) and 2-butanon (Aldrich). Suspensions of PDIPs were mixed with component A – stabilized zinc-rich masterbatch (HZO Farbenzinkstaub, Norzinco GmbH) in epoxy (Epoxidharz CHS141, bisphenol-A epoxy resin, Prochema). Component B – poly(amido amine) cross-linking agent (Durepoxy H15VP, USNER) was added to the suspensions, stirred to homogenize and diluted to get desired wet-concentration. Primer coatings were roll-blade casted on steel panels in 90 μm wet thickness (35±5 μm dry thickness) and let cure for a week at room temperature. Top-coatings were layered on hardened epoxy in an overall wet thickness of 120 μm (80±5 μm dry thickness) with Macrynal SM 2810/75BAC hydroxyacrylic resin (CYTEC Industries Inc.) cured with aliphatic polyisocyanate resin: hexamethylene diisocyanate (Desmodur® N75MPA/X, Bayer Material-Science LLC) which was dissolved in the mixture of n-butyl acetate, xylene and isobutanol (2:2:1). Coatings were let dry and cure for three weeks at room temperature. Scribes were made by a blade as X-cuts.

Schematic representation of preparation of PDIPs and hybrid paint coatings are illustrated in Figure 1. In addition, Table 2 summarizes the composition of paint coatings.

CHARACTERIZATION OF POLYPYRROLE DEPOSITED ALUMINA/MWCNT PARTICLES

Elemental Analysis

Elemental analysis was performed by a vario EL *III* micro and macro CHNOS elemental analyzer (Elementar Analysensysteme GmbH, Germany) with about 4 mg sample.

Electrokinetic Potential

Zeta-potential characterization of the LS and HS NC7000 nanotube samples were performed in water by Malvern Zetasizer Nano ZS equipment, at dispersed phase concentration of 4×10^{-2}. Electrophoretic mobility of the particles was measured by Laser Doppler Velocimetry then Henry's equation was used to determine electrokinetic potentials based on the Smoluchowski approximation. The values of density (2300 kg m^{-3}) and refractive index (2.000) were used for MWCNTs, whereas refractive index (1.330), density (998.21 kg m^{-3}) and dynamic viscosity (1.002 Pas, at 20 °C) of water were used for calculations.

Table 3. Volume fraction (φ) of the rheology investigated dispersions of the nano-particles suspended in dissolved alkyd resin at solid contents of 3.846 wt.%. Thixotropy index and yield stress determined by constant rate and force mode measurements are summarised

Particles	Alumina (φ)	PPy (φ)	MWCNT (φ)	Thixotropy (Pa s–1)	σ (Pa)
alumina	1.67×10^{-2}	-	-	1535	~0.5
p1	1.60×10^{-2}	1.63×10^{-3}	-	5024	9.7
p2	1.62×10^{-2}	1.27×10^{-3}	-	34591	3.4
p3	1.21×10^{-2}	1.40×10^{-3}	3.30×10^{-3}	634615	6.4
p4	1.31×10^{-2}	2.28×10^{-3}	2.34×10^{-3}	4340	1.5
p5	1.45×10^{-2}	2.96×10^{-3}	3.0×10^{-4}	72864	1.3
p6	1.44×10^{-2}	2.14×10^{-3}	6.0×10^{-4}	81720	18.6

Measurements indicated electrokinetic potentials of –24.7 and –31.4 mV of the low and the highly sulfonated nanotube samples at pH of 6.

Transmission Electron Microscopy

PDIP samples were crushed under ethanol and deposited onto copper grids covered by carbon supporting films. Transmission electron microscopy (TEM) data were acquired with a FEI MORGAGNI 268(D) microscope (FEI Co., The Netherlands) operated at 100 kV, with tungsten filament (top-entry; point-resolution = 0.5 nm).

Rheology

Measurements were done by a HAAKE RheoSress RS1 rotational rheometer (Thermo Fisher Scientific Inc., USA) equipped with a cone/plate sensor (2° angle, Ø 35 mm) used at a temperature of 25±0.3 °C. After three days swelling, PDIPs were dispersed by half an hour mortar grinding in of *n*-butanol and xylene (7:93) dissolved alkyd resin (distilled tall oil, SYLVATAL™ 25/30LT, Arizona Chemical, Gersthofen, Germany). Composition of the colloid mixtures are presented in Table 3.

Fourier-Transform Infrared Spectroscopy

PPy structure in *p1*–*p6* PDIP samples was investigated by Fourier transform infrared (FTIR) spectroscopy using a Varian Scimitar 2000 FT-IR spectrometer (Agilent Technologies, Inc., USA) equipped with a broadband Mercury Cadmium Telluride (MCT) detector and a 'Golden Gate' attenuated total reflection (ATR) unit with diamond ATR element. 128 scans were co-added with a spectral resolution of 4 cm^{-1}. All spectra were ATR corrected using the Varian Resolution Pro 4.0 spectral handling software package.

Cyclic Voltammetry

Three-electrode cell configuration was applied for the electrochemical measurements. A platinum spiral ($A \approx 13$ cm^2) and saturated calomel electrode (SCE) were used as counter and reference electrodes, respectively. A platinum disc working electrode (diameter of 13 mm) was incorporated in a Teflon tube. Pressed pellets of PDIPs were attached to the working electrode by a Teflon bracket and wetted with 0.1 M acetonitrile solution of sodium perchlorate (Aldrich) used as electrolyte for the measurements. PPy contents of the pellets were reduced at –0.4 V for 30 min, then voltammetry scanning was performed at a rate of 10 mV s^{-1}. Current transients are normalized to unit mass of the PDIPs and the surface of working electrode. Prior to repeatability of the measurements and to avoid limitations in ionic mass transport, around 0.15 g of the samples was pressed with a maximum pressure of 250 kg cm^{-2}. Half an hour dwell-time preceded voltammetry investigation to ensure proper wetting and swelling of the sample pellets.

CHARACTERIZATION OF PAINT COATINGS

Electrochemical Impedance Spectroscopy

A platinum spiral, SCE and glass-tube-fitted paint coatings covered steel panels (A=5.3 cm^2) were used as counter, reference and working electrodes, respectively. The electrochemical cell and the Zahner IM6eX potentiostat–frequency analyzer (Zahner-elektrik GmbH and Co., Germany) were placed into two grounded Faraday cages wired to each other. OCPs were measured in an open cell configuration. Then impedance measurements were carried out in the frequency range of 10 kHz and 20 mHz applying 20 mV sinusoidal perturbation according to the investigation principles of high resistance paint coatings [39].

Radio-Frequency Glow-Discharge Optical-Emission-Spectroscopy

RF-GD-OES analysis was started with sample preparation including the removal of urethane top coat by gentle polishing, leaving the epoxy primer intact (in a 40 μm thickness). Elemental depth profiling experiments were performed using a GD–Profiler 2™ radio-frequency glow-discharge optical-emission-spectrometer (RF-GD-OES) instrument (HORIBA Jobin Yvon S.A.S., France) equipped with a Paschen–Runge type polychromator (with 46 channels) optical system (focus distance: 0.50 m, diffraction grate: 2400 mm^{-1}, optical resolution: 18–25 pm) and a Cherny–Turner type monochromator (HR640, focus distance: 0.64 m) kept under nitrogen. Sample area was investigated under argon atmosphere at constant pressure of 500 Pa. Prior to high sensitivity detection of the corrosive analytes and coating elements, sputtering was performed at constant power (50 W) with voltage settings for the module (8.5 V) and the phase (3.0 V). Flushing time was 20 s at radio frequency plasma excitation of 13.56 MHz with an anode diameter of 4 mm. Depth profile data are presented over 40 μm of the epoxy primer coatings with an additional 20 μm sputtered crater in the steel substrates on average.

CHARACTERIZATION OF IMMERSION TESTED COLD-ROLLED STEEL SPECIMENS

X–Ray Photoelectron Spectroscopy

Immersion tested paint coatings on the steel panels were removed by three times of 5 min ultrasonication in the mixture of dichloromethane, methanol, toluene and tetrahydrofurane (6:1:1:1). XPS spectra were recorded from the steel surface on a Kratos XSAM 800 spectrometer operated at fixed analyzer transmission mode using Mg $K\alpha_{1,2}$ (1253.6 eV) excitation. Pressure of analysis chamber was lower than 1×10^{-7} Pa. Wide scan spectra were recorded in the 100–1300 eV kinetic energy range, applying pass-energy (80 eV) in 0.5 eV steps with 0.5 s dwell time. High-resolution spectra of characteristic photoelectron lines of constituent elements, and the C1s region for carbon-containing layers were recorded at pass-energy (40 eV) by 0.1 eV steps with 1 s dwell time. Chemical states of the constituents were determined by XPS database [40,41]. Kratos Vision 2000 software was used for quantifications with experimentally determined sensitivity factors. Omicron EA 125 type electron-spectrometer was employed with an excitation energy of AlKα (1486.6 eV) in the fixed analyzer transmission mode. Regions of interest were analyzed with the pass energy of 30 eV. CasaXPS [42] was applied to fit measured data and XPSMultiQuant [43,44] used for quantitative evaluation presuming homogeneous depth distribution of each component. Spectra were referenced to the energy of C1s line of polymeric or hydrocarbon type adventitious carbon set at 285.0 eV binding energy, and if it was applicable to the Fermi edge (0.0 eV binding energy) of the iron substrate.

Immersion and Salt-Spray Chamber Corrosion Tests

Immersion tests were carried out for 254 days with 1 M sodium chloride solution at room temperature, changing the electrolyte in a 42 day period. Corrosion tests were performed by a spray/prohesion/humidity cabinet (SF/MP/AB100, C+W Specialist Equipment Ltd., UK) installed with a Jun-Air (OF302, Gast Manufacturing Inc., USA) oil-free compressor. The 45 min cycles were divided into two phases. In the first 25 min, samples were exposed to salt fog combined with prohesion.

Dosage rate of the sodium chloride solution (5 wt.%) was at 7 cm^3 min^{-1}. Temperature of the chamber was set to 35 ^{o}C and the salt solution was heat up to 45±3 ^{o}C. In the second phase, the temperature was held at 35 ^{o}C for 20 min without application of salt mist. Test cycles were repeated for 142 days.

Results and Discussion

Investigation of Ppy-Deposited Alumina/Mwcnt Particles

Transmission Electron Microscopy

Compared to the unmodified alumina shown in Figure 2a, enhanced dispersity of the water-solvent-prepared *p1* sample presented in Figure 2b was experienced when it was casted from polar protic solvent like ethanol.

At higher resolution, the PPy-deposition caused coalescence of low number of the modified alumina particles at nano-scale is observed and pointed by arrow A on the microscopy image of Figure 2c. On the other hand, bridging type aggregation resulted in extensive, various sub-micron-scale aggregation of the PPy modified particles, signed by arrow B and C in the image. The thin layer of nearly uniform PPy coverage on the nano-size alumina carrier was also seen. In contrast, the aqueous-ethanol-solvent-prepared *p2* presented in Figure 2d exhibited partially globular polymer structure within the range of 30–70 nm with high micro-scale dispersity.

Observations revealed the all around surface coverage of alumina by PPy as it is seen in Figure 2e.

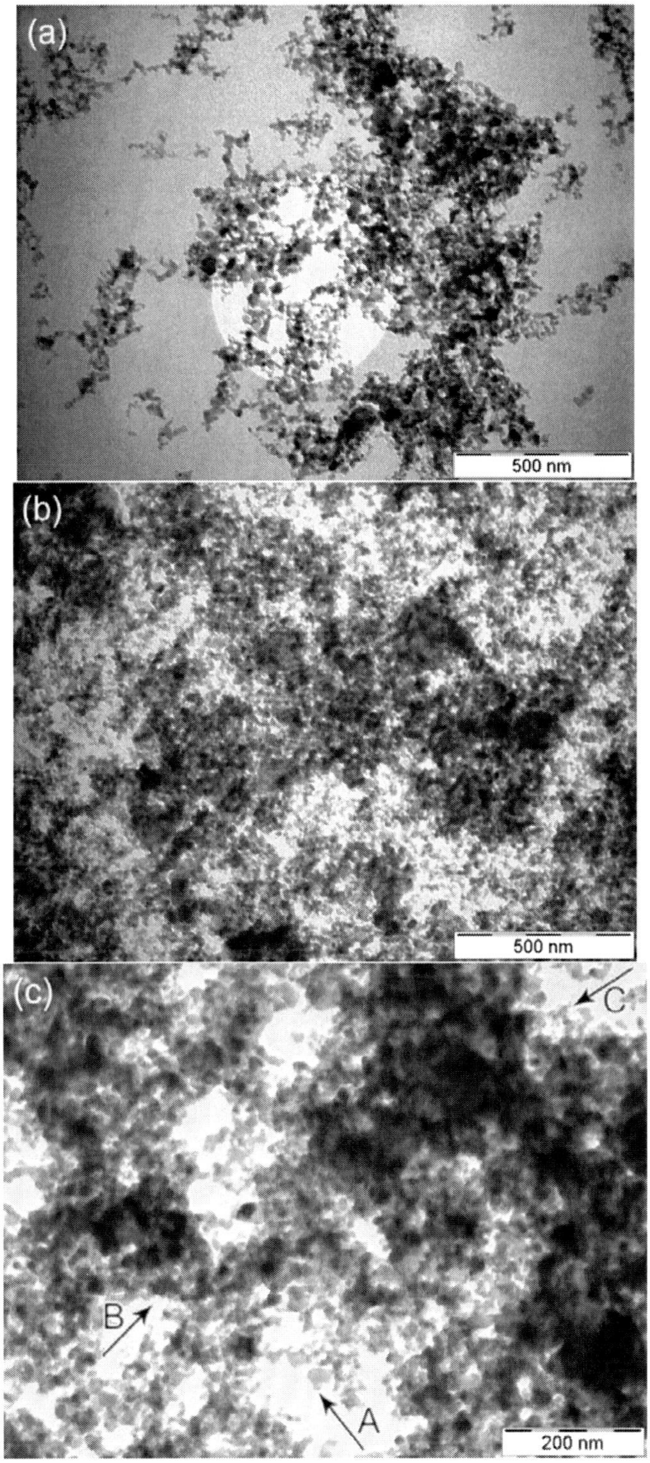

Figure 2. (Continued)

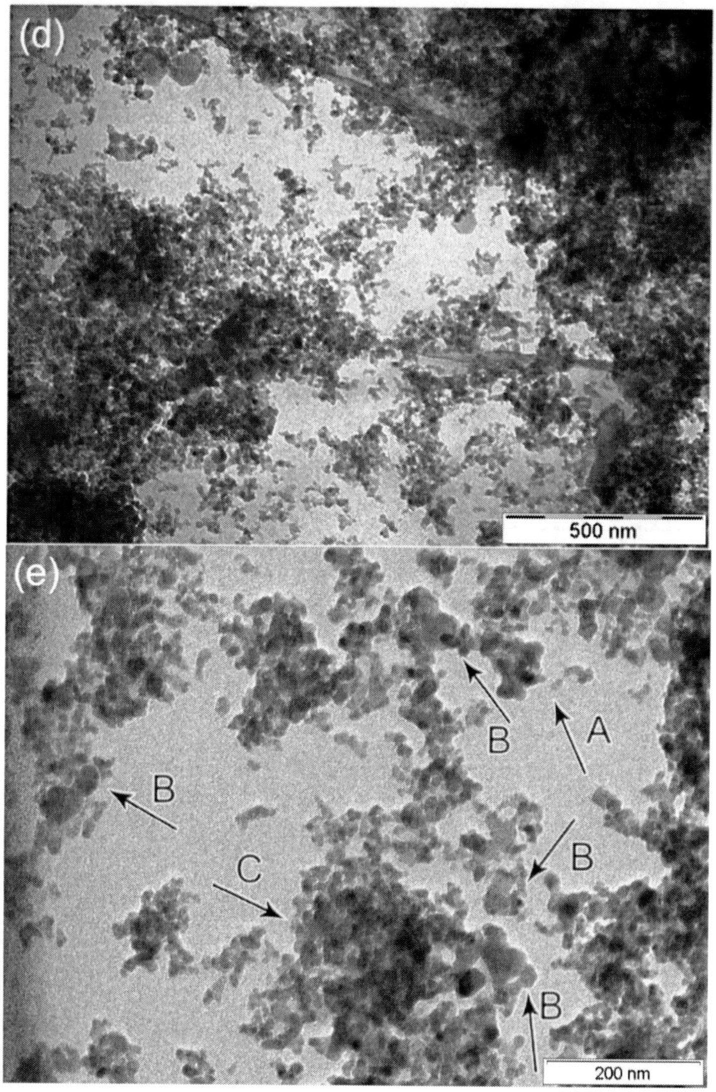

Figure 2. TEM images of (a) the nano-size boehmite alumina and the PPy-deposited alumina inhibitor particles: (b and c) *p1* prepared in water and (d and e) *p2* in aqueous ethanol.

The almost homodispersed sub-micron-scale structure of the *p2* contained individually dispersed PPy-modified particles in minor quantity (signed by arrow A) besides the majority of polymer coalescence caused moderately aggregated particles (noted by arrow B) and the 100 nm size agglomerated chunks (noted by arrow C). Increased dispersity of the PPy-deposited alumina is a result of homopolymer modification (organophilic modifier ensuring a certain range of entropy hindrance based steric repulsion) causing 'dispersant-effect' to prevent aggregation of the alumina carrier thereby supporting sufficient stabilization of highly dispersed composite systems. However, the moderate level of coalescence and polymer bridging type loose aggregation of the particles were manifested in both samples, according to the imperfect deposition and incomplete polymer surface coverage. Nonetheless, bridging type aggregation should be easily overcome by agitation and applied strain or shear-stress.

Figure 3a depicts the polyelectrolyte modified nanotube embedded *p3* sample indicating almost homogeneous sub-micron structure with perfect random distribution of the PPy-modified alumina around the disentangled nanotubes. This is a result of the application of PSS polyelectrolyte and the SDS surfactant ensured high exfoliation of the nanotubes in the preparation mixture. Nonetheless, the more pronounced coalescence and bridging aggregation of the modified alumina were also observed which was resulted by the co-deposition of oppositely charged polymers, i.e. PPy with its PSS dopant. On the other hand, a markedly improved quality of PPy deposition is observed on alumina with a nearly complete polymer-film formation on the modified nanotubes. Figure 3b shows completely coherent microstructure of the modified carriers in *p4* containing large amount of PPy on alumina and the low functionalized NC7000 nanotubes.

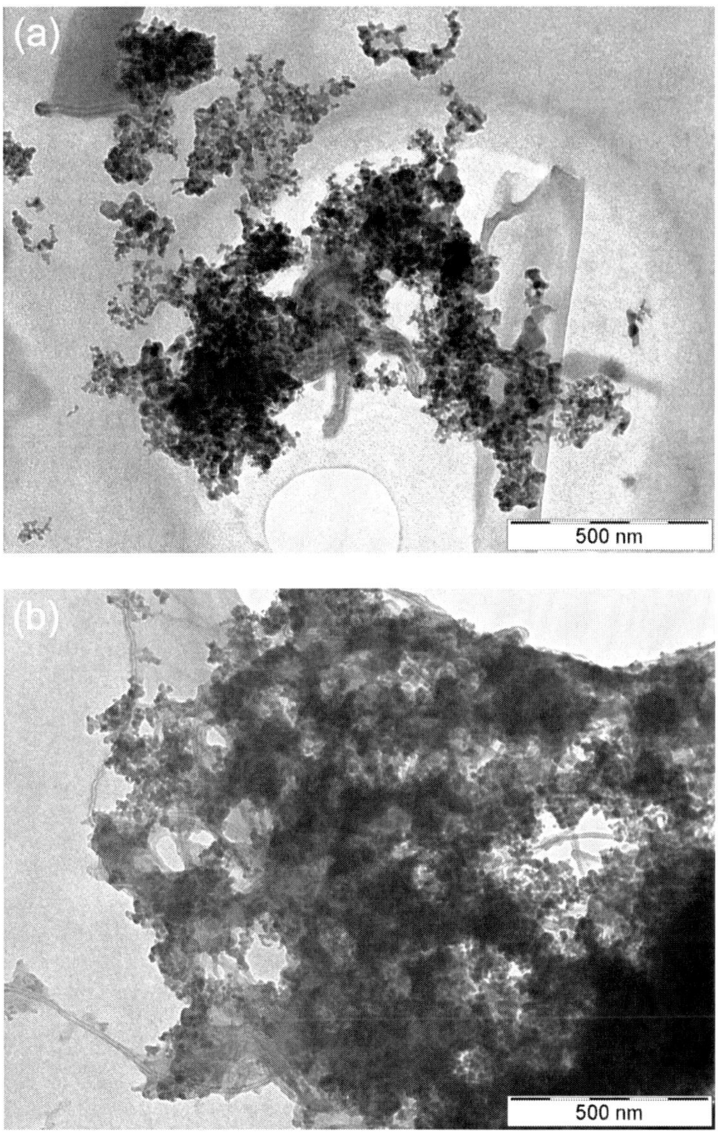

Figure 3. (Continued)

Figure 3. TEM images of the nanotube-embedded inhibitor particles embedded with: (a) *p3* PSS-modified MWCNT, (b and c) *p4* and *p5* low sulfonated MWCNT and (d) *p6* highly sulfonated MWCNT.

This sample indicated massive coalescence type aggregation of the modified particles. However, beyond the homogeneous micro-scale distribution of the carriers, the entire nanotube surface was coated with PPy, entailing in emerge of a more uniform properties of the inhibitor particles. The large amount of PPy and the polyelectrolyte modified small quantity of nanotubes embedded *p5* sample presented in Figure 3c exhibited voluminous nanotube-supported filaments with profound polymer bridging and moderate coalescence aggregation of the particles. Positive interparticle interactions within the bulky filaments rely on the even distribution of thin film co-deposited PPy and PSS on the carriers. Oxidative etching and sulfonate functionalization provided acidic character to the nanotube surface, contributing disentanglements and the formation of homogeneous PPy coverage along with the polyelectrolyte adsorption assisted PPy-anchoring on alumina. In the case of *p6* sample depicted in Figure 3d, homogeneous polymer coverage was obtained both on the alumina and the nanotube carriers. However, the intense coalescence related aggregation of the particles was caused by the effective co-deposition of increased relative amounts of the polymers (Table 1) taking place through primary adsorption of the negatively charged PSS and SDS on alumina with a subsequent anchoring of the oxidative polymerized and positively charged PPy both on the on alumina and the functionalized nanotubes. This procedure has lead to massive coalescence and interparticle bridging of the particles with a strongly flocculated microstructure based on the closely interacted complexes by the synergistic role of 'double flocculants'. Otherwise, beyond the distribution of sulfonated nanotubes in *p4–p6*, dispersity of the PPy-modified particles was found experimentally not comparable to *p1* and *p2,* despite the reversible dispersibility of such types of aggregation and flocculation. Hence, there is a compromise in the overall dispersity when nanotubes are incorporated along with other types of nano-size carriers to form such intimate structures. The preparation routine to obtain *p2* involved using ethanol (13 vol.%) to stabilize the sol of freely dispersed alumina by varying the polarity of dispersant [45] and to assist the formation of dense polymer structure by lowering surface tension of the solvent and increasing the solvation of PPy. Despite the slower polymerization rate and lower deposition yield of PPy, this method was applied because of the increased solubilizing ability allowed the formation of globular polymer structure as it was observed in *p4–p6*. The amount of PPy was set to limit the thickness of superficial polymer film in less than 20 nm on average to minimize the occupied volume in paint dispersions.

The hardship to synthesize high performance anodic inhibitor particles, e.g. fully covered alumina with thin and compact PPy layer, arises from the fact that pyrrole does not adsorb on alumina in moderately solubilizing mixtures and adsolubilized poorly upon the application of SDS. Furthermore, pyrrole may decrease the adsorption ability of SDS on alumina (atypical behaviour) [46]. Thus, beyond the necessity of using SDS and PSS to ensure highly dispersed colloids in order to obtain acceptable degree of nanotube distribution in the PDIPs, PSS was used as an adsorption promoter and immobile complexing polymer to facilitate PPy deposition. However, the use of ethanol as co-solvent should be avoided in that case because of the presence of polyelectrolyte and ionic surfactant. SDS surfactant as a high affinity competitive adsorbing anion assisted well the early stage of PPy deposition on alumina via self-assembly mechanism featured 'Region II' admicelle type solloid adsorption with the formation of interfacial colloidal aggregates through charging and growth [47].

Rheology

Rheology characterization of dispersions of the inhibitor particles was performed to investigate solid-particle-interactions which depend on the concentration, dispersity, extent of interacting surface area and interfacial chemical nature of the filler phase, altogether governing the interfacial bonding with the matrix. Aspect ratio (P) of rod-like particles defined as $P \approx L/D$ (where L and D are the length and diameter) highly affects the viscoelastic behaviour of dispersions [48]. In our case, this was around 100 for both types of the nanotubes with ~20 and 50 nm diameters, on average. In the coatings of *C3* and *C4,* volume fraction of the nanotubes in the epoxy resin reached geometrical semi-dilute range (higher than 5×10^{-3}, Table 2). Similarly, nanotube volume fractions were around that critical value in the rheology tested dispersions of the *p3* and *p4* PDIP samples (summarised in Table 3), at the same solid particle contents of 3.846 wt.%. On the other hand, dilute regime was valid for the *C5–C6* coatings and the rheology examined dispersions of *p5–p6* samples where volume fractions of the anisotropic particles remained under 5×10^{-3} (determined by the ratio of 2(D/4L)) [49]. However, the contents of alumina supported PPy along the part of the nanotube supported polymer in the epoxy resin of all hybrid paint coatings were around the threshold of geometrical semi-dilute range. This is because of the nano-size dimension of the particles related significantly decreased range of geometrical percolation, apart from the additional beneficial effects of the inhomogeneous particle distribution such as aggregation and flocculation of the alumina supported PPy causing lowered rheological and electrical percolation thresholds. Hence, explicit interaction of the nanotubes and their filaments are expected to be negligible in the *C5* and *C6,* whereas regarded as a long-range interaction in the semi-dilute type *C3* and *C4* paint coatings. According to the scope of this investigation, basic rheology study was performed to reveal the microstructure and interaction of the inhibitor particles. To draw appropriate conclusion about the corrosion resistivity of the *C1–C6* coatings containing various types of functional particles, solid contents of the rheology tested dispersions were chosen around the lower range of corrosion tested paint formulations.

To compare flow behaviour of the dispersions of the inhibitor particles, dissolved alkyd resin was also characterized. Almost ideal Newtonian flow behaviour of dissolved alkyd polymer and its dispersion with bare alumina was reflected by the wide-frequency range invariant complex viscosity, linear ramp of storage and loss moduli as well as the little changes in the phase angle (Figure 4a). Viscous flow character of the dispersion is associated with the absence of coherent microstructure.

High surface/volume ratio of the alumina is accounted for the increased dynamic viscosity of the dispersion at low solid content as a result of the high volume exclusion effect of the nano-size filler. Such character changed enormously by some nm thick water-solvent-prepared PPy layer was deposited on alumina and dispersed at the same solid concentration as bare alumina.

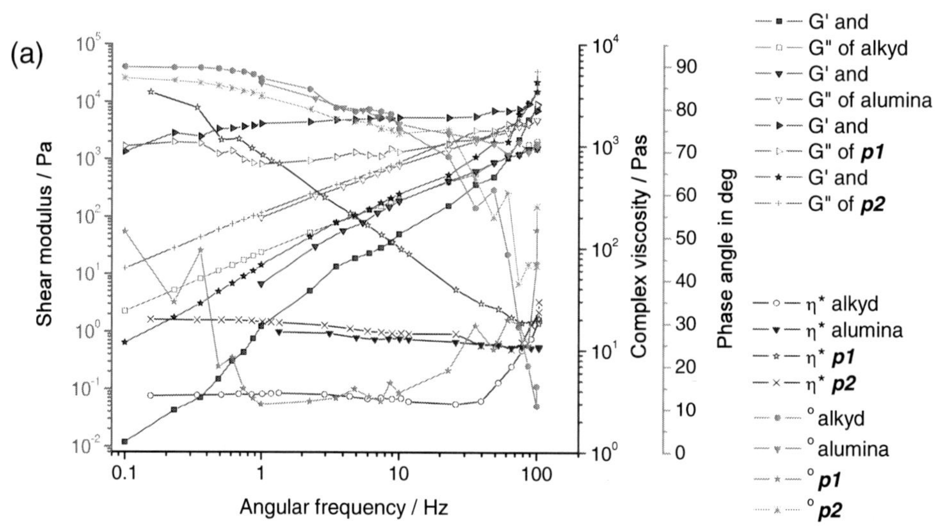

Figure 4. (Continued)

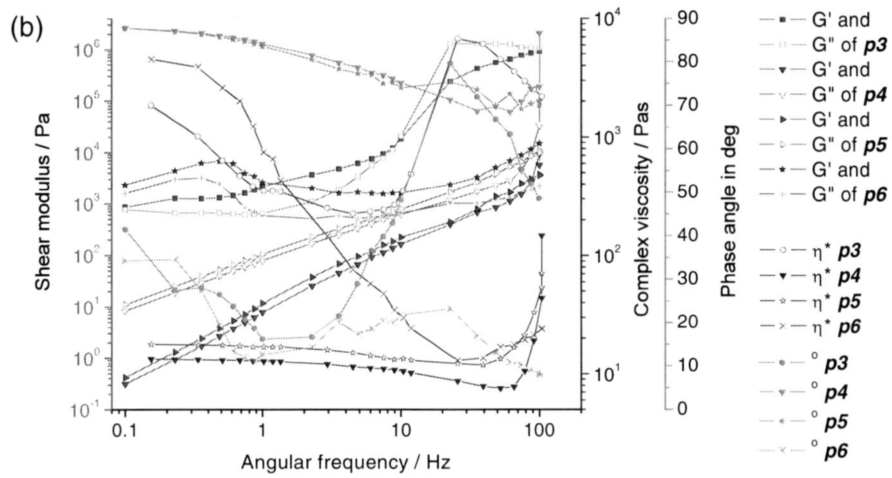

Figure 4. Frequency sweep type dynamic measurements of (a) dissolved alkyd resin, alkyd dispersion of boehmite alumina, *p1* and *p2*, and the nanotube-embedded (b) *p3–p6* particles at solid concentration of 3.846 wt.%, at a stress of 1 Pa.

The strong viscoelastic response of the pseudo-solid like colloid structure of *p1* dispersion is manifested by the entire frequency-range higher storage and lower loss moduli. Non-linear characteristic of the dispersion arranged by week attractive interparticle interaction through the dissolved polymer is reflected by the gradually decreasing complex viscosity, which is attributed to the shear-induced reorientation, nearly complete disruption of the highly associated three-dimensional system at high strain-rates.

In contrast, *p2* dispersion exhibited nearly strain-rate independent viscous characteristic (with higher loss and lower storage moduli due to inefficient load-transfer) which supports the suggestion of an absence of substantial interparticle interaction among the solid particles.

The lower scattered data of *p3–p6* dispersions confirm a good micro-scale dispersity of the nanotubes. However, in the co-dispersion of PPy-deposited alumina and MWCNT, profound effects of the various degree of aggregation, different distribution of the particles were experienced. Frequency-sweep test revealed strongly interacted colloid microstructure of *p3* dispersion by the intersecting storage and loss moduli at 10 Hz with the related elastic-viscous transition (critical stress-level) cross-over (Figure 4b). This was followed by steeply increasing moduli and complex viscosity indicating a more viscous flow behaviour (reflected by the high loss factor). The gradual ramp of higher loss and smaller storage moduli with the resultant high phase-angle reflect on less coherent structure of the *p4* dispersion, similarly to the considerably lower nanotube quantity comprised *p5* colloid sol. Coherent but less extensively associated flocculated colloid dispersion of *p6* were ascertained by the more elastic than viscous character (low phase shift) without a cross-over over the whole frequency range. These are associated with a strongly interacted, lower density nanotube-supported filamentous structure, which is in accordance with the TEM observations. Amplitude sweep measurement indicated stress-induced deterioration of the *p1* colloid structure with an elastic-viscous cross-over at 2.5 Pa (Figure 5a) which is a consequence of the increased interaction of the particles related to the thin film deposited PPy. On the other hand, the stress-ramp invariant viscous – fluid-like characteristic of *p2* sol containing partially globular PPy is connected to a far less associated and interacted three-dimensional structure. An elastic-viscous transition was noted to take place at 2 Pa in the *p3* dispersion with shear stress induced significant rupture of the colloid system (Figure 5b). No similar mechanical activation of the sol structure was observed in the *p4,* featuring predominantly viscous characteristic. Decreased complex viscosity of the *p5* sol (Figure 5c) compared to *p1*–*p2* is connected to the almost diminished inter-floc interactions among the nanotube-supported filaments arranged with aggregated PPy-modified alumina. Two-stages of the considerably high stress-activated elastic-viscous transition of the flocculated *p6* dispersion (reflected by the phase-shift at 5 Pa) confirm strong inter-floc association of the filaments in the colloid network.

Small hysteresis loop and the low monotonic shear-thinning at low strain rates (Figure 4a) suggest marginal structural viscosity of the dissolved alkyd matrix similarly to the alumina contained monodisperse system which manifested ignorable degree of association – floc formation of the randomly distributed particles. The Newtonian plateau of the viscosity over high strain-rates means a hydrodynamically controlled dispersion of the freely moving alumina particles. Aside from the slight bridging aggregation, the low stress required (Figure 6a) and the small thixotropy featured macroscopic flow behaviour of *p1* suspension (Table 3) are associated with rapid breakdown response and micro-scale association recovery of the thin-layer polymer deposited isotropic particles built highly interacted but less coherent medium-size flocs.

The apparent dynamic viscosity at high shear-rates was nearly the same as obtained with the bare alumina sol (Figure 6b), reflecting on an ignorable hydrodynamic effect of the interfacial PPy layer when complete collapse of the microgel flocs occurred.

The partially globular PPy contained suspension of *p2* indicated higher thixotropy as a result of the gradually disrupting, lower-site-number interacting, coherent larger size flocs with reorientation of the particles in the not fully collapsing micro-gels (even when durable straining performed at 300 s^{-1}).

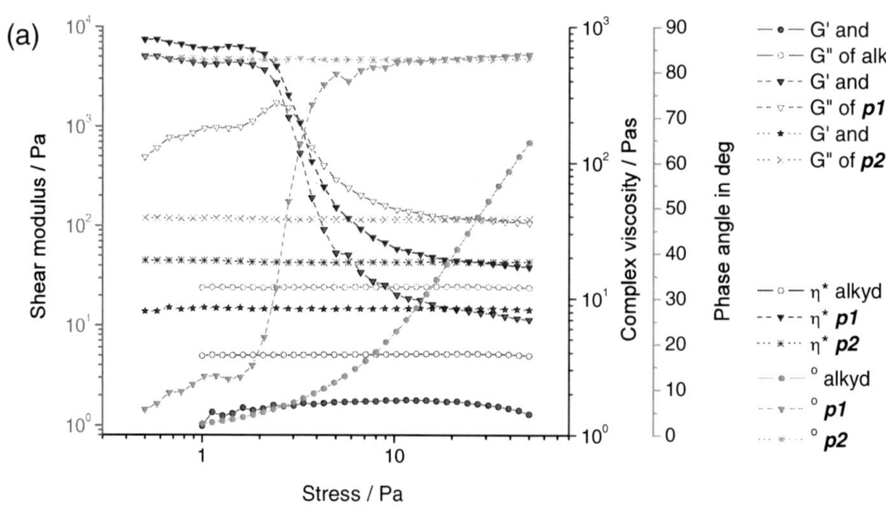

Figure 5. (Continued)

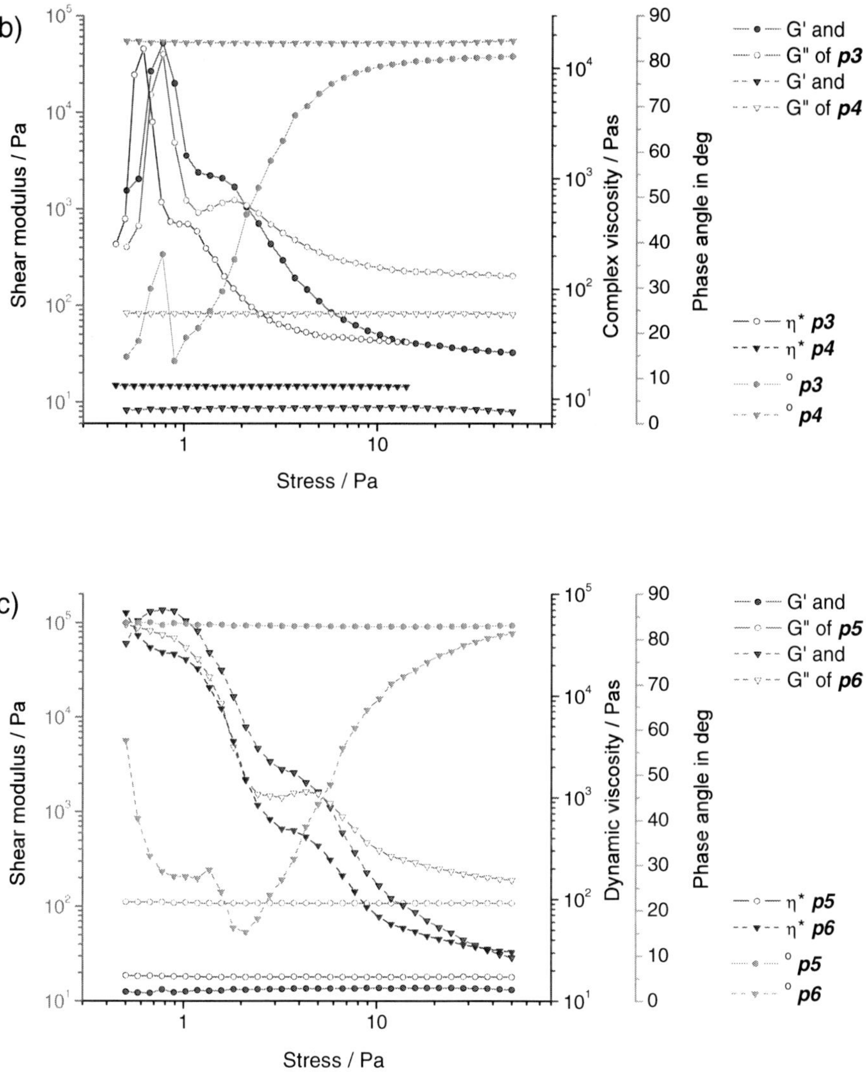

Figure 5. Amplitude sweep type dynamic measurements of (a) dissolved alkyd resin, alkyd dispersion of *p1* and *p2*, and the nanotube-embedded (b) *p3*–*p6* particles at solid concentration of 3.846 wt.%, amplitude sweep at 1 Hz.

The lower pseudoplasticity with the largest stress-required straining (highest apparent dynamic viscosity) and the medium high yield-stress of the *p3* dispersion (Table 3) reveal altogether highly associated and coherent voluminous filaments built floc system featuring medium degree of three-dimensional inter-floc linkage. Strong thixotropy character of the gel-like dispersion is a result of the nanotube interactions around the rheology percolation related entanglement regime [50] allowing slower recovery of the gradually disrupted flocs. In contrast, at similar MWCNT volume fraction, pseudoplastic behaviour of the *p4* dispersion is connected to the fast response of preferential alignment–orientation of the nanotubes parallel to strain direction.

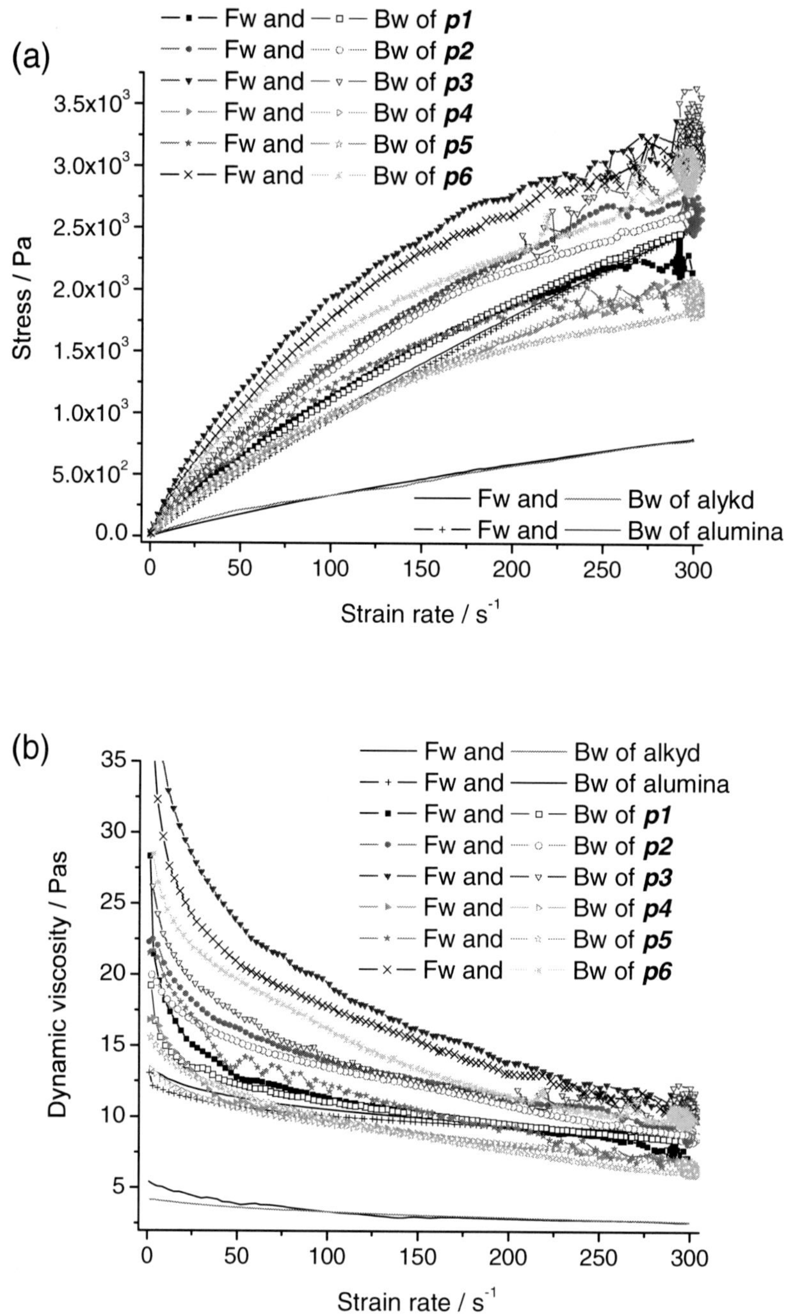

Figure 6. Steady measurements with upward (uw.) and downward (dw.) test results of (a) shear stress and (b) dynamic viscosity in the constant strain rate range of 0–300 s^{-1}, (c) constant force measurements in the stress range of 0–200 Pa, at solid concentration of 3.846 wt.%.

Consequently, the quite low apparent dynamic viscosity, thixotropy and yield-stress are attributed to a small density of poor inter-floc linkage, the absence of effective three-dimensional floc interaction, which is in accordance with the TEM observations. At lower PPy and nanotube contents, *p5* dispersion indicated increased dynamic viscosity, high thixotropy and very low yield-stress with pronounced shear-thinning effect which are connected to the lower extent of inter-floc linkage and stress-induced gradual attrition of the medium coherent flocs. This is a consequence of the double-flocculants modification resulted coalescence of the modified particles. The higher extent and strong aggregation of the particles and the related lower dispersity (effective volume fraction) of the system are revealed by the convergence of dynamic viscosity at high shear-rates where low hydrodynamic flow resistance was measured. Pronounced shear-thinning and the high dynamic viscosity of the *p6* dispersion are due to the extensively inter-linked microgel structure with strain-rate induced gradual but partial attrition of the nanotube-supported filamentous flocs. This was obviously accompanied by the release of entrapped dissolved polymer, leading to decreased effective hydrodynamic size of the remaining flocs at higher strain rates. The high yield-stress and thixotropy confirm a strongly inter-linked floc system, i.e. highly flocculated network based pseudo-solid-like structure featuring sluggish reorientation of disrupted flocs and aggregated modified particles.

Despite the poor packing effect and increased effective volume fraction of the rod-like particles [51,52], higher apparent viscosity was only measured to the *p6* dispersion over the entire strain-rate where volume fraction of the modified nanotubes was well under the experimentally determined rheological percolation regime related entanglement threshold of the bare MWCNTs ($\sim 7\times 10^{-3}$) [50]. This is because of the high dispersity and coherent three-dimensional structure of the inhibitor particles in spite of the decreased dispersibility of the polymer complex modified particles. On the other hand, the lack of ability to build up effective hydrogen bonding between alumina and the dispersant, leads to an inefficient enthalpy gain leveraging unpreferred entropy. Hence, hydrated alumina repels much of the lipophile matrix, so highly dispersed composite is not supported. Therefore, insignificant rheology control is connected to the weak van der Waals forces between the solid and the vehicle. On the other hand, PPy is in favour of interacting with the rheology tested lipohile alkyd and the paint formulated epoxy matrix via van der Waals forces and some hydrogen bonding. This is due to the similar Hansen solubility parameters of pyrrole ($\delta_{D,P,H}$ = 19.2, 7.4, 6.7) describing its dispersion, polar and hydrogen bonding ability with ethanol (15.8, 8.8, 19.4), epoxy (~15.6, 8, 4.6) and phenol (18, 5.9, 14.9) [53] segments contained structural polymers. Thus, layer type deposition of PPy resulted in increased interparticle interactions among the adjacent particles, which must be enhanced by the polar and hydrogen bonding ability of the polyelectrolyte and functionalized nanotubes. As a conclusion, basic rheology study was employed as representative macro-scale appraisal of the particle dispersions corroborating the suggestions made by TEM observations.

FT-Infrared Spectroscopy

The FTIR spectra obtained for the PDIP samples are depicted in Figure 7. The selected spectral region of 1800-1000 cm^{-1} implies the characteristic C–C/C=C and C–N stretching vibration bands of the polypyrrole ring [54,55]. The band at 1569 cm^{-1} in the spectra of water (*p1*) and aqueous-ethanol-prepared (*p2*) particles belongs to the combination of conjugated C–C and C=C stretching modes of the PPy rings.

This band, however, is blue-shifted compared to the one of the free-standing polymer situated at 1530 cm^{-1}. The bands at 1634 and 1340 cm^{-1} can be assigned to the deformations of protonated species like NH_2^+. For the *p2,* beside the strongest band at 1635 cm^{-1}, the medium intense N–H deformation band of terminal –NH_2 groups at 1110 cm^{-1} suggests a mediated oxidation – doping of the polymer, which might lead to partial cleavage of some pyrrole rings [56].

For the nanotube embedded samples (*p3–p6*), PSS bands were also present in the spectra (marked with asterisks in the figure). The in plane deformation bands of =C–H groups detected at 1294 and 1047 cm^{-1} are due to the greater PSS polyelectrolyte proportion in the *p3* and *p5* samples. Spectral effects of the SDS can be neglected because of the thorough consecutive washing process. There are marked differences in the typical spectral features of the PPy. The characteristic bands around 1560/1570, 1170 and 1050 cm^{-1} proved a good quality of the PPy in *p3–p6* without overoxidation, confirming an increased stability/chemical durability of the nanotube-supported PPy. Despite the strong spectral masking effect of the sulphoxide group, the relative high band intensities at 1170 cm^{-1} (N–C stretching of pyrrole rings) and 1050 cm^{-1} (=C–H in-plane deformation) can be related to a compact PPy structure with increased electroactivity and electrical conductivity based on electron hopping near the N–C bond [57].

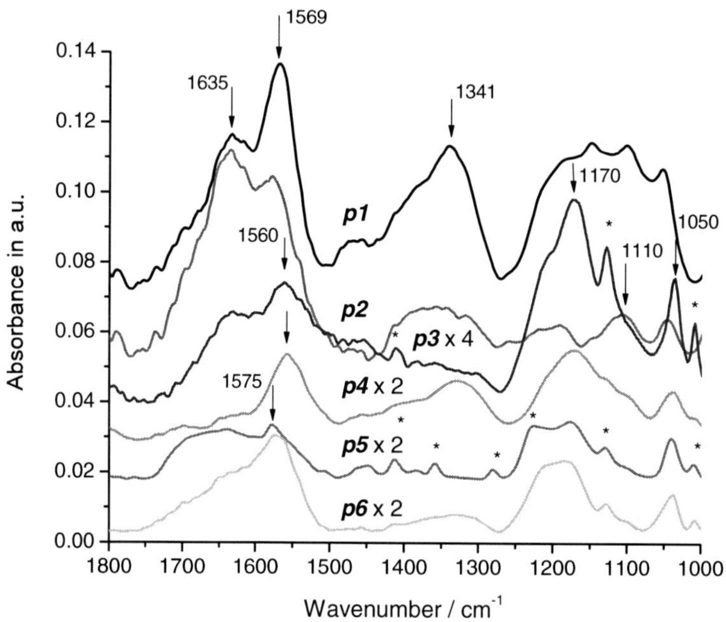

Figure 7. FTIR spectra of PDIP samples: water- and aqueous-ethanol-prepared PPy-deposited alumina particles (*p1* and *p2,* respectively), comprising PSS-modified MWCNT (*p3*), low sulfonated MWCNT (*p4* and *p5*) and highly sulfonated MWCNT (*p6*). Asterisks in the spectra indicate characteristic bands of PSS polyelectrolyte.

Suppressed δC–H deformations indicated by the bands at 1294 and 1047 cm^{-1} suggest stronger co-planar characteristic of the PPy segments throughout the polymer backbone. The relative high band intensities at 1170 cm^{-1} (C–N stretching) and 1038 cm^{-1} (=C–H in-plane bending) in the spectrum of *p1*, *p3* and *p4* suggest a chain-like, co-planar PPy structure. For the low functionalized nanotube-alumina contained *p4* sample, PPy spectrum resembles to those of bare alumina supported polymer obtained in *p1* and *p2* samples. Hence, the spectrum suggests a more compact, mostly film-like PPy structure in *p4*. The broad band around 1330 cm^{-1} assigned to the N–H bending of NH^+ groups suggests a partially doped, protonated PPy. The suppressed C–N stretching and C–H bending intensities compared to the one of the conjugated C–C/C=C stretching of pyrrole rings point to the fact that the deposited polymer might mainly be in a three-dimensional structure. In the case of *p4*, reduced concentration of SDS and PSS allowed the formation of moderately protonated – doped PPy with medium degree of deposition. Hence, highly conductivity and electroactive inhibitor sample was obtained as it is clearly indicated by cyclic voltammetry.

Collective vibration of C=C/C–C inter and intra-rings appears at 1560/1570 cm^{-1} compared to 1530 cm^{-1} of the neat PPy. This blue shift can be explained by an enhanced electron delocalization in the polymer backbone [58] and was found to be higher for the sulfonated nanotube containing *p5–p6* samples (around 1575 cm^{-1}). The same blue-shift accompanied with the rising of a new band around 1220 cm^{-1} in the C–H and N–H in-plane deformation region is connected to the changes in the doped – protonated and base forms of PPy. The spectrum of *p6* PDIP containing highly sulfonated MWCNT and alumina supported PPy is somewhat different owing to the polymer partial interaction with the highly functionalised nanotubes and the complexing PSS polyelectrolyte. On the other hand, PPy content indicated film-like and globular character with the bands at 1160, 1570 cm^{-1} and the wide-range around 1350, 1660 cm^{-1}, respectively. Therefore, the high wavenumber of collective C–C/C=C vibrations at 1573 cm^{-1} together with the C–H/N–H in-plane deformation around 1210 cm^{-1} suggest the presence of a base form of the PPy. Hence, it is obvious that PPy is doped predominantly by the highly sulfonated nanotubes whereas alumina supported polymer could be complexed with small portion of PSS. The relative high band intensities at 1170 cm^{-1} (N–C stretching of pyrrole rings) and 1050 cm^{-1} (=C–H in-plane deformation) detected in *p3* and *p4* are associated with compact polymer structure with increased electrical conductivity. In addition, the collective vibration of C=C/C–C inter and intra-rings at 1560/1570 cm^{-1} (compared to 1530 cm^{-1} of neat PPy). The blue shift associated with an enhanced electron delocalization in the backbone [67,68] is typical to the sulfonated MWCNT-embedded *p5* and *p6* (1575 cm^{-1}). These observations regarding the structure of the PDIP samples corroborate cyclic voltammetry results of increased conductivity and enhanced redox-activity of the nanotube-embedded inhibitor particles.

Cyclic Voltammetry

Electroactivity and conductivity are important properties having strong influence on the corrosion inhibition function of the PPy regardless it is used as stand-alone polymer films or carrier-supported forms in composite coatings.

Taking into account that PPy contents of *p1* and *p2* (as conducting network) were around the percolation threshold [59], valuable coulombic efficiency of the samples was ascertained (Table 1) by the transient charging voltammetry currents which are presented in Figure 8.

Highly reversible current responses were exhibited, confirming moderate electrical conductivity and redox activity of the Ppy contents of the inhibitor particles.

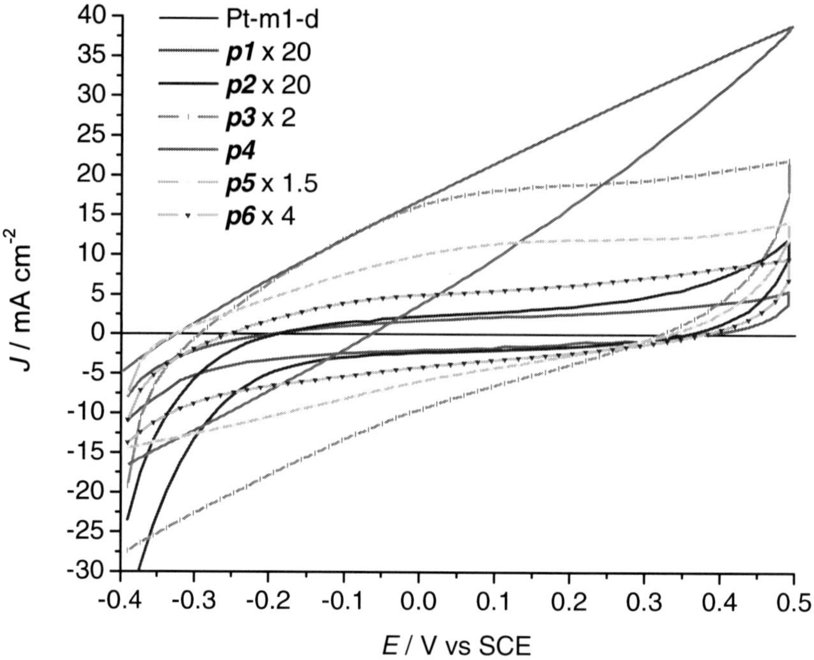

Figure 8. Cyclic voltammograms of water- and aqueous-ethanol-prepared PPy-deposited alumina particles (*p1* and *p2*), with PSS-modified MWCNT content in *p3*, low sulfonated MWCNT in *p4* and *p5*, and highly sulfonated MWCNT in *p6*, referenced to the base current of platinum electrode. Sweep rate is 10 mV s^{-1}.

Similar charging efficacy and the highly reversible current transient of the *p2* comprising lower amount of PPy, suggest an altogether increased mobility of charges in the conducting polymer network compared to *p1*. Current transient of the *p3* indicated medium degree of electrochemical reversibility with improved conductivity and redox ability which are related to the greater current response. The almost ideal resistance like voltammetry current response of the *p4* sample manifests enhanced reversibility of the PPy and charge mobility in the polymer backbone, which are due to the enhanced interaction between the PPy and the nanotubes based increased electrical conductivity of the modified particles. The highly dispersed but low amount of nanotube-incorporated strongly aggregated filamentous structure comprised less interlinked PPy network in *p5* allowed the evolution of relatively high current transients with medium fast kinetics related appropriate reversibility since moderate coulombic efficiency was obtained (containing greater amount of surfactant similarly to *p3*). The current image was also resulted by the greater specific polyelectrolyte content of *p5* which is responsible for the somewhat restrained mobility of the charges in PPy backbone. This is a consequence of the closely interacted complexes of the PPy and PSS molecules, allowing limited conjugation in the partially doped polymer network by the larger distortion and conformational adjustments of the Ppy segments. The current response related enhanced reversibility and improved charge mobility (despite the medium PPy and low nanotube contents) resulted in considerable coulombic efficiency of the PPy in the *p6* sample. This is connected to the highly interconnected, homogeneously distributed nanotube-supported, thin film deposited PPy network with closer interaction and efficient electron transfer between the nanotube carrier and the deposited PPy. Although, PPy charging efficiencies were around 4%, high electroactivity and the moderate electrical conductivity must contribute to the interaction of sacrificial zinc pigments. Increased efficiencies of the galvanic function of the hybrid paint coatings are expected based on the rheology indicated nanotube interactions at a certain solid concentration. Therefore, advanced galvanic protection function of the hybrids is expected to be comparable to the traditional type ZRPs.

INVESTIGATION OF IMMERSION TESTED ALKYD PAINT COATINGS AND STEEL SUBSTRATES

Open-Circuit Potential and Electrochemical Impedance Spectroscopy

The markedly fluctuated, positive polarized OCPs of the salt-solution tested pigment-free epoxy coated steel sample (*E*) are presented in Figure 9a. The measured impedance spectra shown in Figure 9b exhibited almost ideal capacitive behaviour up to 152 days, confirming the firm passive protective function, very low electrolytic conductivity of the coating. Then, the low rate of electrolyte infiltration was disclosed by the stabilized OCPs and the little increasing low-frequency phase-angle minimum in the impedance spectra. The positive mixed potentials suggest notable shrinkage in the area of anodic processes, giving evidence about the suppressed rate of iron dissolution. However, around 225 days the anodic polarization slowly decreased due to the acceleration of anodic processes at the steel surface. At the end of the immersion test, steel substrate was corrosion-free but the primer completely delaminated. Evaluation immersion test results are summarised in Table 4.

The conventional type zinc-rich *Z* provided strong and remarkably durable cathodic polarization – thermodynamic immunization of the steel substrate for 127 days (Figure 9a) which was in accordance with the intense sacrificial action of zinc pigments. In the meantime, deteriorating barrier profile of the coating was reflected by acquired EIS data featuring small and decreasing impedance moduli with high-frequency-range shifted breakpoint-frequencies depicted in Figure 9c.

Table 4. Evaluation of hydrochloric acid solution tested paint coatings, corresponding to the EN ISO 4628-8:2005 standard

Coatings	Degree of blistering size and density		Degree of delamination
E	1	1	complete
Z	2	2	severe
ZA	2	4	moderate
C1	1	3	none
C2	-	0	none
C3	-	0	none
C4	-	0	none
C5	-	0	none
C6	-	0	none

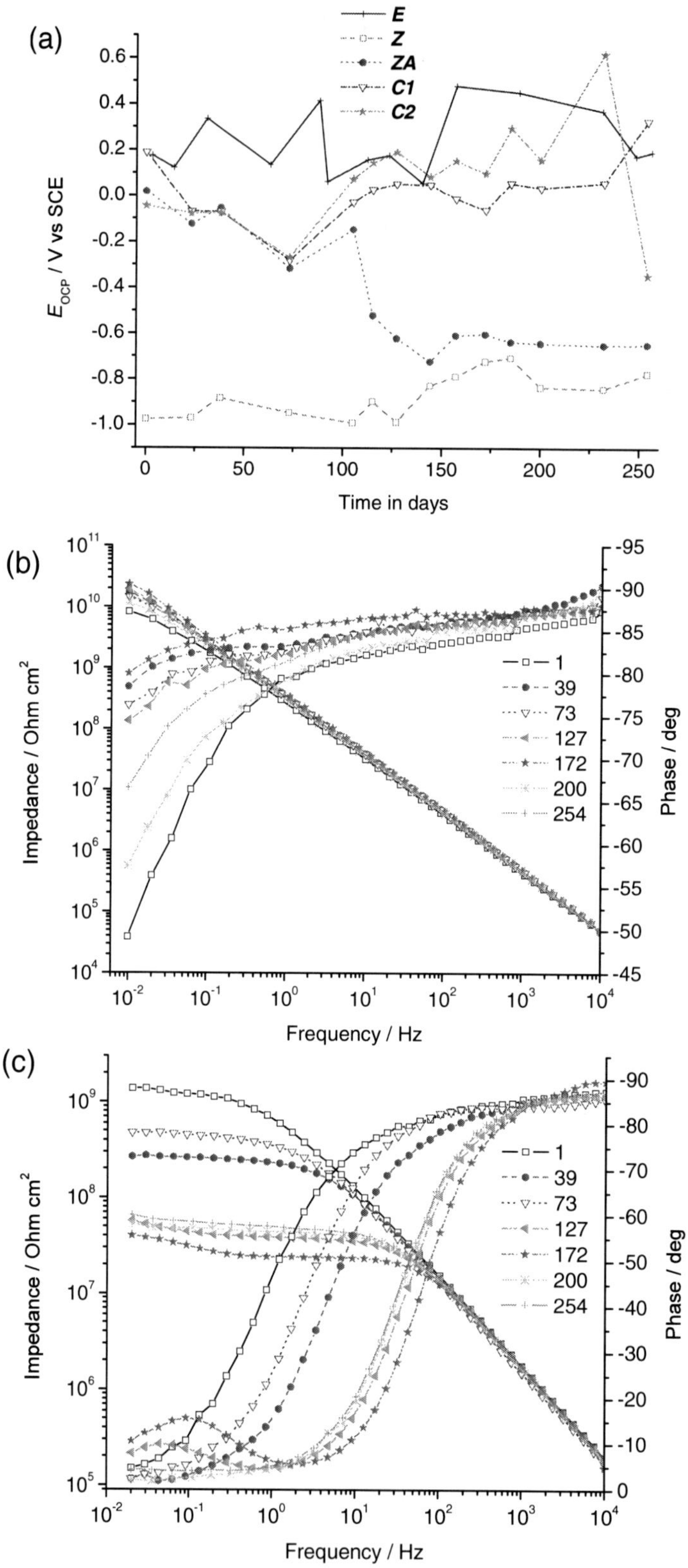

Figure 9. (Continued)

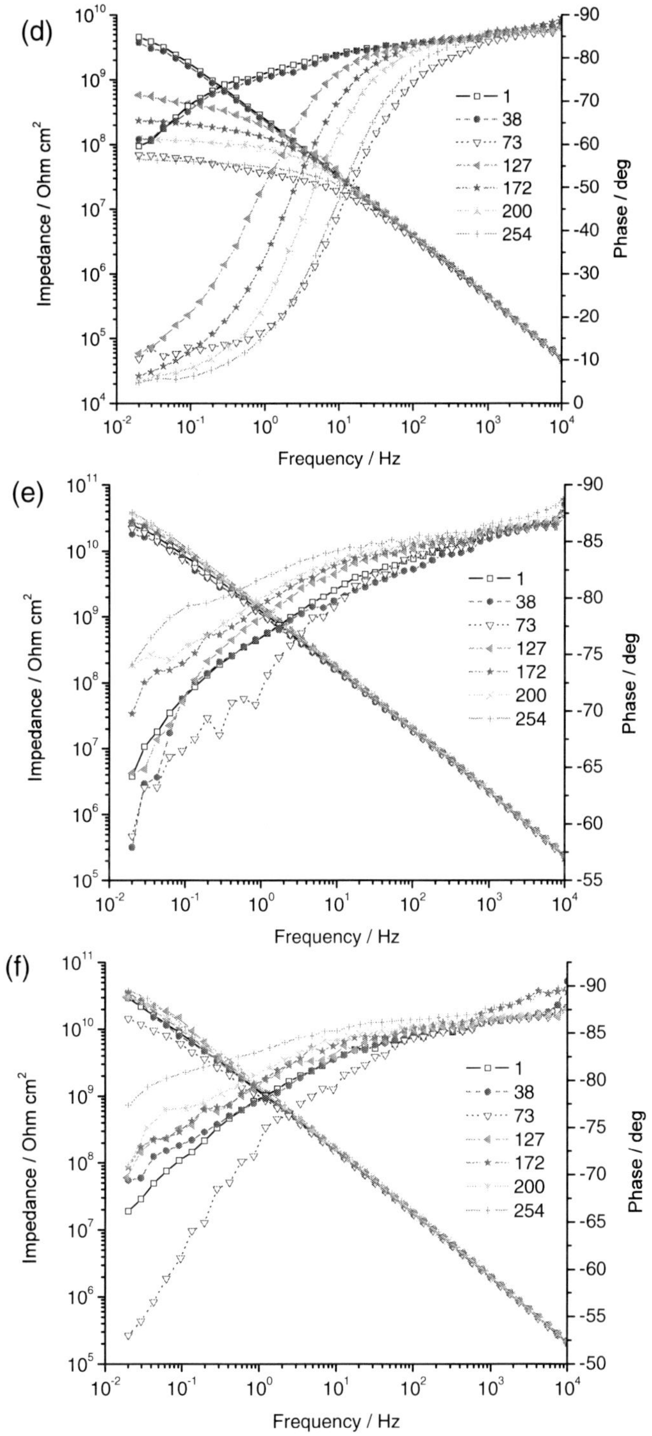

Figure 9. Open-circuit potentials (a) and electrochemical impedance spectra (Bode plots) of steel samples protected with paint coatings of (b) neat epoxy (*E*), (c) conventional type zinc-rich epoxy (*Z*), (d) zinc-rich epoxy with alumina (*ZA*), (e) *C1* and (f) *C2* hybrid paint coatings containing water- and aqueous-ethanol-prepared *p1* and *p2* particles, immersion tested with 1 M sodium chloride solution for 254 days.

The notable drops in the impedance moduli measured at 38 and 172 days correlated with the deactivation and reactivation of cathodic galvanic function of the primer by the increasing and decreasing extent of electrolyte ingress. These processes were followed by moderate recoveries in the barrier nature around 73 and 200 days, which connected to the lessened electrolyte penetration and lower rate of dissolution of the accumulated zinc corrosion products. Evolved second time constant in the phase diagram is attributed to a greater double-layer capacitance. The interfacial galvanic cell at the steel surface is connected to severe deterioration and de-adhesion of the primer coating (containing partly heavily corroded metallic pigments) with its diminishing all around protective function. In this case, enhanced sacrificial function of zinc is propelled by the great deal of electrolyte infiltration and the possible high rate of oxygen reduction reaction (ORR) accompanied with the subsequent evolution of strongly alkaline milieu as well. Despite the severe delamination of coating *Z* which is in accordance with the severe oxidative degradation [60,61] and saponification of the epoxy binder, no corrosion spots were detected on the steel surface.

Except for a positive potential shift around 100 days (Figure 9a) followed by a short-term recovered barrier nature of the *ZA* (Figure 9d), steadily decreasing OCPs with the appearance of breakpoint-frequency and the high low-frequency phase-angle minimum measured until 73 days, suggest an altogether low efficiency galvanic function of the zinc-rich/alumina primer coating along with the steady electrolyte infiltration associated continuously degrading barrier behaviour which became pronounced from 127 days. Stable OCPs around –0.6 V vs. SCE are connected to steel corrosion whereas decreasing low-frequency impedance moduli with the migration of breakpoint- and saddle-frequencies towards the high-frequency domain unveiled gradually increasing electrolytic conductivity. At the end of immersion test, the coating was moderately blistered and slightly delaminated. The steel substrate was somewhat corroded with considerable quantity of precipitated zinc corrosion products on the surface. Consequently, paint coatings based on the composition of highly dispersed alumina and spherical zinc pigments at contents of 3.21 and 70 wt.% provides neither efficient galvanic nor firm barrier protective function to steel substrates against corrosion.

The similarly changing anodic shifted OCPs manifested generally low activation of the galvanic function of the *C1* and *C2* coatings (Figure 9a) by electrolyte ingress due to the firm barrier behaviour of the coatings. The latter is reflected by the high impedance moduli and the quite low phase-angle minimum in the low-frequency domain as it is shown in Figures 9e and f, allowing very limited electrolyte access into the primers and to the anodic pigments. Such nature of the coatings changed around 73 days when OCPs shifted cathodically coinciding with the lower impedance amplitudes and higher low-frequency-range phase-angle minimum associated increased electrolytic conductivity. The anodic polarized OCPs with improved capacitive features at 100 days denote effective recovery of the corrosion resistance of the coatings as a result of the accumulation of small amount of zinc corrosion products. The absence of valuable cathodic galvanic action of the hybrid primers is connected to the complex function of the highly dispersed PDIPs supporting lower extent of electrolyte ingress and inhibition of the sacrificial and self-corrosion of the anodic zinc pigments. During the second half time-period of the test, *C1* and *C2* coatings afforded stable corrosion prevention which was noted by the ignorable changes in the OCPs and the impedance data giving evidence about the hindered electrolyte infiltration into the coatings. Apart from the small number and size of blisters on the surface of *C1* in conjunction with its moderate galvanic function and the osmosis propelled blistering, no de-adhesion or delamination of the hybrid paints and steel corrosion were experienced.

Within the first 50 days of immersion, the large amount of polyelectrolyte modified nanotubes contained coating *C3* manifested slow activation of its hybrid primer by the strong anodic polarized mixed potentials shown in Figure 10a with effective barrier functionality reflected by the impedance data presented in Figure 10b. Anodically shifted OCPs are caused by the shrinkage in the effective anodic surface and the inhibited sacrificial action of the zinc pigments, leading to comparable areas of the galvanic coupled anodic and cathodic reactions. The main reaction of oxygen reduction took place obviously both on the steel surface and zinc pigments along with the low intensity galvanic action of the primer, resulting in low potential gradient in the coating/steel interface. The little increased electrolyte penetration through the *C3* was marked by the cathodic shifted OCPs at 100 days which occurred 25 days later than in the case of *ZA*, *C1* and *C2* coatings. This was accompanied by an ignorable impedance modulus drop and insignificant increase in the low-frequency phase-angle minimum.

However, the 50 days period of continuous activation of the *C3* and the more intense anodic action of its zinc content was followed by efficient pore-filling recovery. This was indicated by the stabilized positive OCPs and the more capacitive coating behaviour associated improved electrolytic barrier function which remained stable to the end of immersion test. All these features of the hybrid paints are mainly attributed to the beneficial action of the *p3* type anodic inhibitor particles having high dispersity and extensively interconnected microstructure with well conducting and redox active nanotube-supported PPy network in the epoxy vehicle.

The slightly lower electrolytic barrier nature of the low sulfonated nanotube contained coating *C4* was reflected by the insignificant cathodic polarization of the OCPs (Figure 10a) and the EIS data indicated breakpoint-frequency in the low frequency domain (Figure 10c). The evolution of negative polarized OCPs within the first 20 days reflects on moderate activation rate of the coating by the penetrating corrosive electrolyte.

The following positive potential shift at 40 days coincided with the greater pore resistance connected enhanced barrier profile (more capacitive character) revealing fast recovery of the corrosion resistance of the *C4* resulted by the efficient accumulation of corrosion products.

This process took place repeatedly at a longer time-scale until 120 days when more stable passive protection function developed and remained stable to the end which was reflected by the slow negative polarization tendency in the OCPs and the unchanged impedance data.

Potential fluctuations and the greater ionic conductivity of the *C4* are resulted by the strongly aggregated and less interconnected microstructure of the *p4* particles in epoxy around the bulky zinc pigments.

Besides the negatively and positively shifted OCPs, impedance spectra presented in Figure 10d with the almost unchanged impedance parameters manifested gradual but an overall slightly increased ionic conductivity of the *C5* up to 172 days of immersion. Similarly to the *C4*, the low quantity of sulfonated nanotubes in *C5* with high relative amount of the polyelectrolyte indicated two–stages of electrolyte ingress related minor activation and blocking.

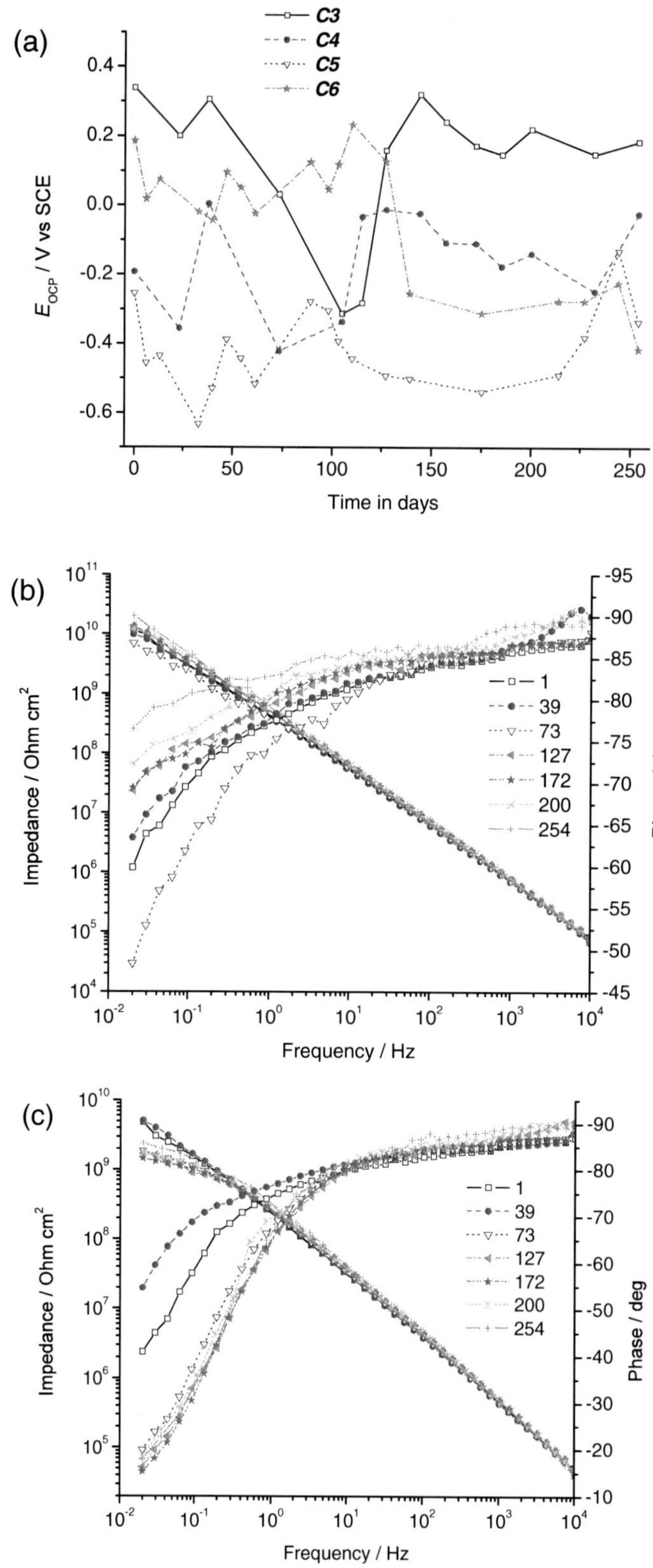

Figure 10. (Continued)

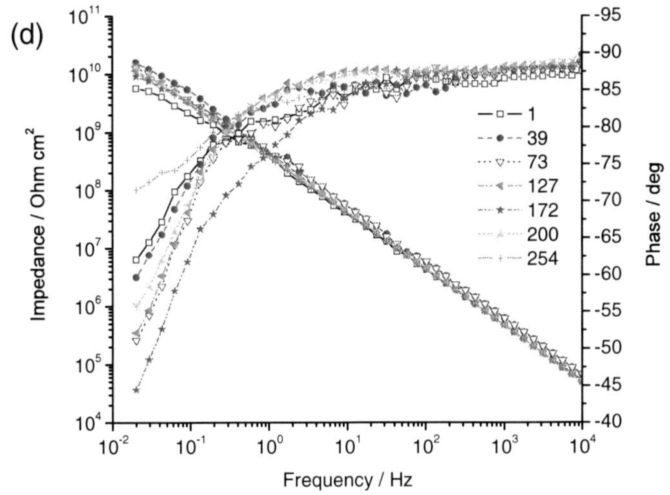

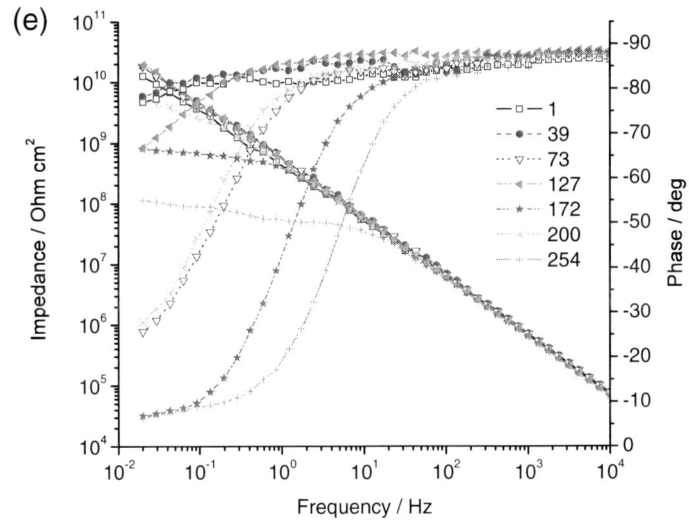

Figure 10. Open-circuit potentials (a) and electrochemical impedance spectra (Bode plots) of steel samples protected with hybrid paint coatings containing inhibitor particles embedded with (b) PSS-modified MWCNT (*C3*), low sulfonated MWCNT (*C4* and *C5*) and highly sulfonated MWCNT (*C6*), immersion tested with 1 M sodium chloride solution for 254 days.

The little increased pore-resistance is related to the corrosion resistance recovery of the *C5* occurred around 50 and 100 days. From 200 days, OCPs became more anodic as a result of the mixed potentials of the varied surface area related half electrode reactions connected to the suppressed zinc dissolution and the galvanic coupled ORRs on the steel surface. The improved barrier nature of the primer was noted by the continuously decreasing low-frequency phase-angle minimum. These changes are connected to the slow development of the long-term passive protection phase of the hybrid paints (similarly to the conventional ZRPs). The less anodically shifted potentials with firm barrier functionality of the coating are attributed to the lower inhibition of zinc corrosion at low extent of electrolyte penetration which was a consequence of the effect of the highly aggregated and least coherent microstructure of the *p5* particles in the epoxy resin.

In the case of coating *C6*, the frequently fluctuating anodically polarized OCPs measured in the first half of the immersion test were followed by cathodic polarization and stabilization of the mixed potentials (Figure 10a) as a consequence of the altered corrosion preventing behaviour of the coating. This was accompanied by the changing barrier nature of the hybrid primer coating between its activated and recovered forms signified by the migration of the breakpoint-frequency towards the high frequency domain and the increasing low-frequency phase-angle minimum in the impedance spectra shown in Figure 10e. Thus, the highly sulfonated nanotube-embedded *p6* sample with medium quantity of polyelectrolyte and the densely interconnected flocculated microstructure of the particles around the zinc pigments contributed to a less durable barrier functionality of the *C6* coating with higher zinc corrosion.

Radio-Frequency Glow-Discharge Optical Emission Spectroscopy

Hybrid paints were investigated by RF-GD-OES to draw sound conclusion about the electrolytic conductivity of the primer coatings based on the comparison of the composition of the initial and the immersion tested of paint coatings, detecting elements in ppm quantities by this high resolution depth-profiling technique. Changes in the quantities of corrosive species and coating elements are presented in relative intensities (normalized to the intensity of carbon) as a function of sputtered depth.

By this method, the EIS data suggested low ionic conductivity–permeability of the nano-size alumina contained zinc-rich *ZA* was confirmed detecting unchanged quantities of the corrosive species in the whole cross-section of the primer (Figures 11a and b). However, aside from the unaffected sum of the anions, the 6% increment in the oxygen content in the region of epoxy-steel surface suggest moderate barrier profile and the related low efficacy of the intense sacrificial and self-corrosion of zinc (galvanic function of the primer). This must have been the result of the absence of any valuable corrosion inhibition of the zinc pigments which was available to hybrid coatings afforded by the PPy-modified inhibitor particles. Almost the same quantities of the corrosive analytes like sodium, potassium and chlorine were detected through the cross-section of epoxy primer in *C1* after the corrosion propagation (Figures 12a and b), corroborating the impedance data suggested largely hindered electrolyte ingress. Potassium concentration in the hybrid is estimated about 4.5×10^{-2} mol dm^{-3} and only little affected by the immersion test. Minor relative intensity changes of the coating elements are related to the cross-sectional and spatial fluctuations in the local pigment/binder contents. In contrast, sodium and potassium were detected in higher quantities in the solution-primer interface of the immersion tested *C2*, as it is shown in Figures 13a and b. As a result of the slow electrolyte infiltration, sodium and potassium accumulated in the epoxy at concentrations of 2×10^{-3} and 1×10^{-1} mol dm^{-3}, respectively. However, the unchanged chlorine concentration in *C2* suggests very low electrolytic conductivity of the hybrid primer to the anionic species. This reflects the ignorable effect of the *p2* type particles on the ionic permeability and the corrosion resistive character of such type of hybrid formulation. In addition, detected potassium in the epoxy near the solution interface is due to the some percentage of potassium impurity in the sodium chloride and the smaller hydrodynamic radius resulted higher mobility, faster diffusion and migration in the organic coatings [62].

Moreover, the greater oxygen content in the primer-steel interface (~20–40 μm) and the slightly increased quantities of the cationic species in the 0–20 μm depth-range hint on the viable cathodic–galvanic function of the *C2* hybrid.

Nonetheless, the nanotube contained hybrid coatings manifested more advanced passive protection functionalities which were revealed by the comparison of corrosive species. No remarkable accumulation of any corrosive analyte was found in the cross-section of the *C3* (Figures 14a and b) which is accordance with the impedance data suggested massive and stable barrier nature of the coating.

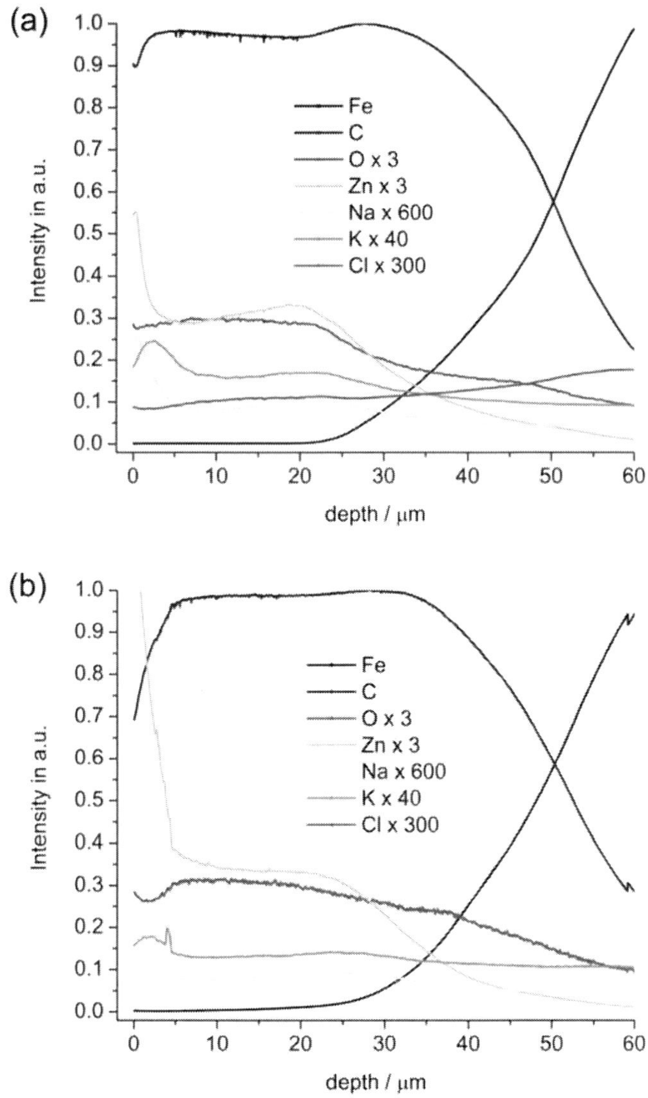

Figure 11. GD-OES depth profiled coating elements and corrosive analytes in the epoxy primer of *ZA*: (a) before and (b) after the 254 days of immersion test.

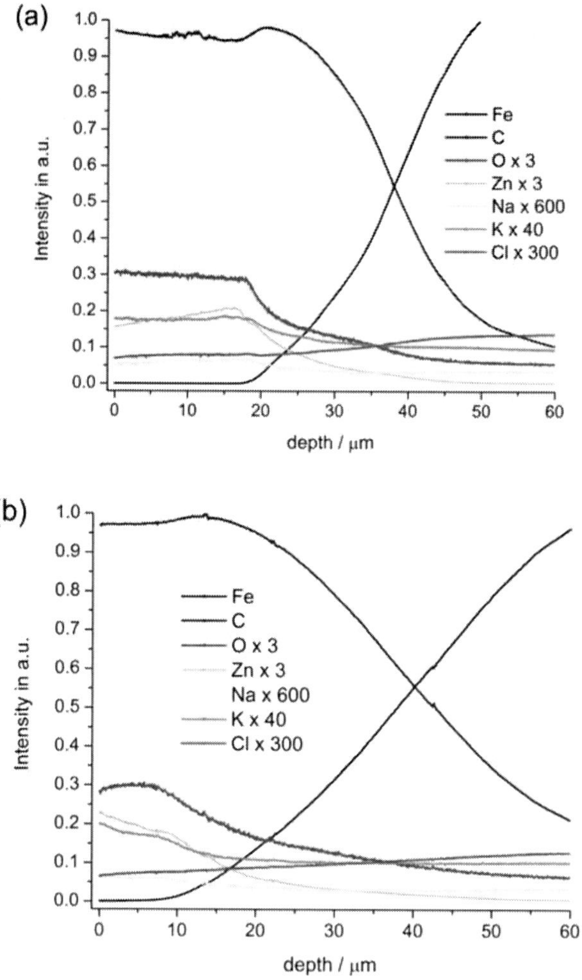

Figure 12. GD-OES depth profiled coating elements and corrosive analytes in the epoxy primer of *C1*: (a) before and (b) after the 254 days of immersion test.

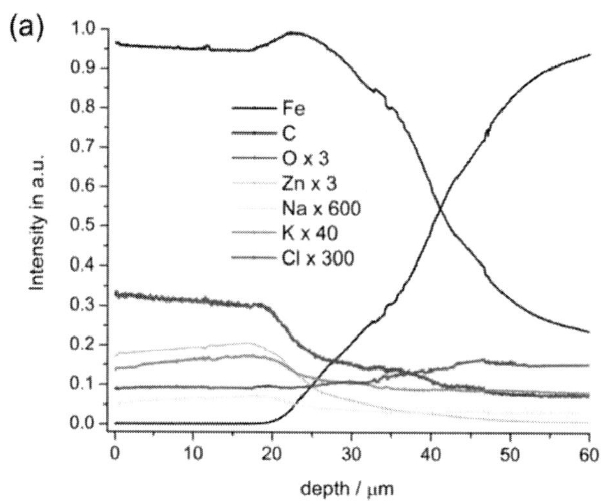

Figure 13. (Continued)

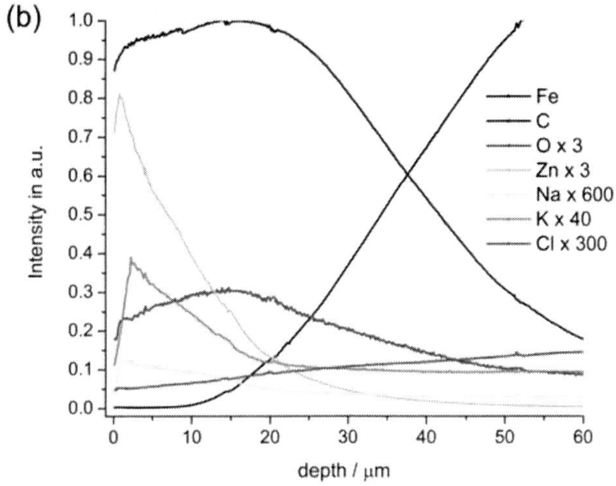

Figure 13. GD-OES depth profiled coating elements and corrosive analytes in the epoxy primer of *C2*: (a) before and (b) after the 254 days of immersion test.

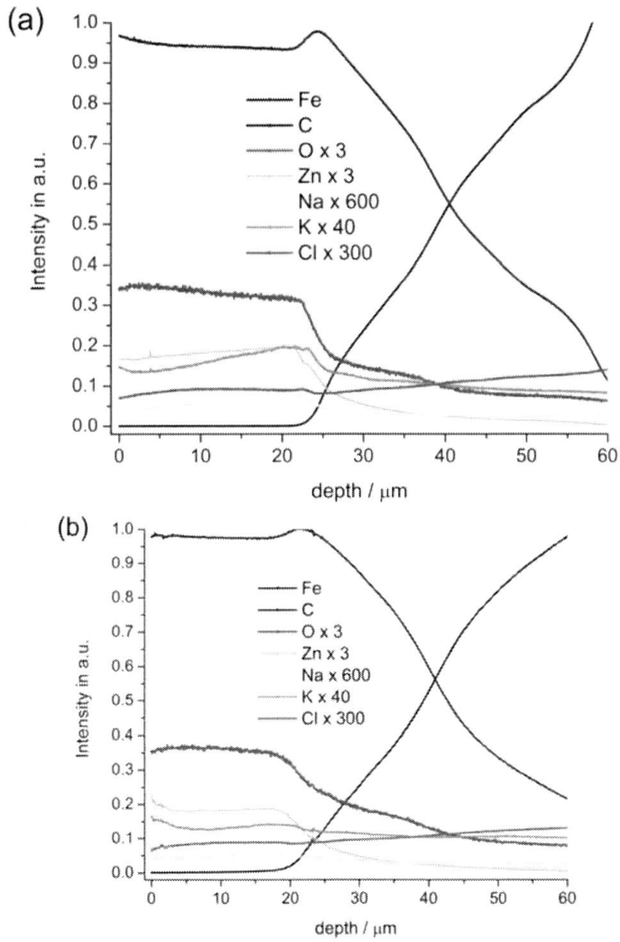

Figure 14. GD-OES depth profiled coating elements and corrosive analytes in the epoxy primer of *C3*: (a) before and (b) after the 254 days of immersion test.

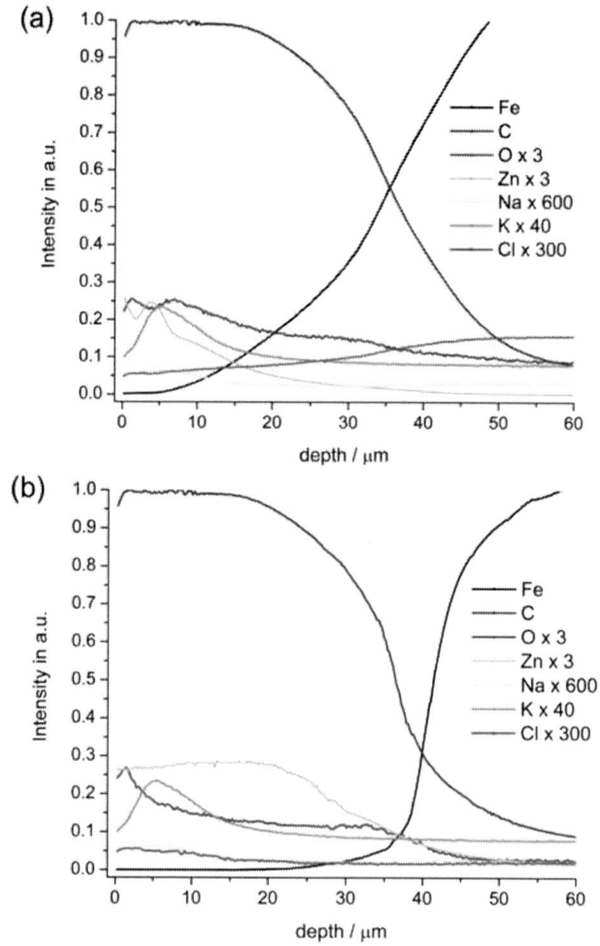

Figure 15. GD-OES depth profiled coating elements and corrosive analytes in the epoxy primer of *C4*: (a) before and (b) after the 254 days of immersion test.

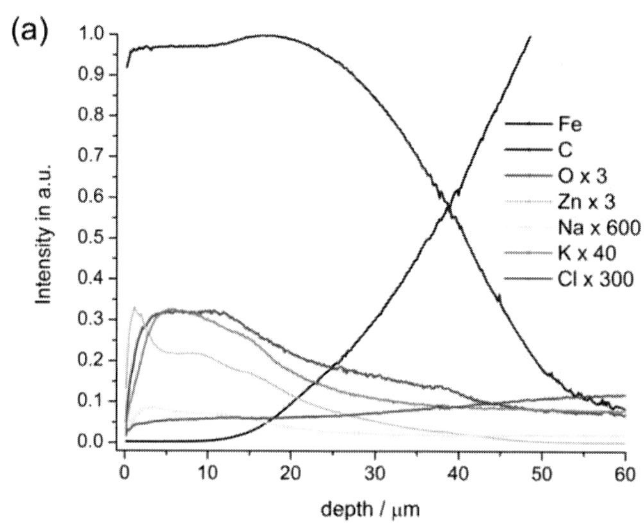

Figure 16. (Continued)

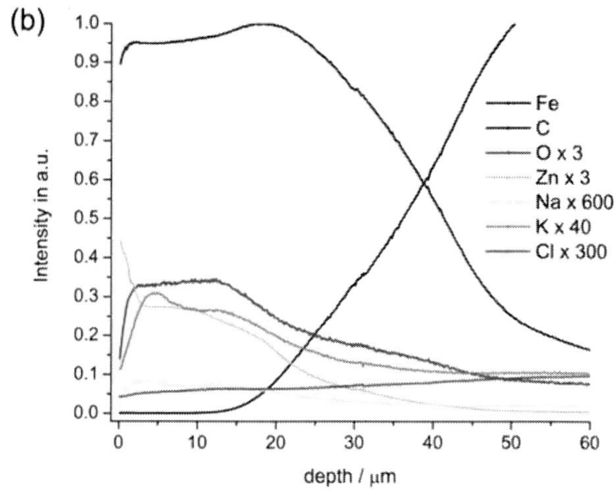

Figure 16. GD-OES depth profiled coating elements and corrosive analytes in the epoxy primer of *C5*: (a) before and (b) after the 254 days of immersion test.

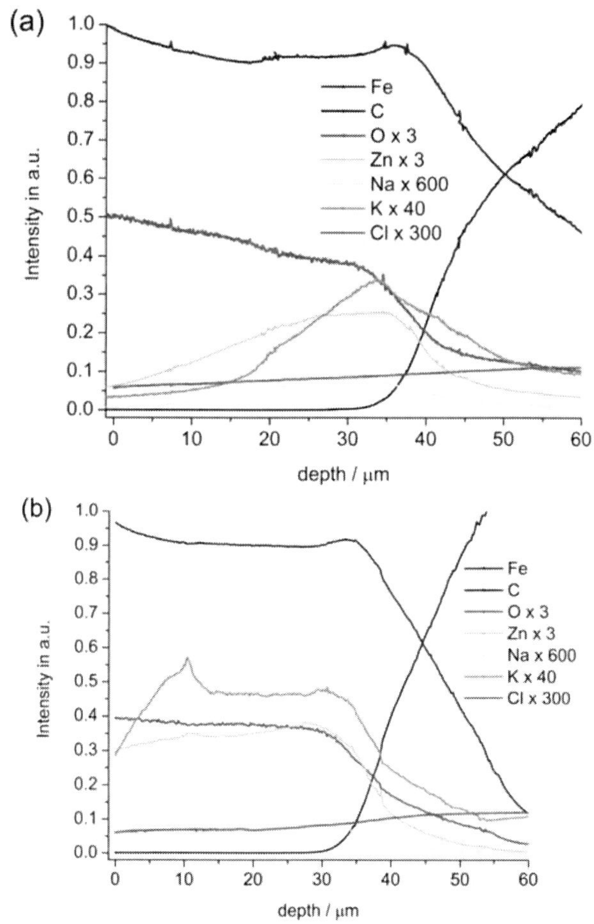

Figure 17. GD-OES depth profiled coating elements and corrosive analytes in the epoxy primer of *C6:* (a) before and (b) after the 254 days of immersion test.

However, the sensibly increased zinc bound oxygen measured in 20 μm thickness of the epoxy next to the steel surface (which was considerably lower than in the case of *ZA*) is connected to the lower dispersity of the *p3* (compared to *p1* and *p2*) and the less comprehensive inhibition of zinc corrosion. Similar phenomenon was observed when ionic species were depth-profiled in coating *C4* as it is shown in Figures 15a and b. The only exception was noted as an undetectable change in the oxygen content, suggesting not only effective barrier nature but also limited galvanic function of the hybrid primer (partially as a consequence of the low activation of anodic zinc) loaded with completely incoherent flocs which are composed of strongly aggregated nanotube-supported filaments. Insignificant increments in the contents of oxygen and the ionic species in the *C5* were disclosed by the measured depth profile data presented in Figures 16a and b, remaining within the ranges of paint composition related fluctuations in the specific contents of the components. The impedance suggested decreasing efficacy of electrolytic barrier ability of the *C6* during the immersion test is corroborated by the detection of greater amount of potassium (by 3.3%) marked by the intensity difference near the solution interface shown in Figures 17a and b. This is accounted for the effect of the polyelectrolyte and the highly sulfonated nanotube contents of the *p6* PDIP sample with its strongly coherent microstructure resulting in the formation of well interconnected zinc pigments in the epoxy primer. These properties lead to the evolution of moderately intense galvanic characteristic which was indicated by the slightly cathodic OCPs and the decreasing amplitudes featured impedance spectra vs. time exposure.

X–Ray Photoelectron Spectroscopy

To characterize the paint-steel interface of the corrosion tested samples, XPS measurements were performed after removing paint coatings by sonication in appropriate solvents. The experiments gave information on the amount and the chemical state of interfacial iron oxides as well as those of paint residues.

Figure 18a depicts the Fe 2p spectra of steel samples with the reference spectra of Fe(II)-rich iron oxide and Fe_2O_3 (containing Fe(III) states). Binding energy of the lowest energy structure, the Fe $2p_{3/2}$ peak is around 711.0 eV and the Fe 2p doublet is accompanied by satellites separated from the main peaks by 8.4 eV. These characteristics indicate the dominant form of iron in Fe(III) ionic state in the interface, although the weak shoulders shifted from the main peaks by 5-6 eV indicate the presence of Fe(II) ions as well. In addition, a weak feature around 707.0 eV indicates the presence of metallic iron, suggesting that thickness of the oxide layer is not enough to completely attenuate the signal arising from the underlying substrate. In fact, these features are characteristic for native iron oxides formed under ambient conditions [63]. Therefore, the oxide layer on the surface corresponds to the one formed before the application of paint coating, demonstrating the corrosion protection abilities of the paint formulations. Samples contained somewhat varying amounts of sodium as it is shown in Figure 18b (stronger peak corresponds to higher Na content), with *C6* being the most sodium-rich. Assignment of the sodium contribution is not particularly straightforward, as a significant part of Na compounds ranging from metallic Na to halides, phosphates or carbonates has a 1s binding energy around 1072.0 eV.

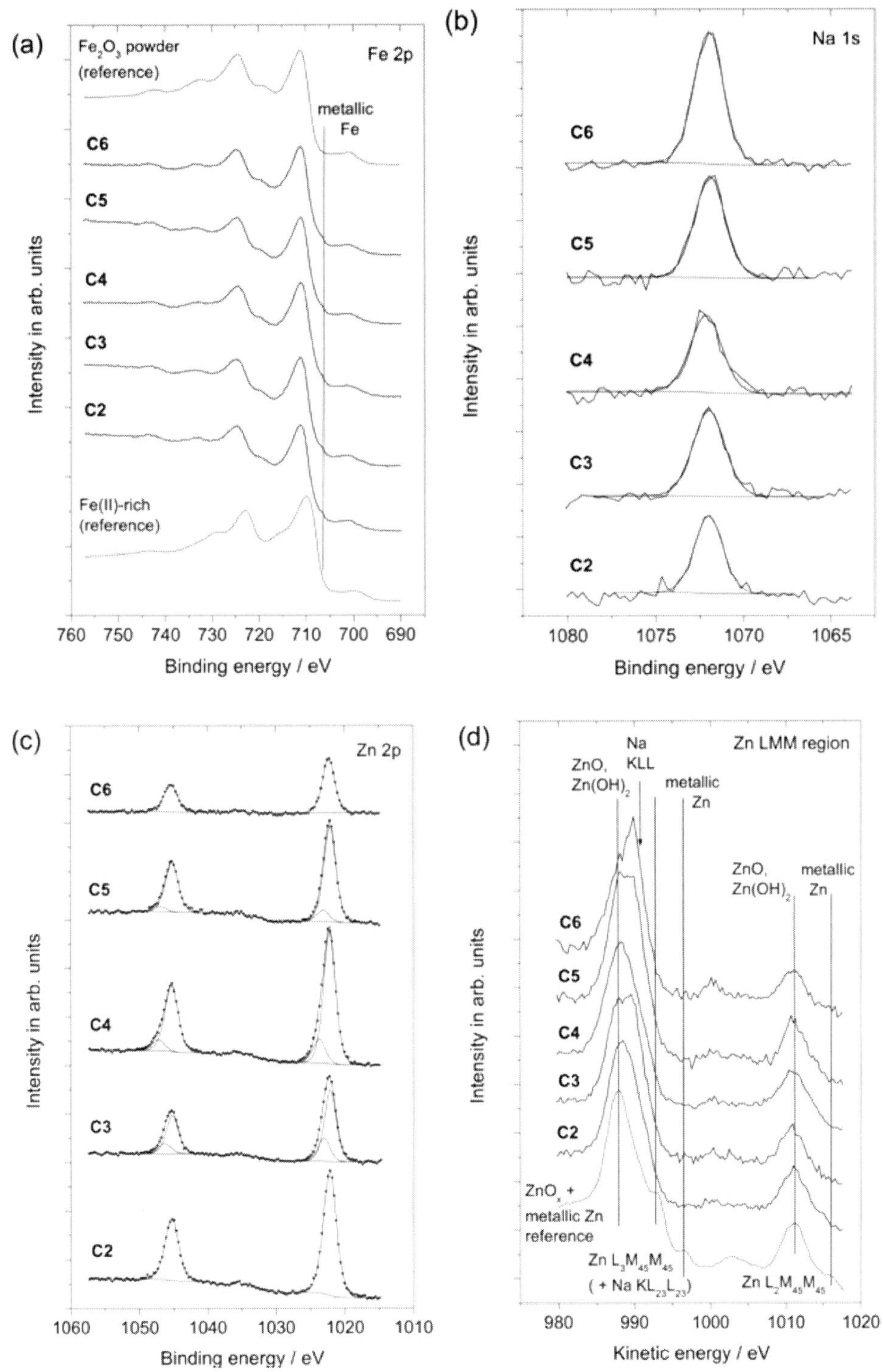

Figure 18. XPS spectra of particular regions measured in the paint-steel interface of the immersion tested paint coatings: (a) Fe 2p with the reference spectra of Fe_2O_3 powder (rich in Fe(III) states) and the Fe(II) rich evaporated thin film, (b) Na 1s, (c) Zn 2p and (d) the X-ray excited Auger Zn LMM spectra, where intensity of the peaks correlate with the element contents of the samples. A spectrum of a Zn oxide/hydroxide containing layer with some metallic Zn contribution is also shown as reference.

Hence, the same value is found in this work at the paint-steel surface. Nevertheless, because of the low chlorine level, NaCl is not a particularly probable interfacial compound in this case.

The intensity related contents of zinc presented in Figure 18c also varied from sample to sample. Under the coatings of *C2* and *C6*, rather symmetric Zn $2p_{3/2}$ line was detected around 1022.1-1022.2 eV binding energy, while in the case of coatings *C3*, *C4* and *C5* more asymmetric peak shapes were obtained. In the latter cases fitting was accomplished by two components, with the minor contribution around or above 1023 eV. As far as the identification of the compounds contributing to the main zinc line is concerned in accordance with the literature data [64], metallic zinc as well as the range of zinc oxides and certain halides exhibit a $2p_{3/2}$ binding energy around 1022 eV. Although the Zn LMM Auger line positions show pronounced variations in these materials [64], the exact overlap of the most intense Zn Auger peak and the Na KLL Auger line still complicates the analysis of the chemical state of Zn in the present samples. Nevertheless, data obtained on the Zn LMM Auger peak region are presented in Figure 18d. To assist peak assignment, a reference spectrum taken from zinc oxide/hydroxide surface is shown with some metallic zinc. It is obvious that the Zn $L_3M_{45}M_{45}$ structure measured at the paint-steel interface exhibits significant spectral intensity around the kinetic energy of 988 eV, which is characteristic for the zinc oxides. On the other hand, contributions from metallic zinc around 992-993 eV and 996-997 eV are missing, indicating that metallic Zn in the paint-steel interfaces is negligible. Still, there is a considerable spectral contribution around 990 eV, which is missing from the reference spectrum and whose intensity varies in accordance with the sodium contents of the samples. This peak can be thus assigned to the Na $KL_{23}L_{23}$ Auger line. Its kinetic energy around 990 eV is in agreement with the possible assignments given in the table. The minor Zn $2p_{3/2}$ peak observed in the samples coated with *C3*, *C4* and *C5* can be assigned to a range of corrosion products such as $ZnCl_2$, zinc hydroxide or $Zn(OH)_x(CO_3)_y$ like compounds [65]. Potassium contents of the coatings are at or below the detection level of the experiment. The carbon content characterizes the amount of paint residues remaining on the surface after the solvent treatment, which varied a little from sample to sample. Nevertheless, the relative amounts of the oxidized carbon species showed only small variations, confirming the lack of severe oxidative degradation of the organic matrix in the interfacial region. Similarly, no strong variation in the O 1s line shape was observed, where the strongest component was always the metal-oxide related contribution. Qualitative evaluation of the XPS results based on the relative quantities of the main components provided important information on the corrosion resistive abilities of the paint coatings. These evaluation results are summarized in Table 5. As a comparison between the hydrophilic alumina contained zinc-rich *ZA* and the traditional type zinc-rich *Z*, the high relative proportion of zinc hydroxide compared to the contents of zinc oxide, iron oxide/hydroxide and carbon confirms an intense but low efficiency galvanic function of the *ZA* despite the EIS characterized low electrolytic permeability. This is signified by the comparison with the greater porosity featured but more efficiently functioning *Z*, exhibiting the accumulation of low quantity of zinc hydroxides along with the formation of similar relative amounts of iron oxide/hydroxide. Almost the same oxidative conversion, especially the carboxyl yield of epoxy suggests intense electron-transfer-reaction related galvanic function of the primers and the viable cathodic polarization driven ORRs, leading to severe degradation of the vehicle in the coating-steel interfacial region of the *ZA* and *Z*.

Table 5. The ratio of particular species based on the XPS quantification data acquired from the surface of immersion tested steel specimens

Forms of elements	*E*	*Z*	*ZA*	*C1*	*C2*	*C3*	*C4*	*C5*	*C6*
$ZnO_x(OH)_y$/ carbon content	-	0.23	0.46	0.05	0.08	0.04	0.10	0.04	0.04
$Zn(OH)_2$/ carbon content	-	0.01	0.05	0.01	<0.01	<0.01	0.01	<0.01	<0.01
$Zn(OH)_2$/ ZnO	-	0.05	0.12	0.21	~0.10	0.30	0.16	0.11	~0.10
$ZnO_x(OH)_y$/ $FeO_x(OH)_y$	-	0.43	0.79	0.22	0.18	0.13	0.21	0.15	0.07
$FeO_x(OH)_y$/ carbon content	-	0.52	0.58	0.25	0.46	0.30	0.46	0.28	0.57
COOH yield of epoxy (%)	2	11	12	8	7	6	3	6	5

Corresponding to the high porosity of zinc-rich *Z,* increased rate of sacrificial and self-corrosion of zinc is indicated by the amounts of carbon normalized zinc corrosion products, exceeding the ratio obtained for the all hybrid coatings. On the other hand, zinc corrosion products composed of not passivating hydroxides were detected around the same or even lower relative quantities on the steel surface coated with hybrid paints. Beyond the higher barrier nature of the *C1–C6* hybrid coatings, the low amount of greater proportion of zinc hydroxides reflect on the active galvanic function of the primers with greater efficiency steel protection. The ratio of oxidized species of zinc and iron also showed a similar tendency. So, the relative quantities of the metallic iron compared to its oxidized forms became dominant for the *Z* and *ZA* coatings compared to other hybrid coatings except for the *C6.* In addition, the altered ratio of oxidized iron and the carbon contents is minor affected by the higher portion of paint residues on the surface of steel panels. This is in agreement with various degree of oxidative degradation – carboxyl yield of the epoxy binder and the related extent of de-adhesion, delamination of the coatings. Nonetheless, increased hydroxyl and carbonyl contents in *C4* and *C6* as well as the carboxyl conversion related oxidative degradation of epoxy in *C3* and *C5* are connected to the greater electrical bulk conductivity and galvanic function of the nanotube-loaded hybrid primers with the implied catalytic activity of the nanotubes for ORRs [77-79].

Salt-Spray Chamber Test

The wet and dry cycled salt mist test provided harsh enough environment to characterize all around corrosion resistive function of the paint coatings. Figures 19 and 20 depict coating samples from which the photos were taken right after drying and removing salt traces remained from the chamber test.

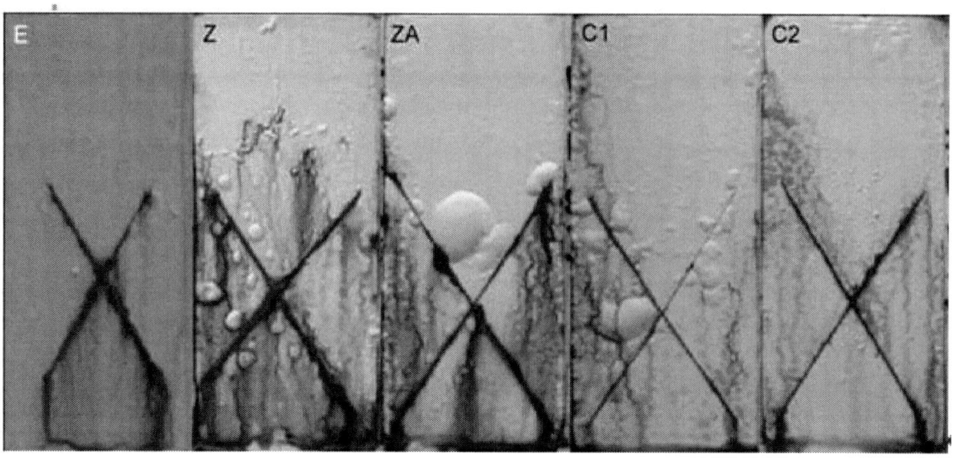

Figure 19. Photographs of cyclic salt-spray chamber tested paint coatings, after a 142 day period, except for the neat epoxy paint coating which was tested for 16 days.

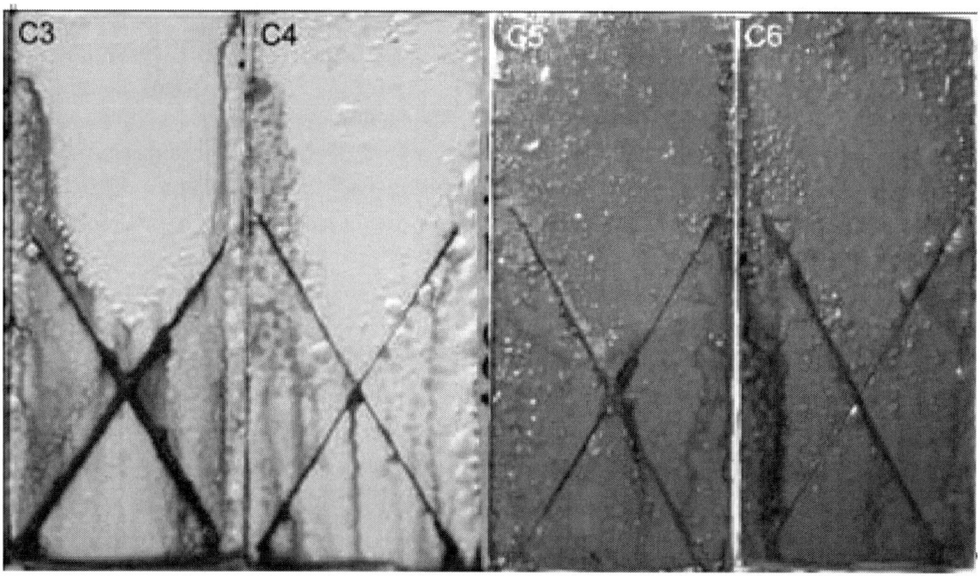

Figure 20. Photographs of cyclic salt-spray chamber tested paint coatings, after a 142 day period.

The neat epoxy (*E*) did not prevent rapid rust formation but exhibited fine condition at intact areas (Figure 19), according to the excellent passive protection ability of the pigment and particle-free paint coatings. Therefore, chamber test was terminated within 16 days. The alumina contained zinc-rich *ZA* indicated severe coating damages and stain formation around the X-cut which are attributed to the low electrical conductivity related poor galvanic function of the primer. However, the better condition of *ZA* compared to *Z* at the buried surface is connected to the highly dispersed nano-size alumina filler caused decreased porosity of the primer. Although efficient galvanic action of the *Z* was reflected by the slightly migrated delamination front around the cuts, the low resistivity against blistering and the intensity of rusting support the idea of only moderate overall protection capability. In the case of hybrid coatings, the highly dispersed PPy-deposited alumina particles loaded *C1* indicated excellent rust inhibition but failed at coating delamination, which suggest intense but less efficient galvanic function of the primer. This is probably due to a somewhat decreased barrier profile of the coating. In contrast, *C2* exhibited very good state regarding all aspects of the usual measures. On the other hand, polyelectrolyte modified nanotube contained *C3* afforded moderate rust formation inhibition but excellent resistivity against delamination and blistering (Figure 20). Furthermore, the coatings of *C4* and *C5* provided superior steel protection reflected by the insignificant intensity steel rusting and strong resistance against delamination. However, the apparently low susceptibility to blistering was noted at the buried areas. In addition, the moderate protection ability of the *C6* is concluded based on the density and size of blistered spots and the intensity of steel rusting, besides the valuable active function against delamination (Table 6).

Table 6. Evaluation of salt-spray chamber test results corresponding to the EN ISO 4628-8:2005 standard, regarding degree of delamination and blistering of paint coatings and steel rusting

Coatings	Blistering size and density		Rusting	Delamination
E	2	1	Ri 4	considerable
Z	3	2	Ri 3	slight
ZA	2	2	Ri 4	severe
C1	2	2	Ri 2	considerable
C2	2	2	Ri 2	very slight
C3	2	3	Ri 3	very slight
C4	3	2	Ri 3	slight
C5	3	3	Ri 3	slight
C6	3	2	Ri 2	slight

As for the repeatability of the experimental results, multiple coating samples were formulated for parallel corrosion test procedures. Immersion tests were repeated at the other surface area of the same coating and on another sample, indicating less than 10 relative percent differences in the state and lifetime of the coatings measured by the extent of scattering in the open-circuit potentials and the impedance data over the whole test periods. Quality assessment of the parallel samples was performed to comply with the ISO standard to gain overview about the repeatability of the all around protection characteristics of the paint coatings over certain time periods. The ongoing visual observation of the intensity, surface density and homogeneity of steel rusting and coating blistering were completed with checking the adhesion of the primers at the end of tests.

Minor differences in the corrosion resistive function are connected to the changes in the sub-micron dispersity of the inhibitor particles and their distribution in the epoxy around the zinc pigments, with some local fluctuation in the concentration of the highly anisotropic nanotube-embedded particles. These factors are mainly connected to the repeatability of the paint formulation and film formation procedures (effect of grinding the particles, mixing and homogenizing paint dispersions).

Given the fact that electrical percolation is contributed by the higher nanotube contents than rheological percolation because effective restriction of the mobility of polymer chains becomes pronounced at lower nanotube concentration than the formation of extensive three-dimensional network for electron conduction [66-68] in epoxy around 0.5 wt.% [69,70]. Therefore, owing to the immersion and salt-spray chamber test results, it is highly obvious that nanotube-embedded anodic PDIPs have an influence on the galvanic efficiency, cathodic corrosion preventing function of the zinc-rich hybrid coatings at zinc content of 70 wt.% (25.7% by volume) compared to the conventional liquid ZRPs with 90 wt.% zinc contents (55.8% by volume). In hybrid paints, it is to be noted that concentration of the nanotubes was around or lower the theoretically and experimentally defined geometrical and rheological percolation – entanglement regime of the nanotubes. Due to the various degree of polymer-complex assembly caused aggregation and flocculation of the PDIPs, rheological percolation – effective interparticle interaction should not be experienced by rheology characterization. On the other hand, despite the low nanotube contents, effective contribution to the electrical contacts between the zinc pigments is expected to be affected by the application of higher proportion of auxiliary electrically conducting particles – nano-size alumina supported PPy whose concentration was around the percolation threshold in the inhibitor particles [59].

In addition, due to the XPS and RF-GD-OES indicated low intensity of zinc corrosion, high performance and durable corrosion resistance of the hybrid coatings are surmised to take advantage of additional function of the highly dispersed PDIPs beyond the lower rate of zinc corrosion, such as mediating the interfacial electron transfer reactions (allowing moderate cathodic polarization).

Regarding the remarkably durable adhesion of the hybrid paints, PDIPs is accounted for scavenging oxidative radicals [71] and eliminating strong nucleophile agents [72-74] taking a vital role in preventing chemical corrosion processes. These functionalities are able to moderate degradation rate of the epoxy vehicle, accumulation and leaching of the corrosion products [75,76] as well as delaying cohesive and adhesive failure of the binder, thereby suppressing the processes which ultimately lead to cathodic delamination phenomena. All these aspects reflect on the positive influence of the PPy-modified carbon nanotubes on the corrosion preventive performance of the zinc-rich hybrid paint formulations over the traditional type ZRPs.

However, utilization of the nanotubes must meet some relevant requirements prior to successful adoption into real-life applications. These are the high exfoliation and even distribution of the particles to promote acceptable micro-scale homogeneity in the paint coatings, as well as suppressing catalytic activity towards oxygen reduction reactions [77-79]. In the present case, the latter is ensured by the uniform and compact PPy layer deposition helping to retain preferred physical and mask undesired chemical nature of the nanotubes. Thus, appropriate nano-particle compositions and paint formulations with the nanotubes are expected to initiate substantial improvement in the corrosion preventing function of the metal-rich primer coatings, thereby opening new vistas to multiple ranges of corrosion protection applications.

CONCLUSION

Although aqueous-ethanol-prepared PPy-deposited alumina particles contribute effectively to the corrosion preventive function of the zinc-rich epoxy primers, massive protection performance improvement is gained by the use of intimately structured anodic inhibitor particles composed of nano-size alumina and nanotube carriers supported PPy. This approach provides viable solution to the yet unresolved percolation and porosity related phenomena of the ZRPs, despite the filamentous-like aggregated microstructure caused lower micro-scale dispersity of the nanotube-incorporated particles.

As a result of well-balanced active–passive protection functionalities, optimal corrosion resistance is afforded by the hybrid paint coatings composited with polyelectrolyte-modified MWCNT (at 7.1×10^{-3} volume fraction) embedded particles containing polypyrrole in 3.56 wt.%. Furthermore, stable steel protection is provided against corrosion by the hybrid coatings with 5.0×10^{-3} and 7.0×10^{-4} volume fractions of the low sulfonated nanotubes at 5.79 and 7.54 wt.% polypyrrole contents in the inhibitor particles. Zinc-rich hybrid formulation with highly sulfonated MWCNT-embedded inhibitor particles provide intense galvanic function with somewhat lower efficiency barrier profile of the paint coatings at polypyrrole and nanotube volume fractions of 4.6×10^{-3} and 1.3×10^{-3}, respectively.

Firm protection ability of the hybrid coatings is connected to the enhanced redox properties and conductivity of the nanotube-supported and unexchangeable counter-ion doped PPy. Moreover, the electrical conductivity between the zinc pigments is effectively contributed since the efficacy of galvanic function of the zinc-rich hybrid primers is increased along with the inhibition of zinc corrosion and the improved barrier nature of the coatings. Moderate galvanic action of the hybrids takes longer time to develop but the evolved passive function of the coatings more durable than in the case of traditional liquid ZRPs. This is due to the enhanced barrier behaviour reinforced by the PPy-modified alumina/carbon nanotube inhibitor particles and the accumulation, long-term pore-filling corrosion products.

Owing to the limited exploitation of hybrid paints, zinc-rich primers containing polypyrrole modified nanotubes at low amounts are recognized as worthy candidates to be utilized in cost effective and environmentally friendly organic coating applications.

ACKNOWLEDGMENTS

This work was partially carried out as part of the TÁMOP-4.2.1.B-10/2/KONV-2010-0001 project in the framework of the New Hungarian Development Plan. The realization of this project is supported by the European Union, co-financed by the European Social Fund. Péter Németh is gratefully acknowledged for the TEM observations.

REFERENCES

[1] C. M. Abreu, M. Izquier, M. Keddam, X. R. Nóvoa and H. Takenouti, *Electrochim. Acta* 41, 240 (1996).

[2] V. Zivica, *Bull. Mater. Sci.* 25, 371 (2002).

[3] S. G. Real, A. C. Elias, J. R. Vilche, C. A. Gervasi and A. Di Sarli, *Electrochim. Acta* 38, 2029 (1993).
[4] D. Pereira, J. D. Scantlebury, M. G. S. Ferreira and M. E. Almeida, *Corros. Sci.* 30, 1135 (1990).
[5] C. M. Abreu, M. Izquierdo, P. Merino, X. R. Nóvoa and C. Pérez, *Corrosion* 55, 1173 (1999).
[6] R. N. Jagtap, R. Nambiar, S. Z. Hassan and V. C. Malshe, *Prog. Org. Coat.* 58, 253 (2007).
[7] J. R. Vilche, E. C. Bucharsky and C. A. Giúdice, *Corros. Sci.* 44, 1287 (2002).
[8] S. Shreepathi, P. Bajaj and B. P. Mallik, *Electrochim. Acta* 55, 5129 (2010).
[9] M. T. Rodríguez, J. J. Gracenea, J. J. Saura and J. J. Suay, *Prog. Org. Coat.* 50, 68 (2004).
[10] H. Shi, F. Liu and E.-H. Han, *Surf. Coat. Technol.* 205, 4532 (2011).
[11] R. N. Jagtap, P. P. Patil and S. Z. Hassan, *Prog. Org. Coat.* 63, 389 (2008).
[12] C. A. Gervasi, A. R. Di Sarli, E. Cavalcanti, O. Ferraz, E. C. Ducharsky, S. G. Real and J. R. Vilche, *Corros. Sci.* 36, 1963 (1994),
[13] H. Marchebois, M. Keddam, C. Savall, J. Bernard and S. Touzain, *Electrochim. Acta* 49, 1719 (2004).
[14] A. Meroufel and S. Touzain, *Prog. Org. Coat.* 59, 197 (2007).
[15] D. Wang and G. P. Bierwagen, *Prog. Org. Coat.* 64, 327 (2009).
[16] O. Øystein Knudsen, U. Steinsmo and M. Bjordal, *Prog. Org. Coat.* 54, 224 (2005).
[17] Y. Shao, C. Jia, G. Meng, T. Zhang and F. Wang, *Corros. Sci.* 51, 371 (2009).
[18] J. S. Hammond, J. W. Holubka, J. E. deVries and R. A. Dickie, *Corros. Sci.* 21, 239 (1981).
[19] M. A. Hernandez, F. Galliano and D. Landolt, *Corros. Sci.* 46, 2281 (2004).
[20] E. Armelin, M. Martí, F. Liesa, J. I. Iribarren and C. Alemán, *Prog. Org. Coat.* 69, 26 (2010).
[21] A. Meroufel, C. Deslouis and S. Touzain, *Electrochim. Acta* 53, 2331 (2008).
[22] Steven K. Geer and Todd R. Hawkins, *United States Patent* No. 7,595,009, Sept 29, 2009.
[23] Sze Cheng Yang and Wenguang Li Yang, *United States Patent* No. 6,803,446, Oct. 10, 2004.
[24] W. Wunderlich and M. Takahashi, Ceramic Transactions Vol. 133 Ed. M. Matsui, et al., Am. Ceram. Soc. (2002) pp 189–194.
[25] E. Akbarinezhad, M. Ebrahimi, F. Sharif, M. M. Attar and H. R. Faridi, Prog. Org. Coat. 70, 39 (2011).
[26] R. N. Rothon, *Adv. Polym. Sci.* 139, 67 (1999).
[27] J. I. Martins, T. C. Reis, M. Bazzaoui, E. A. Bazzaoui and L. Martins, *Corros. Sci.* 46, 2361 (2004).
[28] D. R. Paul and L. M. Robeson, *Polymer.* 49, 3187 (2008).
[29] G. Paliwoda-Porebska, M. Stratmann, M. Rohwerder and K. Potje-Kamloth, *Corros. Sci.* 47, 3216 (2005).
[30] J. Bonastre, P. Garcés, F. Huerta, C. Quijada, L. G. Andión and F. Cases, *Corros. Sci.* 48, 1122 (2006).
[31] P. Ocón, A. B. Cristobal, P. Herrasti and E. Fatas, *Corros. Sci.* 47, 649 (2005).

[32] B. N. Grgur, N. V. Krstajic, M. V. Vojnovic, C. Lacnjevac and Lj. Gajic-Krstajic, *Prog. Org. Coat.* 33, 1 (1998).
[33] M. Tiitu, A. Talo, O. Forsén and O. Ikkal, *Polymer.* 46, 6855 (2005).
[34] R. Partch, S. G. Gangolly, E. Matijevic, W. Cai and S. Arajs, *J. Coll. Interface Sci.* 144, 27 (1991).
[35] H. Chriswanto, H. Ge and G. G. Wallace, *Chromatographia* 37, 423 (1993).
[36] S. Maeda and S. P. Armes, *J. Mat. Chem.* 4, 935 (1994).
[37] A. Malinauskas, *Polymer* 42, 3957 (2001).
[38] M. Ionita and A. Pruna, *Prog. Org. Coat.* 72, 647 (2011).
[39] E. Akbarinezhad, F. Rezaei and J. Neshati, *Prog. Org. Coat.* 61, 45 (2008).
[40] J. F. Moulder W. F. Stickle, P. E. Sobol and K. D. Bomben, Handbook of X-Ray Photoelectron Spectroscopy, Perkin-Elmer Corp., Eden Prairie, Minnesota, USA (1992).
[41] C. D. Wagner, A. V. Naumkin, A. Kraut-Vass, J. W. Allison, C. J. Powell and J. R. Rumble Jr., NIST X-ray Photoelectron Spectroscopy Database, Version 3.4, National Institute of Standards and Technology, Gaithersburg, MD 2003, http://srdata.nist.gov/xps/
[42] N. Fairley, www.casaxps.com/
[43] M. Mohai, *Surf. Interface Anal.* 36, 828 (2004).
[44] M. Mohai, XPS MultiQuant: Multi-model X-ray photoelectron spectroscopy quantification program, Version 3.00.16 (2003), http://www.chemres.hu/aki/XMQpages/XMQhome.htm/
[45] S. Krishnakumar and P. Somasundaran, *Colloids Surf.* 117, 37 (1996).
[46] G. P. Funkhomer, M. P. Arévalo, D. T. Glatzhofer and E. A. O'Reart, *Langmuir.* 11, 1443 (1996).
[47] P. Somasundaran, *J. Phys. Chem.* 70, 90 (1966).
[48] D. Wu, W. Liang and Z. Ming, *J. Polym. Sci. B* 45, 2239 (2007).
[49] E. S. G. Shaqfeh and G. H. Fredrickson, *Phys. Fluids* A 2, 7 (1990).
[50] M. S. P. Shaffer, X. Fan and A. H. Windle, *Carbon* 36, 1603 (1998).
[51] M. Poorteman, M. Traianidis, G. Bister and F. Cambier, *J. Eur. Ceram. Soc.* 29, 669 (2009).
[52] M. Poorteman, P. Descamps, F. Cambier and A. Leriche, *J. Eur. Ceram. Soc.* 12, 103 (1993).
[53] C. M. Hansen, Hansen solubility parameters: a user's handbook CRC Press, 2007.
[54] G. Socrates, Infrared and Raman Characteristic Group Frequencies, 3rd edition, 2001, John Wiley and Sons, Ltd. Baffins Lane, Chichester, England, 182.
[55] B. Tian and G. Zerbi, *J. Chem. Phys.* 92, 3892 (1990).
[56] A. Gergely, É. Pfeifer, I. Bertóti, T. Török, E. Kálmán, Corr. Sci. 53, 3486 (2011).
[57] O. Patil, A. J. Heeger and F. Wudl, *Chem. Rev.* 88, 183 (1988).
[58] L. Liu, C. Zhao, Y. Zhao, N. Jia, Q. Zhou, M. Yan and Z. Jiang, *Europ. Polym. J.* 41, 2117 (2005).
[59] G. Cho, D. T. Glatzhofer, B. M. Fung, W.-L. Yuan and E. A. O'Rear, *Langmuir.* 16, 4424 (2000).
[60] P. A. Sorensen, K. Dam-Johansen, C. E. Einell and S. Kill, *Prog. Org. Coat.* 68, 283 (2010).
[61] T. Nguyen, J. B. Hubbard and J. M. Pommersheim, *J. Coat. Technol.* 68, 45 (1996).

[62] F. Deflorian and S. Rossi, *Electrochim. Acta* 51, 1736 (2006).
[63] G. Bhargava, I. Gouzman, C. M. Chun, T. A. Ramanarayanan and S. L. Bernasek, *Appl. Surf. Sci.* 253, 4322 (2007).
[64] S. Schürz, G. H. Luckeneder, M. Fleischanderl, P. Mack, H. Gsaller, A. C. Kneissl and G. Mori, *Corros. Sci.* 52, 3271 (2009).
[65] B. Zhang, H.-B. Zhou, E.-H. Han and W. Ke, *Electrochim. Acta* 54, 6598 (2009).
[66] Q. Zhang, S. Rastogi, D. Chen, D. Lippits and P. J. Lemstra, *Carbon* 44, 778 (2006).
[67] J. Sumfleth, S. T. Buschhorn and K. Schulte, *J. Mater. Sci.* 46, 659 (2011).
[68] J. Zhu, S. Wei, A. Yadav and Z. Guo, *Polymer.* 51, 2643 (2010).
[69] Y. S. Song and J. R. Youn, *Carbon* 43, 1378 (2005).
[70] Y. J. Kim, T. S. Shin, H. D. Choi, J. H. Kwon, Y. C. Chung and H. G. Yoon, *Carbon* 43, 23 (2005).
[71] M. N. Ismail, M. S. Ibrahim and M. A. A. El-Ghaffar, *Polym. Degrad. Stab.* 62, 337 (1998).
[72] T. Osaka, T. Fukuda, K. Ouchi and T. Momma, *Thin Solid Films* 215, 200 (1992).
[73] O. Inganäs, R. Erlandsson, C. Nylander and I. Lundström, *J. Phys. Chem. Solids* 45, 427 (1984).
[74] F. Beck, P. Braun and F. Schloten, *J. Electroanal. Chem.* 267, 141 (1989).
[75] G. Grundmeier, W. Schmidt and M. Stratmann, *Electrochim. Acta* 45, 2515 (2000).
[76] G. Williams, R. J. Holness, D. A. Worsley and H. N. McMurray, *Electrochem. Commun.* 6, 549 (2004).
[77] X. Y. Zhu, S. M. Lee, Y. H. Lee and T. Frauenheim, *Phys. Rev. Lett.* 85, 2757 (2000).
[78] G. Jürmann and K. Tammeveski, *J. Electroanal. Chem.* 597, 119 (2006).
[79] I. Kruusenberg, N. Alexeyeva and K. Tammeveski, *Carbon* 47, 651 (2009).

In: Applications of Carbon Nanotubes
Editor: Ajay Kumar Mishra
ISBN: 978-1-62081-918-0

Chapter 11

CARBON NANOTUBES - BASED SENSORS

A. Maity[1*], Ch. V. V. Ramana[2] and M. K. Moodley[2]
[1]Smart Polymers Group, Polymers and Composites,
Materials Science and Manufacturing, Council for Scientific and
Industrial Research, Pretoria, South Africa
[2]Department of Physics, School of Chemistry and Physics,
University of Kwa-Zulu Natal, Westville Campus, Durban, South Africa

ABSTRACT

Carbon nanotubes (CNT), made by layer of graphite rolled-up into cylinder, are the subject of one of the most important areas of research in nanotechnology. Their commercial applications ranging from electronics to chemical process control have meant that an enormous amount of efforts have been undertaken on the investigation of CNTs in the past decades. Despite this high level of research activity, very little attention has been paid to potential applications in the sensors with CNTs. Sensors are devices that detect or measure physical and chemical quantities such as temperature, pressure, sound, and concentration. The measurands are converted into an electrical signal. The main requirements of a good sensor are high sensitivity, fast response, low cost, high volume production, and high reliability. Sensors continue to make significant impact in everyday life with applications ranging from biomedical to automotive industry. This has led to intensive research activities across the world in developing new sensing materials and technologies. However, since the discovery of CNTs by Iijima, CNTs has generated keen interest among researchers to develop CNTs-based sensors for many applications. The application of CNTs in next-generation of sensors has the potential of revolutionizing the sensor industry due to their inherent properties such as small size, high strength, high electrical and thermal conductivity, and high surface area. CNTs can be introduced in devices in two ways: (1) either by depositing previously grown nanotubes material from a liquid suspension or by transferring from a solid source or (2) the nanotubes may be grown in location by chemical vapour deposition (CVD). While deposition from a liquid may be combined with surface chemistry to attach pre-treated or selected nanotubes to specific areas of the chip, the latter technique allows for the direct growth of nanotubes at the desired location determined by lithographically defined catalyst patterns without

[*] E-mail: amaity@csir.co.za.

further treatment of the CNTs. For mass production, accurate control of these methods will be crucial. Although the use of CNTs for sensing applications is promising, they are also associated with certain limitations. The objectives of this chapter are: (i) to describe the synthesis, purification, and properties of CNTs; (ii) to perform an extensive survey to highlight the areas where CNTs-based sensors can be applied; (iii) to outline the potential challenges in commercializing CNTs-based sensors; and (iv) to suggest possible research areas for addressing technical challenges associated with CNTs-based devices.

1. Introduction to Sensor Technology

A sensor is a device which detects a physical quantity and converts into a signal which can be interpreted as a good representation of the quantity being "sensed". Clearly, this is a very broad and general description of a sensor due to the fact we find sensors of many different types and function occupying virtually every sphere of our lives.

In the human body, the recognized five senses of sight (ophthalmoception), hearing (audioception), taste (gustaoception), smell (olfacoception or olfacception), and touch (tactioception). Further to this, other senses include temperature (thermoception), kinesthetic sense (proprioception), pain (nociception), balance (equilibrioception) and acceleration (kinesthesioception). The human body (Figure 1) has five sensing capabilities namely smell (chemical-electrical), touch (mechanical), sight (optical), hearing (mechanical) and taste (chemical).

All of these physiological engineered "devices" are more advanced in function than any man-made device. It is the aim of the sensor designer to emulate the capabilities of our naturally occurring sensors and apply them to situations where humans are not required or where in extreme and dangerous environmental conditions where, it is difficult for humans to function or provide reliable information. Furthermore, it is not practical to have a human just to operate as sensor.

In order to understand the world around us, it is important by means of develop and ways of gathering information. This information can be very useful in understanding how the Universe works and also help us predict disasters either man-made or natural and ultimately make the world a better place to live in.

In all these, sensors play an important role. It is therefore important that the sensors we develop provide reliable information, remain very sensitive, low cost and have a long life-cycle.

In recent years, more efforts in improving sensor technology were made due to the discovery of new materials and new forms of old materials which exhibit properties that could be exploited for superior sensing devices.

The carbon nanotube, a tubular form of a graphene sheet, took the science scene by storm in the early 1990's. Research on this material followed the discovery of another common form of carbon: the C_{60} molecule commonly referred to as the Bucky ball for which the 1995 Nobel Prize was awarded to Smalley et al. [1-2]. Unlike C_{60}, the single walled carbon nanotube (SWCNT) was shown to be material of great promise and potential and a candidate replacement of silicon in integrated circuits [3].

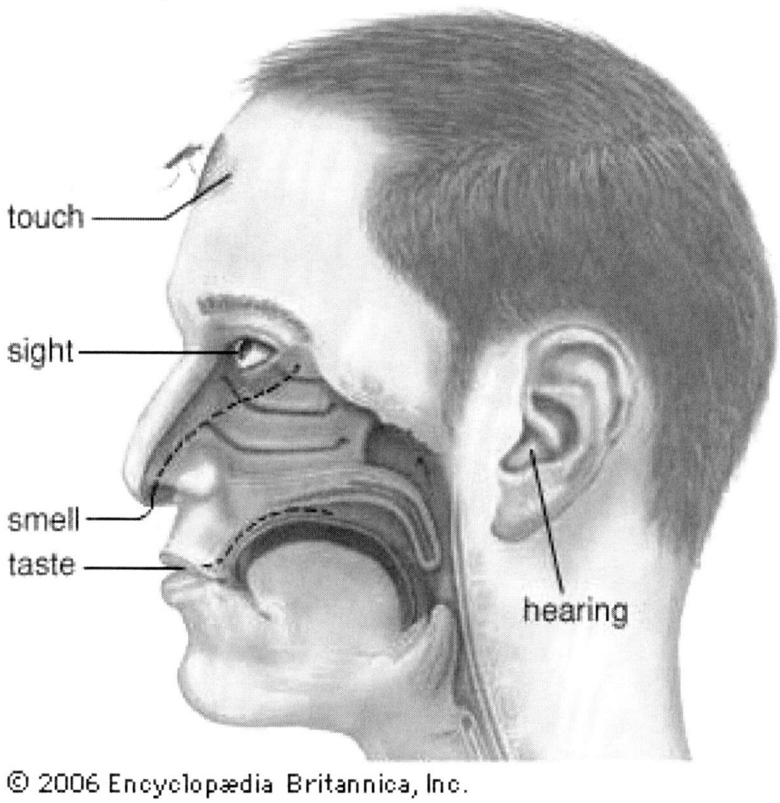

Figure 1. The five senses [Encyclopedia Britannica].

However, much works still needs to accomplished regarding the controlled synthesis and the exact positioning of nanotubes into a functional IC device and also scalable to mass production. Nevertheless in other applications, carbon nanotubes such as in sports equipment [4-5], nanotube inks [6], fibre composites [7] and other specialized materials sold commercially for research purposes. Slowly but surely, the direct use of carbon nanotubes in advanced such as in sensors will become more widespread such as those produce by the company Nanomix [7]. It is the intention of this chapter to highlight current and future research undertakings utilizing carbon nanotubes in sensor devices.

1.1. Sensor Characteristic and Mechanisms of Operation

The basic operation of a sensing device is that material which is sensitive to a change or stimulus which provides a signal that can be measured. Coupled to the sensing material is a transducer, which converts the stimulus signal into a measurable quantity. Together, they form the heart of the sensor. The dynamics of how fast a change in the stimulus signal can be measured, the rate of change in the stimulus and the transducers response determine whether the sensor device behaviour is static or dynamic in nature. These are most important attributes expanded below [8-9]:

1.1.1. Static Characteristics of a Sensor

- *Accuracy*: when the measured response by the transducer, correctly reflects the actual stimulant. Availability of reference standard increases the reliability and accuracy of the measurement.
- *Error*: The difference between the transducers measurement from true value of the stimulant. A reference standard or calibration will be important to have in critical applications.
- *Sensitivity*: The ratio of a change in the transducers response to a change stimulant. A sensitive detector would imply a linear relation between the two quantities.
- *Resolution*: The sensitive to the transducers response is to the slightest/minimalist changes in the stimulants value.
- *Selectivity*: A sensor that is able to measure a single stimulant in the presence other different stimuli describes is sensitivity. In other words sensor discriminates between two similar but different stimuli?
- *Repeatability*: Reflects the ability of the sensor to produce identical measurements under the same conditions.
- *Reproducibility*: The ability of the sensor device to produce identical measurements under identical conditions after the numerous on/off cycles of the device.
- *Stability*: The ability of the device to produce the same measurement over an extended period of time when the input stimulant is also constant in time?
- *Response time*: The time taken by the transducer to reach a stable measurement (95%) of its final value in response to a stimulant
- *Recovery time*: The time taken by the transducer to return to its baseline value upon the removal of the stimulant.

1.1.2. Dynamic Characteristics of a Sensor

The sensor response to a variable input is different from the exhibited, when the input signals are constant. The dynamic characteristics are due to energy-storing elements within the sensor device such as masses, inductances, capacitances etc. Typically, these type of devices have to be modeled by zero-order, first-order and second-order mathematical relations that link the input stimuli to the output response or measurement. Typically, in zero-order type sensors, the sensors only amplify the input signal and also do not have any energy-storing elements within the device. First-order type sensors have one element that stores energy and one that dissipates it. An example of this system would be a standard mercury thermometer immersed in a liquid. A second-order type sensors, will have the components of heat capacity and thermal resistance characterizing the devices response to a stimulus change [8-9]. For example for this type is Thermometer with heat protection.

1.2. Basics of a Sensor Device

Sensors are designed to measure physical and chemical phenomena and the principle mechanisms behind each of these are different. In this chapter, we discuss the use of carbon nanotubes as a sensor and the physical or chemical mechanism involved in this new material which make an excellent material for sensor applications. The most common transduction mechanism involved in chemical and biological sensors are the conductometric and capacitive effect. This is due largely to their ease and low cost of fabrication. Figure 2 shows a simple schematic a sensor device measuring changes in the capacitance or conductance when a sensor material is exposed to some stimulus e.g. a poisonous gas. In this device, the sensing material is coated between two electrodes connected with an applied voltage. As the sensing material interacts with the stimulant or a gas molecule, a change in the conductance is registered. In this way, given a known concentration of gas, the change in the conductance could be calibrated. In the case of capacitance, a charge build-up on one of the electrodes causes an electric field between the electrodes. The sensing material behaves as a dielectric and therefore affects the electric field. As with conductance, changes in the capacitance can be calibrated when the device is exposed to a known value of the stimulant. In most cases, in the measurement of conductance and capacitance, the electrodes take the shape of inter-digital transducers, shown in Figure 3. Variations or modifications in Figure 2, is the inclusion of a third electrode, giving rise to what is known as the field effect transistor. This is shown in Figure 4. This design provides several advantages over the two electrode design. These are: better control of the biasing voltage, faster recovery times, and higher sensitivities [10]. The drawback is that it is harder to fabricate this type of device.

1.3. CNT Chemical and Physical Sensors

We will focus our chapter on the use of carbon nanotube in chemical and physical sensors where the use of CNTs has several advantages over traditional materials such as metal oxides:

- High surface area in the region of 1000 m^2 g^{-1}[11] which creates the space for function chemical groups to be grafted onto the walls of the nanotube.
- High aspect ratio means that the nanotube appears as nanowire.
- Large absorption capacity.
- High strength and high chemical stability.
- High electrical conductivity and high current capacity.

These properties make this material attractive in advanced device applications.

2. Synthesis, Functionalization and Characterization of Carbon Nanotubes

2.1. Methods for the Synthesis of SWCNTs

There are numerous reviews and books in which the synthesis of carbon nanotubes have been described [12-16]. While there have been a plethora of methods used to synthesize carbon nanotubes, this section only highlights the most common methods used to synthesize SWCNTs.

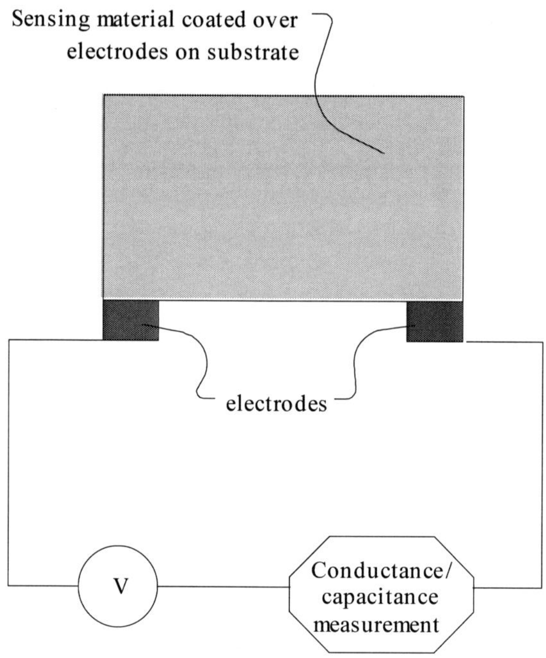

Figure 2. A typical schematic design of a sensing device to measure changes in conductance or capacitance when the device is exposed to a stimulus.

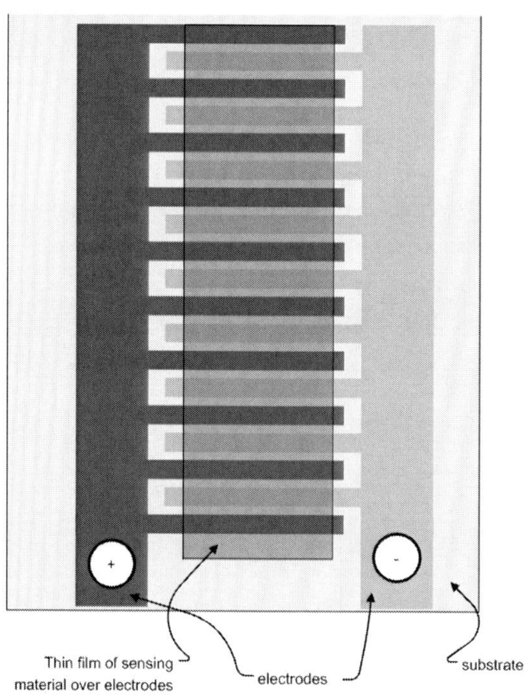

Figure 3. Inter-digital electrodes fabricated onto a substrate. Sensing material is then deposited on this finger like electrodes.

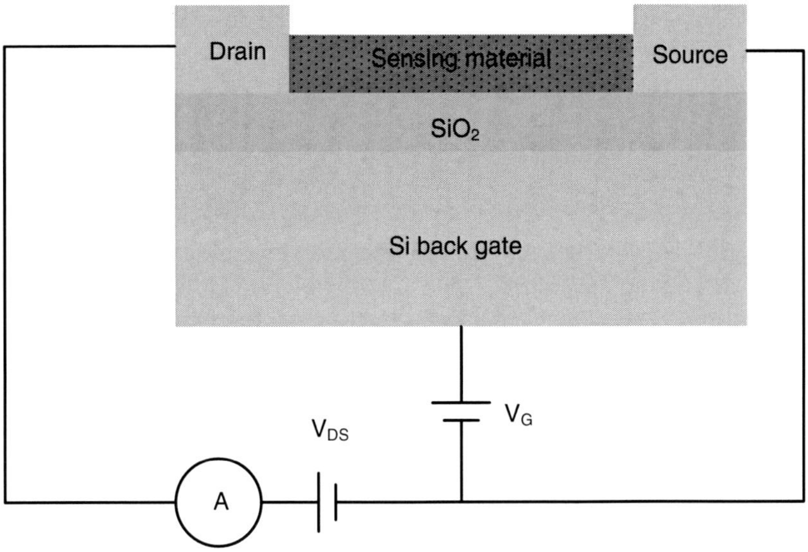

Figure 4. A schematic of a field effect transistor device.

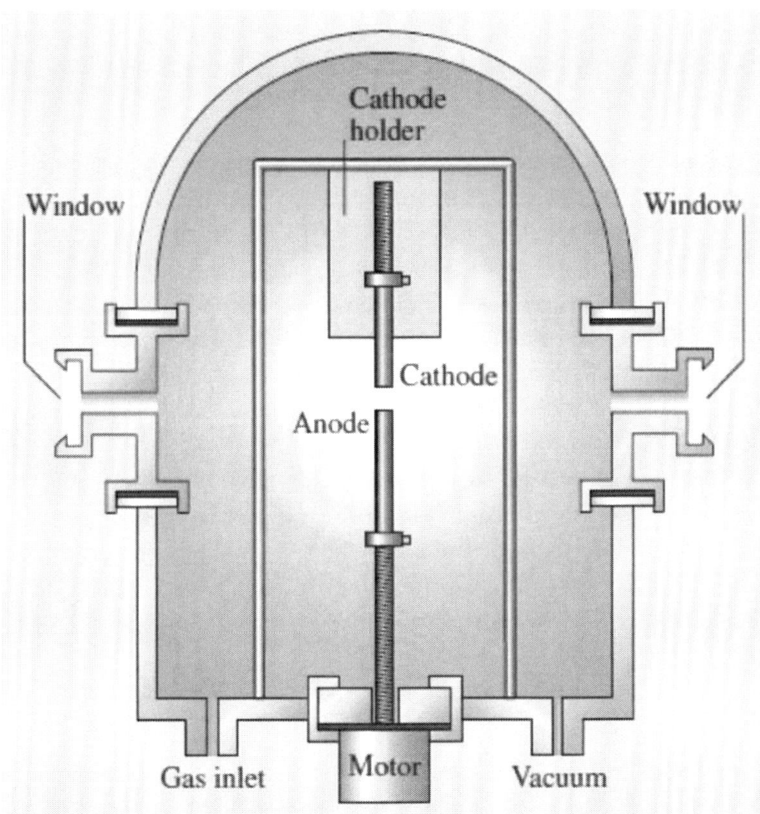

Figure 5. A schematic of an electric-arc reactor used to synthesize SWCNTs [12].

Whether the synthesis of SWCNTs occurs near the focus of a high-powered laser, or between two arcing graphite electrodes, in a hot furnace full of hydrocarbon gas, or even in the middle of a flame, the common factors and basic prerequisites are an active catalyst, a source of carbon, and adequate energy.

2.2. Arc Discharge

Arc discharge was the first recognized method for producing both SWCNTs and MWCNTs, and has been optimized to be able to produce gram quantities of the material. Arc-discharge synthesis use a low-voltage (~12 to 25 V), and high-current (50 to 120 amps) power supply such as an arc welder. An arc is generated across a 1-mm gap between two graphite electrodes 5 to 20 mm in diameter. An inert gas such as He or Ar is used as the atmosphere for the reaction, at a pressure of 100 to 1000 Torr. Iijima produced the first MWCNTs by this method [17].

He found that nanotubes formed on the cathode, along with soot and fullerenes. Iijima and Ichihashi [18] and Bethune et al. [19] were the first to report on the production of SWCNTs. Figure **5** shows schematics of an arc-discharge apparatus [12]. Both Iijima and Bethune found that SWCNTs could only be formed by adding metal catalyst's to the anode [20-21].

2.3. CVD

Chemical vapour deposition is a method in which thermal energy provided by a furnace or a plasma source is used to dissociate an organic gas or vaporize an organic chemical rich in carbon. In the CVD method, metals such as Ni, Fe or Co etc. are used as catalysts. Mo and Ru are sometimes added as promoters to the catalyst to render the feedstock more active for the formation of CNTs. In 1996 a CO-based CVD process to produce SWCNTs was developed [22] that popularized the CVD method to produce SWCNTs. The CVD process encompasses a wide range of synthesis techniques, from the gram-quantity bulk formation of nanotube material to the formation of individual aligned SWCNTs on SiO_2. Figure 6(a)-(c) shows a typical schematic of an experimental setup for a CVD process [23]. In Figure 6(a), well aligned SWCNT can be grown. This method is particular good for use in advanced electronic applications was the properties of SWCNTs can be exploited [6(a)]. Patented methods for producing SWCNTs in large quantities such as the HiPco (high pressure disproportion of CO)[24] are shown in Figure 6(b). A combination of a plasma source and CVD, called plasma enhanced CVD (PECVD) (Figure 6(c) is also a very good method for growing aligned SWCNT for advanced electronic applications [25].

2.4. Laser Ablation Synthesis

The first laser was built in 1960. Thereafter, physicists immediately made use of it as a means of concentrating/depositing large quantities of energy inside a very small volume in a relatively short time span. During the interaction between the laser beam and a material, numerous phenomena occur and each of these processes are sensitive to parameters such as the characteristics of the laser beam, the incoming power density, the nature of the target. Depending on the laser power incident on the target, the material can heat up, melt or vaporize.

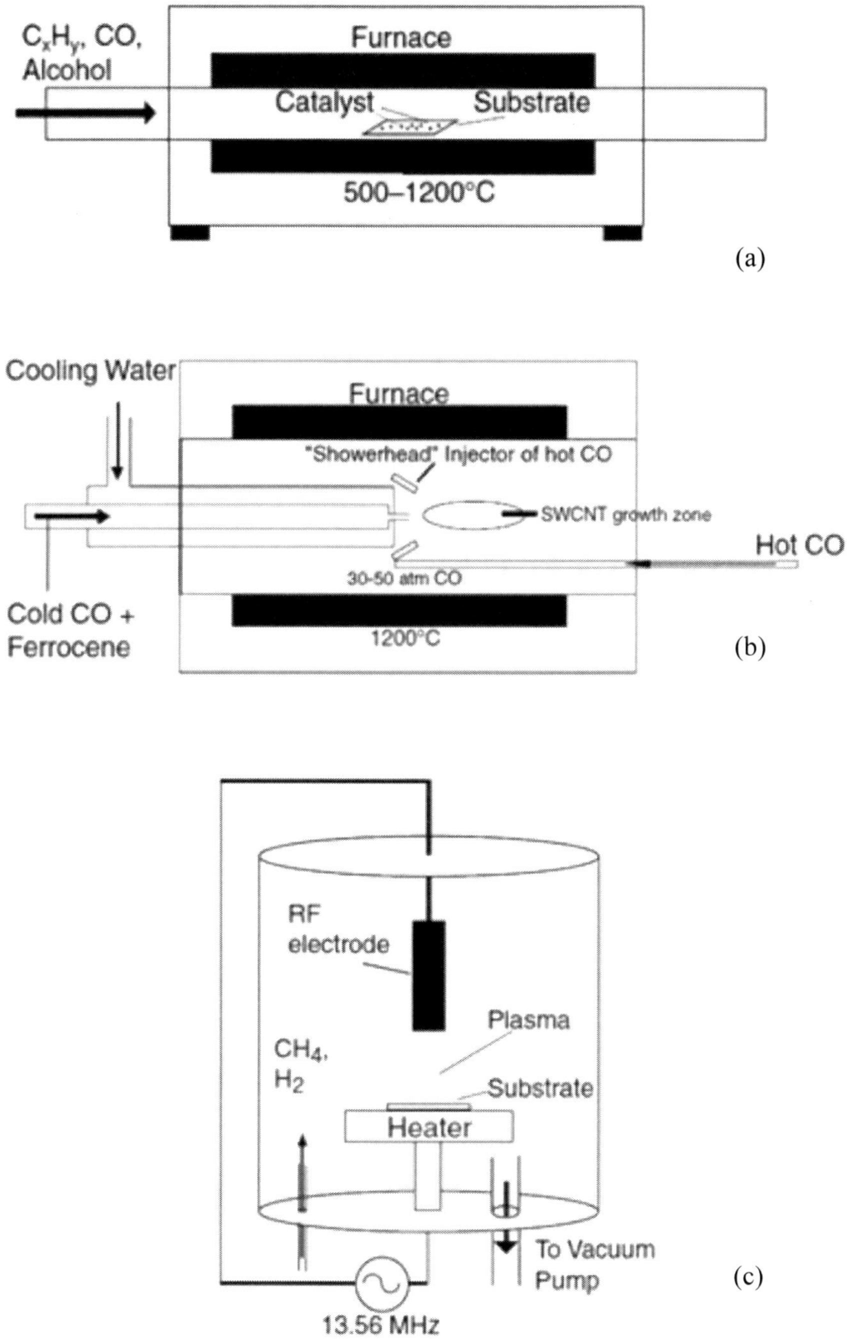

Figure 6. (a)-(c). shows a typical schematic of an experimental setup for a CVD process [23].

While a laser was successfully used to synthesize fullerene-related structures for the very first time in 1985 [1], the synthesis of SWCNTs by laser ablation took another ten years of research [26-28]. Figure 7 shows the schematic of experiments involving the use of lasers to synthesize SWCNTs.

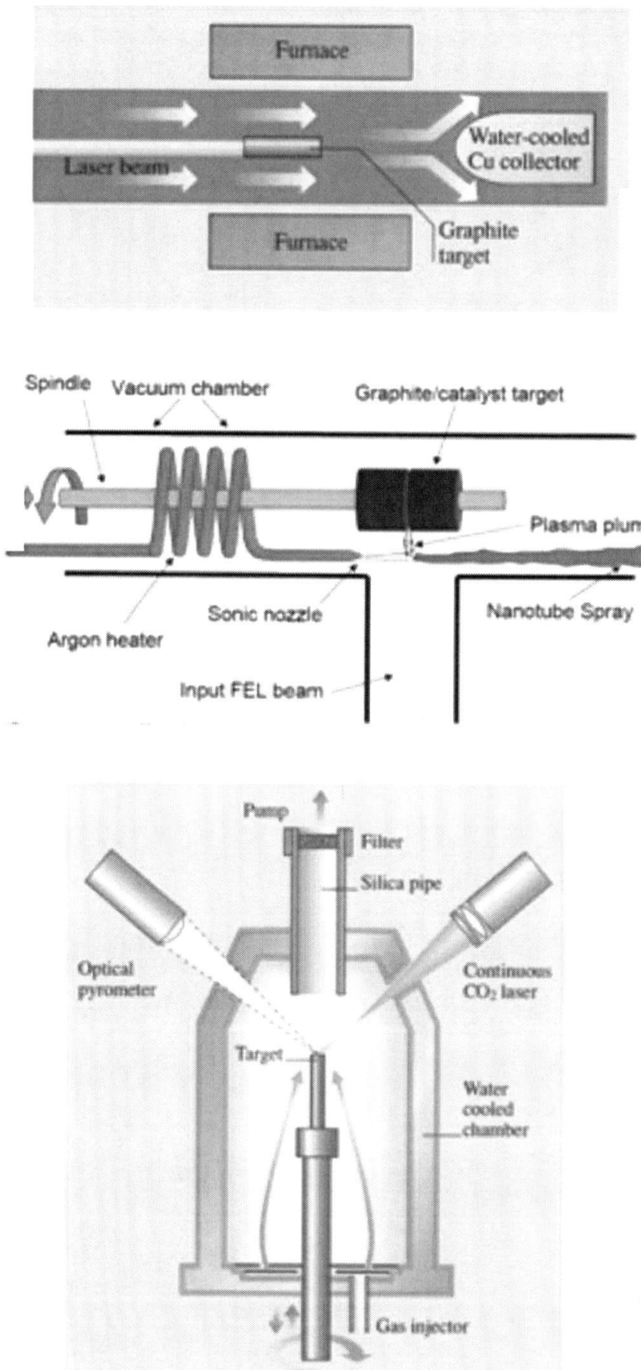

(a)

(b)

(c)

Figure 7. Schematic of experiments involving the use of lasers to synthesize SWCNTs [12, 23].

2.5. Functionalization

Single–walled carbon nanotubes (SWCNTs) are one of the most striking recent discoveries in chemistry and materials science because of their unique structural, electronic and outstanding mechanical, optical properties [29]. They find several applications in the emerging fields of nanoscience and nanotechnology including optical, microelectronic devices, sensors, catalysts, drug delivery, energy storage systems, electro chromic display and light emitting devices etc. [30]. However, the studies and applications of SWCNTs are hindered by processing and manipulation difficulties owing to their insolubility or poor dispersion in common solvents and polymer matrixes [31]. During the past few years, significant research has been carried out on the functionalization and subsequent dissolution of SWCNTs in various solvents which mainly includes covalent [32] and non-covalent [33] fictionalisations. The covalent functionalization perturbs the conjugated π-structure of the SWCNTs and results in adverse changes to its optical and electronic properties. On the other hand, non covalent functionalization enables one to tailor their properties and hence attracts particular research interest. Because of this reason, non-covalent attachment on the surface of the CNTs with numerous aromatic molecules, oligomers, polymers has been investigated [34].

All functionalization methods of CNTs can be divided into two major techniques (*Scheme 1A-E*):

Exohedral Functionalization (Scheme 1A-D):

This technique can be subdivided into three subgroups based on the mechanism of attachment of different groups or compounds to the sidewall of the CNTs:

- Covalent functionalization by attaching functional groups to the nanotube ends or defects [35-36].
- Covalent functionalization through "sidewall functionalization" [35-36].
- Noncovalent exohedral functionalization, for example, wrapping nanotubes by polymers [35].

Covalent functionalization can be realized by either modification of carboxylic acid groups on the surface of the SWCNTs or direct addition of reagents to the side wall. Carboxylic acid groups are introduced at the defects sites of SWCNTs by refluxing with HNO_3 and followed by addition of amine group/alcohol group have been employed by amidation/esterification reactions *(Scheme 2).*

Alternatively, covalent functionalization was also achieved by the addition of reagents to the side walls includes carbenes, fluorine, aryl radicals, aryl cations, nitrenes, radicals and 1,3 dipoles [38]. In the functionalization of SWCNTs through carbenes, $CHCl_3$-NaOH interaction [39], phenyl (bromodichloromethyl) mercury [40] and dipyridyl imidazolium [41] systems were utilized and was found to link to the surface of the CNTs. A different degree of fluorination can be accomplished by treating the SWCNTs with molecular fluorine at temperature between 150-600°C for dissolution in alcohol [42], although the fluorine atoms can be further substituted by alkyl groups to achieve better dissolution [43].

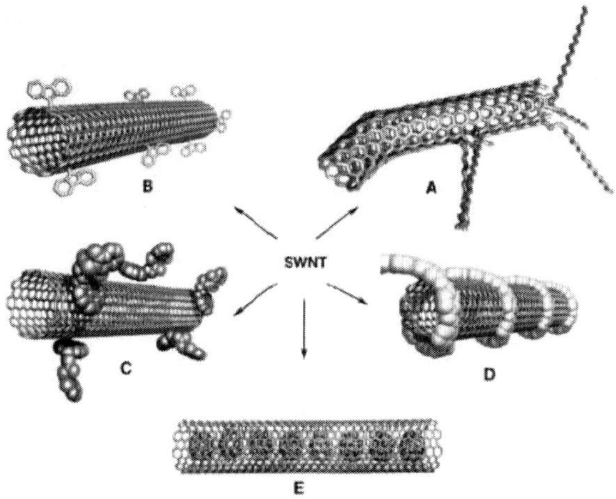

Scheme 1. Functionalization possibilities for SWNTs: (A) defect-group functionalization, (B) covalent sidewall functionalization, (C) non-covalent functionalization with surfactants, (D) noncovalent exohedral functionalization with polymers and (E) endohedral functionalization.

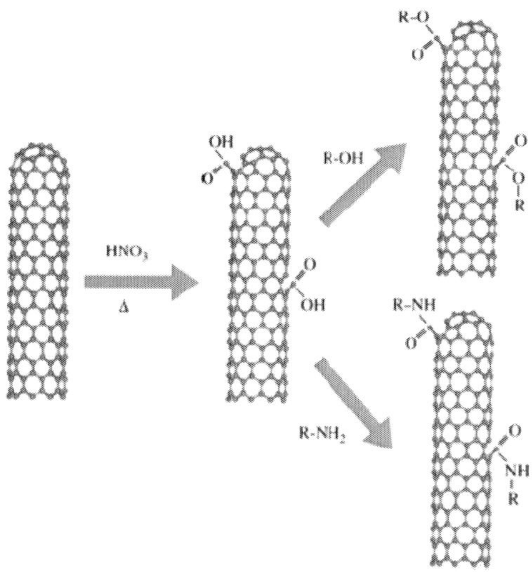

Scheme 2. Schematic representation of the chemical modification of CNTs through oxidation followed by the esterification or amidization of the carboxyl groups [37].

Covalent sidewall functionalization of carbon nanotubes yields material with increased solubility and better dispersibility in polymer matrices was performed by diazonium chemistry [44]. First, derivatization of SWCNTs was achieved by electrochemical reduction or in situ generated diazonium salts.

In addition to the electrochemically reductive coupling of aryl diazonium salts to CNTs that resulted in C-C bond formation and oxidative coupling of aromatic amines resulted in the attachment of amines directly to the surface of CNTs. In recent report, a molecule of chloroform was found to add to the side walls of CNTs upon mechanochemical reaction in the presence of $AlCl_3$, Lewis acid [45]. It is worth reporting that the solubility of SWCNTs, 50 mg/mL in $CHCl_2$ or $CHCl_3$, was achieved by the powerful methodology of 1,3-dipolar cyclo-addition reaction with azomethine ylides [46].

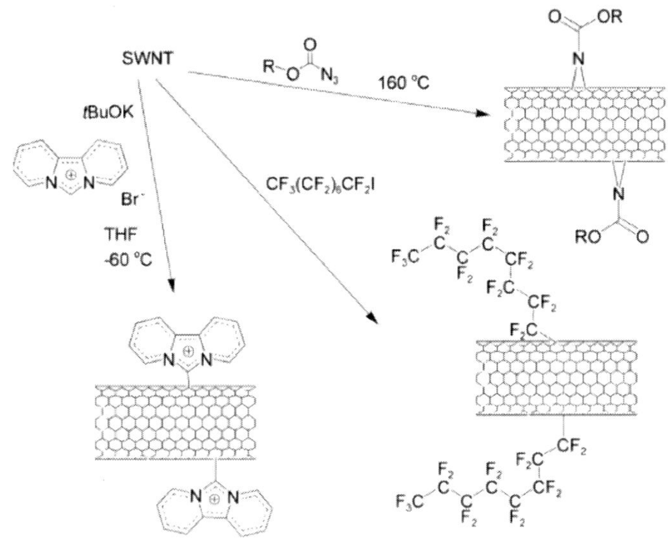

Scheme 3. Functionalization of SWCNTs with nitrenes, radicals and carbene reactions [41].

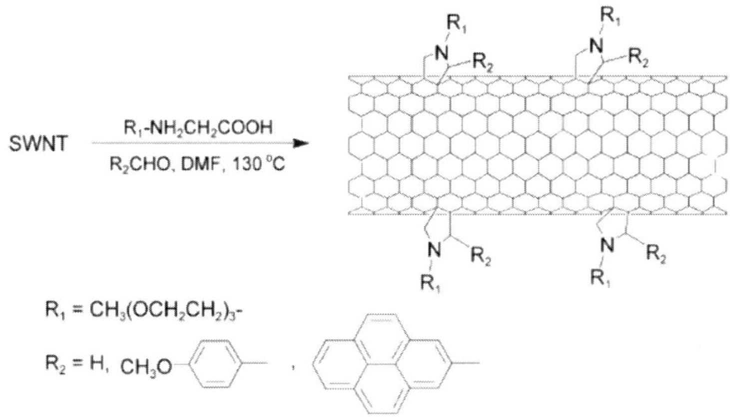

Scheme 4. Functionalization of SWCNTs by 1,3 dipolar cycloaddition reaction.

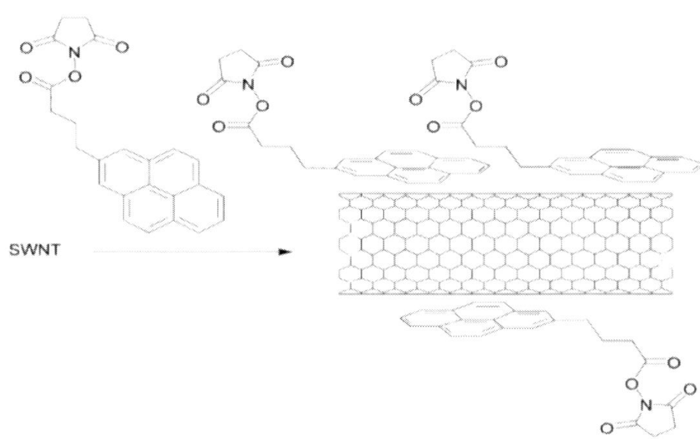

Scheme 5. Non-covalent functionalization of SWCNTs with pyrene moieties [41].

Non-covalent chemistry includes surfactant modification, polymer wrapping and polymer absorption in which the polymers were produced by an in situ or emulsion polymerisation [38]. This type of functionalization is particularly attractive because of the possibility of attaching various groups without disturbing the π system of the graphene sheets. Recently, noncovalent anchoring molecules to the side walls of SWCNTs have been achieved using a planar pyrene moiety that could absorb through strong π-π stacking interactions. Because of pyrene moiety, a variety of molecules (DNA, proteins and other smaller bio molecules) with intriguing properties could be further coupled via amide linkage [41].

In a different approach for the non-covalent functionalization of CNTs wrapped with poly*N*-vinylcarbazole conjugated polymer were synthesized by in situ solid state polymerisation technique [47]. FE-SEM and HR-TEM studies clearly indicated that the polymer wraps itself around the CNTs as a result of π-π stacking interaction.

Endohedral Functionalization (Scheme 1E):

In this techniques, SWCNTs are functionalised by filling them with different nanoparticles, this can be achieved either by

- Exploiting the phenomenon of spontaneous penetration when nanotubes are filled with colloidal suspensions followed by evaporation of the carrier liquid; or by
- Wet chemistry, when the nanotubes are filled with some compounds, which react under particular thermal or chemical conditions and produce nanoparticles. These nanoparticles then become trapped in the nanotubes.

2.6. Characterization of Nanotubes

The extraordinary physical properties of single-walled carbon nanotube (SWCNTs) and the exploitation of this material for use in advanced technological devices means that this material has to be properly characterized and exhibit uniform physical characteristics such that it can be used in consumer products. The financial success that can be derived from the exploitation of such products will depend not only on their novelty, but more importantly on consistency in performance and long-term reliability. Thus, a complete material characterization will be required before SWCNTs could be used in such products. No single experimental method has been developed to describe a SWCNT completely. Since the discovery of SWCNTs, techniques have been used to describe the properties of SWCNTs and carbon materials such as Raman spectroscopy, IR spectroscopy, UV-VIS absorption spectroscopy, electron microscopies, electron spin resonance, X-Ray diffraction, X-Ray photo-electron spectroscopy, nuclear magnetic resonance, thermal gravimetric methods etc [48-49]. But, in reference to SWCNTs in particular, a class of characterization methods has been referred to as carbon nanotube metrology [50]. These methods have the ability to distinguish between individual SWCNTs can be divided into three main groups:

- optical microscopies
- scanning probe microscopies
- electron microscopies

Each method provides a different set of information which contributes to the complete characterization of the material. Under optical, we have Raman, photoluminescence and absorption spectroscopies. Under scanning probe microscopies, we have atomic force microscopy (AFM) and scanning tunneling microscopy (STM) which is the most widely used methods. Electron microscopies, due to its nature of observing matter which are too small for the optical microscope were the first to observe SWCNTs. The above techniques provide an array of measurements that allow for a comprehensive description of carbon nanotubes. The key parameters that need to be determined on carbon nanotubes for it to be used in advanced devices are:

- conducting type i.e. either semi or metallic
- the diameter and chirality distribution
- length
- alignment

Methods to remove impurities such as metallic catalysts, which are remnants from the synthesis process, usually involve the use of acids to oxidize the metals. During the purification process, damage can occur to the nanotubes. Therefore, the nanotube material needs to be continuously monitored to prevent irreparable damage to the nanotubes. In these instances, optical spectroscopy methods are versatile and non-destructive which can be applied to detect the presence of carbon nanotubes, chemical modifications and to identifying the chiral index (n, m).

In this chapter, the optical methods of Raman and photoluminescence spectroscopy to characterize carbon nanotubes are described. These are the two most widely used optical methods currently in use to characterize carbon nanotubes.

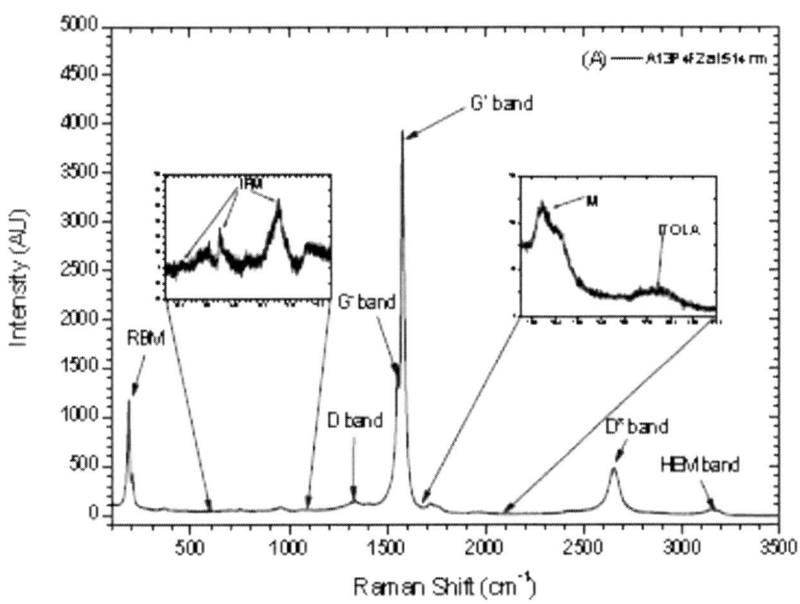

Figure 8. A typical Raman Spectrum of as-prepared SWCNTs [65(a)].

2.6.1. Raman Spectroscopy (RS)

Raman spectroscopy is the measurement of the Raman "effect" and is named after Indian Physicist C.V. Raman (born 1888, died 1970) [51]. The "effect" which is the inelastic scattering of light by a molecule or system of molecules was predicted by Russian physicist Adolf Smekal in 1923.

However, it was only in 1928 that experiments performed by Raman and Krishnan, was this effect discovered [52]. Figure 8 shows a typical Raman spectrum measurement made on a film of as-prepared SWCNTs. This sample consisted of individual and bundles of SWCNTs together with metal catalysts and amorphous carbon. The first-order and second-order features are visible in the spectrum. In a typical analysis, first order features usually suffice but for more detailed analysis, it is necessary to study all features of the spectrum which improves the physical description and understanding of phonon processes in SWCNTs [53-54]. Figure 9 shows a laser prepared sample, excited by 647 nm and 488 nm lasers showing the Raman spectrum SWCNTs bundles and the splitting of the G bands (a) metallic nanotubes (b) semi-conducting.

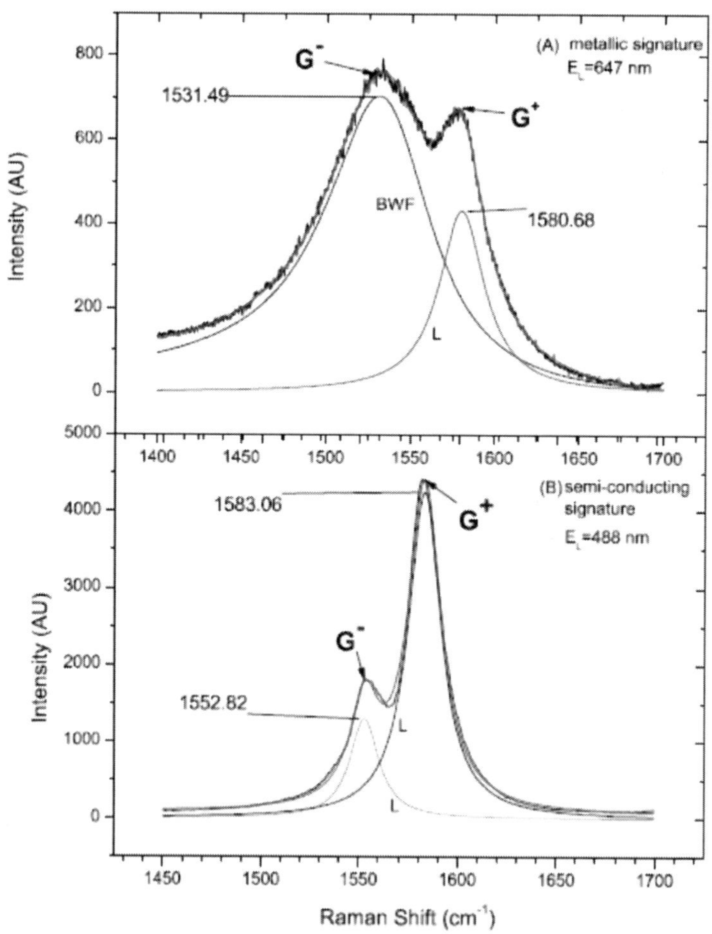

Figure 9. A laser prepared sample, from this study, excited by 647 nm and 488 nm lasers showing the Raman spectrum SWCNTs bundles and the splitting of the G bands (a) metallic nanotubes (b) semi-conducting [65(a)].

2.6.2. Photoluminescence Spectroscopy (PL)

The term photoluminescence (PL) describes any process in which light is absorbed by a medium to generate an excited state, and then re-emission of light of lower frequency upon recombination to a ground state. SWCNTs have a huge potential of being applied in technological advanced applications. A major obstacle to such efforts has been the diversity of tube diameters, chiral angles, lengths and aggregation in nanotube samples obtained from current synthesis techniques.

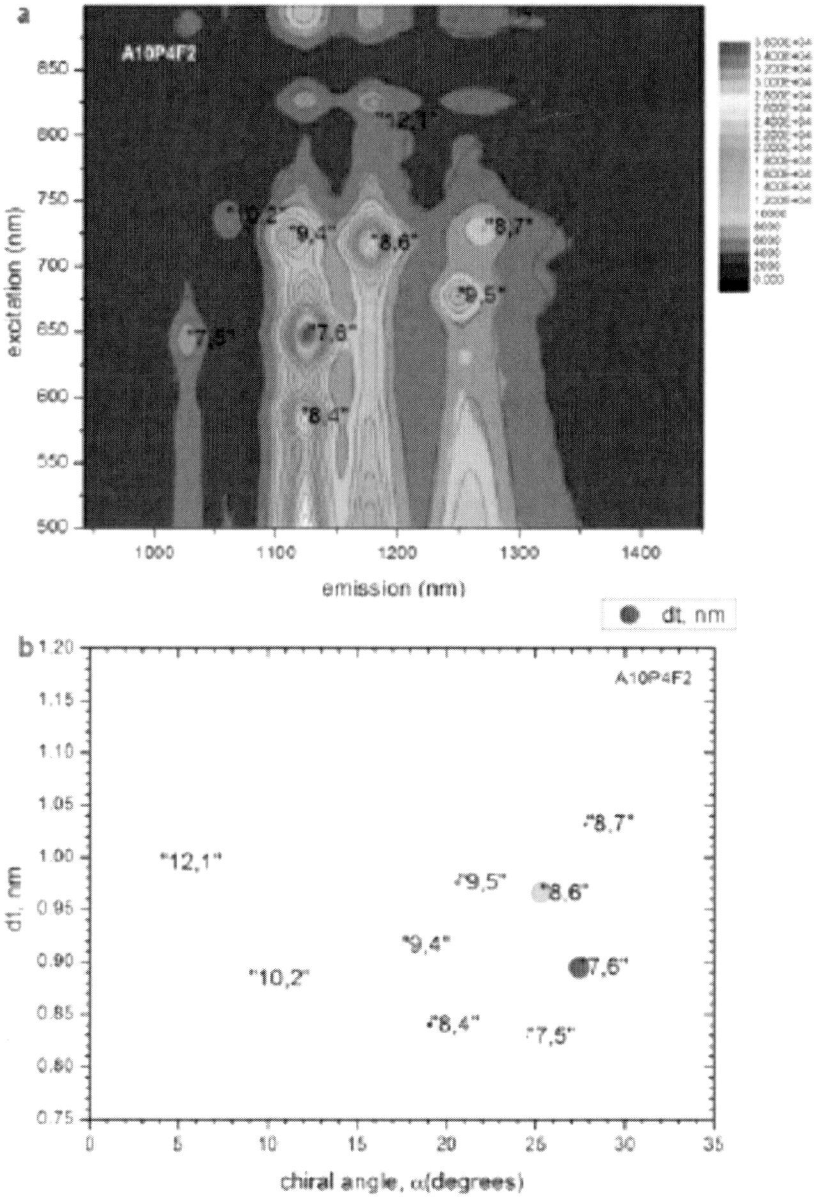

Figure 10. (a) A photoluminescence spectrum of a laser synthesized sample dispersed and suspended in D2O containing 2% by weight of sodium cholate and (b) a plot of the diameter vs chiral angle [65(a)].

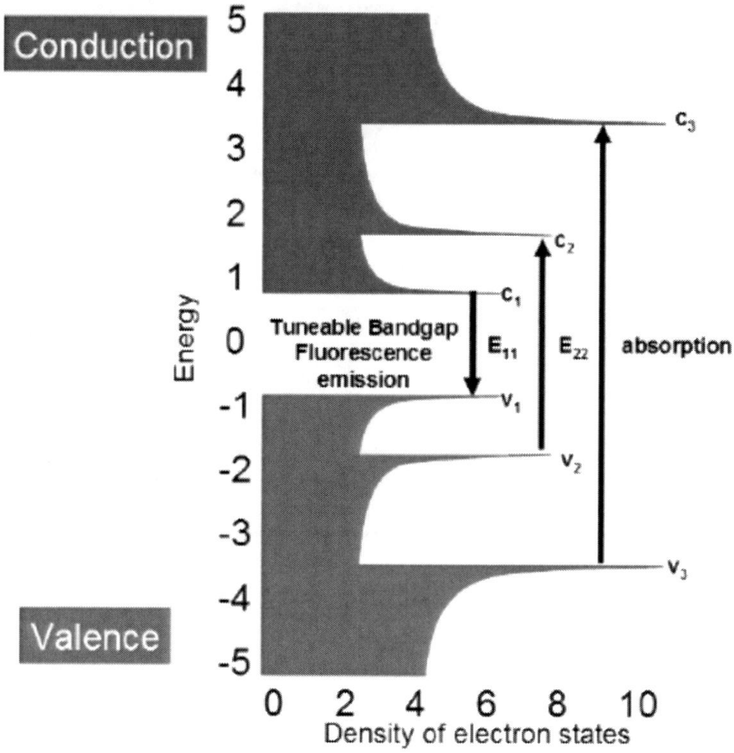

Figure 11. A Schematic illustrating the density of states (DOS) for SWCNT. Solid arrows depict the optical excitation and emission transitions while dashed arrows denote non radioactive relaxation of the electron, in the conduction band, and a hole, in the valence band before emission [55].

PL has become an important tool in measuring a fraction of the total population where this fraction represents all semi-conducting tubes. Bachilo et al. [55], laid the ground work for use of PL to determine chiral indices (n, m) of semi-conducting nanotubes and became the basis on which the software program called Nanosizer [56]. Figure 10 shows (a) A photoluminescence spectrum of a laser synthesized sample dispersed and suspended in D_2O containing 2% by weight of sodium cholate and (b) a plot of the diameter vs chiral angle. Figure 11 represents a schematic illustrating the density of states (DOS) for SWCNT. Solid arrows depict the optical excitation and emission transitions while dashed arrows denote non radioactive relaxation of the electron, in the conduction band, and a hole, in the valence band before emission [55].

2.6.3. Scanning Electron Microscopy (SEM)

SEM is the most widely used electron microscopic technique because of versatility and ease of use. Samples can be quickly analyzed since sample preparation need not be elaborate. Accelerating voltages up to 30 kV are available which means a very wide range of magnification is possible, enabling very good resolution. SEM analysis can yield image resolutions in the range of 0.6 nm but these resolutions are only possible for metallic samples. Figure 12 shows a sample SEM image of an as-prepared SWCNT material. Only bundles are visible. Generally with current SEMs, it is not possible to resolve individual SWCNTs.

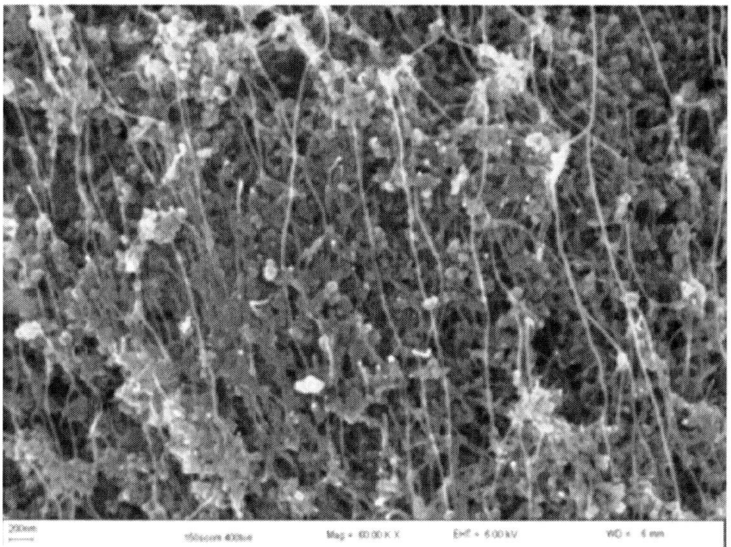

Figure 12. A sample SEM image of an as-prepared SWCNT material [65(a)].

Since, the amount of secondary electrons produced from a low atomic number element such as carbon is not sufficient to resolve the nanotubes with diameter of less than ± 1nm. However, high resolution SEM is now a routine analytical tool.

This technique provides an important high throughout characterization tool since "seeing is believing". It is important to know the dimensions of the structures fabricated and the materials prepared when characterizing device structures. Thus a SEM has become an indispensable tool in nanometrology [57].

2.6.4. Transmission Electron Microscopy (TEM)

Among the various analytical techniques that have been used to study carbon nanotubes, it was in a transmission electron microscopy (TEM) discovered by Iijima [17] in 1991. Compared to SEM, a TEM principal of operation is different. TEM requires a dedicated and expertly trained materials scientist to operate the TEM to its full potential. In TEM, the transmitted electrons are used to create an image of the sample. Scattering occurs when the electron beam interacts with matter. Scattering can be elastic (no energy change) or inelastic (energy change). Elastic scattering can be both coherent and incoherent (with and without phase relationship). Elastic scattering occurring from well-ordered arrangements of atoms as in a crystal, results in coherent scattering, giving spot patterns. This can be in the form of rings in the case of a polycrystalline material. However, inelastic scattering also occurs, which also gives regular patterns as in the case of Kikuchi patterns [58]. Inelastic processes give characteristic absorption or emission, specific to the compound or element or chemical structure. Because of all these diverse processes, a transmission electron microscope is akin to a complete laboratory. The nature of bonding in a nanotube is described by orbital hybridization. The chemical bonding of nanotubes is composed entirely of sp^2 bonds, similar to those of graphite. This bonding structure, which is stronger than the sp^3 bonds found in diamonds, provides the molecules with their unique strength. Nanotubes naturally align themselves into "ropes" held together by Van derWaals forces. Figure 13 shows the schematics and accompanying TEM images of a (a) SWCNT, (b) DWCNT and (c) MWCNT.

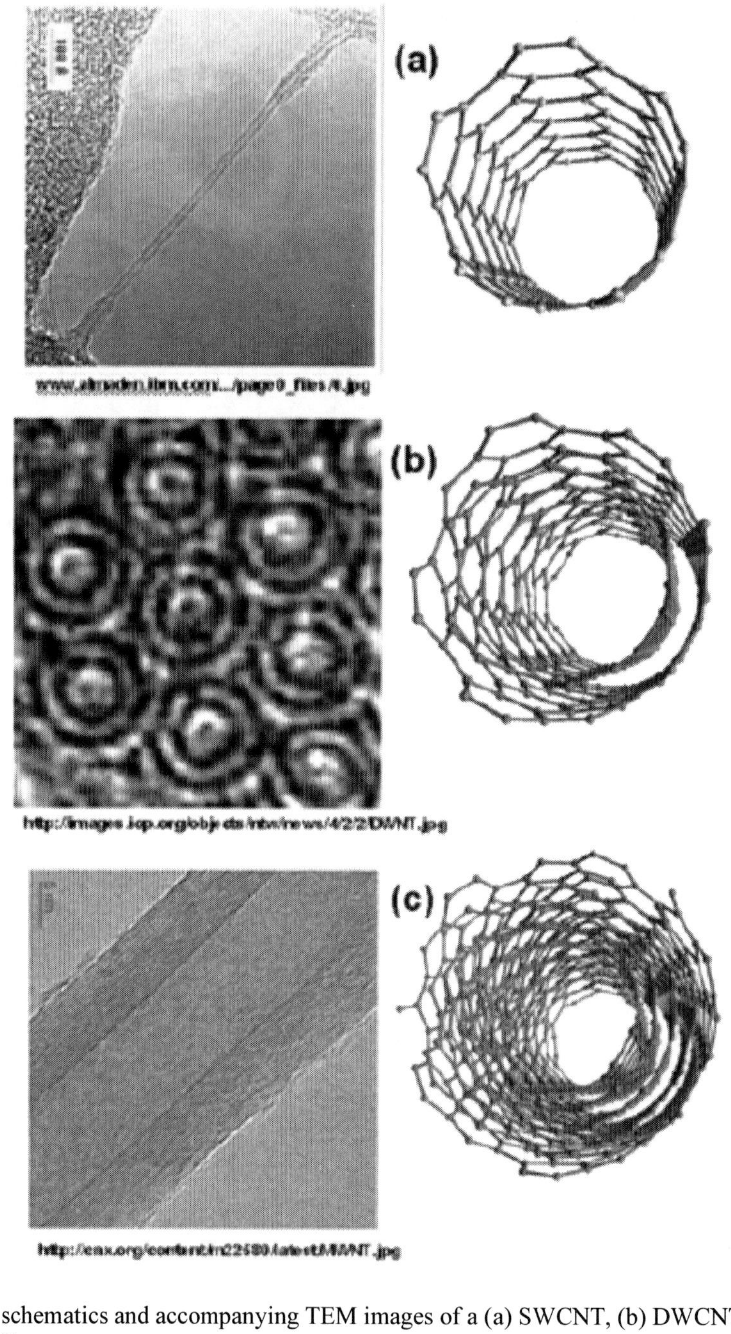

Figure 13. The schematics and accompanying TEM images of a (a) SWCNT, (b) DWCNT and (c) MWCNT [65(a)].

3. Carbon Nanotubes: Structure and Properties

Carbon nanotubes (CNTs; also known as buckytubes) are allotropes of carbon with a cylindrical nanostructure. These cylindrical carbon molecules have novel properties that make them potentially useful in many applications in nanotechnology, electronics, optics and other fields of materials science. They exhibit extraordinary strength and unique electrical properties, and are efficient thermal conductors. The ends of a nanotube can be capped with a hemisphere of the buckyball structure.

3.1. Structure

To imagine a single-wall carbon nanotube (SWCNT), consider a perfect graphene sheet of sp^2 -hybridized carbon atoms arranged in hexagons and rolled into a cylinder so that the hexagonal rings connect coherently. The tips of the tube can be are sealed by two caps, each cap being a hemi-fullerene of the appropriate diameter. Geometrically, there is no restriction to the diameters of a tube. However, calculations have shown that collapsing a single-wall tube into a flattened two-layer ribbon is energetically more favorable than maintaining the tubular morphology beyond a diameter value of ≈ 2.5 nm [59].

On the other hand, it is easy to grasp intuitively that the shorter the radius of curvature, the higher the stress and the energetic cost to maintaining a tubular structure. However, SWCNTs with diameters as low as 0.4 nm have been synthesized successfully [60]. A suitable energetic compromise is therefore reached for tubes with diameters of ≈ 1.3 nm, the most frequent diameter encountered, regardless of the synthesis technique. There is no such restriction on the nanotube length, which only depends on the limitations of the preparation method and the specific conditions used for the synthesis (thermal gradients, residence time and so on).

Experimental data are consistent with these statements, since SWCNTs wider than 2.5 nm are only rarely reported in the literature.

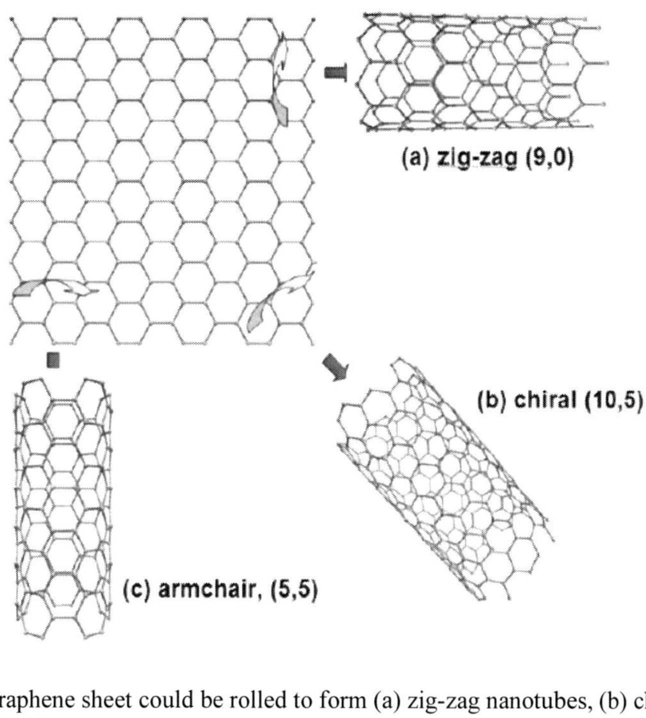

Figure 14. How a graphene sheet could be rolled to form (a) zig-zag nanotubes, (b) chiral nanotubes and (c) armchair [65(a)].

These features make single-wall carbon nanotubes a unique example of single molecules with huge aspect ratios. As illustrated in Figure 14, there are many ways to roll a sheet of graphene into a single-wall nanotube.

Some of the resulting nanotubes possess planes of symmetry both parallel and perpendicular to the nanotube axis (such as the SWCNTs generated in Figure 14(a) and 14(c), while others does not, such as that of Figure 14(b). Similar to the terms used for molecules, the latter are commonly called.chiral, nanotubes, since they are unable to be superimposed on their own image in a mirror.

3.2. Properties

3.2.1. General Properties

The diameters of SWCNTs fall in the nanometer regime, commonly between 0.8 nm to 1.6 nm but can be hundreds of micrometers long. SWCNTs are stable up to 750°C in air but may suffer damage before this temperature is reached due to oxidation mechanisms at the atomic scale, as demonstrated by the fact that this is one of the methods of filling SWCNTs with other elements or molecules. They are stable up to between 150°-1800°C in an inert atmosphere, beyond which they transform into regular, polyaromatic solids. The properties of a SWCNT, like any molecule, are heavily influenced by the way that its atoms are arranged. The physical and chemical behavior of a SWCNT is therefore related to its unique structural features [12].

3.2.2. Electronic and Optical Properties

The electronic and optical properties of the tubes are considerably influenced by the environment. The electronic transition energies are in the infrared and visible spectral range [25]. The one-dimensional Van Hove singularities have a large influence on the optical properties of SWCNTs. Visible light is selectively and strongly absorbed by SWCNTs. Photoluminescence can be observed in individual SWCNT aqueous suspensions stabilized by the addition of surfactants. Detailed photo-excitation maps provide information about the helicity (chirality) dependent transition energies and the electronic band structures of SWCNTs. Agglomeration of tubes into ropes or bundles influences the electronic states of SWCNTs. Photoluminescence signals are quenched for agglomerated tubes [61].

3.2.3. Mechanical Properties

The tensile strength of SWCNTs can be 20 times that of steel and has been measured as to be as high as 45 GPa [62]. Very high tensile strength values are also expected for ideal (defect-free) MWCNTs, since combining perfect tubes concentrically is not detrimental to the overall tube strength, provided the tube ends are well capped (otherwise, concentric tubes could glide relative to each other, inducing high strain). Tensile strength values as high as 150 GPa have actually been measured for perfect MWCNTs prepared from an electric arc [63].

3.2.4. Reactivity

The chemical reactivity of SWCNTs is believed to derive mainly from the caps, since they contain six pentagons each, as opposed to the tube body, which supposedly only contains hexagons. Indeed, applying oxidizing treatments to carbon nanotubes (air oxidation, wet-chemistry oxidation) selectively opens the nanotube tips [64]. SWCNTs can be opened by oxidation methods and then filled with foreign molecules such as fullerenes. This suggests the occurrence of side defects. The defect sites have been proposed to be due to Stone-Wales mis-arrangements which may occur every 5 nm along the tube length and involves about 2% of the carbon atoms in a regular (10,10) SWCNT [65].

4. FABRICATION OF CNT SENSOR DEVICES

Sensors are a class of devices that have found widespread use, ranging from the detection of gas molecules to the real time tracking of chemical signals in biological cells. In general, a sensor comprises an active sensing element and a signal transducer, and produces an electrical, optical, thermal or magnetic output signal. While the sensing element is responsible for the selective detection of the analyte, the transducer converts a chemical event into an appropriate signal that can be used with or without amplification to determine the analyte concentration in a given test sample. One key to advanced technology emerging over the past decade has been nonmaterial's. The aggressive advance of smart materials, solid state devices, and biomimetic technologies and the concurrent push towards miniaturization are making the understanding and development of materials on the nanometer level critical and are encouraging the design of nanoscale structure and functionality into materials systems. The focus on nano structuring of materials systems has been further sharpened by the need to develop materials having novel and/or enhanced properties without resorting to new synthetic chemistries with the associated environmental and cost issues. Over the last 40 years, feature sizes in complementary metal oxide semiconductor (CMOS) technologies have been scaled from 3 μm to the current sub-50 nm using the "top-down" scaling techniques [66]. This "scaling" has resulted in an increased processing power and transistor density while reducing the cost per transistor [67]. These classical methods employ a sequence of deposition, pattern definition, doping, lithographic and etching steps to build solid-state semiconductor devices and integrated circuits. As the process technologies scale beyond the sub-10nm feature sizes, above fabrication methods result in increased process costs, variability and longer fabrication turnaround times. To push the CMOS technology to its limits and to reap the benefits of scaling, non-traditional alternatives are needed while fabricating devices. One such approach is the use of "bottom-up" nanotechnologies or even a combination of the bottom up and the "top-down" fabrication methodologies [68]. In the bottom-up approach, analogous to the biological systems, atoms or organic molecules are self assembled to build electronic structures with novel electronic, optical, or magnetic properties. Integrated circuits obtained with this approach of molecular-level control of material composition and structure may lead to devices and fabrication strategies not possible with top-down methods [69]. Increasingly technology relies on the ability to control materials on very small length scales. This has been accomplished, traditionally, through what is known as "top-down" techniques, i.e. using tools to whittle away undesired material to leave the final product. This approach has been used for countless years; examples include: fashioning an arrowhead from a rock, building a table from lumber, assembling a car, and fabricating a microchip. New "bottom-up" techniques are being explored as a compliment to traditional "top-down" methods. In contrast to removing the materials we don't need, "bottom-up" techniques simply construct the desired features from fundamental building blocks, usually spontaneously through self-assembly, without the need for patterning. The rapid advance in electronics technology and proliferation has brought great changes to our life and society in recent years. On the other hand, it has also exerted adverse effects on the natural environment. Current production processes require large-scale production lines as well as many plants and equipment, and their top down technologies result in a substantial waste of materials. Concerns about global warming due to increasing carbon dioxide emission and declining natural resources have led to demands for the replacement of these high-cost, environmentally burdening processes by methods that are more efficient in energy and material consumption. All these have heightened interest in bottom-up processes for electronic device production with soluble organic semiconductors [70-71].

4.1. Foundry Technology

The most significant innovations in CMOS fabrication since the inception of silicon VLSI. High K metal gate (HKMG) enables a revival in transistor scaling that had stalled with poly SiON gate technology. GLOBALFOUNDRIES is driving the global standard for High K Metal Gate (HKMG) with seven world-leading semiconductor product companies including free scale, IBM, Infineon, NEC, ST, Samsung, and Toshiba. This HKMG solution is far superior to that currently pursued by the other leading foundries, in both scalability (die size, design compatibility, performance) and manufacturability. This solution is a “Gate-First” approach that shares the process flow, design flexibility, design elements and benefits of all previous nodes based upon poly SiON gates. The approach of GLOBALFOUNDRIES enables full scaling from 40nm to 28nm, where the approach of the other leading foundry suffers a 10% or much greater hit in die size, depending on the user’s standard cell library.

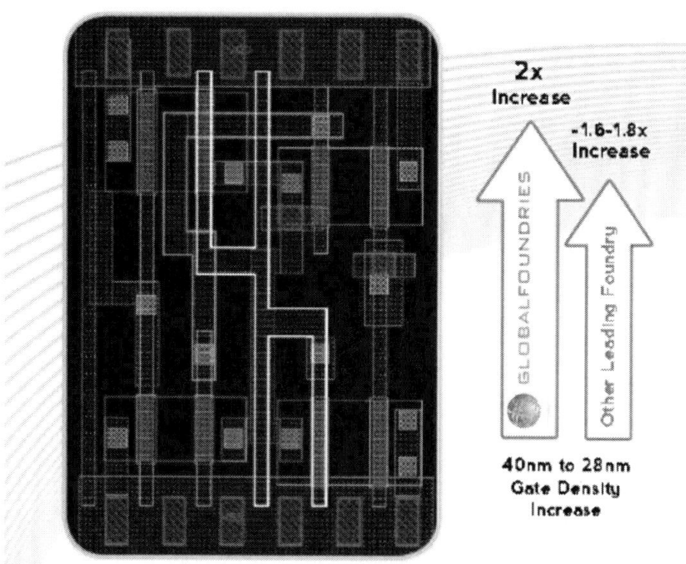

Figure 15. Dense routing is enabled by GLOBALFOUNDRIES’ HKMG. Layout schematic shows Metal 1 outlined in brown. Poly interconnect, outlined in white, enables flexible connection and substantially smaller die size of ~10-20% or more depending on user’s standard cell library.

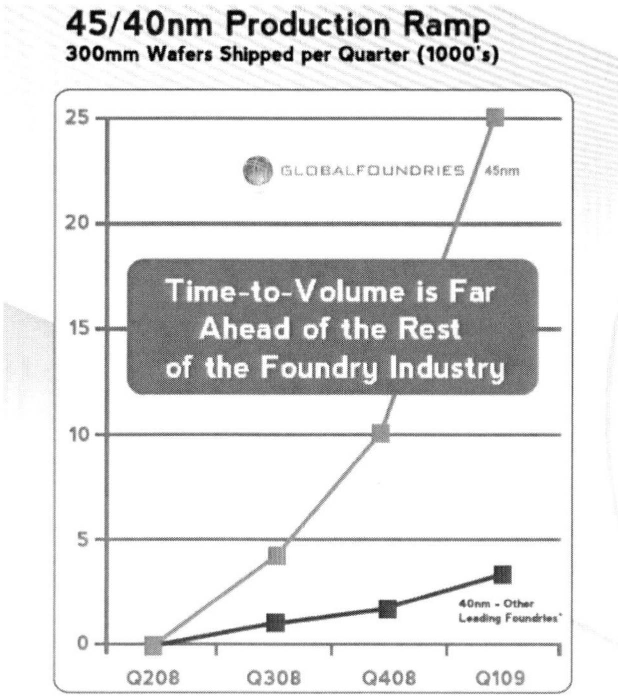

Figure 16. 45/40nm foundry production ramp. Similar leadership by GLOBALFOUNDRIES is anticipated for the 32/28nm foundry industry production ramp of HKMG-based technologies in 2010-2011.

The cell layout and routing advantages of GLOBALFOUNDRIES' design rules that allow tighter pitch; dual orientation poly gates and flexible poly interconnect, in contrast to the single orientation poly restrictions by the other leading foundry shown in Figure 15. In order to mitigate the penalties imposed by these poly restrictions, an additional metal layer must be utilized to prevent the cell from blowing up 10-20% or more. GLOBALFOUNDRIES also enables >50% SRAM area scaling – featuring a 0.120um^2 high density bit cell. This is the smallest foundry bit cell offered for the 28nm node. Leakage emerged as a major yield limiter at the 45/40nm node, preventing other leading foundries from ramping production is shown in Figure 16.

Example:

Y.Zhou et al [72] reported the concept of the CMOS integrated SWNTs. A CMOS CNT growth method that bridges the gap between CVD and CMOS in which micro fabrication technology is used to create micro heaters that can locally generate hot spots for SWNT synthesis while leaving surrounding areas at room temperature. The Figure 17 shows the concept of the CMOS integrated SWNTs.

4.2. Soft lithography

Lithography is a process in which the substrate to be patterned is covered with a layer of radiation-sensitive polymer, a photoresist.

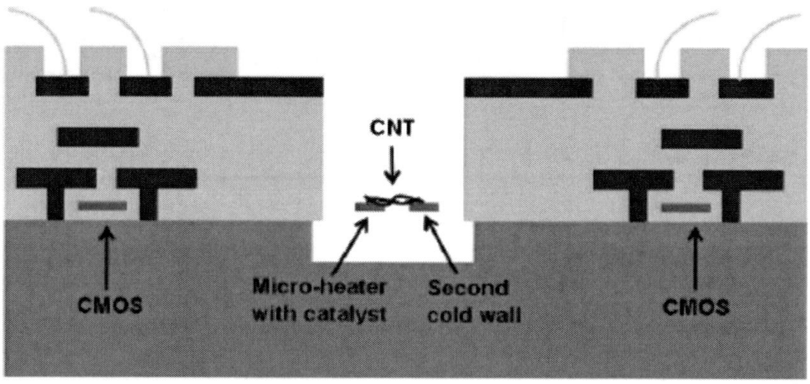

Figure 17. The concept of the CMOS integrated SWNTs.

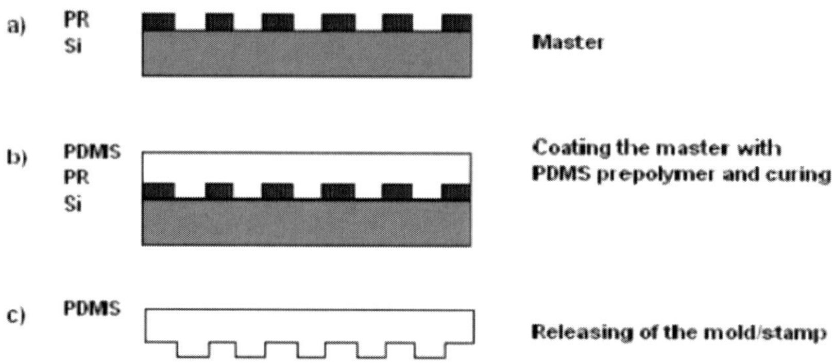

Figure 18. The principle of fabrication of polymer stamp or mold for soft lithography.

The resist is locally exposed to radiation, which modifies the structure of the polymer chain to change its solubility properties. After exposure, the resist is developed in a suitable solution, and the desired pattern is created onto the substrate. Conventional lithography includes photolithography and beam (electron or ion) lithography. Eeva Läänilläinen [73] in his master's thesis reported that the Soft lithography is based on pattern transfer using a mold that is deforming or patterning the substrate material. Most of the soft lithographic techniques use mold formed of a master template. The fabrication of an elastomeric mold is presented in Figure 18. The mold replicates the inverse of master features, and it can be used in turn to deform or pattern the next material.

The process begins with the master fabrication, usually made of silicon having topographical features like trenches and ridges as shown in Figure 18a. The mold is fabricated by casting the liquid precursor, usually a pre polymer mixture into the master (Figure 18b). The mold is then cured to make it solid, and removed from the master (Figure 18c). This mold will then be used in subsequent patterning steps. The fabrication of the master is usually done with photolithography or when very small features are needed, by electron beam lithography. The master has to have smooth sidewalls and bottom surfaces and it can be coated with a release layer to ease the removal of the mold.

In master fabrication soft lithography is not independent of clean room facilities and high-cost equipment. Its strength is rather in replication: once the master is created with the expensive or low-throughput equipment, it can be used multiple times, with no need to have the slow EBL step done each time a pattern is created, thus overcoming the limits of slow throughput with scanning beam lithography and making soft lithography an attractive alternative in nanoscale patterning. Variety of materials can be used, and some of these techniques allow the patterning of curved non-planar surfaces and the replication of three dimensional structures.

Example:

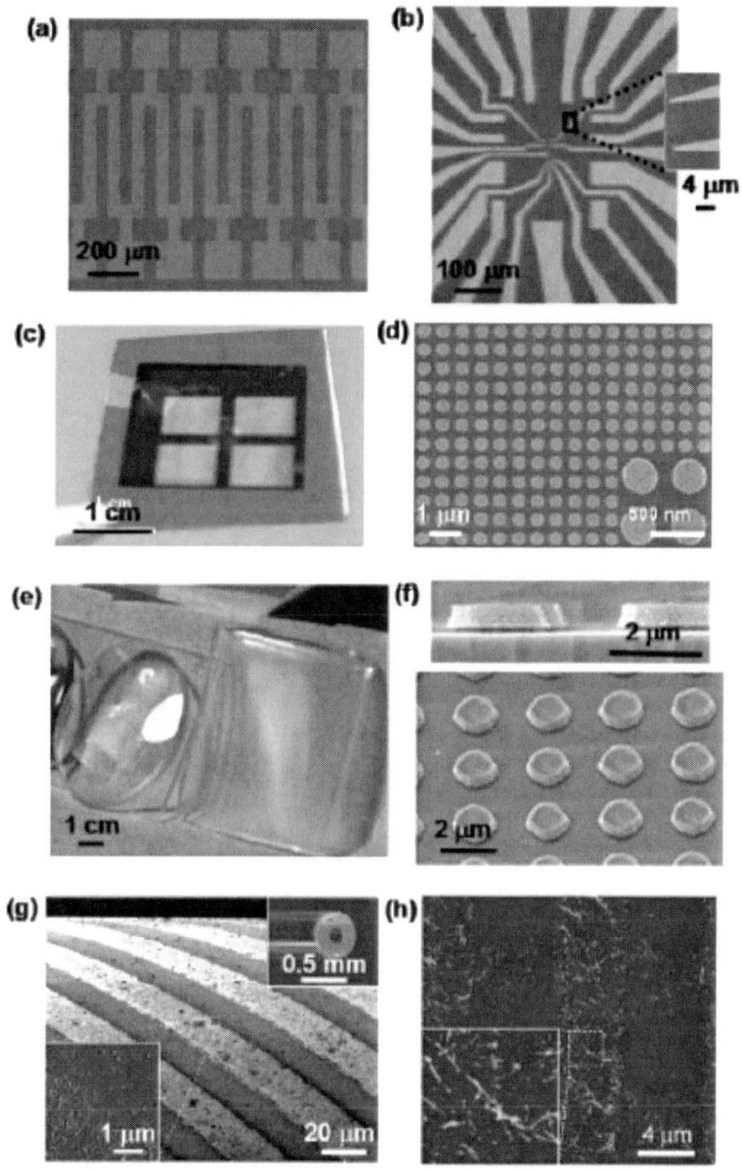

Figure 19. Optical images, SEMs, and AFMs of patterns of metals (a-d), dielectrics (e,f), and SWNTs (g,h) formed by soft-lithographic transfer printing with solid inks of these materials [74-76].

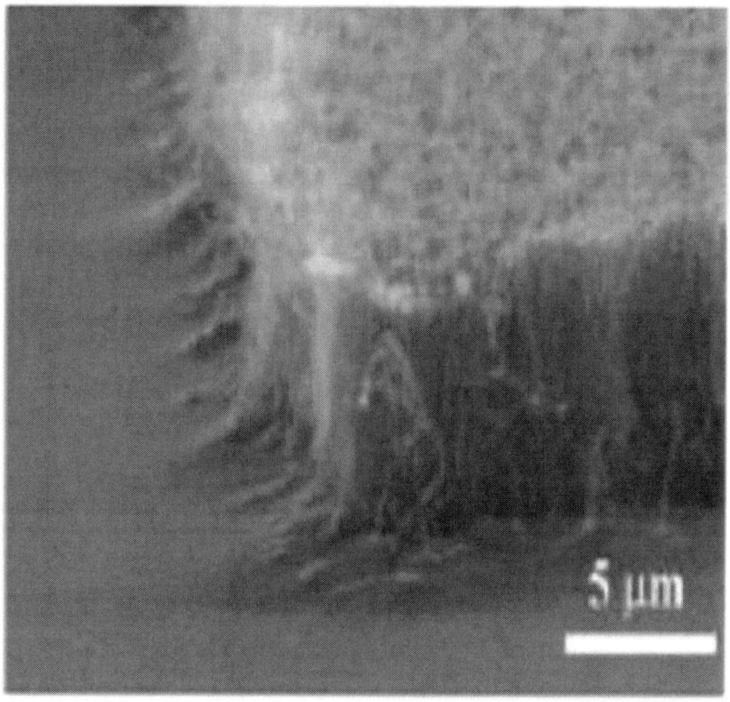

Figure 20. Schematics for and CNT films resulting from CVD [79].

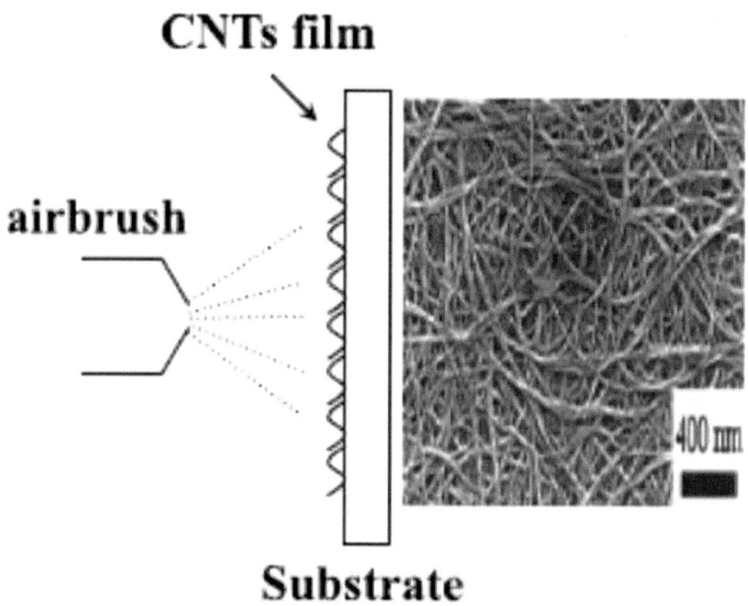

Figure 21. Schematics for and CNT films resulting from spray coating [82].

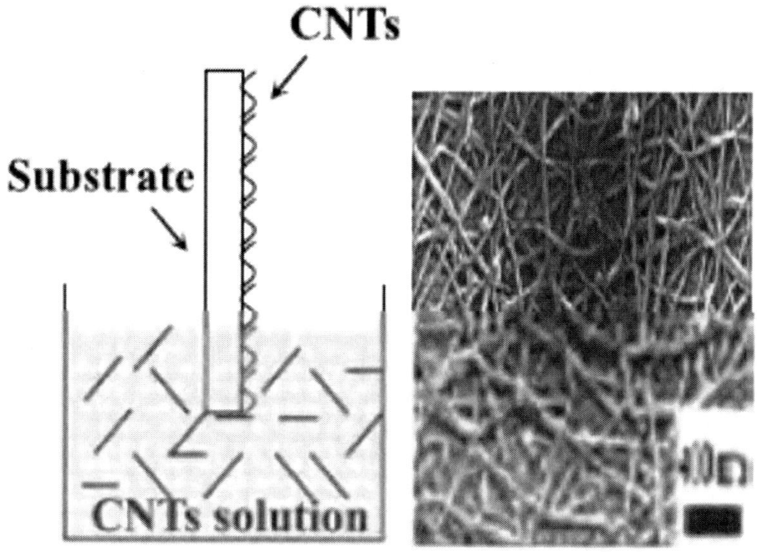

Figure 22. Schematics for and CNT films resulting from dip coating [82].

Many applications of CNTs are in the form of thin film. For example, the transparent CNT film is seen as having great potential to replace the indium tin oxide (ITO) film as transparent electrodes in the future because the price of indium is keeping increase. Thus, fabrication of CNT thin film draws much research attention. There are several methods for fabricating a CNT thin film:

CVD: This method [77-81] is suitable for the fabrication large area CNT films. A catalyst of alloys of Fe, Co and Ni is coated on the substrate which is then placed into a high-temperature chamber, to which hydrocarbon gas is added. Figure 20 shows the SEM image of the CNT film created by CVD.

Spray coating: The CNT solution is sprayed on the substrate with an airbrush and heated to dry the CNT droplets. This simple, low-cost method [82-83] is also suitable for the fabrication large area CNT films. Figure 21 shows the schematic of the spray coating method and the SEM image of the CNT film spray coating on the substrate.

Dip-coating: The substrate is immersed into a uniform CNT solution. When removed from the substrate, the CNT solution is adhered to the substrate. As the adsorption CNT solution dries, the CNT network is formed on the substrate. This method [84-86] has advantages in the fabrication of large area CNT films, but high conductivity requires the substrate to be dipped several times and the film thickness does not correlate linearly with the dipping times. Figure 22 is the schematic of the dip-coating method and the SEM image of the CNT film dip-coating on the substrate.

Filtration: The uniform CNT solution is prepared and passed through a nano porous filter, leaving a uniform CNT film which can then be transferred to the desired substrate. This method [87-91] has the advantages in the homogeneity of the resulting CNT film but the size of the CNT film has limited by the size of the filter. Figure 23 shows the schematic of the filtration method and the SEM image of CNT film filtration on the nano filter.

Imprint: This method involves printing CNT film with a standard office laser printer [92], inkjet printer [93-94] or contact printer [95-97]. In a laser printer, CNT powder replaces the toner. The CNTs could be patterned on polymer film. Figure 24 shows the schematic of the laser-printer method and the CNT film laser-printed on the substrate. In inkjet printing, the CNT solution replaces the ink. Figure 24 shows the CNT film inkjet on the substrate. This method has an advantage in that the patterns are easily edited by a graphic or word processing software on a computer. The contact printing method is popular for transferring CNT film.

The growth or network CNT film is laid on the target substrate, and the transfer takes place during the contact printing process. Figure 24 shows the schematic of the contact printing method and the SEM image of the CNT film contact printing on the substrate.

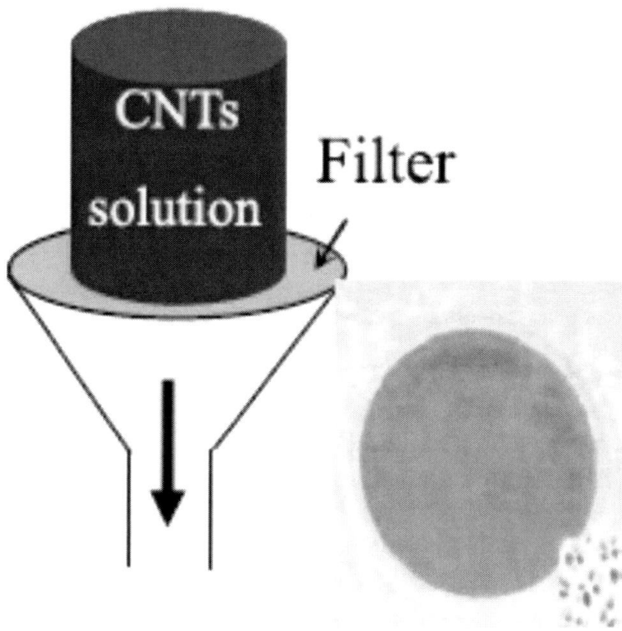

Figure 23. Schematics for and CNT films resulting from filtration [89].

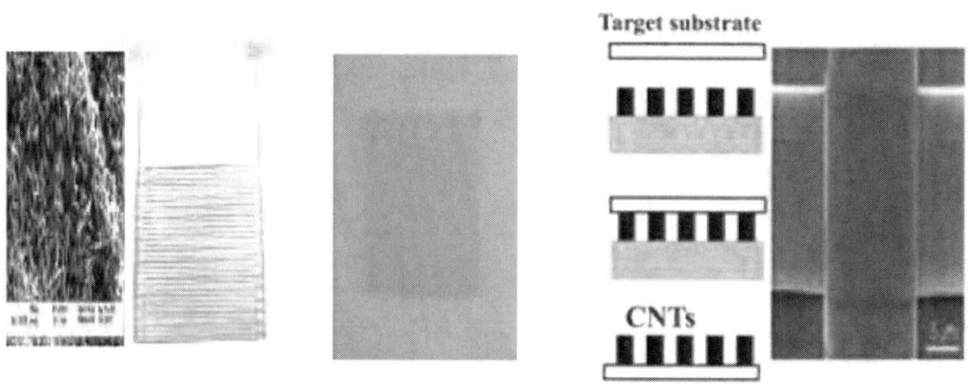

Figure 24. Schematics for and CNT films resulting from laser-printer, ink-jet and contact printing [92, 93, 97].

Electro phoretic deposition: A uniform CNT solution is prepared in the sink. Two metals are connected to direct current (DC) and are used as the anode and the cathode. The CNT film is then deposited on the anode surface. This method [98-100] is well-suited to the fabrication of large-size CNT films, and produces films with high homogeneity and proper surface roughness. However, the substrate needs a conductive anode and must be immersed in the solution. Figure 25 shows the schematic of the electro phoretic deposition method and the SEM image of CNT film electro phoretic deposition on the substrate.

Laser transfer: This method can produce large CNTs films [103] and transfer the film to a selected position [101, 102, 103-105]. A uniform CNTs solution is placed on a transparent substrate. After the CNTs self-assemble on the transparent substrate, the mask is placed in front of the substrate. High intensity irradiation forms a thick CNT film.

The mask is removed and we apply a low laser intensity to irradiate the thick pattern of CNTs. The transparent substrate has a thin residual CNTs film. Figure 26 shows the schematic of the laser transfer method and an image of the CNT film deposition on the substrate.

5. Commercialization of Carbon Nanotubes Sensors

Currently most of the CNT sensor devices demonstrated are at laboratory level or at pre-commercialization phase.

Some devices are already being developed for deployment in space for the exploration of planets [106]. The problem in the slow commercializing devices based on carbon nanotubes is the reproducibility of the performance of the device.

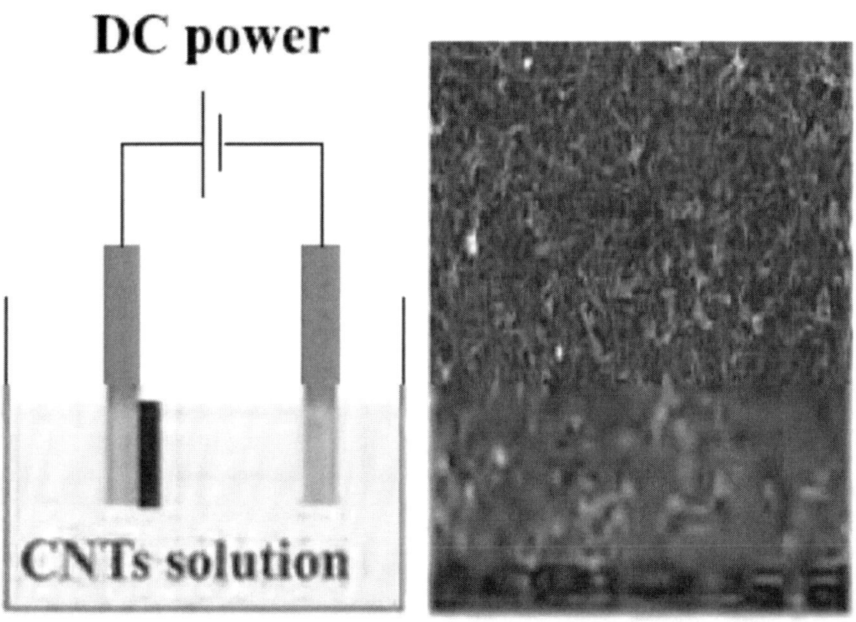

Figure 25. Schematics for and CNT films resulting from electro phoretic deposition [99].

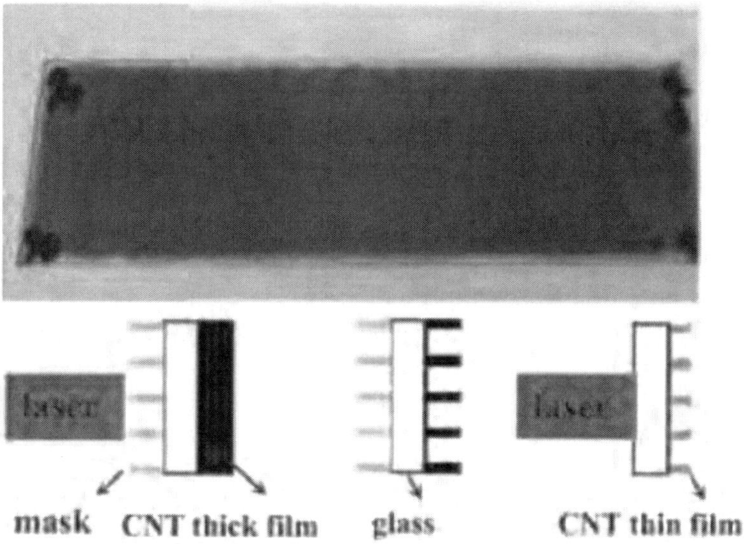

Figure 26. Schematics for and CNT films resulting from laser transfer [103].

Thus rigorous testing and improving the reproducibility of the processes involved in fabricating a device is required. Furthermore, the reliable synthesis of specific nanotubes has plagued the applications area of this material though this has been shown to lesser of a problem in gas sensor devices [107]. To advance the use of nanotubes in gas sensing devices, Nanōmix [reference: www.nano.com] is a leading nanoelectronic detection company have developed detection devices based on its *Sensation*™ technology. These devices based on carbon nanotubes have been functionalized for respiratory and bio detection applications.

CONCLUSION

The nanometric structure and properties of carbon nanotubes lend itself to being an excellent advanced material for exploitation in device applications. While in its early years it was strongly suggested that carbon nanotubes would eventually replace silicon, in recent times it has found more feasible applications such as in sensors devices. To this end, more amenable technologies such as soft lithography technologies have been employed to fabricate cost-effective devices. The rich physical and chemical structure of a carbon nanotubes, allows a skilled chemist to graft specific molecules on the nanotubes scaffold to achieve high functionality to enhance sensing capabilities and target particular analytes or stimulants in sensing applications. However, laboratory successes have not as yet led to the market for commercial products. Issues such as reliable synthesis of nanotubes, repeatable fiction-alization and the reliable deposition of the material onto prefabricated device are still areas which require refinement. Furthermore, in the area the gas sensor, the understanding of the charge transfer mechanism between stimulant and the nanotubes still require much more work. Nevertheless, there is much to gain from continuing research in this area of sensor technology involving carbon nanotubes, since the successful commercialization of such devices to prove to exhibit performances far exceeding available current devices.

REFERENCES

[1] H. W. Kroto, J. R. Heath, S. C. O. Brien, R. F. Curl and R. E. Smalley, *Nature* 318, 162 (1985).
[2] H. Kroto, Swedish Academy of Sciences, Stockholm, Sweden, 44 (1996).
[3] P. Avouris, and R. Martel, *MRS Bulletin*, 306 (2010).
[4] Http://www.eastonbellsports.com/.
[5] Http://www.zyvextech.com/.
[6] Http://www.swentnano.com/.
[7] Http://www.nanocomptech.com/.
[8] E. Comini, C. Baratto, G. Faglia, M. Ferroni, A. Vomiero and G. Sberveglieri, *Prog. Mater. Sci.* 54, 167 (2009).
[9] K. Kalantar-Zadeh, and B. Fry, Springer, New York, USA, (2008).
[10] T. Someya, J. Small, P. Kim, C. Nuckolls and J. Yardley, *Nano Letters,* 3, 877 (2003).
[11] Http://www.azonano.com/article.aspx?ArticleID=980.
[12] M. Monthioux, P. Serp, E. Flahaut, M. Raza.nimanana, C. Laurent, A. Peigney, W. Bacsa J.-M. Broto, Springer, 2007, Ch. 3, p. 43.
[13] M. Meyyappan, CRC Press, Washington, D.C., (2005).
[14] J. Lui, S. Fan and H. Dai, *MRS Bulletin* 29, 244 (2004).
[15] P. Harris, *Carbon,* 45, 229 (2007).
[16] C. Rao, and A. Govindaraj, RSC Publishing, Cambridge, (2005).
[17] S. Iijima, *Nature,* 307, 56 (1991).
[18] S. Iijima and T. Ichihashi, *Nature,* 363, 603 (1993).
[19] D. Bethune, C. Kiang, M. de Vries, G. Gorman, R. Savoy, J. Vazquez and R. Beyers, *Nature,* 363, 605 (1993).
[20] S. Farhat, M. de La Chapelle, A. Loiseau, C. Scott, S. Lefrant, C. Journet and P. Bernier, *J. Chem. Phys.* 115, 6752 (2001).
[21] C. Journet, W. Maser, P. Bernier, A. Loiseau, M. DelaChapelle, S. Lefrant, P. Deniard, R. Lee and J. Fischer, *Nature*, 388, 756 (1997).
[22] H. Dai, A. Rinzler, P. Nikolaev, A. Thess, D. Colbert and R. Smalley, *Chem.Phys. Lett.* 260, 471(1996).
[23] D. Mann, Taylor and Francis, Ch. 2, p.19 (2006). [23 (a)] P. Avouris and R. Martel, *MRS Bulletin*, p. 306 (2010).
[24] P. Nikolaev, M. Bronikowski, R. K. Bradley, F. Rohmund, D. Colbert, K. Smith and R. Smalley, *Chem. Phys. Lett.* 313, 91 (1999).
[25] P. Avouris, Carbon nanotube electronics and photonics, *Physics Today*, 34, 62 (2009).
[26] T. Guo, P. Nikolaev, A. Thess, D. Colbert and R. Smalley, *Chem. Phys. Lett.* 243, 49 (1995).
[27] P. Eklund, B. Pradhan, U. Kim, Q. Xiong, J. Fischer, A. Friedman, B. Holloway, K. Jordan and M. Smith, *Nano Letters*, 2, 561 (2002).
[28] A. A. Puretzky, G. Eres, C. M. Rouleau, I. N. Ivanov, D. B. Geohegan, *Nanotechnology*, 19, 055605 (2008).
[29] M. S. Dresselhaus, G. Dresslhaus, and P. Avouris, Springer-Verlag Press: Heidelberg, 2001.

[30] K. P. Lee, A. I. Gopalan, P. Santhosh, S. H. Lee and Y. C. Nho, *Composites Science and Technology*, 67, 811 (2007).

[31] R. H. Baughman, A. A. Zakhidov and W. A. de Heer, *Science*, 297, 787 (2002).

[32] H. M. Li, F. O. Cheng, A. M. Duft and A. Adronov, *J. Am. Chem. Soc.* 127, 14518 (2007).

[33] M. Zheng, A. Jagota, E. D. Semke, B. A. Diner, R. S. Mclean, S. R. Lustig, R. E. Richardson and N. G. Tassi, *Nature Materials*, 2, 338 (2003).

[34] C. Vijayakumar, B. Balan, M. J. Kim, and M. Takeuchi, *J. Phys. Chem. C* 115, 4533 (2011).

[35] A. Hirsch, *Angewandte Chemie-International Edition,* 41, 1853 (2002).

[36] S. Banerjee, T. Hemraj-Benny and S. S. Wong, *Advanced Materials*, 17**,** 17 (2005).

[37] K. Balasubramanian and M. Burghard, *Small*, 1, 180 (2005).

[38] S. Qin, D. Qin, W. T. Ford, D. E. Resasco and J. E. Herrera, *Macromolecules,* 37, 752 (2004) and references cited therein.

[39] Y. Chen, R. C. Haddon, S. Fang, M. A. Rao, P. C. Eklund, W. H. Lee, E. C. Dickey, E. A. Grulke, J. C. Pendergrass, A. Chavan, B. E. Haley and R. E. Smalley, *J. Mater. Res.*13, 2423 (1998).

[40] J. Chen, M. A. Hamon, H. Hu, Y. Chen, M. A. Rao, P. C. Eklund and R. C. Haddon, *Science*, 282, 95 (1998).

[41] D. Tasis, N. Tagmatarchis, V. Georgakilas and M. Prato, *Chem.A- Europ. J.* 9, 4000 (2003).

[42] E. T. Mickelson, C. B. Huffman, A. G. Rinzler, R. E. Smalley, R. H. Hauge and J. L. Margrave, *Chem. Phys. Lett.* 296, 188 (1998).

[43] P. J. Boul, J. Liu, E. T. Mickelson, C. B. Huffman, I. W. Chiang, K. A. Smith, D. T. Colbert, R. H. Hauge, J. L. Margrave and R. E. Smalley, *Chem. Phys. Lett.* 310, 367 (1999).

[44] J. L. Bahr, J. Yang, D. V. Kosynkin, M. J. Bronikowski, R. E. Smalley and J. M. Tour, *J. Am. Chem. Soc.* 123, 6536 (2001).

[45] N. Tagmatarchis, V. Georgakilas, M. Prato and H. Shinohara, *Chem. Commun.* 2010 (2002).

[46] V. Georgakilas,K. Kordatos, M. Prato, D. M. Guldi, M. Holzinger and A. Hirsch, *J. Am. Chem. Soc.* 124, 760 (2002).

[47] A. Maity, S. Sinha Ray and S. K. Pillai, *Macromol. Rapid Commun*. 28, 224 (2007).

[48] H. Czichos, and T. Saito, L. S. (Eds.), Spinger, Heidelberg, (2006).

[49] C. R. Brundle, C. A. Evans, and S. Wilson, Sotneham, USA, (1992).

[50] A. Jorio, E. Kauppinen, A. Hassanien, Springer-Verlag, *Ch. Carbon-Nanotube Metrology,* p. 63 (2008).

[51] R. Krishnan, Vol. 1, Marcel Dekker, New York, Netherlands, p. 32 (1971).

[52] C. Raman and K. Krishnan, *Nature*, 121, 501(1928).

[53] M. Dresselhaus, G. Dresselhaus, R. Saito, and A. Jorio, *Physics Reports,* 409, 47 (2005).

[54] C. Hierold, Vol. 8, Wiley- VCH, Ch. 8, p. 125 (2008).

[55] S. Bachilo, M. Strano, C. Kittrell, R. Hauge, R. Smalley and R. Weisman, *Science*, 298, 2361(2002).

[56] F. Division, Nanosizer Version 2.0, Horiba (2008).

[57] T. Pradeep, Tata McGraw-Hill Publishing Company Limited, New Delhi, (2007).

[58] B. Fultz, and J. Howe,Springer, Heidelberg, (2008).

[59] J. Tersoff and R. Ruoff, *Phys. Rev. Lett.*73, 676 (1994).

[60] N. Wang, Z. K. Tang, G. D. Li and J. S. Chen,*Nature*, 408, 50 (2000).

[61] M. J. OConnell, S. Bachilo, C. Human, V. Moore, M. Strano, E. Haroz, K. Rialon, P. Boul, W. Noon, C. Kittrell, J. Ma, R. Hauge, R. Weisman and R. Smalley, *Science,* 297, 593 (2002).

[62] Y. Saito, M. Okuda, N. Fujimoto, T. Yoshikawa, M. Tomita and T. Hayashi, *Japan. J. Appl. Phys.* 33, L526 (1994).

[63] J. Gavillet, A. Loiseau, F. Ducastelle, S. Thair, P. Bernier, O. Stet'phan, and J.-C. C. J. Thibault, *Carbon*, 40, 1649 (2002).

[64] A. Gorbunov, R. Friedlein, O. Jost, M. Golden, J. Fink and W. Pompe, *Appl.Phys. A: Mater.Sci. Proc.* 69, S593 (1999).

[65] D. Geohegan, H. Schittenhelm, X. Fan, S. Pennycook, A. Puretzky and M. Guillorn, *Appl. Phys. Lett.* 78, 3307 (2001). [65(a)] Mathew Kisten Moodley, Ph.D thesis, University of Witwatersrand, Johannesburg, South Africa, December (2010).

[66] Nowak, E. J. *IBM J. Res. and Dev.* Vol. 46 No. 2/3 March/May, 169-180 (2002).

[67] C. Li, B. Lei, W. Fan, D. Zhang, M. Meyyappan and C. Zhou, *J. Nano Sci. Nanotechnol.* 7, 138 (2007).

[68] L. Wei and M. Charles Lieber, *Nature Materials*, 6, 841(2007).

[69] A, Sathyapalan, A. Lohani, S. Santra, S. Goyal, M. Ravikanth, S. Mukherjee and V. R. Rao, *Australian Journal of Chemistry*, 58, 810 (2005).

[70] H. Sirringhaus, T. Kawase, R. H. Friend, T. Shimoda, M. Inbasekaran, W. Wu and E. P. Woo, *Science*, 290, 2123 (2000).

[71] E. C. P. Smits, S. G. J. Mathijssen, P. A. van Hal, S. Setayesh, T. C. T. Geuns, K. A. H. A. Mutsaers, E. Cantatore, H. J. Wondergem, O. Werzer, R. Resel, M. Kemerink, S. Kirchmeyer, A. M. Muzafarov, S. A. Ponomarenko, B. de Boer, P. W. M. Blom and D. M. de Leeuw, *Nature*, 455, 956 (2008).

[72] Y. Zhou, J. Johnson, L. Wu, S. Maley, A. Ural, and H. Xie, Proceedings of the 2008 IEEE International Conference on Nanotechnology (8th) Held in Arlington, TX on August 18-21, (2008).

[73] Eeva Läänilänen, Master's Thesis (Master of Science in Technology), Department of Electrical and Communications Engineering, Helsinki University of Technology, Espoo, 29 June (2006).

[74] K. Felmet, *Appl. Phys. Lett.* 85(15), 3316 (2004).

[75] Y.-L. Loo, *J. Am. Chem. Soc.*124(26), 7654 (2002).

[76] M. A. Meitl, *Nano Lett.* 4(9), 1643 (2004).

[77] J. I. Sohn, S. Lee, Y. H. Song, S. Y. Choi, K. I. Cho and K. S. Nam, *Appl. Phys. Lett.* 78, 901 (2001).

[78] G. Pirio, P. Legagneux, D. Pribat, K. B. K. Teo, M. Chhowalla, G. A. J. Amaratunga and W. I. Milne, *Nanotechnology*, 13, 1 (2002).

[79] T. Y. Tsai, C. Y. Lee, N. H. Tai and W. H. Tuan, *Appl. Phys. Lett.* 95, 013107 (2009).
[80] A. Mathur, S.S. Roy, Kiran S. Hazra, D.S. Misra and J.A. McLaughlin, *Diamond and Related Materials*, 19, 914 (2010).
[81] M. Mann, Y. Zhang, K.B.K. Teo, T. Wells, M.M. El Gomati and W.I. Milne, *Microelectronic Engineering*, 87,1491 (2010).
[82] Y. I. Song, C. M. Yang, D. Y. Kim, H. Kanoh and K. Kaneko, *J. Colloid Inter. Sci.* 318, 365 (2008).
[83] S. Kim, J. Yim, X. Wang, D. D.C. Bradley, S. Lee and J. C. deMello, *Adv.Funct.Mater.* 20, 2310 (2010).
[84] Y. I. Song, G. Y. Kim, H. K. Choi, H. J. Jeong, K. K. Kim, C. M. Yang, S. C. Lim, K. H. An, K. T. Jung and Y. H. Lee, *Chem. Vap. Deposition* 12, 375 (2006).
[85] A. Rahy, P. Bajaj, I. H. Musselman, S. H. Hong, Y. P. Sun and D. J. Yang, *Appl. Surf. Sci.* 255, 7084 (2009).
[86] Y. M. Liu, Y. C. Fan, Y. C. Chen, Y. Sung and M. D. Ger, *Appl. Surf. Sci.* 256,1731 (2010).
[87] Z. Wu, Z. Chen, X. Du, J. Sippel, M. Nikolou, K. Kamaras, J. R. Reynolds, D. B. Tanner and A. G. Rinzler, *Science*, 305,1273 (2004).
[88] C. S. Woo, C. H. Lim, C. W. Cho, B. Park, H. Ju, D. H. Min, C. J. Lee and S. B. Lee, *Microelectronic Engineering*, *84*, 1610 (2007).
[89] D. Zhang, K. Ryu, X. Liu, E. Polikarpov, J. L., Mark E. Tompson and C. Zhou, *Nano Lett.* 6, 1880 (2006).
[90] J. H. Shin, D. W. Shin, S. P. Patole, J. H. Lee, S. M. Park and J. B. Yoo, *J.Phys. D: Appl. Phys.* 42, 045305 (2009).
[91] M. Qian, T. Feng, K. Wang,H. Ding, Y. Chen,Q. and Li,Z. Sun, *Physica E,* 43, 462 (2010).
[92] N. Saran, K. Parikh, D. S. Suh, E. Munoz, H. Kolla and S. K. Manohar, *J. Am. Chem. Soc.*126, 4462 (2004).
[93] J. W. Song, Y. S. Kim, Y. H. Yoon, E. S. Lee, C. S. Han, Y. Cho, D. Kim, J. Kim, N. Lee, Y. G. Ko, H. T. Jung and S. H. Kim, *Physica E*, 41, 1513 (2009).
[94] H. Okimoto, T. Takenobu, K. Yanagi, Y. Miyata, H. Shimotani, H. Kataura and Y. Iwasa, *Adv. Mater.* 22, 3981 (2010).
[95] A. Kumar, V. L. Pushparaj, S. Kar, O. Nalamasu, and P. M. Ajayan, *Appl. Phys. Lett.* 89, 163120 (2006).
[96] H. Liu, D. Takagi, S. Chiashi and Y. Homma, *J. Am. Chem. Soc.* 4, 933 (2010).
[97] C. L. Pint, Y. Q. Xu, S. Moghazy, T. Cherukuri, N. T. Alvarez, E. H. Haroz, S. Mahzooni, S. K. Doorn, J. Kono, M. Pasquali and R. H. Hauge, *J. Am. Chem. Soc.*4, 1131 (2010).
[98] L. Valentini, D. Bagnis, R. Cagnoli, F. Meloni, A. Mucci, L. Schenetti and J. Kenny, *Diamond and Related Materials*, 17, 1573 (2008).
[99] S. Pei, J. Du, Y. Zeng, C. Liu and H. M. Cheng, *Nanotechnology* 20, 235707 (2009).
[100] S. B. Bon, L. Valentini, J. M. Kenny, L. Peponi, R. Verdejo and M. A. Lopez-Manchado, *Phys. Status Solidi. A* 207, 2461 (2010).
[101] Shiang-Kuo, Chang-Jian, Jeng-Rong Ho, J-W John Cheng and Cheng-Kuo Sung, *Nanotechnology* 17, 1184–1187 (2006).

[102] Shiang-Kuo Chang-Jian, Jeng-Rong Ho and J.-W. John Cheng, *Microelectronic Engineering*, 87, 1973 (2010).

[103] Shiang-Kuo Chang-Jian, J. R. Ho and J. W. John Cheng, *Solid State Communication*, 150, 666 (2010).

[104] A. A. Stramel, M. C. Gupta, H. R. Lee, J. Yu and W. C. Edwards, *Optics and lasers in Engineering,* 48, 1291 (2010).

[105] C. Boutopoulos, C. Pandis, K. Giannakopoulos, P. Pissis, and I. Zergioti, *Appl. Phys. Lett.* 96, 041104 (2010).

[106] ww.nasa.gov.

[107] Q. Cao and J. A. Rogers, *Adv. Mater.* 21, 29 (2009).

In: Applications of Carbon Nanotubes
Editor: Ajay Kumar Mishra

ISBN: 978-1-62081-918-0

Chapter 12

SYNTHESIS OF MULTI-WALLED CARBON NANOTUBES BY FLOATING CATALYST CHEMICAL VAPOR DEPOSITION AND THEIR UTILITY FOR HYDROGEN SENSING AT ROOM TEMPERATURE

***Darío A. Zilli*[1,2], *Pablo R. Bonelli*[1,2] *and Ana L. Cukierman*[1,2,3*]**

[1]Programa de Investigación y Desarrollo de Fuentes Alternativas de Materias Primas y Energía (PINMATE) - Depto. de Industrias,
Facultad de Ciencias Exactas y Naturales, Universidad de Buenos Aires,
Buenos Aires, Argentina
[2]Consejo Nacional de Investigaciones Científicas y Técnicas (CONICET),
Buenos Aires, Argentina
[3]Cátedra de Farmacotecnia II. Depto. de Tecnología Farmacéutica,
Facultad de Farmacia y Bioquímica, Universidad de Buenos Aires,
Buenos Aires, Argentina

ABSTRACT

Self-aligned muti-walled carbon nanotubes (CNTs) assemblies are synthesized by floating catalyst chemical vapor deposition. Iron phthalocyanine, as a precursor, different reaction temperatures (880-1050 °C), total gas flow rates (10-100 cm^3/min), and H_2 molar fractions (0.25-0.75) are used. Morphological characteristics of the CNTs are assessed by scanning and transmission electronic microscopy. Different multivariate analysis techniques are applied to relate the operating variables investigated and the morphological characteristics, the latter including inner and outer diameters of the individual CNTs building the assemblies, length of the CNTs, aspect ratio, and volume of the individual CNTs. It is found that the temperature is the operating variable which exerts the highest influence on the morphological characteristics, with values increasing with temperature raise. Lower effects are inferred for the gas flow rate and H_2 composition, both in the same direction but in opposition to that determined for the

* E-mail: analea@di.fcen.uba.ar; anacuki@ffyb.uba.ar.

reaction temperature. Besides, for certain sets of operating conditions, the synthesized CNTs assemblies evidence large scale occurrence of naturally grown Y-junctions-like CNTs, that are relevant to potential applications in nano-electronics. The orientation of the concavity of inner structures of the nanotubes indicates that the Y-junctions are formed by the merging of metal catalyst nanoparticles during a tip-growth scenario of the tubes. Formation and growth of these nanostructures are explained by an own previously reported mechanism. On the other hand, the CNTs synthesized under pre-established operating conditions, without further treatment and subjected to a three-stage purification protocol are successfully processed in the form of thin films (buckypapers). Their morphology and electrical behavior are characterized. Purification consists of gas-phase oxidation of the metallic nanoparticles under strictly controlled conditions, subsequent removal of metal oxide nanoparticles by acid treatment, and final dispersion in ethanol by ultrasonication. The buckypapers show bidimensional structures constituted by randomly oriented CNTs, room temperature resistivity of around 10^{-3} $\Omega \cdot m$, and semiconductor-type behaviour. Finally, nano-composite films composed of the pristine and purified CNTs and Pd nanoparticles are obtained by a simple process involving the reduction of palladium in solution and the subsequent filtering of the aqueous suspension of the decorated nanotubes. The nano-composite films show lower values of room temperature resistivity than the buckypapers, pointing to an increase in the density of electrical contact points in the structure of the films. The effect is more pronounced for the films based on the oxidized CNTs, showing the lowest resistivity value (3.2×10^{-4} Ω m). The viability of the nano-composite films as room temperature H_2 sensors is evaluated from the change in their electrical resistance in H_2 atmosphere. They all exhibit a reversible response towards H_2 at low concentration, with repeatable characteristics. However, the sensing capacity of the Pd-decorated oxidized CNTs films is twice larger than that for the films based on the pristine or completely purified CNTs, likely because the surface functionalities generated during oxidation could act as anchor seeds for Pd nanoparticles deposition. The sensing behavior of the former is further examined for H_2 average concentrations in the range 70-2100 ppm. The sensor magnitude of response at room temperature reaches a saturation-like state with a maximum associated value of 2.15% for an average H_2 concentration above 350 ppm. The maximum value is 0.6% for an average H2 concentration of 70 ppm. Accordingly, the films based on the Pd-decorated oxidized CNTs could be used as nano-sensors of H_2 at room temperature.

1. Introduction

In recent years, the field of nanoscience and nanotechnology has undergone an unprecedented growth related to the possibility of controlling materials' structure at molecular or atomic scale. In this scenario, carbon nanotubes (CNTs) currently represent the most relevant and researched class of nanostructured materials. The exceptional properties of CNTs, arising from their unique crystalline, tubular nanometric structures have situated them at the forefront of emerging technologies [1-2]. Since their discovery [3], carbon nanotubes have received great attention as promising candidate materials in a widespread range of fields and applications due to their potentialities for technological innovations. They include use of CNTs in nano-electronics, nano-robotics, gas storage, nano-medicine, catalysis, nano-biotechnology, and environmental monitoring, among other areas [4-13].

Carbon nanotubes are structures of one or more graphite-like sheets rolled up into seamless cylinders of nanometric diameter and micrometric length [14-15]. The type of CNTs (single-, double-, or muti-walled CNTs) and their morphological characteristics (diameter, length, alignment), that influence their properties, and, therefore, potential applications, depend on the synthesis process and experimental conditions employed. Among the different existing processes for the synthesis of CNTs, chemical vapor deposition (CVD) is commonly applied. Various types of CVD have been developed for the synthesis of well-aligned CNT arrays, including thermal CVD, plasma-enhanced CVD, and floating catalyst CVD [16-18]. In particular, in floating catalyst CVD, the catalyst is produced *in situ* through the simultaneous introduction of both the metal catalyst and carbon source required for nucleation and growth of CNTs, leading to the continuous production of CNTs arrays. Compared to other CVD techniques, floating catalyst CVD is attractive for bulk-scale production of aligned CNTs due to its low cost, simplicity, versatility, and scalability [11, 19-20]. However, in spite of considerable progress, understanding of the effect of process variables on properties of CNTs, that is relevant to the successful, reliable scale-up, has not been thoroughly attained. While numerous investigations have attempted to correlate process parameters with CNTs characteristics, selectivity and yield, the effect of process variables still remains unclear when the data are analyzed collectively. A key reason for discrepancies is attributed to the one-factor at a time approach adopted in most studies where the effect of parametric interactions is neglected despite their relevance [21-23].

Furthermore, with the continuous miniaturization of electron devices, development of carbon nanotubes with branched structures has received particular attention in nano-electronics. Due to their special three-terminal tubular structure, CNTs with Y-junctions have potential applications in nano-scale integrated circuits and nanoelectronic devices, such as rectifiers, switches, and logic devices. Other types of junctions, like T-junctions or H-junctions, could also be useful in the prospect of organizing CNTs into complex networks to make nano-electronics circuits [24-26]. However, branched CNTs are more difficult to obtain than conventional CNTs with the disadvantage that their synthesis may require use of templates or additives, which can make the process more complex and/or introduce impurities, in turn affecting properties of Y-junction CNTs [27]. In some instances, however, branched CNTs have been naturally synthesized [24].

On the other hand, in the search of novel materials to serve as small, user-friendly and reliable gas sensing devices, CNTs have also attracted much attention as a feasible gas sensor material [28-31]. Gas sensors are widely used for environmental analysis in industry, medical diagnosis, and several other applications. Common gas sensors are based on metal oxide semiconductors such as tin oxide, zinc oxide, titanium oxide and aluminium oxide. However, problems encountered with these sensors include lack of flexibility, poor response time, and operation at elevated temperature [32]. In recent years, macroscopic forms of CNTs, such as arrays and films, have been reported making the macroscopic manipulation of CNTs practical for gas sensing applications [33]. Carbon nanotube paper, which is a paper-like CNTs film, known as buckypaper, is expected to have the same properties as CNTs. Up to the present, the most popular way of preparing buckypapers has involved the procedures of dispersion and filtration of funcionalized CNTs [34]. Regardless of the form of CNTs in devices, nanotube-based sensors have unique advantages over other materials, and different proposals have been explored. Single- and multi-wall CNTs arrays have been used for detection of ammonia gas, water vapour, volatile organic compounds, and other common gases [2, 30, 35]. Recently, multi-wall carbon nanotube networks of randomly entangled pure and oxidized nanotubes have also shown potentialities as sensing elements for sensitive and selective organic vapor switches [36]. In some cases, CNTs mixed with different metal oxides have been used to fabricate gas sensors. Modification of CNTs with functional groups, metal nanoparticles, oxides, and polymers has been proposed to change their electronic properties and to enhance the selectivity and response to specific gases through the interaction of the target molecules with the functional groups or additives [37-38]. Detection of different gases has been reported modifying single-walled CNTs, using coatings of catalytic noble metals and mixtures of metal oxide semiconductors or polymers [39-40]. Metals exhibit a broad range of electronic, chemical and physical properties that are often highly sensitive to changes in their chemical environment [41]. They are mechanically and chemically robust and stable, and can operate at high temperature and in harsh environment.

In particular, a variety of hydrogen sensors based on CNTs have been developed using different detection principles based on thermal conductivity, electrical conductivity, catalytic combustion, optical properties, and electrochemistry. H_2 gas sensors are necessary to ensure safe operation of hydrogen-based energy equipment and infrastructures, such as H_2 gas stations or fuel cell vehicles, accounting for the crucial role expected for hydrogen as an energy source in the near future. Some attempts to detect hydrogen include CNTs-based gas sensors, which are functionalized by catalytic palladium or platinum [34, 42-46].

Within this context, the present chapter deals with the synthesis of self-aligned muti-walled carbon nanotubes (CNTs) assemblies by the floating catalyst chemical vapor deposition process using iron phthalocyanine, as a precursor, and different reaction temperatures (880-1050 °C), total gas flow rates (10-100 cm^3/min) and H_2 molar fractions (0.25-0.75). Morphology of the CNTs is assessed by scanning and transmission electronic microscopy. The influence of the operating variables on CNTs morphological characteristics is examined by different multivariate analysis techniques. Besides, naturally grown Y-junctions-like CNTs in the synthesized assemblies are investigated. Formation and growth of this kind of nanostructures are explained in the framework of an own previously reported mechanism. On the other hand, the CNTs synthesized under pre-established operating conditions, without treatment and further subjected to purification, are processed in the form of thin films (buckypapers), and their morphology and electrical behavior are characterized. Nano-composite thin films are additionally prepared by a simple and practical filtration method of Pd-decorated CNTs, pristine and further subjected to purification. Their ability as room temperature hydrogen sensor material is evaluated from measurements of the change in electrical resistance when they are exposed to H_2 environment.

2. Experimental Section

2.1. Synthesis of Carbon Nanotubes and Operating Conditions

The floating catalyst chemical vapor deposition process was applied for the synthesis of self-aligned carbon nanotubes assemblies, operating under a flow of Ar and H_2. Analytical grade (99.99 %) iron (II) phtalocyanine ($C_{32}H_{16}N_8Fe$) purchased from Sigma Aldrich was used as organo-metallic precursor.

The set-up and general experimental procedure used for the synthesis of the CNTs were similar to those earlier reported [19, 47-48]. Briefly, the set-up consisted of a semi-continuous tubular reactor (2.2 cm inner diameter, 100 cm length) of horizontal configuration. The reactor, built in quartz, was externally heated by a two-zone furnace commanded by independent programmable temperature controllers, which provided control of the heating rate, temperature, and heating time. It was operated under an accurately controlled flow of H_2 (99.99 % v/v) and Ar (99.99 % v/v), supplied from gas cylinders. Gas flow rates were conveniently regulated to attain the desired H_2 concentration. Gas flow-meters located upstream the reactor inlet allowed to measure and control continuously gaseous streams.

The synthesis procedure was performed as follows. A quartz boat containing a weighted amount of iron phtalocyanine (ca. 200 mg) was placed near the reactor inlet, in the region heated by the first furnace (precursor zone). In turn, quartz substrates were conveniently placed in the reactor region heated by the second furnace (deposition zone). Inert conditions in the reactor were attained by purging air through circulation of a flow of Ar. Oxygen concentration at the reactor outlet was measured periodically by gas chromatography with a Shimadzu instrument equipped with a TCD detector, until lack of O_2 was verified. Preliminary experiments were carried out to assess optimal operating conditions for O_2 analyses. Once inert conditions were ensured, a mixture of Ar and H_2 in equal proportions was subsequently introduced into the reactor by regulating the individual gas flow rates. The reactor was heated by operating the double furnace in such a way that the precursor zone was kept at 650 ºC for 10 min, after the deposition zone was first held at the reaction temperature, above 800 ºC. During this stage, vaporization of iron (II) phtalocyanine and transport of the vaporized molecules by the flowing gas take place simultaneously. Nanoparticles of iron are formed and preferentially deposited in the deposition zone, due to presence of H_2 and the operation mode used. After this stage, the temperature of the first furnace was raised up to the selected reaction temperature in order to enable a uniform, homogeneous temperature throughout the reactor. Further fragmentation of the organic species presumably predominates in the deposition zone in this stage, thus providing the carbon source required for CNTs growth. After 15 min, the furnace was switched off and the reactor was cooled down to room temperature, keeping circulation of the Ar flow. CNTs were deposited on the reactor walls and on the quartz substrates in the deposition zone, near the reactor's outlet. They were easily separated by scratching the surfaces of the reactor's walls and/ or substrates. The synthesis process was carried out for different reaction temperatures (880-1050 °C), total gas flow rates (10-100 cm^3/min) and H_2 molar fractions (0.25-0.75), keeping constant the total reaction time for 60 or 90 min.

2.2. Purification of the Synthesized Carbon Nanotubes

Oxidation of the synthesized CNTs was firstly performed according to the procedure earlier reported [19]. It consisted of treating a batch of the as-synthesized CNTs (CNT-P, P for pristine) in an O_2 atmosphere (O_2 molar fraction of 0.10) at 375 ºC for 90 min, using the same set-up employed for the synthesis. The resulting oxidized CNTs are denoted as CNT-O (O for oxidized).

Part of the oxidized sample was further separated for the preparation of macroscopic CNTs thin films, while the rest was subsequently subjected to acid treatment, to remove oxide nanoparticles. The acid treatment consisted in contacting the oxidized sample with HCl aqueous solution (50 wt%), and gently mixing at room temperature. Afterwards, the CNTs were rinsed repeatedly with distilled water until neutral pH in wash water was attained. Then, the sample was dispersed in ethanol and the dilute dispersion was ultrasonicated at room temperature for 120 min. Ethanol was eliminated by evaporation and the sample was oven-dried overnight at 100 ºC up to constant weight. The so-treated CNTs are denoted as CNT-A (A for acid treated).

2.3. Characterization of the Pristine and Purified CNTs

Pristine arrays of the CNTs deposited on the quartz substrates as well as those subjected to further sequential treatments were examined by scanning and transmission electronic microscopy (SEM and TEM, respectively). SEM images were determined using a Phillips 500 microscope equipped with an energy-dispersive X-ray spectrometer at an accelerating voltage of 25 kV. Prior to SEM observations, the samples were sputter coated with Au under vacuum.

TEM analysis was performed on a Phillips CM-200 instrument operated at 100 kV. For TEM observations, drops of a suspension of the CNTs dispersed in ethanol by ultrasonication were placed on commercial grids. Alcohol evaporation was carried out at room temperature.

Around about 30 micrographs of each sample were used for morphological characterization. Moreover, energy dispersive X-ray (EDS) analysis was conducted to confirm the presence of metallic nanoparticles in the CNTs.

2.4. Preparation of CNTs Macroscopic Thin Films

For preparation of the buckypapers and nano-composites films of Pd-decorated CNTs, pristine CNTs synthesized at 895 °C with a total gas flow rate of 30 cm^3/min, H_2 molar fraction of 0.5, and total reaction time of 90 min were used. Also, the CNTs further oxidized in gas phase as well as those additionally subjected to the acid treatment stage for the removal of metal oxide nanoparticles were employed.

Typically, the CNTs, both pristine and further treated, were ultrasonically dispersed in distilled water for 90 min at room temperature. The suspension (0.3 mg/mL) was filtered through a polyamide membrane (Osmonic Inc., pore diameter 0.45 μm) employing a rectangular shape mould, and then dried overnight at room temperature. The resulting rectangular shaped CNTs film was easily peeled off from the polyamide membrane.

The nano-composites were prepared by reduction of the mixtures of both pristine and further treated CNTs and palladium salt following a similar procedure to that described by Li et al. [37]. The carbon nanotubes (0.3 mg/mL) and $PdCl_2$ (1.0 mg/mL) were added to water to prepare a mixture by ultrasonic dispersion (60 min). Then, 5 mL of aqueous solution of $NaBH_4$ (0.01 M) was added dropwise, into 20 mL of the mixture, in a period of 20 min under ultrasonic treatment for the reduction of Pd^{2+}. Finally, the aqueous suspension (25 mL) of the so prepared Pd-decorated CNTs was diluted with distilled water (500 mL) and ultrasonicated for 2 h. Then, it was filtered through the polyamide membrane and repeatedly washed with distilled water using the same filtering device mentioned above.

2.5. Morphological and Electrical Characterization of the Macroscopic CNTs Based Films

Morphological characteristics of the CNTs buckypapers and nano-composite films were observed on the aforementioned Philips 500 scanning electron microscope at an accelerating voltage of 25 kV. For electrical characterization, the rectangular shaped CNTs films were maintained over the polyamide membrane and cut into smaller rectangular probes to maximize the aspect ratio and minimize edge effects. Measurements of electrical behavior of the buckypapers and nano-composite films were carried out in Ar atmosphere using a standard four-probe method, varying the DC current intensity between 0 and 0.175 A. The resistances of the samples were determined by recording electrical voltage and current intensity at 15 seconds interval until steady state was reached. The electrical contacts to the sample were made with silver paste. Simultaneously, the temperature of the samples was measured with a copper-constantan (T-type) thermocouple.

2.6. Characterization of H_2 Sensing Behavior of the Nano-Composite Films

H_2 sensing characteristics of the Pd-decorated CNTs nano-composite films were determined by measuring the change in electrical resistance when they were exposed to environments with hydrogen. All the measurements were performed at room temperature using a self-built system device applying only 100 mW DC power. A diagram of the system device is illustrated in Figure 1.

The testing chamber had a volume of 190 mL and was operated under normal pressure throughout the measurements. Argon was used as carrier gas. It was fed continuously into the chamber at a flow rate of 30 mL/min. In order to obtain low concentrations of H_2 inside the testing chamber, small doses of hydrogen (30 – 500 $\mu L_{(STP)}$) were injected into the Ar flow.

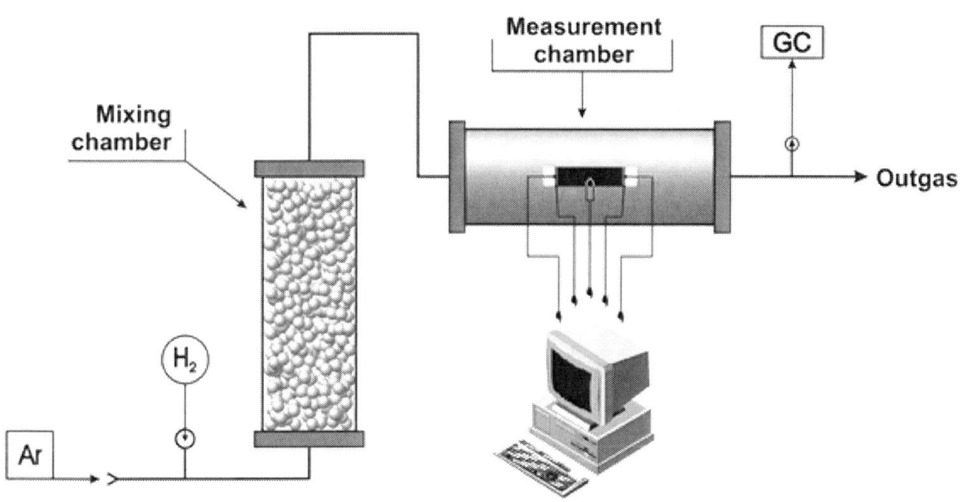

Figure 1. Schematic diagram of the set-up used for characterization of the H2 sensing behavior of the Pd-decorated CNTs nano-composite films [34].

Table I. Representative operating conditions used in the synthesis of CNTs assemblies

Sample	T_R (ºC)	F_T ($cm^3_{STP}\,min^{-1}$)	[H_2]	t_R (min)
CNT-1	1050	10	0.20	60
CNT-2	1050	10	0.25	60
CNT-3	1050	10	0.50	60
CNT-4	1050	30	0.25	60
CNT-5	880	30	0.75	60

TR: Synthesis temperature; FT: total gas flow rate; [H_2]: hydrogen molar fraction; tR: total reaction time.

A mixing chamber (190 mL) was placed between the hydrogen feed and the testing chamber to homogenize H_2 molecules in the carrier gas flow. The instantaneous concentration of H_2 inside the testing chamber was determined by gas chromatography using the aforementioned Shimadzu instrument.

3. Results and Discussion

3.1. Effect of Reaction Temperature, Total Gas Flow Rate and Gas Composition on Morphological Characteristics

The effect of the reaction temperature, total gas flow rate and H_2 molar fraction on morphology of the self-aligned arrays of CNTs, as well as on dimensions of the individual nanotubes building the arrays was examined.

In Table I, some representative examples of the synthesis conditions investigated are shown.

In Figure 2, SEM micrographs of the self-aligned arrays of CNTs synthesized for the conditions detailed in Table I are shown. As reported in own previous studies [19, 47], the synthesis process applied leads to dense mats or "forests" of aligned CNTs composed of near-parallel CNTs that are self-oriented, perpendicular to the substrate surface. The tendency for upward growth of CNTs has been ascribed to steric hindrance and growth competition among CNTs for the carbon source, likely due to the large number of catalytic metal nanoparticles supplied by the precursor. Besides, due to van der Waals interactions between the outermost walls of CNTs, they adhere to each other and form bundles or assemblies [17,19]. Density and degree of alignment of the resulting structures depend on the experimental conditions used, in turn affecting dimensions of the individual carbon nanotubes making up the assemblies.

From SEM images in Figure 2(a-d), it may be inferred that for the same reaction temperature, the operating variables related to the gas flow exert a weak influence on the nanotubes building the arrays. As may be observed in these images, increases in the gas flow rate and H_2 molar fraction led to slightly reduce the length of the self-aligned arrays of CNTs without considerable modifications in the alignment and density of the CNTs in the arrays.

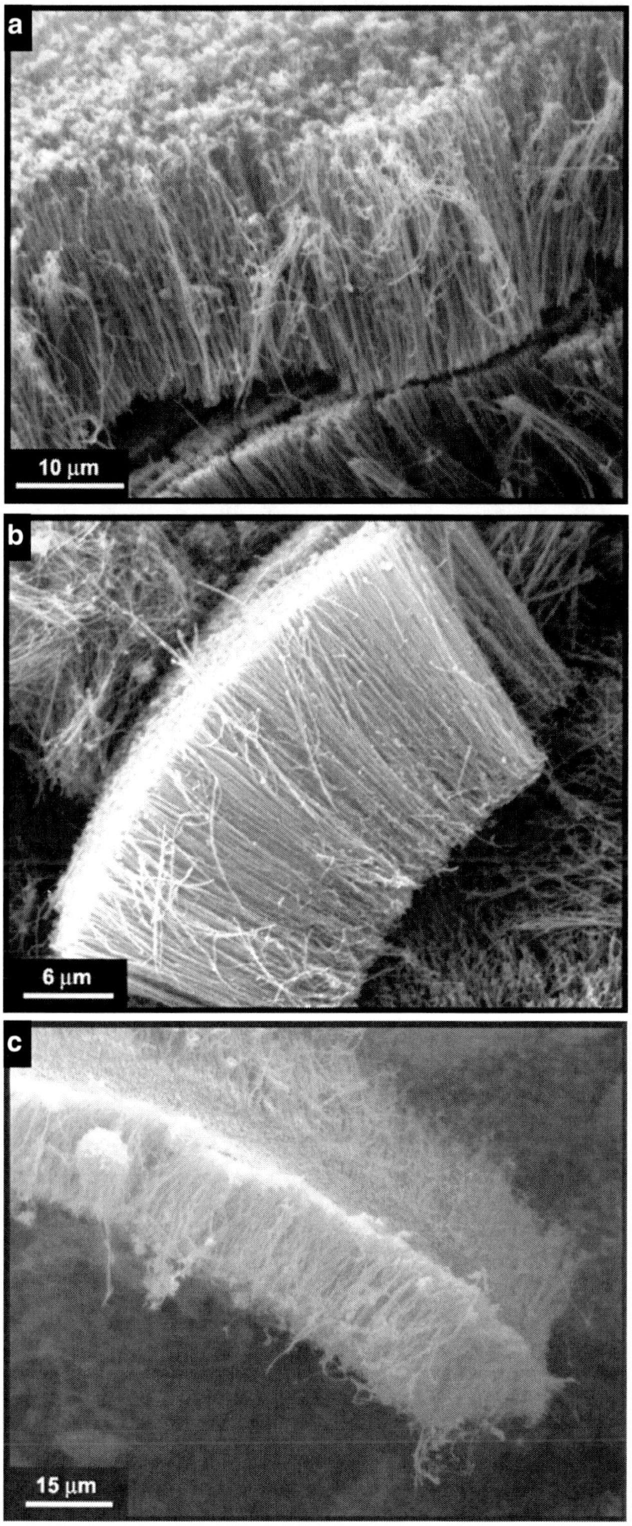

Figure 2. SEM images of the self-aligned arrays of CNTs synthesized for the conditions detailed in Table I. a) CNT-1, b) CNT-2, c) CNT-3, d) CNT-4, and e) CNT-5.

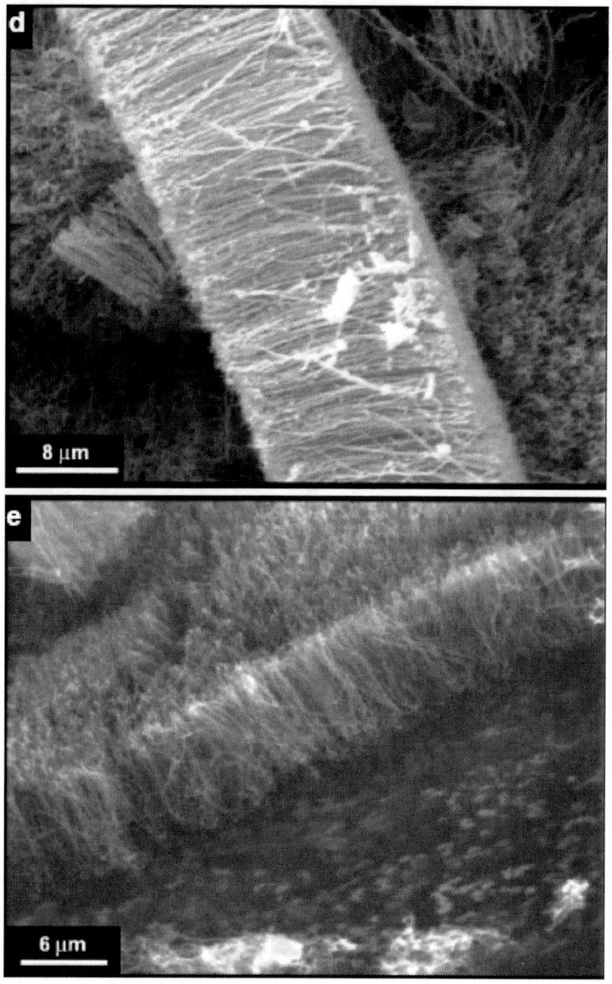

Figure 2. (Cont.). SEM images of the self-aligned arrays of CNTs synthesized for the conditions detailed in Table I: a) CNT-1, b) CNT-2, c) CNT-3, d) CNT-4, and e) CNT-5.

Conversely, the pronounced influence of the temperature on the length of the self-aligned CNTs arrays can be appreciated in Figure 2.e. A marked reduction in the length may be noticed for sample CNT-5 synthesized at the lowest temperature of the investigated range, in accordance with other reported results[49-51]. Higher temperatures promote growth of longer CNTs, with mean length varying from 4 μm to 25 μm. The trend is attributed to a thermal activation process governing CNTs growth rate [19]. Moreover, representative TEM micrographs obtained for the individual nanotubes constituting the self-aligned arrays of CNTs synthesized for the operating conditions detailed in Table I are displayed in Figure 3 (a-e). As seen, the images indicate that the nanotubes possess thick walls built of several graphitic layers, pointing to multi-wall carbon nanotubes and bamboo-shaped nanostructures. Possible mechanisms to explain formation of the latter have been earlier discussed [19]. TEM micrographs were used to measure outer and inner diameters, whose values allowed constructing the histograms which are illustrated in Figures 4 and 5, respectively.

The histograms in Figure 4 show that values of the outer diameters of the individual nanotubes of samples CNT-1 and CNT-2 present a normal distribution (Parts a and b of the figure), that differs substantially from those corresponding to samples CNT-3, CNT-4, and CNT-5 (Parts c to e of the figure). The results for the latter point to a slight effect of the operating variables related to the gas flow on the outer diameter, in comparison with that of the synthesis temperature.

Increase of hydrogen concentration in the gas flow noticeably modifies the shape of outer diameters distribution (Parts a to c of Figure 4), but it slightly reduces the mean value of the outer diameter for which most measured values tend to center around.

Besides, as may be observed in Figure 5 (a-e), none of the histograms constructed with values of the inner diameter show a normal distribution, even though the degree of influence of the operating variables related to the gas flow rate with respect to that of the synthesis temperature is kept. The results for the non-normal distributions of the outer and inner diameters of the individual carbon nanotubes building the self-aligned assemblies are discussed in the next sub-section.

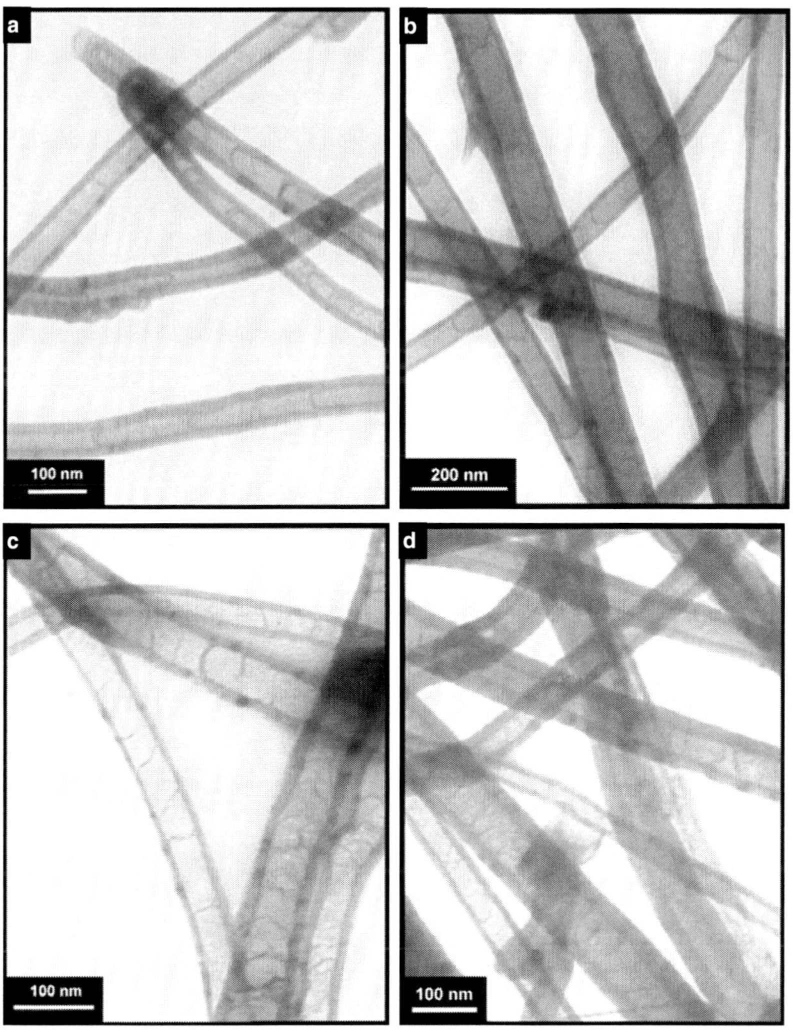

Figure 3. TEM images of the self-aligned arrays of CNTs synthesized for the operating conditions detailed in Table I: a) CNT-1, b) CNT-2, c) CNT-3, d) CNT-4, and e) CNT-5.

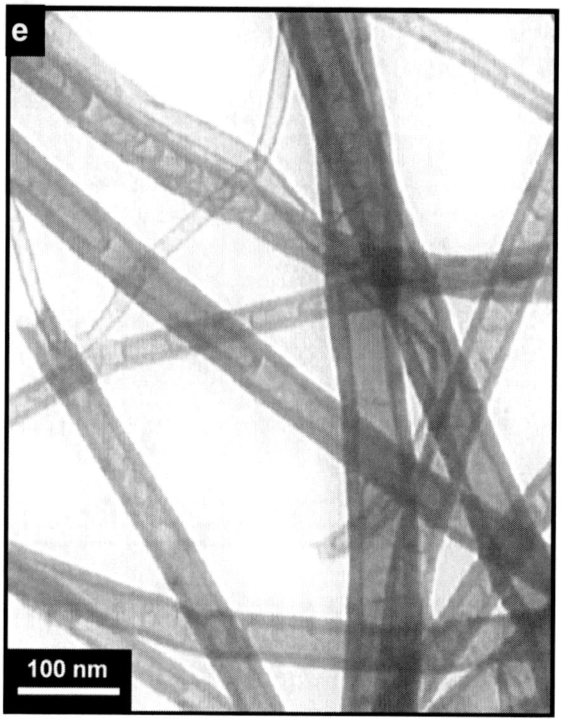

Figure 3. (Cont.). TEM images of the self-aligned arrays of CNTs synthesized for the operating conditions detailed in Table I: a) CNT-1, b) CNT-2, c) CNT-3, d) CNT-4, and e) CNT-5.

3.2. Statistical Analysis between Operating Variables and Morphological Characteristics of the Carbon Nanotubes

Different multivariate analysis techniques (MATs) were applied in order to determine the influence of the operating variables investigated on the following morphological characteristics:

- Inner and outer diameters of the individual CNTs building the assemblies (D_{int} and D_{ext}, respectively),
- Length of the CNTs (L_{CNT}),
- Aspect ratio (R_{CNT}), calculated as L_{CNT}/D_{ext},, and
- Volume of the individual CNTs (V_{CNT}), calculated as $L_{CNT}\cdot(D_{ext}^2-D_{int}^2)$.

The operating variables considered for the analysis were the synthesis temperature (T_R), molar fraction of hydrogen ($[H_2]$), and total gas flow rate (F_T). The results obtained by applying the different techniques to a data set of 2500 elements are discussed below.

Initially, principal component analysis (PCA) was applied [52-53], in an attempt to relate morphological characteristics of the CNTs by a linear combination of the operating variables. Results are shown in Tables II to IV.

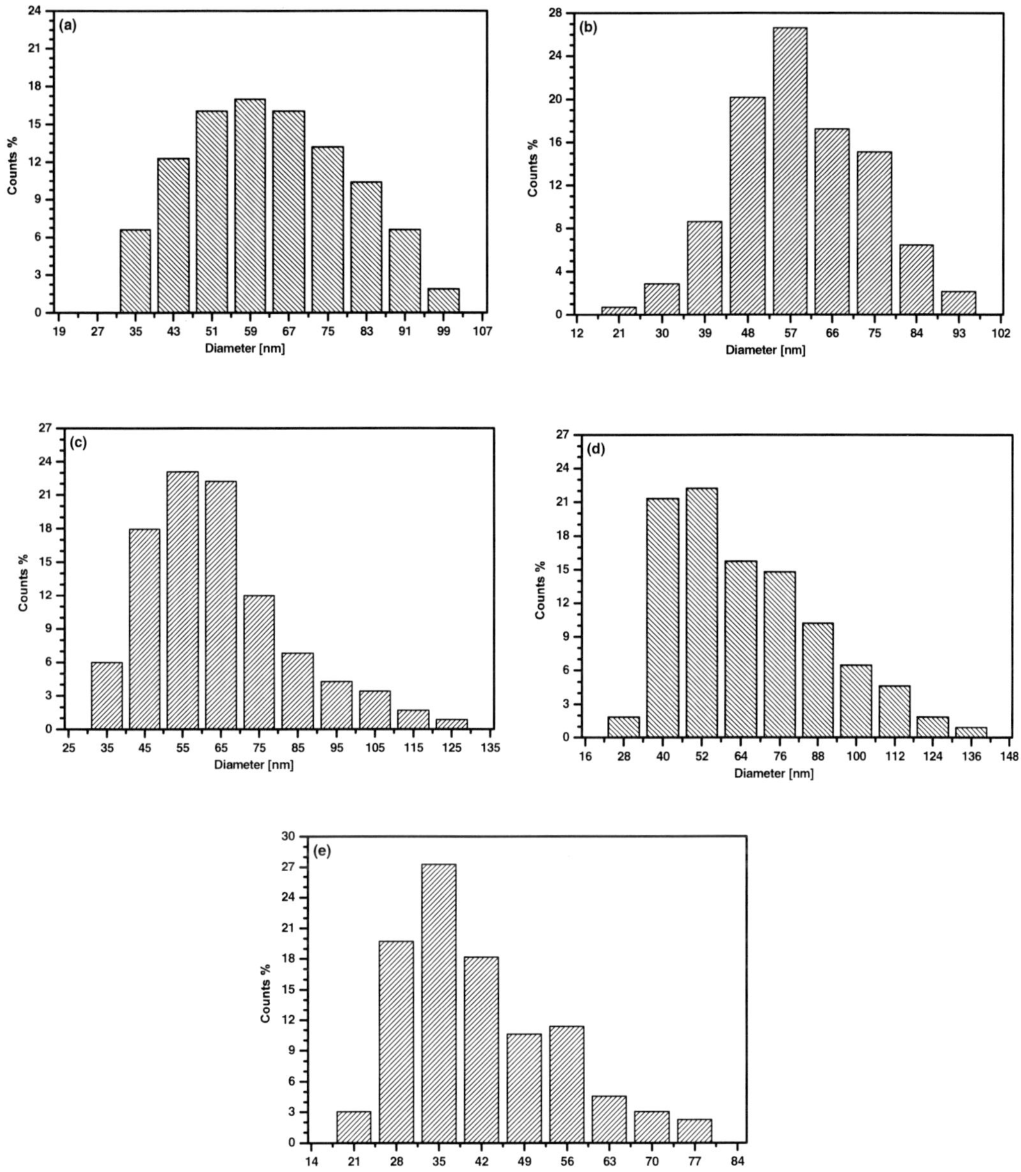

Figure 4. Histograms of outer diameters of the nanotubes of samples: a) CNT-1, b) CNT-2, c) CNT-3, d) CNT-4, and e) CNT-5.

Table II. PCA analysis: Communalities

	With 1 Factor	With 2 Factors	With 3 Factors	Multiple R^2
T_R	0.376477	0.985978	1.000000	0.074615
F_T	0.620863	0.664643	1.000000	0.179637
$[H_2]$	0.564210	0.739134	1.000000	0.157471

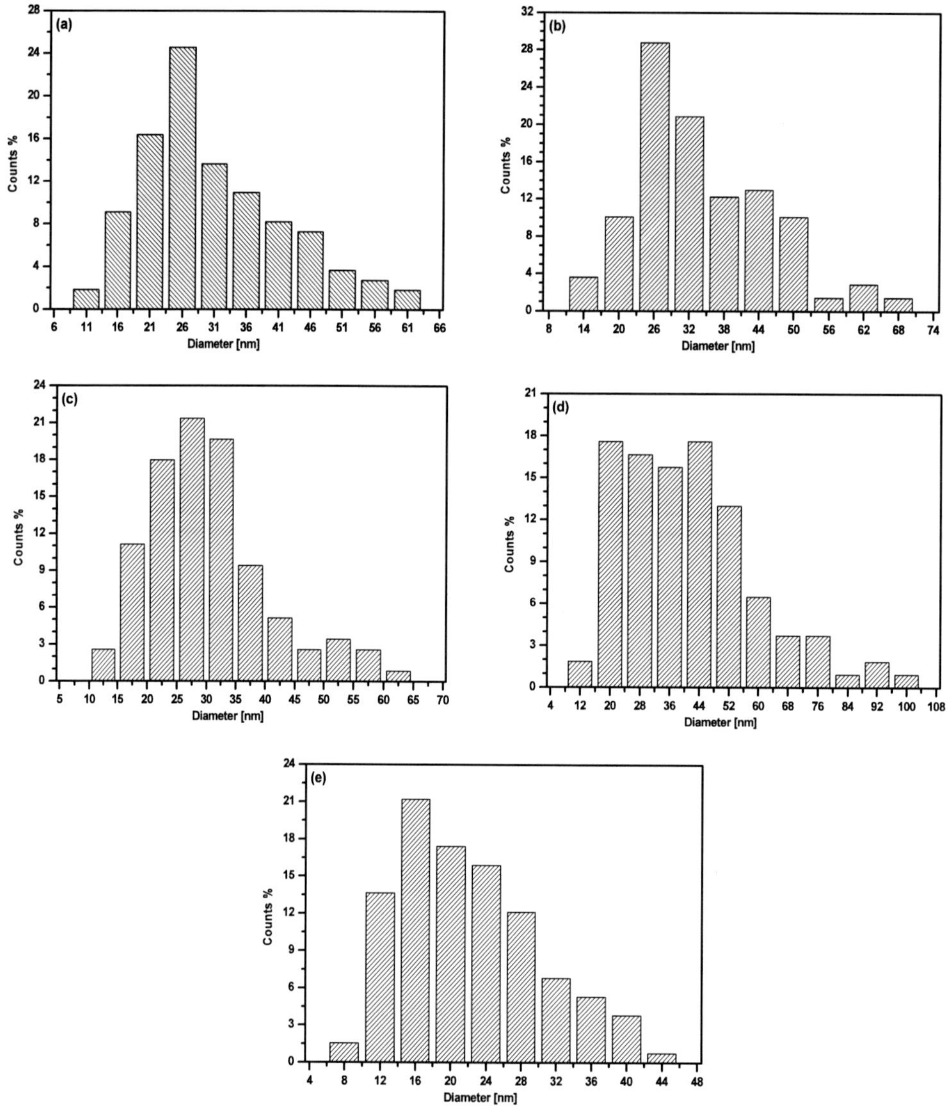

Figure 5. Histograms of inner diameters of nanotubes of samples: a) CNT-1, b) CNT-2, c) CNT-3, d) CNT-4, and e) CNT-5.

Table III. PCA analysis: Factor Loadings

	Factor 1	Factor 2	Factor 3
T_R	- 0.613577	**- 0.780706**	- 0.118414
F_T	**0.787949**	- 0.209237	- 0.579100
$[H_2]$	**0.751139**	- 0.418238	0.510751
Variance Explained	1.561550	0.828205	0.610245
Total Proportion	0.520517	0.276068	0.203415

Table IV. Eigenvalues extracted from PCA analysis

	Eigenvalue	% Total Variance	Cumulative Eigenvalues	Cumulative %
F 1	1.561550	52.05167	1.561550	52.0517
F 2	0.828205	27.60682	2.389755	79.6585
F 3	0.610245	20.34151	3.000000	100.0000

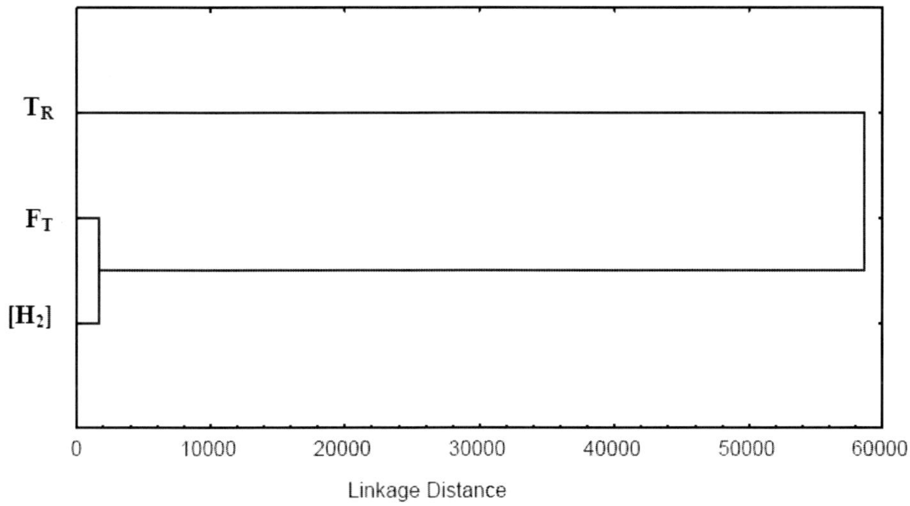

Figure 6. Dendrogram for the three operating variables investigated applying nearest-neighbor clustering.

The results highlighted in bold lettering in Table III correspond to the influence of the variables on each one of the factors where an incidence is observed (components in absolute value > 0.7).

The results highlighted in bold lettering in the second column in Table IV correspond to the proportion of the explained variance for each one of the components. In turn, those of the last column (Cumulative %) show that with the two first components near 80% of the explained variance is obtained. Thus, application of the PCA method to all the experimental data analyzed indicates that more of the 80% of the explained variance is observed only with two of the three necessary factors. Thus, it should be possible to reduce in one the number of initial variables.

On one hand, the total gas flow rate and the hydrogen molar fraction are grouped in one of the factors, whereas the synthesis temperature is independently grouped in the other one.

On the other hand, the experimental data were grouped into clusters by common attributes. Clustering was carried out by applying agglomerative hierarchical classification [53-54]. This method is widely used to describe groups with common, predictive characters or attributes. The results of the analysis by this method show that the synthesis temperature has a different behavior from the two other variables considered ($[H_2]$ and F_T), as may be appreciated in the dendrogram illustrated in Figure 6.

Similarly to the results from the PCA analysis, the dendrogram in Figure 6 shows two well distinguishable groups: one of them is related to the synthesis temperature, while the other one is associated with the total gas flow rate and hydrogen molar fraction. Besides, as aforementioned, values of the inner and outer diameters calculated from TEM micrographs and used to construct the histograms exemplified in Figures 4 and 5, do not strictly show normal distributions.

The Shapiro-Wilk test was used to further check if all the original measured data (D_{int}, D_{ext}, and L_{CNT}) come from a normal distribution population. In accordance with the analysis of TEM micrographs, it was found that the variables do not follow a normal distribution, as exemplified in the Q-Q plot for the length of CNTs shown in Figure 7.

Based on present results and taking into account that criteria of normality were not fulfilled, multiple regression methods to examine the dependence between the morphological characteristics and the operating variables were inapplicable. Therefore, it was necessary to apply non-parametric methods.

In particular, the Spearman's correlation coefficient, denoted as ρ_{Sp}, was applied. It is a non-parametric measurement of the interdependence of two random variables without the intrinsic necessity of a relationship between the variables [55-56]. ρ_{Sp} value indicates the variation direction of a variable with respect to another one, although it does not provide information whether there exists or not a functionality between the variables. Under independence, ρ_{Sp} is zero. It is exactly equal to 1 if there is a monotonic increasing relationship between the variables and to −1 if the association is monotonic decreasing. Moreover, this coefficient allows qualitative comparison of the incidence level that two variables has with respect to a third one. As an example, it is possible to qualitatively compare the incidence that two operation variables has on a given morphological characteristic. The results obtained by this technique are shown in Table V.

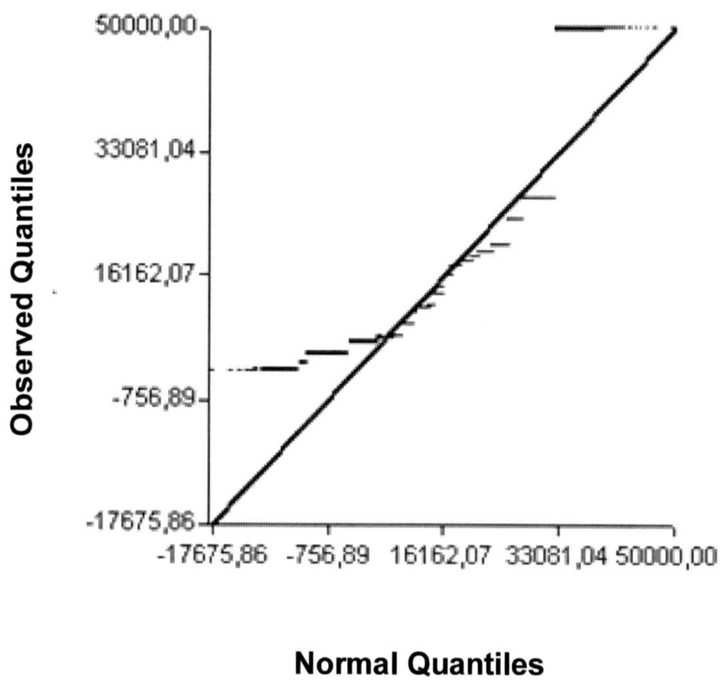

Figure 7. Q-Q plot for the length of the CNTs.

Table V. Spearman's correlation coefficients between the operation variables and the morphological characteristics of CNTs

Morphological Characteristic	T_R	F_T	$[H_2]$
L_{CNT}	0.792	- 0.420	- 0.232
D_{ext}	0.393	- 0.214	- 0.255
D_{int}	0.336	- 0.223	- 0.173
R_{CNT}	0.542	- 0.269	- 0.042
V_{CNT}	0.666	- 0.342	- 0.275

From the results in Table V, it may be inferred that the reaction temperature is the operation variable that has the highest incidence on morphological characteristics, with the value of the latter increasing with temperature raise. In a similar way, it is inferred that incidences of the total gas flow rate and H_2 molar fraction are lower and in the same direction, but both are in opposition to that of the reaction temperature.

Accordingly, considering the techniques applied for the statistical analysis between the operation variables and the morphological characteristics of the CNTs building the assemblies, it arises that those variables related to the gas flow rate present a similar, qualitative incidence that is opposite to the one of the reaction temperature. These results are in line with the analysis reported by See and Harris [57], in the sense that discrepancies in the data concerned with the influence of key variables involved in CNTs synthesis on their properties is due to the fact that the effect of parametric interactions is neglected in the majority of the reported studies.

3.3. Formation of Y-Type Nanostructures in the Self-Aligned Arrays of CNTs

For certain sets of operating conditions, the synthesized CNTs assemblies evidenced large scale occurrence of naturally grown Y-junctions-like CNTs. This kind of nanostructures is relevant to potential applications in nano-electronics for the development of nanometer-sized electronic devices, allowing the possibility of creating junctions that can serve as interconnecting molecular wires. Other potential applications for Y-branched nanotubes may also be envisioned, such as anchors and reinforcement for composite materials [25-27].

Table VI. Operating conditions for which CNTs with Y-junctions structures were observed

Sample	T_R (ºC)	F_T (cm^3_{STP} min^{-1})	$[H_2]$	t_R (min)
CNT-6	1050	10	0.66	60
CNT-7	880	00	0.50	60
CNT-8	880	100	0.66	60
CNT-9	950	10	0.25	60

T_R: Synthesis temperature; F_T: total gas flow rate; $[H_2]$: hydrogen molar fraction; t_R: total reaction time.

Figure 8 (a-d) displays TEM micrographs showing the presence of Y-junctions in the samples synthesized for the different sets of operating conditions reported in Table VI. In these junctions, two thin nanotubes are merged into a thicker nanotube. Metal nanoparticles can be clearly appreciated in the figures. Often, a metal nanoparticle is trapped at the junction (Figure 8d), or close to it in one of the narrower branches (Figure 8a). Arrows in Figure 8d point to the concavity of the inner structure of the nanotubes, giving hints on the way by which these junctions are formed [24].

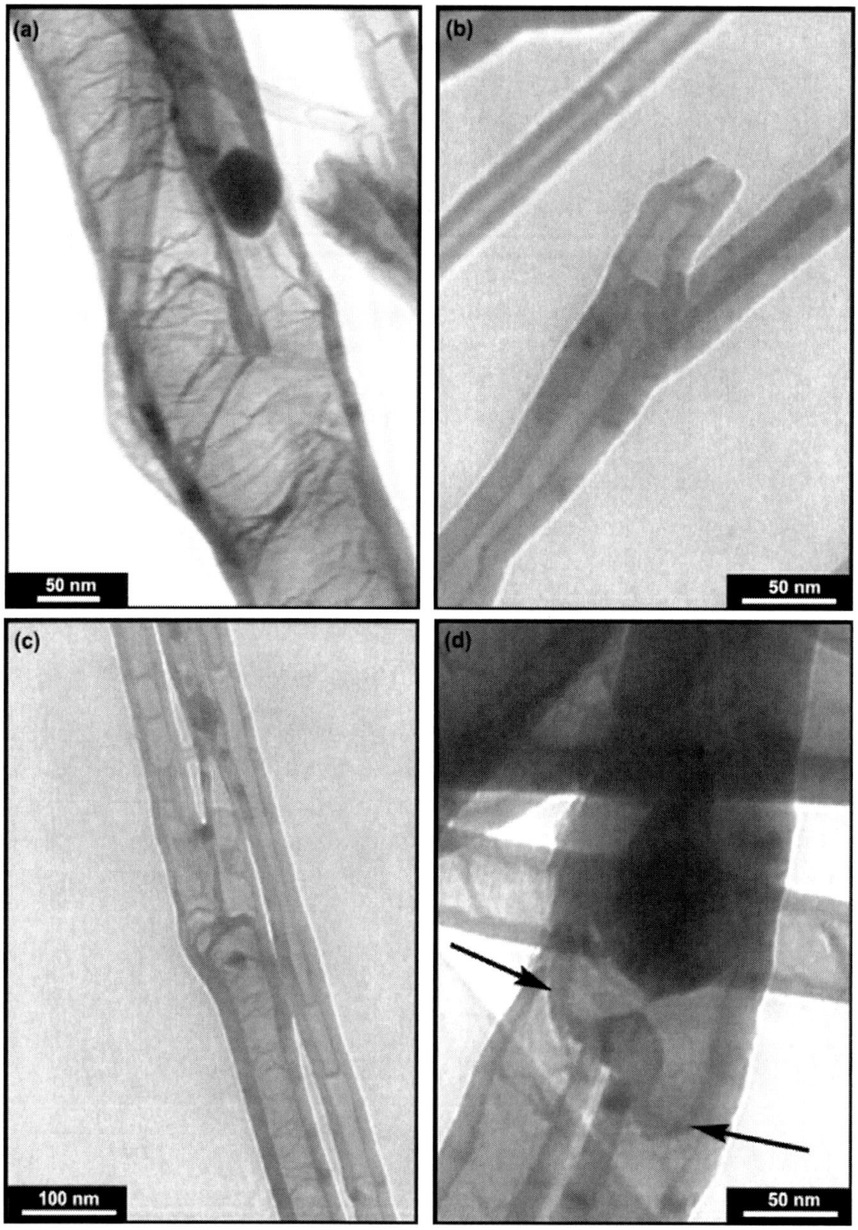

Figure 8. TEM images of some representative examples of carbon nanotubes with Y-like junctions structures observed in samples: a) CNT-6, b) CNT-7, c) CNT-8, and d) CNT-9.

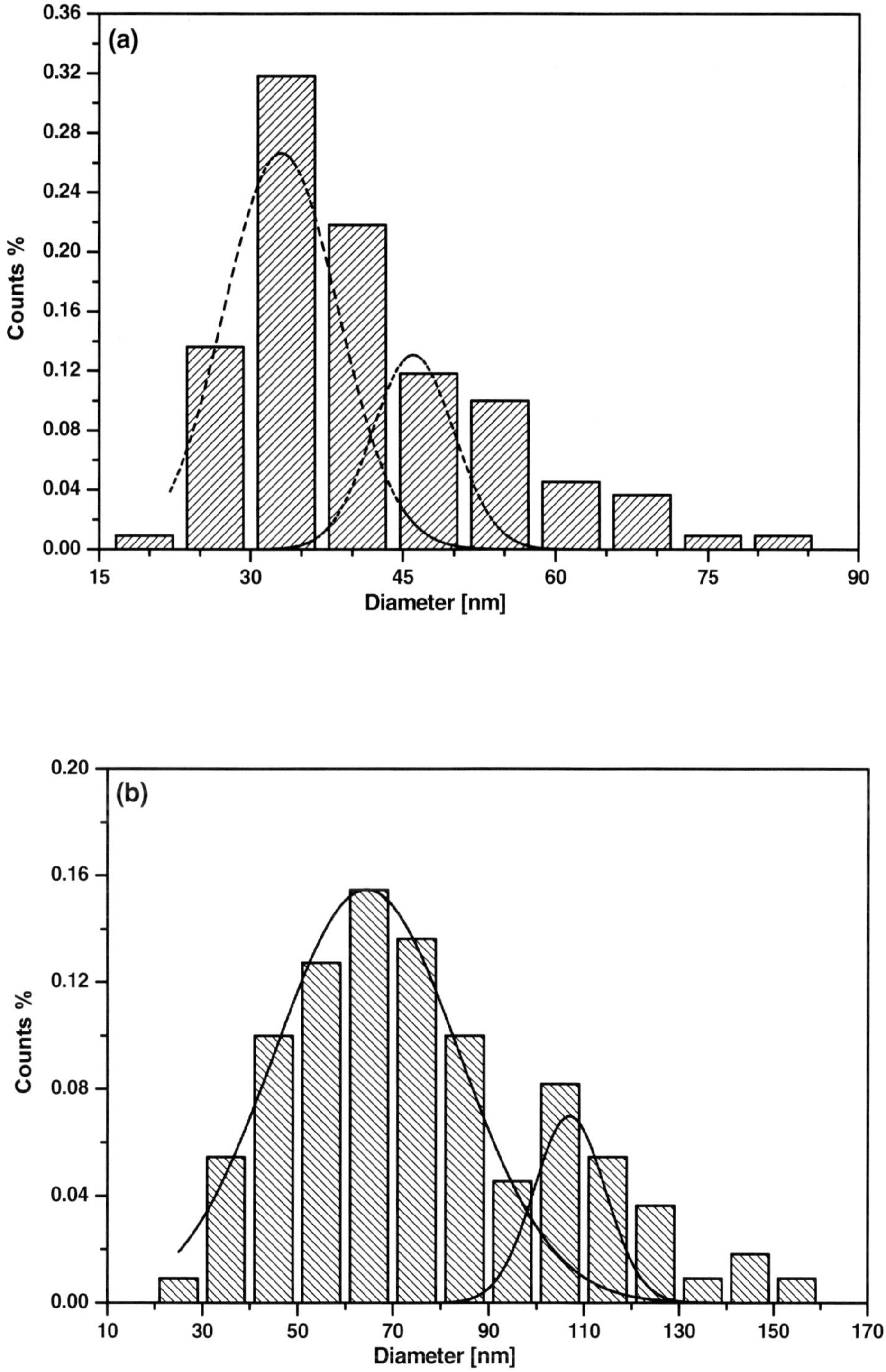

Figure 9. Distributions of outer diameters of the nanotubes in samples: a) CNT-7, and b) CNT-9. Dotted lines represent each one of the Gaussian modes of the bimodal distribution.

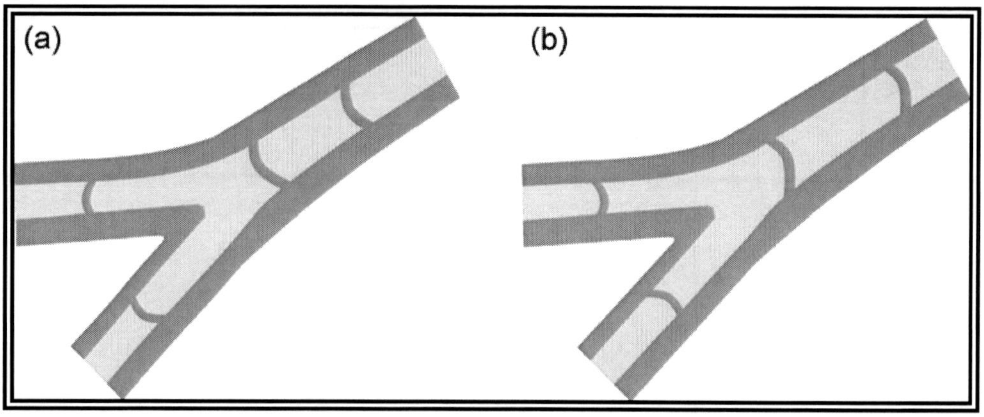

Figure 10. Possible concavities of the inner structures of the nanotubes according to the mechanisms proposed for the formation of Y-junction CNTs in the frame of a tip-growth route: (a) merger of two metal particles into a larger one, and (b) splitting of a metal particle into two smaller particles [24].

Histograms of outer diameters for samples CNT-7 and CNT-9 (corresponding to Figures 8b and 8d) are shown in Figures 9a and 9b, respectively. In both cases, distributions are bimodal.

For sample CNT-7, a first group of nanotubes with a diameter between 25 and 40 nm may be assigned to the two branches of the Y-like junction (e.g. 30 and 32 nm in Figure 8.b), whereas the second group of nanotubes with larger diameters between 40 and 55 nm should correspond to the main branch of the Y-junction (e.g. 45 nm in Figure 8.b). Similarly, the diameter distribution of CNT-9 is also bimodal, with the first group corresponding to the two branches (e.g. 62 and 70 nm in Figure 8.d) and the second one corresponding to the main branch (e.g. 105 nm in Figure 8.d).

The results suggest that among the different mechanisms reported in the literature for CNTs growth, Y-junctions formation should be related to a tip-growth route. In this framework, Zilli et al. [24] proposed that Y-like junctions can occur either from the merger of two small metal particles into a larger one (Figure 10a) or by the splitting of a large metal particle into two smaller ones during the tip-growth of the main branch of the Y (Figure 10b). Accounting for the orientation of the concavity of the inner walls, pointed out by the arrows in Figure 8d, the structure should have been formed by the merger of the two branches of the Y, rather than by the latter.

The mechanism proposed by Zilli et al. [24] to explain the formation of Y-junctions in self-oriented carbon nanotubes arrays considers that all the nanotubes are initially identical; consequently, they grow at the same rate and their metal particles are close to each other. A Y-junction is formed when two particles touch and merge. When the growth continues afterwards, the growth rates are no longer necessarily the same, and the metal particles move therefore apart from each other.

In this merging framework, it was postulated that the occurrence of Y-junctions is proportional to the probability of two particles to meet during the growth of the CNTs. SEM images in Figure 2 along with those displayed in Figure 11 (a-b), the latter showing evolution of CNTs growth process, support the view that the aligned growth of nanotubes maximizes the probability of touching and merging of the metal nanoparticles.

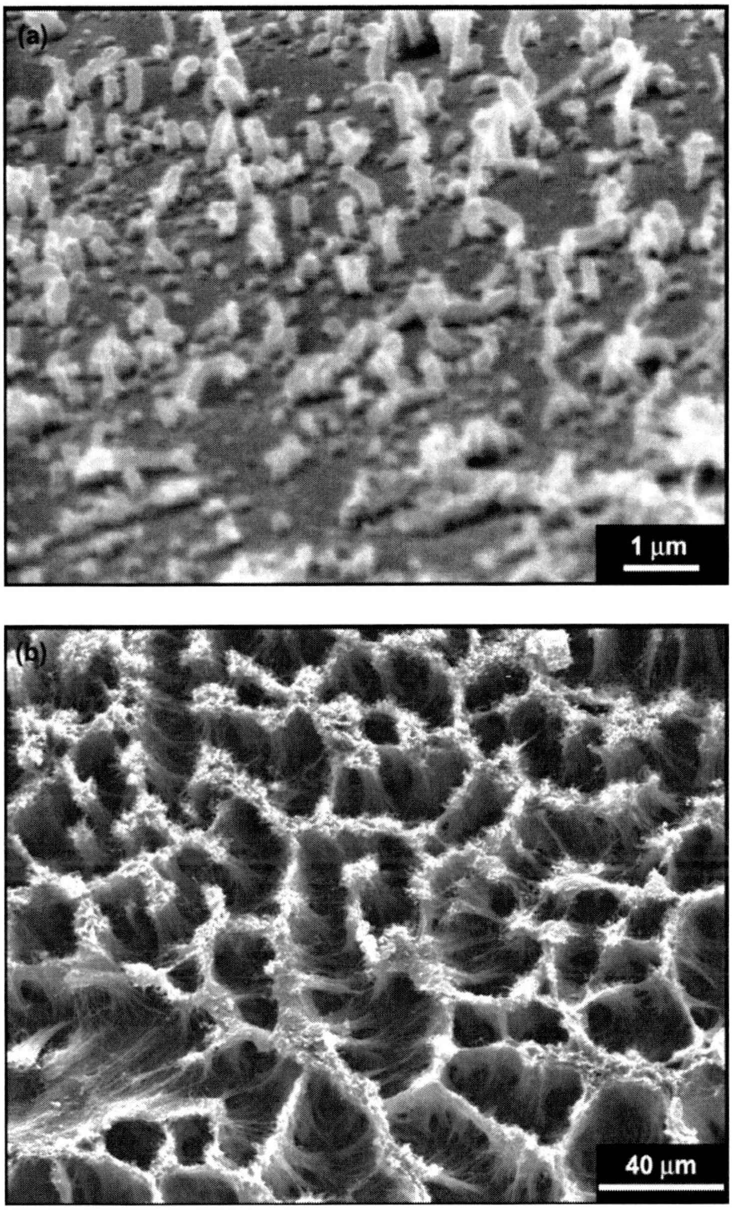

Figure 11. SEM images of carbon nanotubes arrays synthesized at $T_R = 950$ ºC, $F_T = 30$ cm^3/min, $[H_2] = 0.50$, $t_R = 60$ min. (a) A single nanotube grows from a metal catalyst nanoparticle; b) When the nanoparticles are disperse on the quartz substrate the nanotubes bend forming bunches during the growth stage.

3.4. Structure and Electrical Behavior of Macroscopic Thin Films of CNTS

The photograph in Figure 12 shows a buckypaper film obtained from the synthesized pristine CNTs, after being separated from the filter membrane. As may be appreciated in the SEM micrographs in Figure 13(a-d), due to their great elasticity, multi-walled CNTs form bi-dimensional, flexible pseudo-laminated structures interconnected through the same CNTs building the films, allowing to easily separate CNTs films from the filter membrane.

Figure 12. Photograph of a buckypaper developed from the synthesized pristine CNTs after separation from the filtration membrane.

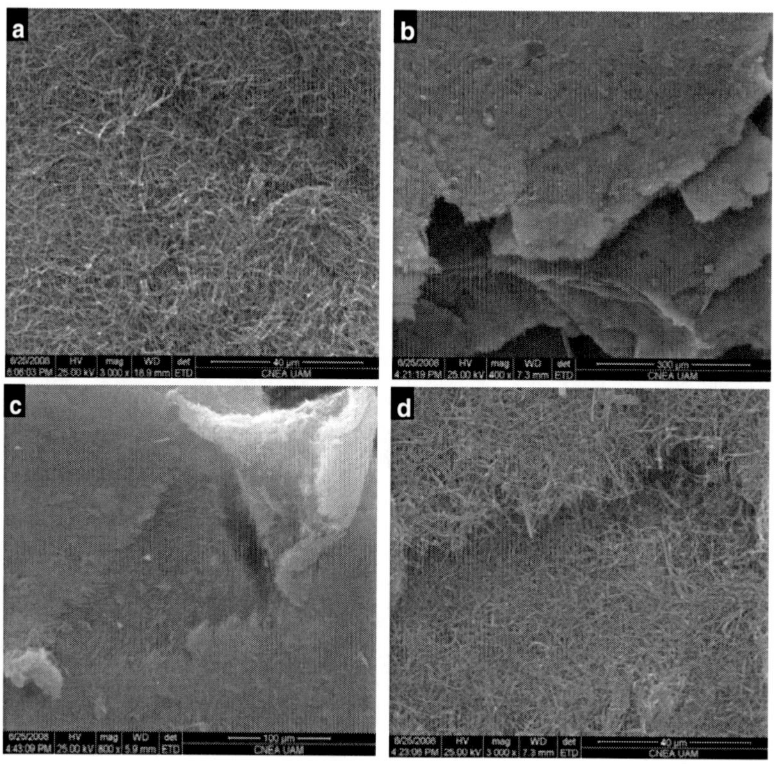

Figure 13. SEM micrographs of the buckypapers developed from the synthesized pristine CNTs, at different magnifications.

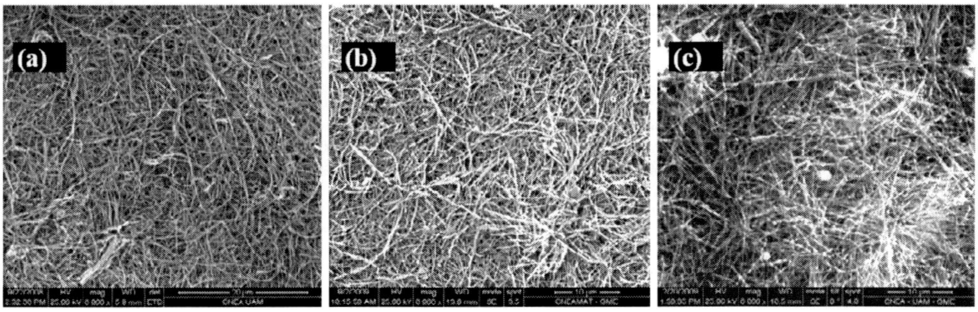

Figure 14. SEM micrographs of the surface of the buckypapers based on: (a) pristine CNTs, (b) oxidized CNTs, and (c) oxidized and acid-treated CNTs.

SEM micrographs of the surface of the CNTs buckypapers films developed from the pristine and further treated nanotubes are comparatively displayed in Figure 14(a-c). It should be stressed that no surfactant was used for their processing and, therefore, high-purity buckypapers were obtained.

As may be noticed in the SEM images in Figure 14, the films are 2D like-structures constituted by randomly oriented CNTs, both pristine (Figure 14.a) and subjected to purification treatments (Figures 14.b and 14.c), parallel to the filter surface. Cross linkages of straight CNTs can be observed, suggesting that good physical contact may be attained among nanotubes. Comparison of the SEM images indicates that the sequence of treatments applied to the CNTs has almost no effect on the morphological conformation of the buckypapers.

Typical electrical responses of the buckypapers obtained from the pristine and further treated CNTs are illustrated in Figure 15.

For the sake of comparison, voltage drops (ΔV_{Norm}) generated between the extreme of the probes by the DC current (I) and normalized by the aspect ratio of each probe, are plotted in Figure 15. Thus, the slope of the curve in the linear region (below 60 mA) represents the room temperature resistivity (ρ_0) of the buckypapers.

As can be appreciated in Figure 15, the treatments applied to the CNTs affect the electrical response of the buckypapers in a different way. In comparison with the CNT-P based buckypapers, the room temperature resistivity of those based on CNT-O is slightly increased, whereas the one corresponding to the CNT-A based buckypapers is slightly decreased.

Values of ρ_0 of the CNTs based buckypapers, expressed in Ω m, are included in the same figure. [35] also reported that the electrical resistivity of single-wall CNTs thin films ($\rho_0 = 3 \times 10^{-5}$ Ω m) increased when the nanotubes were previously subjected to an oxidative treatment with nitric acid. Enhancement in the resistivity of CNT-O based buckypapers could be related to the presence of oxide nanoparticles and oxide groups in the carbon nanotubes.

Besides, during electrical characterization of the buckypapers, their temperature raised due to Joule effect, leading to change their inherent electrical resistivity. Figure 16 shows the temperature dependence of the relative resistivity (ρ/ρ_0), estimated as the ratio between the normalized voltage drop and the DC current intensity, divided by ρ_0. As can be seen, the electrical resistivity of the pristine and further treated CNTs buckypaper films present a strong semiconductor-like dependence with temperature, decreasing its resistivity when the temperature is increased.

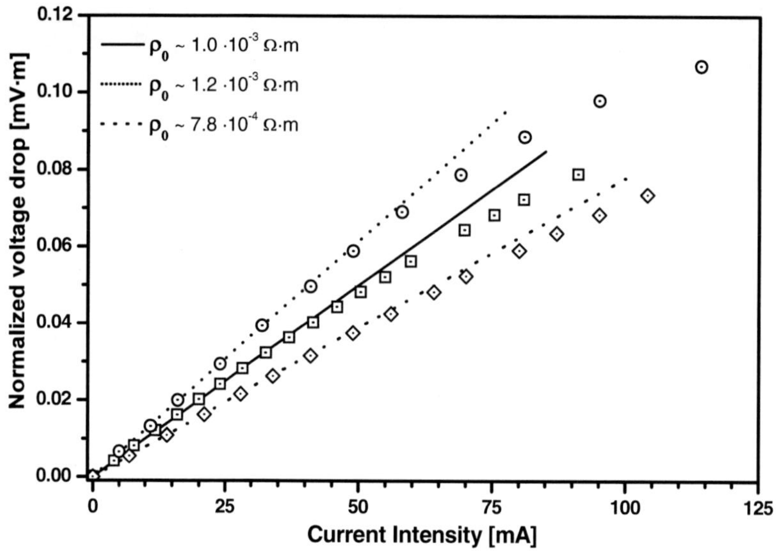

Figure 15. Normalized voltage drop as a function of DC current intensity for the buckypapers based on pristine CNTs (squares), oxidized CNTs (circles) and acid-treated CNTs (rhombus).

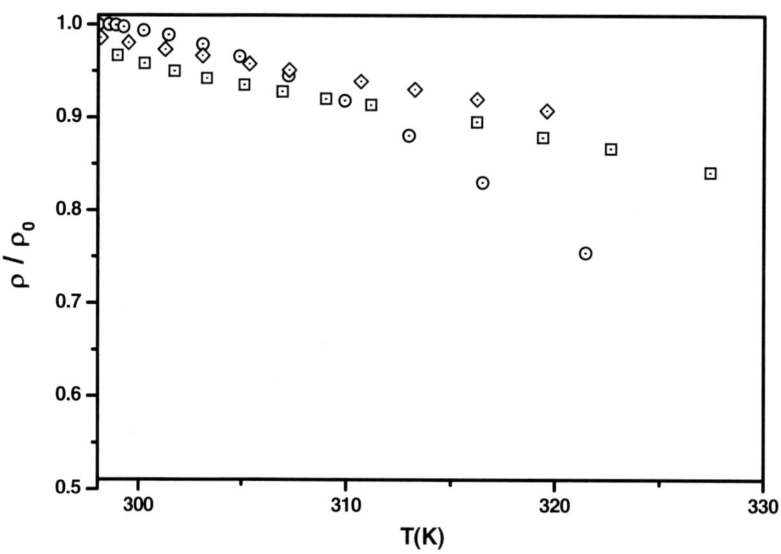

Figure 16. Variation of the relative resistivity (ρ/ρ0) with temperature for the buckypapers based on: the pristine CNTs (squares), oxidized CNTs (circles), and on the CNTs further subjected to acid treatment (rhombus). ρ is normalized to ρ0 at 298 K.

3.5. Structure, Electrical Behavior and H_2 Sensing Ability at Room Temperature of the Pd-Decorated CNTs Nano-Composite Films

The optical photography in Figure 17 exemplifies two sensor nano-composites obtained after processing the films of Pd-decorated CNTs to carry out electrical measurements.

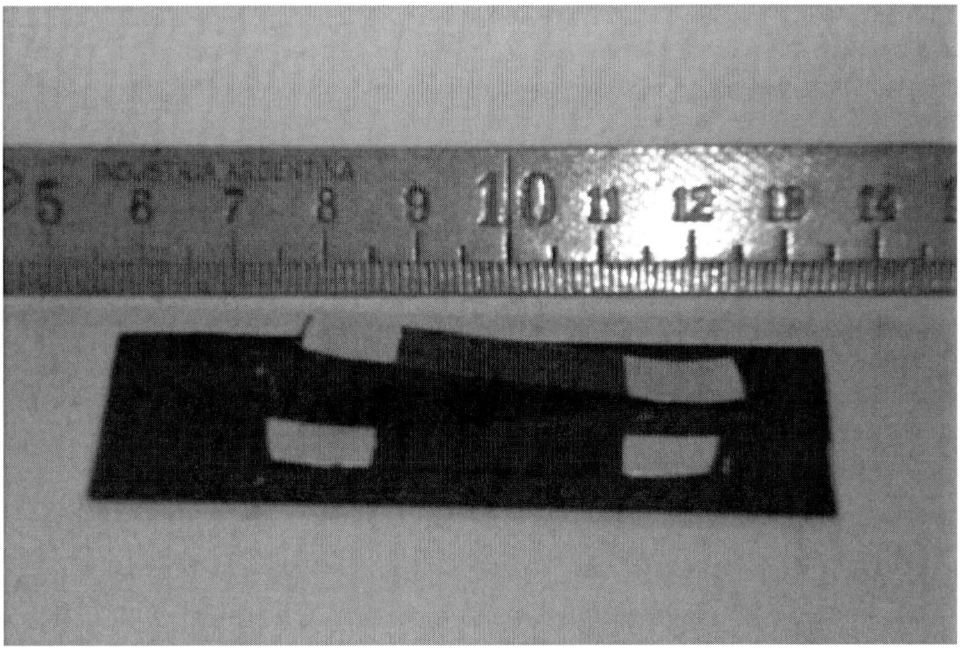

Figure 17. Photography of the nano-composite films based on Pd-decorated carbon nanotubes, after being processed for electrical behavior characterization.

Figure 18. SEM micrographs of the surface of the nano-composite films based on Pd0 nanoparticles and: a) pristine CNTs; b) oxidized CNTs; c) oxidized and acid treated CNTs [34].

Figure 19. Pattern of Pd^0 surface distribution in the nano-composite films based on Pd^0 nanoparticles and: a) pristine CNTs; b) oxidized CNTs; c) oxidized and acid treated CNTs.

In a similar way to the electric assays performed for the buckypapers, edges of the sensor elements were coated with a silver colloidal suspension.

SEM micrographs of the surface of Pd-decorated CNTs nano-composites thin films obtained with both pristine nanotubes and further treated ones are displayed in Figure 18 (a-c). It is apparent that Pd^0 nanoparticles are well spread in the nano-composite films, likely due to ultrasonication prior to the filtration stage. Homogeneous distributions of Pd^0 nanoparticles in the nano-composite films were confirmed by EDS analyses, as shown in Figure 19 (a-c).

CNTs are known to be very hydrophobic and cannot be wetted by some liquids due to their high surface tension, hindering adhesion of most metals [58-59]. Conversely, Pd^0 nanoparticles form clusters well dispersed all over the CNTs surface that can be observed in Figure 18. This should imply that Pd^0 nanoparticles are merged into clusters attached to the surface of the nanotubes in the aqueous suspension, instead of being filtered during the film formation. Figure 18.a shows that Pd^0 nanoparticles clusters in the nano-composite films based on the pristine CNTs (Pd–CNT-P) are larger than those present in the post-treated CNTs based composite films (Pd–CNT-O and Pd–CNT-A, in Figures 18.b and 18.c, respectively). In the latter case, the Pd^0 nanoparticles decorating the surface seem to form nano-clusters partially interconnected, that could lead to improve the electrical contacts among the Pd-decorated nanotubes (Figures 18.b and 18.c). The functional groups generated during the oxidation treatment, such as carboxylic and hydroxyl groups attached to the outer surface of the CNT-O, could act as anchor seeds for Pd nanoparticles deposition [60].

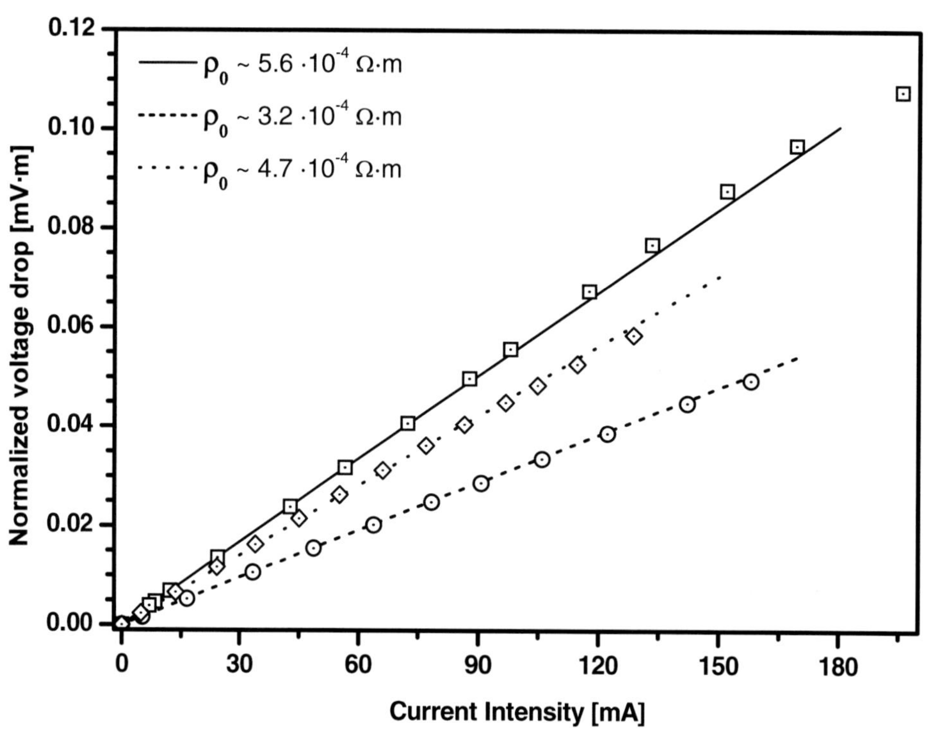

Figure 20. Normalized voltage drops as a function of DC current intensity for the nano-composite films based on Pd nanoparticles and pristine CNTs (squares), oxidized CNTs (circles), and oxidized and acid-treated CNTs (rhombus).

Electrical responses of the nano-composite films based on the pristine and further treated CNTs decorated with Pd^0 nanoparticles are displayed in Figure 20. In the same way as for the buckypapers, voltage drops generated between the extremes of the probes and normalized by the aspect ratio of each probe are plotted as a function of current intensity. Values of the room temperature resistivity (ρ_0) obtained from the slopes in the linear region are detailed in the same figure.

The results in Figure 20 show that the nano-composite sensing elements exhibit a linear behavior in almost all the range of analyzed values.

Only small deviations are noticed at the highest current intensities. From these results, it is inferred that the nano-composite films obtained from the oxidized CNTs present a lower electrical resistivity compared to the two other sensors.

Besides, all the Pd-decorated CNTs nano-composite films show lower values of room temperature resistivity than the CNTs buckypapers, pointing to an increase in the density of electrical contact points in the structure of the films. In addition, the CNT-O based nano-composite shows the lowest value of ρ_0 (3.2×10^{-4} Ω m), in opposition to the CNT-O based buckypaper, which presents the highest value (1.2×10^{-3} Ω m). Compared to the CNT-P based nano-composites, the lower value of ρ_0 should be related with a better distribution of Pd nanoparticles in smaller clusters enhancing the number of electrical contact points inside the film structure, as shown in Figure 18.b.

Similar to the results earlier discussed for the buckypapers, changes in the inherent resistivity of the nano-composite films were determined due to increases in temperature occasioned by the Joule effect. The relative resistivity, estimated as the ratio between the normalized voltage drop and the current intensity, divided by the room temperature resistivity, as a function of the temperature is shown in Figure 21.

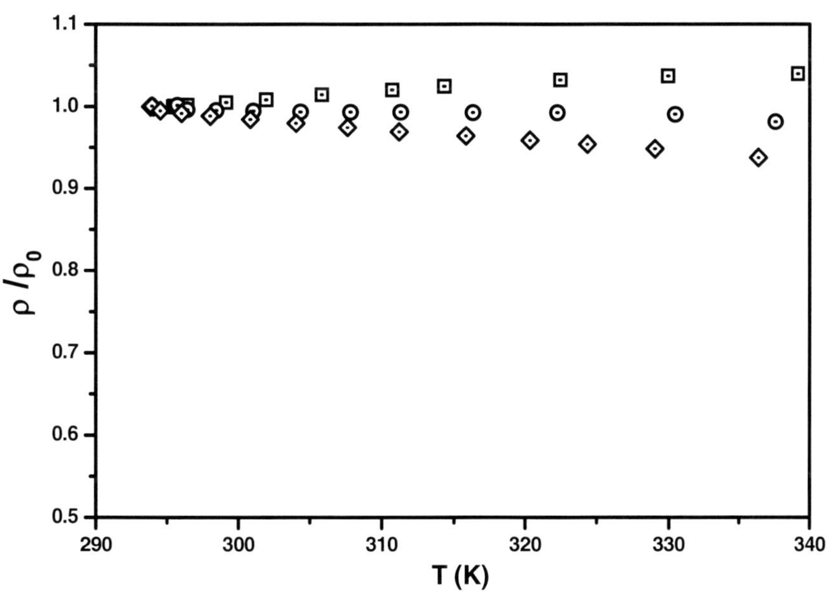

Figure 21. Variation of the relative resistivity with temperature for the nano-composite films developed from Pd^0 nanoparticles and pristine CNTs (squares), oxidized CNTs (circles), and oxidized and further acid-treated CNTs (rhombus).

The results in Figure 21 indicate that the resistivity of Pd–CNT-O and Pd–CNT-A films presents slightly semiconductor-type temperature dependence, whereas the Pd–CNT-P sample exhibited metallic-like type dependence. The latter trend points to an extensive presence of large clusters of Pd nanoparticles contributing to the electrical resistance in addition to cross-linking among the CNTs.

Furthermore, to quantify the sensing behavior of the nano-composite films, the response of the sensors (S) is defined as the percentage of the resistance change due to the exposure to a gas containing H_2 against the resistance measured without the presence of hydrogen, according to:

$$S = 100 \cdot \left(\frac{R_{H_2} - R_{Ar}}{R_{Ar}} \right)$$

where R_{Ar} is the resistance of the film exposed only to the carrier gas (without H_2), and R_{H_2} is the resistance of the films after exposure to a gas containing hydrogen. The response magnitude of the sensors (S_{max}) is defined as the percentage of the maximum resistance change for a gas containing a given concentration of hydrogen. In turn, the response time is defined as the time to reach S_{max}, while the time of recovery is defined as the time to attain the steady state resistance once the atmosphere is free of hydrogen.

To investigate the reproducibility of sensing, the Pd–CNTs nano-composite films were tested for three sensing cycles by injecting doses of H_2 repeatedly. Figure 22 shows the electrical resistance of the Pd–CNT-P based films as a function of time when exposed to H_2 dose of 500 $\mu L_{(STP)}$ for three times. It may be appreciated that the process is completely reversible. The resistance increases with time in the presence of H_2 in the detection chamber, and recovers the initial value once H_2 has been thoroughly displaced by the carrier gas flow. The responses of the Pd-decorated CNTs nano-composite films did not deteriorate after the third cycle, indicating that the sensing characteristics are repeatable.

Competing mechanisms that result in the sensing behavior of the CNTs are attributed to physisorption and chemisorption of hydrogen. Physisorption involves molecular hydrogen adsorption on the CNTs surface, whereas chemisorption involves the adsorption of atomic hydrogen on the surface of CNTs, with Pd playing a catalytic role during the chemisorption process [43]. At room temperature, the adsorbed H_2 molecules on the surface of Pd nanoparticles are dissociated as hydrogen atoms, which dissolve into Pd with high solubility, leading to a decrease in the work function of Pd. This causes electron transfer from Pd to CNTs, leading to a decrease in the nanotube conductance. The process is reversible as dissolved atomic hydrogen in Pd can be combined with O_2 in air to form water leaving the Pd-CNTs system and, enabling to recover the sensor's initial conductance [34,38].

Figure 23 illustrates the measured response of the Pd-decorated nano-composite films based on pristine and further treated CNTs when H_2 dose of 500 $\mu L_{(STP)}$ was injected into the carrier gas flow. All the samples showed a similar behavior, namely an increase in electrical resistance in H_2 atmosphere reaching a maximum value to slightly decrease straight afterwards (saturation-like state); then, the increased resistance suddenly dropped to the original value as the H_2 concentration in the carrier vanished. This behavior is similar to that reported earlier by Kumar and Ramaprabhu [43] who investigated the H_2 sensing performance of thin plastic substrates with Pd-nanoparticles decorated SWCNTs.

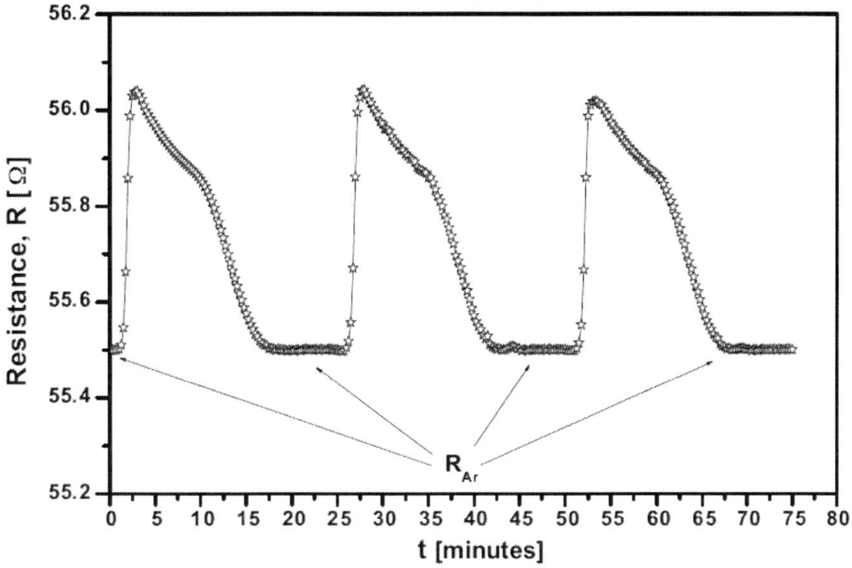

Figure 22. Electrical resistance changes of the nano-composite films based on Pd decorated pristine CNTs as a function of time when exposed to H_2 dose of 500 $\mu L_{(STP)}$ for three cycles [34].

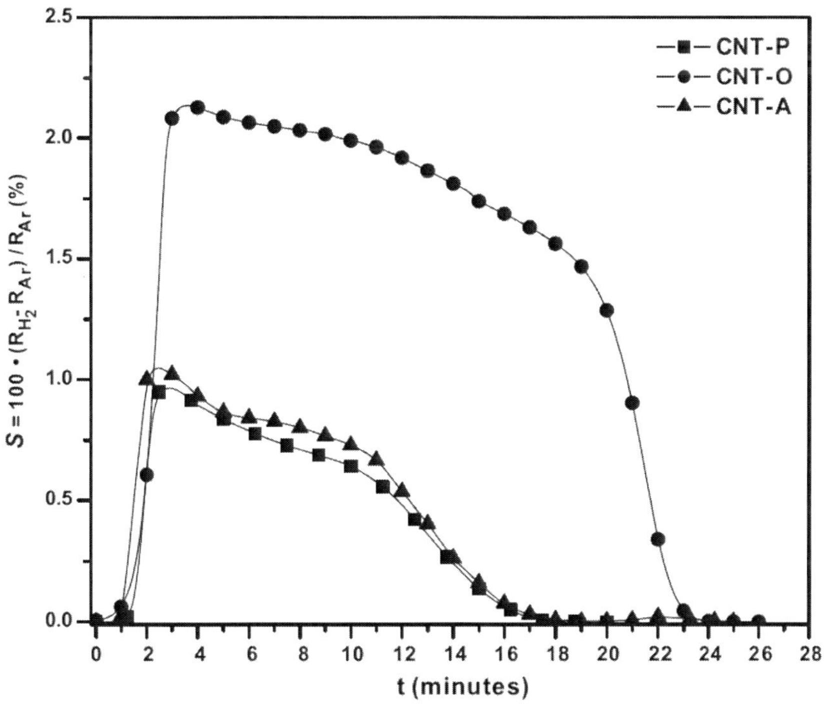

Figure 23. Sensing response of the Pd-decorated CNTs based nano-composite films when H_2 dose of 500 $\mu L_{(STP)}$ was injected into the carrier gas [34].

Comparing the responses for the different sensor elements, it arises that the Pd–CNT-P and Pd–CNT-A based nano-composite films exhibit a similar response towards H_2, but the Pd–CNT-O based film shows a more pronounced one. This suggests that the latter should be more sensitive to the presence of H_2 molecules in the environment, likely due to the distribution of Pd^0 nanoclusters over the CNTs surface (Figure 18.b).

The response magnitude and time of both Pd–CNT-P and Pd–CNT-A films were estimated as ~1.0% and ~150 s, respectively, while the same values associated with the Pd–CNT-O films were ~2.2% and ~200 s. Besides, the time of recovery of all the samples was estimated to be ~210 s.

In order to analyze the differences in the behavior of nano-composite sensors towards H_2, the presence of surface oxide groups and oxide nanoparticles produced, due to the intermediate oxidative stage in the sequential purification treatment, should be taken into account. In CNT-P, most catalytic Fe nanoparticles are encapsulated inside the nanotubes. Nevertheless, during the gas phase oxidation, they should be exposed as their corresponding oxides and oxygenated groups should be generated in active sites over nanotubes surface, but the removal of the oxide nanoparticles occurs during the final acid treatment. Then, iron oxide nanoparticles and oxygenated groups such as –COOH and –OH, present in the MWCNTs, should act as additional anchoring sites for Pd^{+2} ions [61]. Considering these results, the performance of the Pd–CNT-O based sensor was examined varying the dose of H_2 injected into the gas carrier flow.

The measured Pd–CNT-O based sensor response to different doses of H_2 injected (30 and 100 $\mu L_{(STP)}$) in the gas carrier flow is plotted in Fig. 24. For the sake of comparison, the measured response to 500 $\mu L_{(STP)}$ dose of injected H_2 is also included.

The results in Figure 24 shows that the dose of H_2 injected influences the Pd–CNT-O based sensor response. It can be appreciated that the sensor response magnitude reached similar values (S_{max} ~2.2%) when the two higher doses of hydrogen (100 and 500 $\mu L_{(STP)}$) were injected, although the saturation-like state monitored under the former condition holds for a shorter period of time than that measured with the latter.

These two characteristics combined together should indicate the presence of a threshold hydrogen concentration, above which the sensor response attains a maximum value, and remains practically constant as in a saturation-like state, until the concentration of H_2 in the test chamber decreases below it. Nevertheless, after the H_2 dose of 30 $\mu L_{(STP)}$) was injected, the resistance change of the sensor reached a maximum (S_{max} ~0.6%) and decreased to its original value straight afterwards, indicating that the threshold hydrogen concentration was never reached inside the test chamber.

To correlate the response magnitude of the sensor with the effective amount of H_2 inside the measurement chamber, an average hydrogen concentration $\left[H_2\right]_{t_p}$ was estimated, considering the time spent (t_p) until the S_{max} is reached since the dose of H_2 is injected into the carrier ($t_{inj} = 0$), and the following expression:

$$[H_2]_{t_p} = \frac{\int_{t_{inj}=0}^{t_p} Q \cdot [H_2]_{(t)} \cdot dt}{\int_{t_{inj}=0}^{t_p} Q \cdot dt}$$

where $[H_2]_{(t)}$ represents the instantaneous hydrogen concentration inside the measurement chamber, determined by GC since the dose of H_2 is injected (t_{inj}), and Q, the carrier gas flow.

The results of the average H_2 concentration along with the magnitude and time of response of the Pd–CNT-O based sensor (S_{max}), are comparatively listed in Table VII.

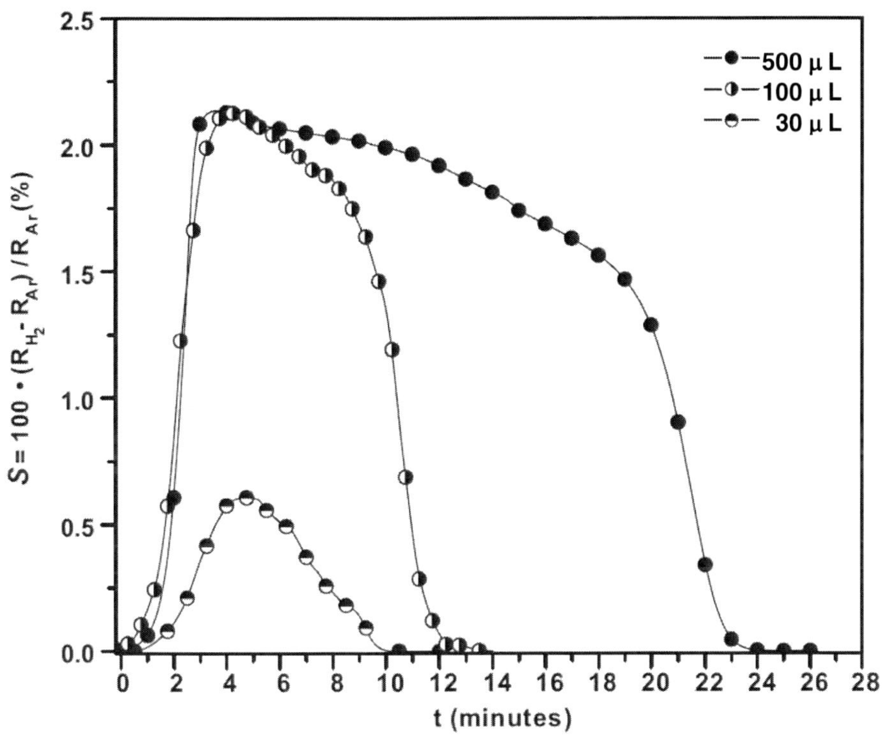

Figure 24. Effect of the H2 dose injected into the carrier flow on the response of the Pd-decorated oxidized CNTs based nano-composite sensor (Pd–CNT-O based sensor) [34].

Table VII. Characteristic values of the nano-composite films based on Pd0 nanoparticles and the oxidized carbon nanotubes

Dose of H_2 (μ L_{STP})	Response time t_p (min)	Average H_2 conc. (ppm)	Response magnitude S_{max} (%)
30	4.5	70	0.60
100	4.0	340	2.10
500	3.3	2100	2.15

They reveal that the Pd-decorated CNT-O based nano-composite films exhibited a reversible response to low concentrations of hydrogen at room temperature. Hence, they show potentialities to be used as low power devices for H_2 sensing at room temperature.

CONCLUSION

Self-oriented multi-wall CNTs arrays have been synthesized by floating catalyst CVD, varying the reaction temperature, the total gas flow rate, and H_2 molar fraction. The effect of these variables on the inner and outer diameters of the individual CNTs building the assemblies, length of the CNTs, aspect ratio, and volume of the individual CNTs has been examined by multivariate analysis techniques. The reaction temperature is the operating variable which exerts the highest influence on the morphological characteristics, their values increasing with temperature raise. The operation variables related to the gas flow, namely the total gas flow rate and H_2 molar fraction, present a qualitatively, similar influence between them, but lower and opposite to that for the reaction temperature.

For certain sets of operating conditions, the synthesized CNTs assemblies evidence large scale occurrence of naturally grown Y-junction-like CNTs. The diameter distributions of the nanotubes in the assemblies are bimodal, with one mode corresponding to the two branches of the Y, and the other one, to the main trunk. The orientation of the concavity of inner structures of the nanotubes indicates that the observed Y-junctions are formed by the merging of metal catalyst nanoparticles during a tip-growth scenario of the tubes. Formation and growth of this kind of nanostructures are explained by an own previously reported mechanism.

On the other hand, the CNTs synthesized under pre-established operating conditions, pristine and subjected to a purification protocol, have been successfully performed in the form of buckypapers. They show 2D like-structures constituted by randomly oriented CNTs, values of room temperature resistivity of around 10^{-3} $\Omega{\cdot}m$, and semiconductor-type behaviour.

Moreover, nano-composite films constituted by the pristine and purified CNTs and Pd^0 nanoparticles have been obtained by a simple process involving the reduction of palladium in solution and the subsequent filtering of the aqueous suspension of the decorated nanotubes. The Pd^0 nanoparticles affect electrical resistivity of the nano-composite films, compared to those developed for the buckypapers. All the Pd-decorated CNTs nano-composite films show lower values of room temperature resistivity than those for the CNTs buckypapers, pointing to an increase in the density of electrical contact points in the structure of the films. The effect is more pronounced for the films obtained from the oxidized CNTs, as evidenced by the lowest resistivity value (3.2×10^{-4} Ω m). The resistivity of the films obtained from Pd^0 nanoparticles and either the oxidized CNTs or the CNTs subjected to the complete purification protocol present a slightly semiconductor-type temperature dependence, whereas those derived from Pd-decorated pristine CNTs exhibit a metallic-like type dependence.

Room temperature H_2 sensing potential of the nano-composite films based on Pd-decorated CNTs, pristine and subjected to purification treatments, has been comparatively examined. The sensing capacity of the nano-composite films has been evaluated from the change in their electrical resistance in H_2 atmosphere. The Pd-decorated CNTs nano-composite films exhibit a reversible response towards hydrogen at low concentrations. Their sensing capacity is affected by the different stages of purification applied to the carbon nanotubes.

The nano-composite films based on Pd-decorated CNTs, either pristine or completely free of catalyst particles, show a similar sensing behavior, while the response of the oxidized Pd–CNTs based films is twice larger. Furthermore, the sensing behavior of the Pd-decorated oxidized CNTs nano-composites is modified by the H_2 concentration. The sensor magnitude of response at room temperature reaches a saturation-like state with a maximum associated value of 2.15% for an average hydrogen concentration above 350 ppm. The maximum value is 0.6% for an average hydrogen concentration of 70 ppm. Accordingly, the results reveal that the nano-composite films prepared with the Pd-decorated oxidized CNTs have potentialities for their use as room temperature H_2 sensors devices.

Acknowledgments

Financial support from Consejo Nacional de Investigaciones Científicas y Técnicas (CONICET), Universidad de Buenos Aires (UBA), and Agencia Nacional de Promoción Científica y Tecnológica - Fondo para la Investigación Científica y Tecnológica (ANPCYT-FONCYT) from Argentina is gratefully acknowledged.

References

[1] Capek, I. 2009. Dispersions, novel nanomaterial sensors and nanoconjugates based on carbon nanotubes. *Adv Coll Interface Sci,* 150, 63-89.

[2] Jiang, Y., Wang, P., Lin, L. 2011. Characterizations of contact and sheet resistances of vertically aligned carbon nanotube forests with intrinsic bottom contacts. *Nanotechnology* 22 365704 (8pp) doi:10.1088/0957-4484/22/36/365704.

[3] Iijima, S. 1991. Helical microtubules of graphitic carbon. *Nature* 354, 56–58.

[4] Iijima, S. 2002. Carbon nanotubes: past, present, and future. *Physica B: Condensed Matter.* 323, 1-4.

[5] O'Connell, M.J. 2006. Carbon Nanotubes: Properties and Applications, CRC Press: Boca Raton, FL, USA.

[6] Klumpp, C., Kostarelos, K., Prato, M., Bianco, A. 2006. Functionalized carbon nanotubes as emerging nanovectors for the delivery of therapeutics. *Biochimica et Biophysica Acta* 1758, 404 – 412.

[7] Dong, L., Subramanian, A., Nelson, B.J. 2007. Carbon nanotubes for nanorobotics. *Nanotoday* 2, 12-21.

[8] Jorio, A., Dresselhaus, G., Dresselhaus, M.S. 2008. Carbon Nanotubes: Synthesis, Structure, Properties and Applications. Springer, Berlin.

[9] Foldvari M., Bagonluri M. 2008. Carbon nanotubes as functional excipients for nanomedicines: I. pharmaceutical properties. *Nanomedicine: Nanotech., Biology and Medicine* 4, 173-182.

[10] Yang, X., Yuan, L., Peterson, V.K., Yin, Y., Minett, A.I., Harris, A.T.2011. Open-ended aligned carbon nanotube arrays produced using CO2-assisted floating-ferrocene chemical vapor deposition. *J. Phys. Chem.* C 115, 1409314097.

[11] Ying, L.S., bin Mohd Salleh, M.A, b. Mohamed Yusoff, H., Rashid, S.B.A., b. Abd. Razak, J. 2011. Continuous production of carbon nanotubes A review. *J. Industrial and Engineering Chemistry* 17, 367376.

[12] Diao, Q., Li, Y., Zhou, M., Xie, G. 2012. A new immunosensor for serum myeloperoxidase based on self-assembly of multi-walled carbon nanotubes/thionine/gold nanoparticles-chitosan. *Advanced Materials Research* 343-344, 1207-1211.

[13] Liu, A. 2012. Preparation and microwave absorption properties of CNTs/HAp composites for use in tumor hyperthermia. *Advanced Materials Research* 345, 280-286.

[14] Dresselhaus, M.S., Dresselhaus, G., Avouris, P. 2001. Carbon Nanotubes: Synthesis, Structure, Properties, and Applications, Springer-Verlag: Berlin Heidelberg

[15] Harris, P.J. 2003. Carbon nanotubes and related structures. New materials for the twenty-first century. Cambridge Press, U.K.

[16] Nxumalo, E.V., Nyamori, V.O., Coville, N.J. 2008. CVD synthesis of nitrogen doped carbon nanotubes using ferrocene/aniline mixtures. *J. Organomet. Chem.* 693, 2942–2948.

[17] Seah, Ch-M, Chai, S-P, Mohamed, A.R. 2011. Synthesis of aligned carbon nanotubes. *Carbon* 49, 4613-4635.

[18] Sun, Z., Qi, L., Shu, Y. 2012. Influence analysis for the controllable growth of carbon nanotubes with CCVD method. *Applied Mechanics and Materials* 101-102, 573-576.

[19] Zilli, D.A., Bonelli, P.R., Cukierman, A.L. 2008a. Effect of synthesis conditions and sequential post-trreatments on morphology, thermal behaviour, and gas adsorption properties of self-aligned multi-walled carbon nantoubes assemblies. In Nanotechnology Research: New nanostructures, nanotubes and nanofibers. X. Huang (Ed.). *Nova Science Publishers. N.Y, USA.* Chapter 5, 195-219.

[20] Atiyah, M.R., Awang Biak, D.R., Ahmadun, F., Ahamad, I.S., Mohd Yasin, F., Yusoff, H.M. 2011. Low temperature growth of vertically aligned carbon nanotubes via floating catalyst chemical vapor deposition method. *J. Mater. Sci. Technol.* 27, 296-300.

[21] See, Ch.H., Dunens, O.M., MacKenzie, K.J., Harris, A.T. 2008. Process Parameter Interaction Effects during Carbon Nanotube Synthesis in Fluidized Beds. *Ind. Eng. Chem. Res.* 47, 7686–7692.

[22] See, Ch.H., MacKenzie, K.J., Dunens, O.M., Harris, A.T. 2009. Multi-parameter optimization of carbon nanotube synthesis in fluidised-beds. *Chemical Engineering Science* 64, 3614 – 3621.

[23] McKee, G.S.B., Deck, Ch.P., Vecchio, K.S. 2009. Dimensional control of multi-walled carbon nanotubes in floating-catalyst CVD synthesis. *Carbon* 47, 2085–2094.

[24] Zilli, D.A., Blacher, S., Cukierman, A.L., Pirard, J.-P., Gommes, C.J. 2008b. Formation mechanism of Y-junctions in self-oriented multi-walled carbon nanotubes arrays. Colloids and Surface A: *Physicochemical and Engineering Aspects* 327, 140-143.

[25] Yousefie, S., Mottahedi, A. A. 2009. Production of Y-junction carbon nanotubes by floating catalyst method with a low cost and highly Y-branched approach. *Int. J. Chem. Tech Research* 1, 53-61.

[26] Yuan, X., Wu, X., An, Y., Hou, Q. 2011. Synthesis of S doped Y-junction carbon nanotubes by CVD method. *Advanced Materials Research* 183-185, 1731-1735.

[27] Fu, D., Zeng, X., Zou, J., Qian, H., Li, X., Xiong, X. 2009. Direct synthesis of Y-junction carbon nanotubes by microwave-assisted pyrolysis of methane. *Materials Chemistry and Physics* 118, 501–505.

[28] Sayago, I., Terrado, E., Lafuente, E., Horrillo, M.C., Maser, W.K., Benito, A.M., Navarro, R., Urriolabeitia, E.P., Martinez, M.T., Gutierrez, J. 2005. Hydrogen sensors based on carbon nanotubes thin films, *Synth. Met.* 148, 15–19.

[29] Sinha, N., Ma, J., Yeow, J.T.W. 2006. Carbon nanotube-based sensors. *J. Nanosci. Nanotechnol.* 6, 573–590.

[30] Rajaputra, S., Mangu, R., Clore, P., Qian, D., Andrews, R., Singh, V.P.. 2008. Multi-walled carbon nanotube arrays for gas sensing applications. *Nanotechnology* 19, 345502–345507.

[31] Fam, D.W.H., Palaniappan, A., Tok, A.I.Y., Liedberg, B., Moochhala, S.M. 2011. A review on technological aspects influencing commercialization of carbon nanotube sensors. *Sensors and Actuators, B: Chemical* 157, 1-7.

[32] Koh, S.K., Jung, H.J., Song, S.K., Choi, W.K., Choi, D., Jeon, J.S. 2000. Sensor having tin oxide thin film for detecting methane gas and propane gas, and process for manufacturing thereof. *US Patent* 6,059,937.

[33] Wang, D., Song, P., Liu, C., Wu, W., Fan, S. 2008. Highly oriented carbon nanotube papers made of aligned carbon nanotubes. *Nanotechnology* 19, 075609–075614.

[34] Zilli, D.A., Bonelli, P.R., Cukierman, A.L. 2011a. Room temperature hydrogen gas sensor nano-composite based on Pd-decorated multi-walled carbon nanotubes thin films. *Sensors and Actuators* B 157, 169-176.

[35] Zhang, T., Mubeen, S., Myung, N.V., Deshusses, M.A. 2008. Recent progress in carbon nanotube-based gas sensors. *Nanotechnology* 19, 332001–332014.

[36] Slobodian, P., Riha, P., Lengalova, A., Svoboda, P., Saha, P. 2011. Multi-wall carbon nanotube networks as potential resistive gas sensors for organic vapor detection. *Carbon* 49, 2499–2507.

[37] Li, Y., Wang, H., Chen, Y., Yang, M. 2008. A multi-walled carbon nanotube/palladium nano-composite prepared by a facile method for the detection of methane at room temperature, *Sens. Actuators* B 132, 155–158.

[38] Zhang, W.D., Zhang, W.H. 2009. Carbon nanotubes as active components for gas sensors, *J. Sens. ID* 160698 16pp.

[39] Star, A., Joshi, V., Skarupo, S., Thomas, D., Gabriel, J.C.P. 2006. Gas sensor array based on metal-decorated carbon nanotubes. *J. Phys. Chem.* B 110, 21014–21020.

[40] Mubeen, S., Zhang, T., Yoo, B., Deshusses, M.A., Myung, N.V. 2007. Palladium nanoparticles decorated single-walled carbon nanotube hydrogen sensor. *J. Phys. Chem.* C 111, 6321-6327.

[41] Ruiz, A., Arbiol, J., Cirera, A., Cornet, A., Morante, J.R. 2002. Surface activation by Pt nanoclusters on titania for gas sensing applications, *Mater. Sci. Eng.* C 19 105–109.

[42] Sippel-Oakley, J., Wang, H.T., Kang, B.S., Wu, Z., Ren, F., Rinzler, A.G., Pearton, S. 2005. Carbon nanotube films for room temperature hydrogen sensing, *Nanotechnology* 16, 2218–2221.

[43] Kumar, K.M., Ramaprabhu, S. 2006. Nanostructured Pt functionlized multiwalled carbon nanotube based hydrogen sensor. *J. Phys. Chem.* B 110, 11291-11298.

[44] Schlecht, U., Balasubramanian, K., Burghard, M., Kern, K. 2007. Electrochemically decorated carbon nanotubes for hydrogen sensing. *Appl. Surf. Sci.* 253, 8394–8397.

[45] Sun, Y., Wang, H.H., Xia, M. 2008. Single-Walled Carbon Nanotubes Modified with Pd Nanoparticles: Unique Building Blocks for High-Performance, Flexible Hydrogen Sensors. *J. Phys. Chem.* C 112, 1250-1259.

[46] La, D-D., Kim, Ch.K., Jun, T.S., Jung, Y., Seong, G. H., Choo, J., Kim, Y.S. 2011. Pt nanoparticle-supported multiwall carbon nanotube electrodes for amperometric hydrogen detection. *Sensors and Actuators* B 155, 191–198.

[47] Zilli, D.A.; Bonelli, P.R., Cukierman, A.L. 2006. Effect of alignment on adsorption characteristics of self-oriented multi-walled carbon nanotubes arrays. *Nanotechnology* 17, 5136–5141.

[48] Zilli, D.A., Bonelli, P.R., Gommes, C.J., Blacher, S., Pirard, J-P, Cukierman, A.L. 2011b. Krypton adsorption as suitable tool for surface characterization of multi-walled CNTs, *Carbon* 49, 980–985.

[49] Lee, Y.T., Kim, N.S., Park, J., Han, J.B., Choi, Y.S., Ryu, H., Lee, H.J. 2003. Temperature dependent growth of carbon nanotubes by pyrolysis of ferrocene and acetylene in the range between 700 and 1000 °C. *Chem. Phys. Lett.* 372, 853-859.

[50] Kim, N.S., Lee, Y.T., Park, J., Han, J.B., Choi, Y.S., Choi, S.Y., Choo, J., Lee, G. H. 2003. Vertically aligned carbon nanotubes grown by pyrolysis of Iron, Cobalt, an nickel phthalocyanines. *J. Phys. Chem.* B 107, 9249–9255.

[51] See, C.H., Harris, A.T. 2007. A review of carbon nanotube synthesis via fluidized-bed chemical vapor deposition. *Ind. Eng. Chem. Res.* 46, 997-1012.

[52] Wherry, R. 1984. Contributions to correlations analysis. Academic Press. N.Y., USA.

[53] Tinsley, H.E.A., Brown, S.D. 2000. Handbook of Applied Multivariate Statistics and Mathematical Modeling. Academic Press. N.Y., USA.

[54] Breiman, L., Friedman, J., Stone, C.J., Losen, R.A. 1984. Classification and Regression Trees. Chapman and Hall. London, U.K.

[55] Hollander, M.; Wolfe, D.A. 1973. Nonparametric statistical methods. *J. Wiley and Sons.* N.Y., USA.

[56] Wilcox, R.R. 2003. Applying contemporary statistical techniques. Academic Press. San Diego, USA.

[57] See, C. H.; Harris, A. T. 2008. A comparison of carbon nanotube synthesis in fixed and fluidised bed reactors. *Chem. Eng. J.* 144, 267-269.

[58] Dujardin, E., Ebbessen, T.W., Hiura, H., Tanigasaki, K. 1994 Capillarity and wetting of carbon nanotubes. *Science* 265, 1850–1852.

[59] Lordi, V., Yao, N., Wei, J. 2001. Method for supporting platinum on single-walled carbon nanotubes for a selective hydrogenation catalyst. *Chem. Mater.* 13, 733–737.

[60] Prasomsri, T., Shi, D., Resasco, D.E. 2010. Anchoring Pd nanoclusters onto pristine and functionalized single-wall carbon nanotubes. *Chem. Phys. Lett.* 497, 103–107.

[61] Zhang, X., Sreekumar, T.V., Liu, T., Kumar, S. 2004. Properties and structure on nitric acid oxidized single wall carbon nanotube films. *J. Phys. Chem.* B 108, 16435–16440.

In: Applications of Carbon Nanotubes
Editor: Ajay Kumar Mishra
ISBN: 978-1-62081-918-0

Chapter 13

DETERMINATION OF PHENOLIC COMPOUNDS BASED ON THE DEVELOPMENT OF BIOSENSORS USING NANO-ESTRUCTURED CARBON MATERIALS

***Mayara Brezolini Rocha*[1], *Bruna Fabrin Somera*[2], *César Ricardo Teixeira Tarley*[2,3]*, *Débora Nobile Clausen*[2] *and Arnaldo César Pereira*[1]**

[1]Universidade de São João del-Rei (UFSJ), Departamento de Ciências Naturais, São João del-Rei, Brazil
[2]Universidade Estadual de Londrina (UEL), Departamento de Química, Londrina, Brazil
[3]Departamento de Química Analítica - Instituto Nacional de Ciência e Tecnologia (INCT) de Bioanalítica, Universidade Estadual de Campinas (UNICAMP), Cidade Universitária Zeferino Vaz s/n, Campinas, Brazil

ABSTRACT

The enzymatic biosensors present some unique characteristics as low cost, simplicity, selectivity and sensibility, which might turn them attractive for the determination of target compounds in complex matrix, for instance, the environmental ones. The objective of this work is to develop amperometric biosensors for the determination of the concentration of phenolic compounds in environmental samples using carbonaceous materials. It was studied three samples of modified xerogels, two with melamine (FRM2B-PN, HFRM2-PN), one with urea (HFRU2-PN ½), besides standard xerogel without modification (Xerogel pH 5.5), carbon nanotubes and graphite. It was also studied the enzymatic activity of the zucchini extract and it was characterized in the prepared paste with the mentioned materials. For the preparation of the electrochemical biosensors, some parameters such as enzymatic extract concentration, hydrogen peroxide concentration, pH and buffer solution were optimized. The higher enzymatic activity was obtained when the enzymatic extract was immobilized onto carbon nanotube. The peroxidase enzyme extracted from several zucchinis (natural

---

* Tel: +55 43 3371 4366. Fax:+55 43 3371 4286. E-mails:ctarleyquim@yahoo.com.br., arnaldocsp@ yahoo.com.br (A.C. Pereira).

source) presented lower activity value than those described in literature and also presented low stability. The amperometric biosensor optimization (work electrode prepared with carbon nanotube paste and enzymatic extract) was evaluated through cyclic voltammetry and it is possible to observe a higher sensibility to the prepared biosensor with 100 μL of enzymatic extract, when the electrochemical measures were made in buffer solution phosphate pH 6.5, containing 10 μmolL-1 of hydrogen peroxide. The activity of the enzymatic extract from the zucchini (natural source) presented low activity, but this extract can still be used in the immobilization of different employed materials. The carbon nanotubes, among the analyzed materials, show the best support for the enzyme immobilization. For the optimization of the biosensor, the result with best sensibility was verified in the electrode prepared with carbon nanotubes paste containing 100 μL of enzymatic extract, when used in a buffer solution phosphate pH 6.5, containing 10 μmolL-1 of hydrogen peroxide.

Keywords: Biosensor; carbon nanotube; peroxidase

1. INTRODUCTION

Nowadays an increasing investment in research in the area of devices for environmental monitoring has been observed due to severe and various pollutants in the environment. In this way, the development of methodologies of determination and quantification of several substances with speed, selectivity and sensibility is a trend in the area of research in environmental chemistry.

All these characteristics are found in biosensors, that present possibility for detection in real-time, preserving at its maximum the real conditions found in nature. In addition, they present low cost of construction and storage, miniaturization potential, easy automation with the construction of simple and portable equipment (Louzada et al. 2004). From the environmental point of view, it is known that phenolic compounds are largely distributed on the environment, from industrial processes such as: plastics, dyes, drugs, pesticides, resins, antioxidants, and specially, paper and cellulose.

The determination of phenolic compounds is made mainly by means of spectro-photometric and chromatographic techniques (Fiamengos et al. 2000, Liu et al. 2004). However, the versatility of electrochemical techniques and the lower limits of detection commonly obtained in comparison to the spectroanalytical techniques, as well as its lower acquisition costs when compared to chromatographic techniques make them to be largely used in the determination of phenolic compounds.

Thus, the electrochemical sensors, due to their unique characteristics, such as: selectivity, low construction and storing costs, miniaturization potential, automation and construction facility of simple and portable equipment for a fast monitoring analysis "on-line" (Somerset et al. 2006, Leonard et al. 2003), represent a promising tool in supplementing the existing techniques. The objective of this work was the preparation of a biosensor using as a support different carbon materials, and the zucchini extract as a natural enzymatic peroxidase font for the determination of the phenolic compound concentration in environmental samples.

2. Literature Review

An enzymatic biosensor is an analytical instrument that combines biological material with a transducer. The biological material needs to be immobilized accordingly in order to be coupled to a transducer. It will monitor the disappearance of a reagent or the appearance of a reaction product of the biological material with the interest substrate, through a signal. This signal may result from a change in the protons concentration, from the liberation or adsorption of gases, such as ammonia or oxygen, emission, adsorption or reflectance of light, heat emission, and in this way, to bring information on the catalyzed reaction by the enzyme. The transducer converts this signal into a measurable response, such as electrical current, electrical potential difference, temperature change or light adsorption through electro-chemical, thermal or optical phenomenon. This signal can also be amplified, processed or stored for further analysis (Louzada et al. 2004). The electrochemical biosensors are the most used due to its practicality, low cost, high sensibility and compatibility to the modern technologies of micro-fabrication, apart from requiring low energy. Among the measurement forms of electrochemical methods – potentiometric, amperometric and spectrometric impedance (impedance, conductance or capacitance) – the amperometric has been the most applied. This technique consists of measuring the current generated in the cell from an applied potential, constant or varied (voltammetric method), on the working electrode (Skoog et al. 2002). The enzymatic biosensors are the main electrochemical biosensors; they depend largely on the process of electron transference. An essential prerequisite for the development of these devices is the high sensibility and fast response, which characterizes a fast and stable electron transference of the biological behavior for the electrode. The mechanism of electron transference in amperometric biosensors may occur by assisted or direct electrical transference. The biosensor generation that uses direct transference (direct communication between enzyme and electrode) is very attractive because it avoids intermediary steps of auto-reactions. The success and performance of this kind of electrode depends largely on the quality of the enzymatic layer immobilization due to possible differences in work conditions (pH, ionic strength, biomolecules stability, work time, etc.) (Habermuller et al. 2000). Phenols can be determined amperometrically by direct electrochemical oxidation. Never-the-less, this kind of procedure has a series of disadvantages, specially, due to a high over voltage. The high applied potential causes a high background current and, consequently, a high level of noise. Besides that, on direct phenol oxidations, a great number of secondary reactions may occur causing mainly a formation of polymeric products which passivates the electrode surface. Amperometric biosensors may minimize these obstacles in the phenol determination since they operate with a low applied potential in relation to the reference electrode. In this way, the current generated by the biocatalyst reaction of oxidation or the reduction of electroactive species on the electrode surface, in general, takes place in potentials close to zero mV versus ECS (Electrode of Saturated Calomel). In this potential, the contribution of interfering species (easily oxidizes or reduced) is minimized (Silva et al. 2006, Freire et al. 2002). The redox enzymes such as tirosinase, lacase and peroxidase are the most used for the construction of amperometric biosensors for phenols. Despite these enzymes have different action mechanisms, their performance in amperometric biosensors for phenolic compound detection have been demonstrated as a sequence of similar reactions, as shown in figure 1. The immobilized enzymes on the surface of the electrode are oxidizes by oxygen (for tirosinase and lacase) or hydrogen peroxide (for peroxidase), and further reduced by phenolic compounds. During this last reaction, the phenols are basically converted in quinones and/or free radicals, and these products, usually eletroactives, can be reduced on the electrode surface in potentials close to 0 (zero) V *vs.* ECS. The measured reduction current is proportional to the phenolic compound concentration in the solution. The detection of consumption of O_2 or H_2O_2 is another possibility of monitoring the concentration of phenolic compounds. Nevertheless, the detection method of phenols based on the reduction of products generated by enzymatic reactions presents advantages such as: manipulation and construction simplicity, electrode protection against the accumulation of secondary polymeric products on its surface (passivation), that are usually observed during direct electrochemical oxidation of phenols; response amplification as a consequence of quinone and/or phenoxy radicals reduction to the initial phenolic compound, besides the work potential band of the electrodes fits the optimum potential interval for electrochemical measures (Freire et al. 2002).

The choice of the biological material of the biosensor depends on the analyte to be studied and must be selective towards substrate. The extraction of enzymes from vegetables has been commonly carried out for building biosensor due to its availability and ease of obtaining.

For instance, the crude extract of peroxidase can be obtained from zucchini (Curcubita pepo) and of polyphenol oxidase from yam (Alocasia maccrorhiza) (Louzada et al. 2004).

Another important variable in the preparation of a biosensor is the support. Among the most used supports for amperometric biosensors construction are the carbonaceous materials, such as glassy carbon, pyrolytic graphite, carbon paste and composite graphite, nanotubes and meso-structered carbons. In addition, the introduction of metals, for example, has been explored for obtaining a larger time life for the sensor (Raoof et al. 2005). Sensors prepared from carbon nanotube show catalytic activity for many electrochemical reactions, including O_2 reduction, which is important for catalytic cathodes in combustible cells.

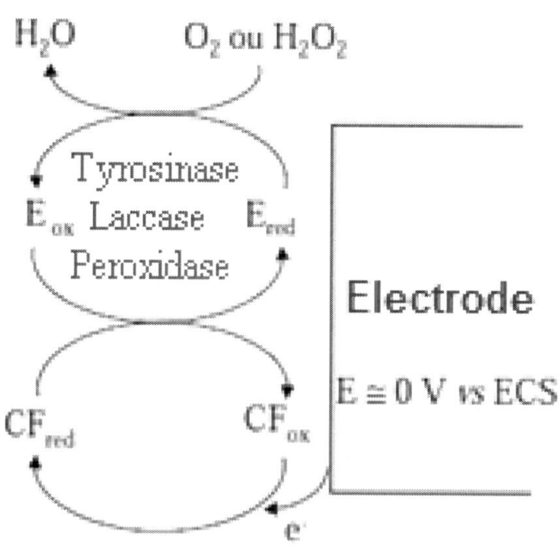

Figure 1. Reaction mechanism in modified electrodes by the enzymes Tirosinase, Lacase and Peroxidase for the determination of phenolic compounds. E_{red} and E_{ox} are oxidizes and reduced forms, respectively of the enzymes Tirosinase, Lacase and Peroxidase; CF = phenolic compound.

Some works on carbon nanotubes associated with carbon paste electrodes or glassy carbon electrodes show excellent electrocatalytic activities for the ascorbic acid oxidation, uric acid, dopamine, 3,4-dihidroxiphenilacetic acid, phenols, tryptophan, carbohydrates, and the hydrogen peroxide reduction and nitrite ions (Katz and Willner 2004).

The sensors can also be prepared using as a support carbon xerogels prepared from the step-growth polymerization of hydroxylated benzenes (phenol, catechol, resorcinol and hydroquinone) with aldehydes (formaldehyde and furfural). The products obtained after drying and pyrolysis of these materials, called carbon xerogel, are mesoporous materials with great applicability.

The carbons xerogels show great applicability as a metal support for catalysis (Figueiredo et al. 2006, Job et al. 2006, Samant et al. 2005). However the acid-based nature of its surface has not yet been explored. These properties can be modified by the introduction of groups as carboxyls, carbonyls, quinones, amines, cyanates and nitriles. The nitrogen groups can introduce basicity and favor the complexation with metals. This property of the surface is of interest in the preparation of catalysts as adsorbents as well as supports of electrochemical sensors (Pastor-Villegas et al. 2002).

The carbon xerogels used in this work (supplied by Profa. Dra. Honoria de Fatima Gorgulho/UFSJ) were modified using urea and melamine as a nitrogen font and are listed in table 1. The characterization data of these materials (superficial area and nitrogen content) are shown in table 2. (Gorgulho et al. 2009).

In this work, three samples of modified xerogels were studied. Two with melamine (FRM2B-PN, HFRM2-PN – which they differentiate from each other in relation to the superficial area) and one with urea (HFRU2-PN ½), standard xerogel with no modification (Xerogel pH 5.5), graphite and carbon nanotube.

Table 1. Xerogels used in the immobilization of biological materials

Standard (no modified)	Modified with melamine	Modified with urea
Xerogel pH = 5.5	FRM2B-PN	HFRU2 - ½ PN
	HFRM2-PN	

Table 2. Superficial area data, and determination of nitrogen content for elementary analysis and photoelectronic spectroscopy of X-rays (XPS)

Samples	% (total mass)[1]	% (total mass)[2]	Superficial Area (m^2g^{-1})
FRM2B - PN	3.5 ± 0.5	1.2	623
HFRM2 - PN	2.7 ± 0.3	2.4	579
HFRU2 – ½ PN	2.5 ± 0.1	1.7	633
Xerogel 5.5	0.0	0.0	720

One studied the enzymatic activity of the zucchini extract and characterized this enzymatic extract immobilized in the paste prepared of the mentioned carbon materials. For the preparation of the electrochemical biosensors, the concentration of the enzymatic extract, the concentration of hydrogen peroxide, pH and the buffer solution were optimized.

3. METHODOLOGY

3. 1. Preparation of the Crude Extract

After washing and drying, 100g of peeled zucchini (Cucurbita pepo) are grounded and homogenized in a blender with 200 mL of phosphate buffer 0,1 mol.L^{-1} (pH 7), containing 10.0 g of protective agent (PVP: polyvinylpyrrolidone) for 5 minutes.

After that, the homogenate is filtered in four layers of gases and centrifuged at 3450 rpm during 20 minutes. The supernatant solution is stored at 4°C and used as enzymatic font of peroxidase.

The polyvinylpyrrolidones (PVPs) are polymers of high molecular mass with outstanding ability for removing phenolic compounds from solutions through the formation of hydrogen bridges to the natural substrates as shown in Figure 2.

3.2. Characterization of the Enzymatic Activity

3.2.1. Enzymes Extraction and Activity Measurement

The enzyme activity is determined by UV/visible spectroscopy measuring the absorbance variation (470 nm) of tetraguaiacol (probe molecule) formed in the enzymatic reaction. Equation (1) represents the reaction between guaiacol and hydrogen peroxide forming tetraguaiacol, catalyzed by peroxidase.

The equipment used was spectrophotometer Varian Cary 50. In this determination it is used 0.2 mL of supernatant solution, 2.7 mL of guaiacol 0.05 molL^{-1} solution prepared in buffer phosphate 0.1 mol L^{-1} pH 7 and 0.1 mL of hydrogen peroxide solution at 25°C. A measurement unit activity (UA) is defined as the quantity of enzymes that causes the addition of 0.001 units of absorbance per minute under the mentioned conditions.

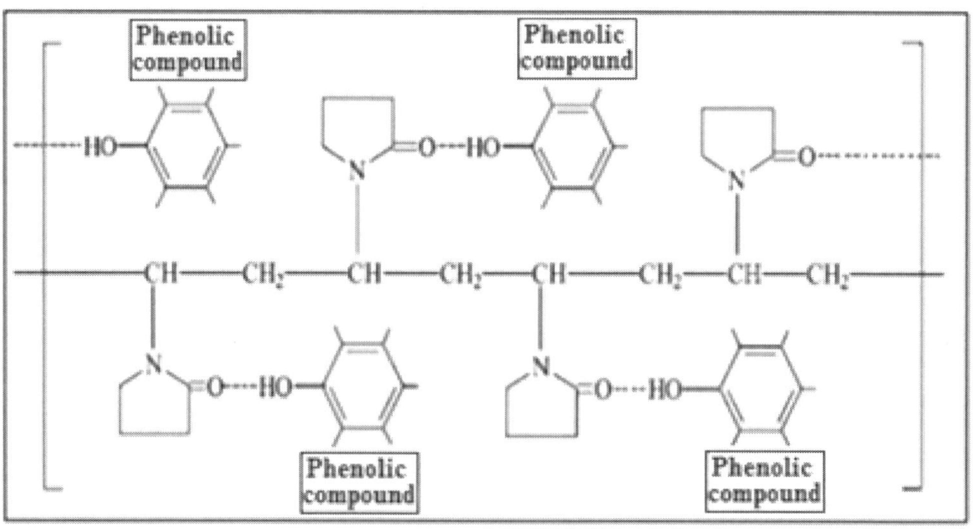

Figure 2. Interaction by hydrogen bridges between PVP and the phenolic compounds.

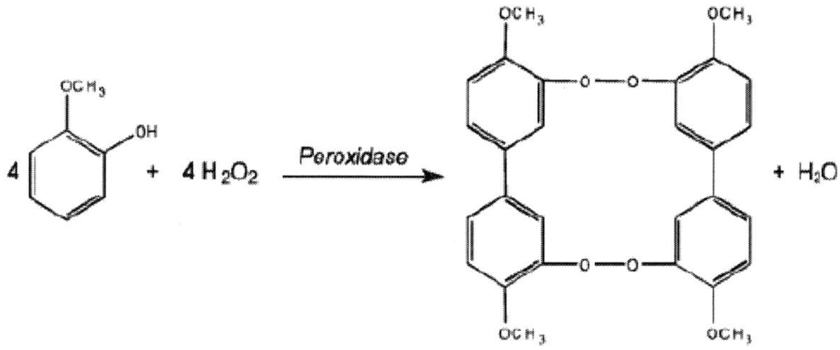

(1)

In this work we propose the results in Units/mL of activity of the crude material (zucchini).

3.2.2. Preparation of Xerogels

The xerogels were synthesized by polycondensation of resorcinol (R) with formaldehyde (F). Sodium carbonate (C) was employed as catalyst of reaction. For the introduction of nitrogen in the xerogel structure, urea (A) at molar ratio R/U = 1 and melamine (M) at molar ratio R/M =2 were used (Gorgulho et. al 2009). The synthesis was accomplished by mixing 9.09 g of resorcinol with sodium carbonate with deionized water. Under stirring, the mixture was heated at 70 °C followed by addition of nitrogen precursors (urea and melamine). In the next step, the mixture was cooled followed by addition of formaldehyde. The mixture was then transferred to an erlenmeyer and kept under heating at 70 °C during 96 h. After this period, a gel was formed. The pH of synthesis was maintained within the range 5.5 to 6.5. Next, the gel solution was filtered under vacuum and dried at 100 °C. Finally, the dried gels were pyrolyzed under flow of nitrogen in a tubular oven; model EDG-10P-S. The xerogel herein defined as xerogel 5.5 was prepared under similar procedure above described, except by addition of precursor of nitrogen (Job et al. 2004).

3.2.3. Immobilization of the Enzymatic Extract in Carbon Samples – Enzyme Activity in the Working Electrode

The biosensors prepared under different carbonaceous materials (carbon nanotube, xerogels and graphite) were prepared homogenizing 10.0 mg of each material and 50 μL of enzymatic extract for 20 minutes. Then 25 μL of mineral oil were added to the mixture and homogenized for 15 minutes. Around 10.0 mg of paste were compressed on the working electrode, as shown in Figure 3.

In the evaluation of the immobilized enzyme activity the electrodes were submerged in a cubet with 2.5 mL of hydrogen peroxide in the concentration of $4x10^{-3}$ $molL^{-1}$ and in 5 minutes intervals 100 μL of hydroquinone solution in the concentration of 0.202 $molL^{-1}$ were added, during 30 minutes. For the blank test the absorbance measurement was carried out in the absence of the immobilized biological material.

In this method, the increase of the absorbance at 470 nm of the p-quinone formed by the reaction of hydroquinone with hydrogen peroxide catalyzed by peroxidase immobilized in the biosensor was measured. The measurements were done in duplicity and realized in a spectrophotometer UV-visible (Varian Cary 50).

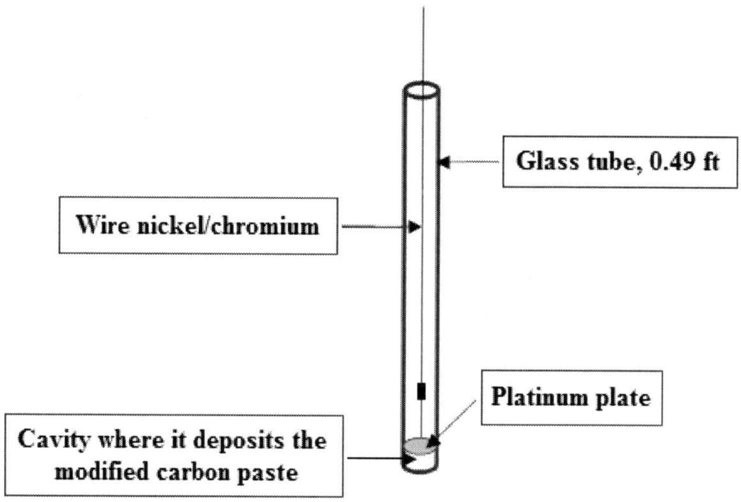

Figure 3. Scheme of the working electrode.

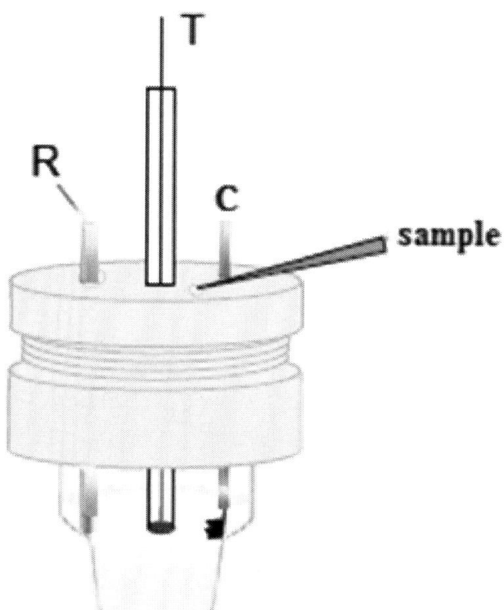

Figure 4. Representation of the electrochemical cell used for the biosensor optimitization, (R) the reference electrode, (T) the working electrode and (C) the auxiliary electrode.

3.3. Optimization of the Biosensor

3.3.1. Preparation of the Biosensors

The biosensors were prepared homogenizing 10.0 mg of nanotube and 100 μL of the enzymatic extract (6.809 units/mL/g) for 20 minutes. Then it is added 30 μL of mineral oil to the mixture and homogenized for 15 minutes. The obtained paste was introduced in a cavity in the working electrode. The scheme is shown in Figure 3.

The electrochemical measurements were performed employing a three-electrode conventional cell: the paste electrode of modified carbon material (nanotube carbon, xerogels and graphite powder) as the working electrode and the electrodes of Ag/AgCl and platinum spiral as reference electrode and auxiliary, respectively, as shown in Figure 4. The equipment used in the electrochemical measurements was the Autolab PGSTAT302N and for data processing the software used was the GPES 4.9.

3.3.2. Evaluation of the Biosensor Sensibility Prepared with Paste with Different Carbon Materials

Pastes with different carbon materials were prepared (modified xerogels with melanine - FRM2B-PN, HFRM2-PN – urea - HFRU2-PN ½ - pattern xerogel - Xerogel pH 5,5 – graphite and carbon nanotube).

Cyclic voltammograms were recorded in the presence of the buffer phosphate 0.1mol/L, in an interval of scan potentials (ΔE = 0.1 to 0.6 V *vs.* Ag/AgCl) at scan rate (ν = 50 mV s^{-1}). It was used as a measurement parameter the cathodic peak current value generated by each pyrocatechol addition. The voltammograms were obtained 10 seconds after the addition of 50 mL of pyrocatechol 0.202 $molL^{-1}$ in the electrochemical cell containing 15 mL of the buffer phosphate solution 0.1 $molL^{-1}$. This procedure was performed for each paste prepared with different carbon materials.

3.3.3. Evaluation of the Biosensor Sensibility Prepared with Carbon Nanotube Paste Containing Different Concentrations of Enzymatic Extracts

The experiment was accomplished by using a modified biosensor with nanotube paste, containing different quantities (Leonard et al. 2003) of immobilized enzymatic extract (30 - 120 μL). These measurements were taken as per the experimental procedure described in item 3. 3. 2.

3.3.4. Study on the Influence of the Hydrogen Peroxide Concentration

In the electrochemical cell (Figure 4), it was employed the modified biosensor with carbon nanotube paste with 100 μL of immobilized enzymatic extract, in buffer phosphate solution 0.1 $molL^{-1}$, with different concentrations (Skoog et al. 2002) of hydrogen peroxide (10 - 100 $\mu molL^{-1}$). These measurements were taken as per the experimental procedure described in item 3. 3. 2.

3.3.5. Study on the Influence of pH

The catalytic action of an enzymatic reaction is usually achieved in very strict pH limits. Each reaction has an optimum pH, that for most of the enzymes is between 4 and 8, and in which the enzyme shows its maximum activity (Belini et al. 2001).

The study was carried out with a biosensor prepared with carbon nanotube paste. It was prepared different buffer phosphate solutions in the pH interval of 5 and 8. These measurements were taken as per the experimental procedure described in item 3. 3. 2.

3.3.6. Study on the Influence of the Buffer

It was used the modified biosensor with carbon nanotube paste with 100 μL of the immobilized enzymatic extract. It was prepared different buffer solutions such as (Phosphate,

Acetate, Tris, Hepes, Pipes and KCl) in pH 6.5. These measurements were taken as per the experimental procedure described in item 3. 3. 2.

4. Results and Discussion

4.1. Characterization of the Enzymatic Activity

4.1.1. Enzyme Extraction and Activity Measurement

The enzymatic extracts were obtained from a natural source (zucchini) in phosphate buffer 0.1 $molL^{-1}$, in the presence of protective agent (PVPs: polyvinylpyrrolidones) used to separate natural phenolic compounds from the enzymatic extract (Fatibello-Filho et al. 2002).

Table 3. Enzymatic activity of the crude extracts, obtained by measurements taken with spectrophotometer UV-visible (Cary), using wavelength of 470nm

Sample	Activity (unit/mL)
Enzymatic extract 1	553
Enzymatic extract 2	701
Enzymatic extract 3	540
Enzymatic extract 4	930
Enzymatic extract 5	1668
Enzymatic extract 6	824
Enzymatic extract 7	4250
Enzymatic extract 8	2044

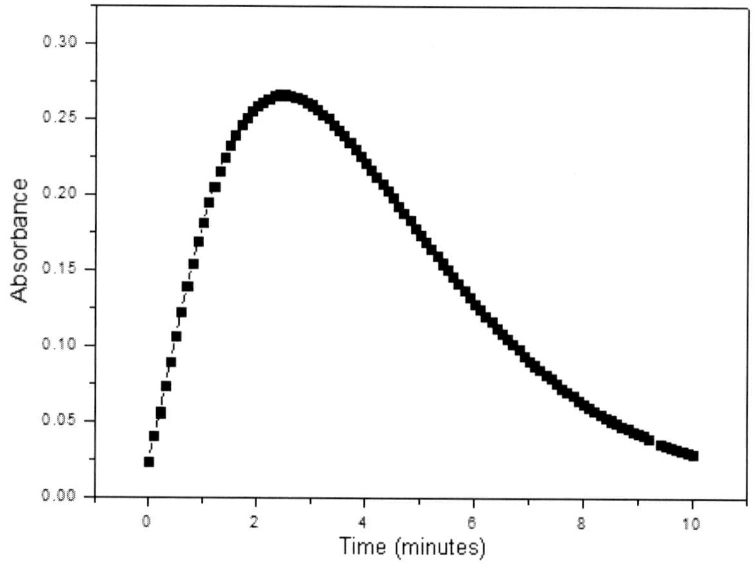

Figure 5. Absorbance curve as a function of time obtained for crude extraction 7 by measurements taken in spectrophotometer UV-visible 7 (Cary), using wavelength of 470nm.

Table 4. Activity of the crude extract

Measurements	Activity (units/mL)
1st day	4250
3rd day	3436
5th day	3227
7th day	2326

Table 5. Activity of the immobilized enzymatic extract, obtained by taken measurements at the spectrophotometer UV-visible, using wavelength at 470nm

Samples	Activity (units/mL)
Carbon nanotube	43.3
FRM2B - PN	20.3
HFRM2 – PN	31.1
HFRU2 – PN ½	35.0
Xerogel pH 5.5	14.6
Graphite	9.8

The activity of the crude enzymatic extract was determined measuring the absorbance variation with the wavelength in 470nm, this measurement was taken with a spectrophotometer UV-visible (Cary). The enzymatic activity was determined from the derived of the obtained absorption curve (Figure 5) and the values are shown in Table 3 (Job et al. 2006).

The natural sources (zucchini) from which the different extracts presented in Table 3 were obtained had different origins and were obtained in different seasons.

The activity of crude extract 7 was monitored for 7 days with the objective of determining its stability. The results are shown in Table 4.

The highest obtained activity was 4250 units/mL, value 4 times smaller than the one found by Fatibello et al (Job et al. 2006). The lowest activity and stability may be explained by the fact that the vegetable tissue used does not have its parameters under control, vegetable type, pesticide use, etc. Because of these facts, probably the peroxidase activity was affected, showing that the vegetable tissue used in the work did not have the appropriate characteristics for obtaining the crude enzymatic extract.

4.1.2. Immobilization of the Carbon Samples – Enzyme Activity on the Working Electrode

The enzyme immobilization was carried out by means of adsorption phenomenon, in which the biological material is adsorbed due to the Van Der Waals forces, hydrogen bonds or hydrophobic forces (Lojou and Bianco 2006). A study of the immobilized enzyme activity was taken to evaluate the best material for the immobilization. Table 5 shows the response for each electrode for the reaction of the hydroquinone oxidation by the hydrogen peroxide catalyzed by the immobilized peroxidase.

With the results presented at Table 5, one can observe that the highest enzymatic activity was obtained when the enzymatic extract was immobilized in the carbon nanotube, due to its optimum absorbance capacity of large molecules such as the enzymes. It is known as well that from literature bibliography (Whitney et al. 1993), this material shows large superficial area, which can contribute for a more effective immobilization of the biological material on its surface.

4.2. Preparation of the Electrochemical Biosensors

4.2.1. Evaluation of the Sensibility of the Biosensor Prepared with Nanotube Paste Containing Different Concentrations of the Enzymatic Extract

For the development of the amperometric biosensor, it was employed carbon nanotubes as a support for the immobilization of the biological material, preparing working electrodes with different enzymatic extract volumes (50 - 200 μL).

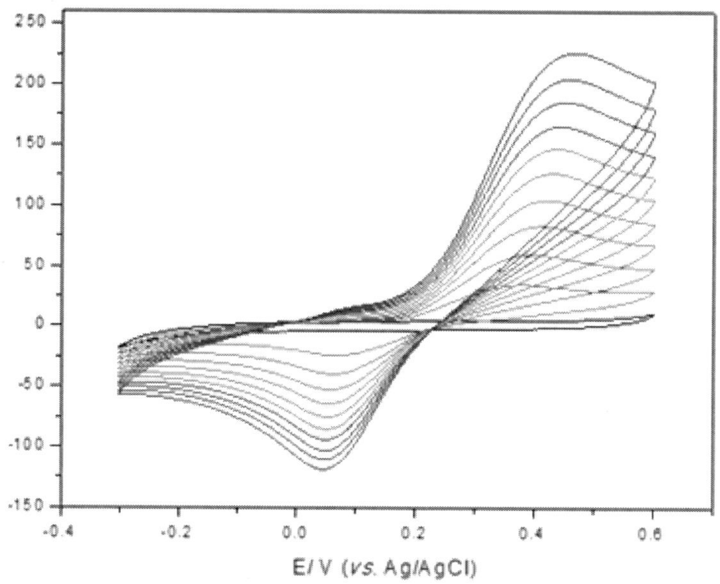

Figure 6. Cyclic voltammograms obtained with the biosensor prepared with carbon nanotube paste, in phosphate buffer 0.1 mol/L, containing 100 μL of the enzymatic extract. ΔE = 0.1 to 0.6 V *vs*. Ag/AgCl with $\nu = 50$ mV s^{-1}.

Table 6. Biosensor response for catechol obtained in different concentrations of hydrogen peroxide

Hydrogen Peroxide Concentration (*μmolL^{-1}*)	Sensibility / μA $mmol^{-1}$ L
05	3.87×10^{-2}
10	6.32×10^{-2}
20	4.03×10^{-2}
40	1.99×10^{-2}
60	2.87×10^{-2}
80	2.60×10^{-2}
100	4.32×10^{-2}

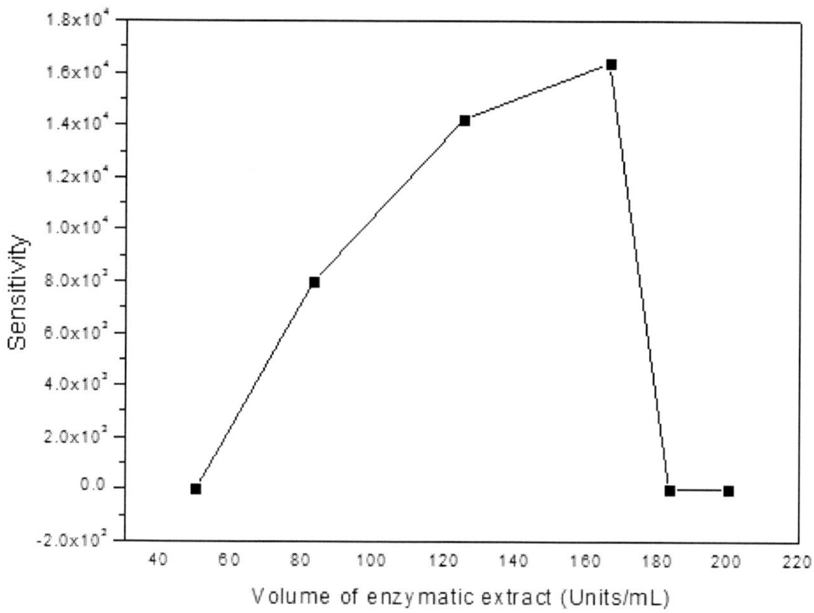

Figure 7. Evaluation of the sensibility of the biosensors prepared with carbon nanotube paste containing enzymatic extract in different volumes.

In this study, we can determine the sensibility of each developed device and these measurements were taken as per the experimental procedure described in item 3.3.3. Figure 6 depicts cyclic voltammograms for pyrocatechol under different concentrations.

Analyzing Figure 7, it is possible to observe a higher sensibility for the biosensor prepared in the presence of 100 μL of enzymatic extract.

Above this concentration, it probably occurs lixiviation of the biological material for the working solution, along with the loss of stability of the developed biosensor.

4.2.2. Study on the Influence of the Hydrogen Peroxide Concentration

For the study on the influence of the hydrogen peroxide concentration, it was prepared solutions of this compound in different concentrations (0.5 $\mu molL^{-1}$ - 100 $\mu molL^{-1}$). These measurements were taken as per the experimental procedure described in item 3.4.2.

The results are presented in Table 6. The result obtained in this analysis shows that the modified working electrode with carbon nanotube is more sensitive when the measurement is taken in the electrolyte prepared with 10 $\mu molL^{-1}$ of hydrogen peroxide.

In concentrations above that, the obtained results may be associated to the inhibition of the enzyme catalytic site, in other words, formation of inactive enzyme in high peroxide concentrations.

4.2.3. Study on the Influence of pH

The main aim of this study was to determine the optimum pH in which occurs the catalytic action of the enzymatic reaction. These measurements were taken as per the experimental procedure described in item 3.4.2. The results are presented in Table 7.

With the obtained results, shown in Table 7, one can observe higher sensibility of the modified electrode when the analysis is done in pH 6,5. This pH value reflects the optimum conditions for the enzymatic catalytic reaction. Fast pH changes may cause the enzyme denaturation.

4.2.4. Study on the Influence of the Buffer Solution

Solutions were prepared with different buffer systems to evaluate the sensibility of the modified working electrode, in the conditions so far optimized, with carbon nanotube in each different buffer solution.

The measurements were taken as per the experimental procedure described in item 3.4.2. The results are presented in Table 8.

According to Table 8, higher sensibility in the electrode in the presence of phosphate buffer solution was observed, in agreement to previous published works for similar systems to the one studied here (Huang et al. 2001, Choi et al. 2007, Tsai et al. 2007).

Table 7. Biosensor response for catechol obtained in different pH values

pH	Sensibility / μA $mmol^{-1}$ L
5.0	$3.441x10^{-2}$
5.5	$3.439x10^{-2}$
6.0	$3.701x10^{-2}$
6.5	$10.493x10^{-2}$
7.0	$5.968x10^{-2}$
7.5	$2.880x10^{-2}$
8.0	$3.922x10^{-2}$

Table 8. Biosensor response for catechol obtained in different buffer solutions

Buffer Solutions	Sensibility / μA $mmol^{-1}$ L
Acetate	$1.28x10^{-2}$
Phosphate	$6.39x10^{-2}$
Hepes	$2.57x10^{-2}$
Tris	$3.31x10^{-2}$
Pipes	$2.82x10^{-2}$
KCl	$3.50x10^{-3}$

CONCLUSION

The enzymatic activity obtained from a natural source (zucchini) showed low activity and little stability compared to previous results in literature, although this extract can still be used in the immobilization over the different employed materials.

The nanotube, among the analyzed materials, shows the best support for enzyme immobilization since it showed higher enzymatic activity, due to its large superficial area, which can contribute to a more effective immobilization of the biological material on its surface.

In the optimization of the working electrode with the obtained sensibility as a parameter of evaluation, the best results occurred when the preparation of the electrode was made with carbon nanotube paste containing 100 μL of the enzymatic extract; furthermore, the best support electrolyte was the phosphate buffer solution pH 6.5, containing 10 $\mu molL^{-1}$ of hydrogen peroxide.

This work demonstrates the efficiency of the carbon nanotube matrix in the development of biosensors based on peroxidase for determining phenolic compounds. Besides that, it shows the possibility of developing economic viable devices, which is proven by the use of enzymatic extract from natural sources.

ACKNOWLEDGMENTS

The authors would like to thank CNPq, FAPESP, FAPEMIG, INCT (Instituto Nacional de Ciência e Tecnologia de Bioanalítica), Fundação Araucária and CAPES for financial support and fellowships. ACP and MBR would like to thank Profa. Dra Honoria de Fatima Gorgulho for supplying of xerogels.

REFERENCES

Belini, A.; Filho, O. F.; Clemente, E.; Braz, J.; *Food Technol.*, 2001, 4, 75.

Choi, H.N.; Han, J.H.; Park, J.A.; Lee, J.M.; Lee, W.Y.; *Electroanalysis*, 2007, 19, 1757.

Fatibello–Filho, O., Vieira, I. C.; *Quim. Nova*, 2002, 25, 455.

Fiamengos, Y.C.; Stalikas, C.D.; Pilidis, G.A.; Karayannis, M.I.; *Anal. Chim. Acta*, 2000, 403, 315.

Figueiredo, J.L.A.; Pereira, M.F.R.A.; Serp, P.B.; Kalck, P.; Samant, P.V.; Fernandes, J.B.; *Carbon*, 2006, 44, 12, 2516.

Freire, R.S.; Thongngamdee, S.; Durán, N.; Wang, J.; Kubota, L.T.; *Analyst,* 2002, 127, 258.

Gorgulho, H.F.; Gonçalves, F.; Pereira, M.F.R.; Figueiredo, J. L.; *Carbon*, 2009, 47, 2032.

Habermuller, K.; Mosbach, M.; Schuhmann, W. Resenius, *J.; Anal. Chem.,* 2000, 366, 560.

Huang, Y.; Duan, X; Wei, Q; Lieber, C.M.; *Science*, 2001, 291, 630.

Job, N.;Pirard, R.; Marein, J.; Pirard, J. P.; *Carbon,* 2004, 42, 619.

Job, N.A.; Pereira, M.F.R.; Lambert, S.; Cabiac, A.; Delahay, G.; Colomer, J.F., Marien, J., Figueiredo, J.L., Pirard, J.P.; *J. Catal.,* 2006, 240, Issue 2, 160.

Katz, E.; Willner, I.; *Chem. Phys. Chem.,* 2004, 5, 1085.

Leonard, P.; Hearty, S.; Brennan, J.; Dunne, L.; Quinn, J.; Chakrborty T.; O'Kennedy, R.; *Enzyme Microb. Tech.,.* 2003, 32, 3.

Liu, R.; Zhou, J.L.; Wilding, A.; *J. Chromatogr. A,* 2004, 1022, 179.

Lojou, E.; Bianco, P.; *J. Eletroceram.,* 2006, 16, 79.

Louzada, E. S.; Luccas, P.O.; Magalhães, C. S.; *Analytica,* 2004, 11, 56.

Pastor-Villegas, J., Duran-Valle, C. J. J., Grzyb, B.; Machnikowska, H.; Weber, J.V.; *Carbon,* 2002, 40, 397.

Raoof, J. B.; Ojani, R.; Nadimi, S. R.; *Electrochim. Acta*, 2005, 50, 4694.

Samant, P.V., Fernandes, J.B., Rangel, C.M., Figueiredo, J.L.; *Cata. Today*, 2005, 102, 173.

Silva, R.D.C.L.; Damos, F.S.; Oliveira, A.B.; Beck, J.; Kubota, L.T.; *Sens. Actuators B, Chem.*, 2006, 117, 274.

Skoog, D. A.; Holler, F. J.; Nieman, T. A.; *Princípios de Análise Instrumental.* 5 ed., *Bookman*, 2002.

Somerset, V.S.; Klink, M.J.; Sekota, M.M.C.; Baker, P.G.L.; Iwuoha, E.I.; *Anal. Chim. Acta*, 2006, 39, 1683.

Tsai, Y.C.; Chen, S.Y.; Lian, H.W.; *Sens Actuators*, 2007, 125, 474.

Whitney, T.M.; Jiang, J.S.; Searson, P.C.; Chien, C.L.; *Science,* 1993, 261, 261.

In: Applications of Carbon Nanotubes
Editor: Ajay Kumar Mishra
ISBN: 978-1-62081-918-0

Chapter 14

CARBON NANOTUBES IN ENVIRONMENTAL REMEDIATION – A GLOBAL PERSPECTIVE

K. Pillay*
Department of Chemical Technology,
University of Johannesburg, South Africa

ABSTRACT

Carbon nanotubes have been identified as potential sorbents for the remediation of various organic and inorganic pollutants over the years. These materials have even demonstrated a superior adsorption capability to that of activated carbon. This chapter highlights some of the key observations made in the use of carbon nanotubes for the extraction of contaminants from aqueous solution over the past decade. The studies conducted to date are critically reviewed where both the advantages and shortcomings of using carbon nanotubes in environmental remediation are examined. Contradictions in the literature are discussed. Some case studies where carbon nanotubes have been used for the removal of key pollutants such as Cr(VI) and Hg(II) are highlighted. Modifications to the nanotube surface for the selective extraction of a target pollutant also features as a key topic in this chapter. In the regard, the effect of functional groups and the introduction of various heteroatoms is discussed. Other modifications such as the effect of acid and base treatment and the dopant concentrations are also examined. The impact of the studies conducted to date on the large-scale industrial usuage of carbon nanotubes in water treatment is critically discussed. Lastly, the chapter concludes with a discussion on aspects that are lacking in the use of carbon nanotubes for environmental remediation and future research perspectives are discussed.

INTRODUCTION

Carbon nanotubes are group of carbon materials which were discovered in 1991 by Iijima [1]. Such materials have been gaining increasing recognition for environmental applications

* E-mail: kriveshinip@uj.ac.za. Tel: +27 11 559-6128. Fax: +27 11 559-6425.

over the past decade. This is mainly due to their well defined porosity and functionality which has granted these materials a superior adsorption capacity to even that of conventional adsorbents like activated carbon [2].

The studies conducted to date on the use of these materials in environmental remediation are therefore reviewed in this chapter. This includes an intensive critical discussion where both the strengths and weaknesses of such studies are highlighted. The chapter also focuses on studies that have contributed significantly to the remediation of key pollutants like Cr(VI) and Hg(II). Much emphasis is placed on parameters such as the effect of contact time, solution pH, competing ions and initial concentrations. Since the costs incurred upon using these materials are expensive [3], some emphasis is also placed on desorption and regeneration.

Selectivity of the nanotubes for a specific pollutant is another important topic of discussion in this chapter. The importance of selectivity cannot be overemphasized as many environmental systems especially wastewater contain a variety of pollutants which can compete for active adsorption sites on the nanotubes. Competitive adsorption can pose problems in environmental remediation as the target pollutant can sometimes be out-competed by other pollutants which are of secondary importance because their toxic effects are not as lethal as that of the target pollutant. Thus, strategies which aim at improving the selectivity of carbon nanotubes for the uptake of a specific pollutant are also discussed. These include functionalisation or heteroatom doping and acid-base treatment.

Laboratory scale applications are of little use if such applications cannot be practically implemented. Strategies in improving the practical usage of these nanotubes for the treatment of industrial effluents are also critically discussed.

Lastly, the shortcomings of using carbon nanotubes in environmental remediation are discussed. This is accompanied by a discussion on recommendations for future studies on the use of carbon nanotubes in environmental remediation.

Studies on the Use of Carbon Nanotubes in Environmental Remediation

Research on the use of carbon nanotubes (CNTs) for the removal of both inorganic and organic pollutants began in 2001 where single-walled carbon nanotubes (SWCNTs) were used for the uptake of Pb(II), dioxins, furans and even fluorides [4-6]. Such studies lead the US Environmental Protection Agency (EPA) to commission the use of these materials in environmental remediation [7]. Other key studies which have contributed to this field are summarized in Table 1.

The studies highlighted in Table 1 show that carbon nanotubes have demonsrated a huge potential to adsorb a variety of inorganic and organic pollutants. A striking feature of these materials is their superior adsorption capability to that of activated carbon It has been proposed that this could be due to the hygroscopic and porous nature of activated carbon where moisture is trapped into the pores and hydrophilic contaminants like Cr(VI) simply dissolve in this moisture [26]. However, apart from this postulation little plausible explanations for this discrepancy in adsorption capabilities has been furnished in the literature.

Table 1. Some Key Studies on the Use of Carbon Nanotubes in Environmental Remediation (from Pillay [8])

Type of nanotube	Pollutant	Main observations	Comments	Reference
Unmodified form and microwave modified CNTs	Bisphenol A	Slightly higher uptake capacity for the modified form at 7^0C (69,93 mg/g as opposed to 60,98 mg/g for the unmodified CNTs).	Selectivity studies were not attempted.	Kuo [9]
SWCNTs	17-α-ethinyl estradiol and Bisphenol A.	Can adsorb rapidly and retain these strongly.	None	Pan et. al. [10]
SWCNTs and MWCNTs purified with a sodium hypochlorite solution	Zn(II).	1. This purification method rendered the CNTs more favourable for the uptake of Zn(II) by making the CNTs more hydrophilic. 2. The adsorption was also pH dependent. 3. Both types of CNTs were better adsorbents than powdered activated carbon (PAC)	1. The concentration range studied was 10-80 mg L^{-1}. 2. Did not conduct any competitive adsorption studies with competing cations that may co-exist with Zn(II) in wastewater. 3. No report on removal mechanism.	Lu and Chiu [2]

Table 1. (Continued).

Type of nanotube	Pollutant	Main observations	Comments	Reference
Oxidised MWCNTs	Ni(II)	Up to 93% Ni(II) could be desorbed from the oxidised MWCNTs at pH< 2 using distilled water as the desorbing agent. 2. Lack of dependence on ionic strength. 3. Equilibrium is attained within 40 min of contact time	Despite proving that these materials can be regenerated by desorption, the authors still argued that costs limit the practical utility of these MWCNTs and did not study the selectivity of these materials.	Chen and Wang[11]
As grown and oxidised MWCNTs	Ni(II)	1. Adsorption capacity of oxidised MWCNTs was superior. 2. A short time was required to attain equilibrium and noted that the adsorption capability of this material was also superior to that of peanut shell	These workers also neglected to study the effect of competing cations and did not conduct any desorption experiments.	Kandeh and Meunier [12]
Oxidised MWCNTs	Cd(II)	1. MWCNTs oxidised with $KMnO_4$ gave the best results. 2. Oxidation also lowered the pH_{pzc} of these materials and increased the quantity of surface functional groups.	Did not examine the effects of competing anions or determine the regenerability of these materials.	Li et. al. [14]

Type of nanotube	Pollutant	Main observations	Comments	Reference
SWCNTs.	Benzene, Toluene and Chlorobenzene	The uptake of was hindered by the presence of oxygen containing functional groups	These authors accounted for this in terms of steric hindrance presented by the functional groups.	Chin et. al. [15]
SWCNTs.	Naphthylamine, 1-Naphtol and 1-Phenol	The uptake was suppressed by the presence of oxygen-containing functional groups.	Authors attributed this to the weakening of π-π interactions and increased electrostatic repulsions by carboxylic groups.	Sheng et. al. [16]
Acid-treated MWCNTs	Pb(II).	1. The adsorption capacity reported was 85 mg g-1 which is higher than that reported for unmodified MWCNTs and activated carbon. 2. The equilibrium adsorption time was shortened to 20 minutes. 3. Provided experimental evidence that chemical interactions between Pb(II) and oxygen containing functional groups accounts for 75,3% of the Pb(II) adsorption while the formation of PbO, $PbCO_3$ and $Pb(OH)_2$ account for only 3,4% of the adsorption	The adsorbed Pb(II) aggregates occur mainly on the ends and defect sites of the MWCNTs.	Wang et. al. [17,18]

Table 1. (Continued).

Type of nanotube	Pollutant	Main observations	Comments	Reference
Oxidised MWCNTs	Ni(II)	1. The presence of polyacrylic acid (PAA) enhanced the uptake of by at low pH. 2. That the adsorption was independent of the addition sequence of PAA and Ni(II). 3. The presence of foreign ions had no effect at low pH.	The strength of the complexes of Ni(II) with PAA adsorbed onto the MWCNTs was higher than the strength between Ni(II) and oxygen containing functional groups.	Yang et. al. [19]
MWCNTs	Cu(II)	Adsorption was positively affected by the presence of humic and fulvic acids at pH>7.5	Adversely affected at pH < 7.5.	Sheng et. al [20]
SWCNTs doped with iron	Benzonitrile	More effective at removal than the undoped materials.	No reports of the ability of Fe-doped CNTs to remove inorganic metal ions.	Aguiar et. al. [21]
High crystalline MWCNTs (HC-MWCNTs)	Volatile organic carbons (VOCs)	Highly selective uptake	No report on regeneration capacities	Sone et. al. [22]
Activated carbon, carbon nanotubes and carbon encapsulated nanoparticles	Co(II) and Cu(II).	Both types of carbon nanomaterials had a superior adsorption capability to that of activated carbon.	The carbon nanomaterials only lost 5-11% of their adsorption efficiency in saline environments where competing ions exist while activated carbon lost 30-50% of its adsorption efficiency.	Pyrzynska and Bystrzejewski [23]
Pristine MWCNTs	2,3-dichlorophenol and pentachlorophenol	1. 94% removal 2. The adsorption was dependent on factors such as pH, temperature, mass of adsorbent and analyte concentration. 3. Rapid extraction from solution	Both studies also did not report on the regenerability and selectivity of the adsorbent.	Salam et. al.[24]; Salam and Burk [25]

A variety of carbon nanotubes have also been used in these studies with oxidized carbon nanotubes featuring as the most widely used materials. Such materials have demonstrated a promising potential for cation uptake and have not only demonstrated a superior adsorption capability to that of activated carbon but also have higher adsorption efficiencies to that of other low cost adsorbents [27]. In this regard oxidized SWCNTs have demonstrated a better efficiency than their multi-walled counterparts since their zeta potentials are more negative and the electrostatic interactions with cations are much greater [27].

Despite all the studies initiated on topic, some inherent flaws in the studies undertaken are evident upon examining Table 1.

The first of these flaws is the shortage of desorption and regeneration studies which are important in light of the costs involved in synthesizing CNTs.

The few studies which have reported on desorption and regeneration have however shown good desorption and regeneration capabilities [18, 27].

Another shortcoming of the studies undertaken thus far is that chemical affinity competition studies are lacking. This is important since complex chemical matrices do prevail in industrial effluents and other environmental samples. Since CNTs have demonstrated a great potential to adsorb a multitude of pollutants, it is of great interest to examine which pollutant is selectively adsorbed in a complex matrix. The topic of selectivity is discussed in detail later on in this chapter but it is of interest to examine how a tailor-made CNT can selectively adsorb a target pollutant. Li et. al. [14] showed that multi-walled carbon nanotubes (MWCNTs) selectively adsorbed Pb(II) over Cd(II) and Cu(II) in both single component and multi-component systems.

Thus, selectivity for a specific pollutant can be achieved. This study also showed that the adsorption was higher in single component systems and this indicated that adsorption in a complex matrix needs improvement.

The wide range of pollutants reported in Table 1 exclude priority pollutants like Cr(VI) and Hg(II). Such pollutants need special attention because of their unique type of toxicities [8]. The next section of this chapter is therefore devoted to studies which have focused on the removal of Cr(VI) and Hg(II) from aqueous solution.

Removal of Cr(VI)

Cr(VI) has become a priority pollutant due to its toxic effects which range from a multitude of cancers to chronic dermatitis [28,29]. The removal of Cr(VI) from aqueous solution has therefore become a topic of much interest. Studies on the removal of Cr(VI) by CNTs have been initiated quite recently. The use of different types of CNTs has also been explored. For instance, Di et. al. [36] showed that ceria nanoparticles supported on aligned CNTs were more efficient at adsorbing Cr(VI) than activated carbon, Al_2O_3 and ball milled CNTs.

These workers also reported a maximum uptake capacity of 30,2 mg per g and observed that the best adsorption activity was achieved over a pH range of 3,0 – 7,4. Similar observations were also made by Qu [3] and Di {30].

Another study by Hu et. al. [31], shed some light on the mechanism of Cr(VI) removal.

Table 2. The Influence of Various Parameters on the Uptake of Cr(VI)

Parameter	Observation	Reference
Contact Time	Increased adsorption with increased contact time up to 12 hours and 280 hours	Hu et. al. [30] Pillay et. al. [26]
Solution pH	Higher adsorption at low pH and lower adsorption at High pH Effective adsorption over the pH range 3,4-7,0	Pillay et. al. [26] Hu et. al. [30]
Competing Anions	Reduced uptake in the presence of competing anions	Pillay et. al. [26]
Initial Concentrations	Reduced Uptake with increasing initial concentrations	Pillay et. al. [26]
Ionic Strength	No Influence	Hu et. al. [30]
MWCNT dosage	Increased removal capacity with increased dosage	Hu et. al. [30]

These workers used oxidized MWCNTs and proposed an adsorption mechanism which occurred via surface complexation. These authors also suggested that Cr(VI) is reduced to Cr(III) on the surface of the nanotubes and that both forms of chromium are retained. This has been experimentally verified by the authors.

The Influence of Selected Parameters on the Removal of Cr(VI)

A number of parameters tend to influence the uptake of Cr(VI). These are summarized in Table 2.

The observation made for the effect of competing anions sheds more light on chemical affinity competition and the selectivity of the CNTs for Cr(VI). The selectivity for Cr(VI) removal is poor if competing anions out-compete the Cr(VI) oxyanions for active adsorption sites on the CNTs namely functionalized and unfunctionalised MWCNTs [26].

There is an apparent contradiction in the literature where Hu et. al. [30] have observed that ionic strength has no influence on the uptake of Cr(VI) when varying concentrations of a $NaClO_4$ solution were used. This suggests that certain oxyanions like ClO_4^- do not pose a threat to the uptake of Cr(VI).

The effect of solution pH is somewhat more complex where this is not only accounted for in terms of the balance between H^+ and OH^- ions and the electrostatic interactions with the Cr(VI) oxyanions but also in terms of its influence on the speciation of chromium. Cr(III) predominates at low pH [26] so it is also possible that this form of chromium is adsorbed at low pH.

There is some agreement in the literature on the influence of contact time and initial concentrations although contact times have differed in individual studies. The effect of initial concentrations has been accounted for in terms of the saturation of active adsorption sites [26].

The observation made by Hu et. al. [30] on the effect of MWCNT dosage is an interesting observation as this sheds some light on the usage of CNTs for the large scale treatment of industrial effluents. This certainly implies that the MWCNTs can be used for large scale industrial applications but the cost implications are not promising as these materials are expensive to synthesise. Thus regeneration and reusability of these materials is important.

Desorption and Regeneration of Cr(VI)

The regeneration and desorption of Cr(VI) once adsorbed onto CNTs has not been extensively explored. Only one study to date has reported that 100% Cr(VI) can be desorbed using distilled water [26]. This is promising in the sense that this could address cost factors to some extent.

Removal of Hg

The removal of Hg by CNTs has been studied to a lesser extent than the removal of Cr(VI). Tawabini et. al. [31] used MWCNTs functionalized with carboxyl, carbonyl and hydroxyl groups for the uptake of Hg(II).

These workers achieved 100% removal at pH 8 using 10 mg MWCNTs and even observed a good adsorption efficiency for concentrations as low as 0,1 mg L^{-1}.

Some important parameters which influence the uptake of Hg(II) are outlined in Table 3.

The information in Table 3 reveals that there is some agreement in the literature concerning the effects of contact time and pH.

Table 3. The Important Parameters Influencing the Uptake of Hg(II)

Parameter	Observations	Reference
Contact Time	Increased adsorption with increased contact time until equilibrium is attained	Tawabini et. al. [32], Pillay [8], Pillay et.al. [33]
Solution pH	Increased uptake at high pH and reduced pH	Tawabini et. al. [32], Pillay [8], Pillay et.al. [33]
Competing Cations	High selectivity for the removal of Hg(II) by sulphur doped MWCNTs (S-MWCNTs)	Pillay [8], Pillay et.al. [33]
Sulphur Content	Linear increase in Hg(II) removal with increasing sulphur content	Pillay [8], Pillay et.al. [33]
Mixing Rate	Increased adsorption with increasing mixing rate	Tawabini et. al. [32],
CNTs dosage	Increased adsorption with increased dosage	Tawabini et. al. [32],

However, the solution pH does again have an impact on the speciation of Hg and it has been observed that the adsorption at very high pH is restricted due to the precipitation of hydroxyl-mercury species [8.32, 33].

An interesting result is the observation that S-MWCNTs are very selective to the uptake of Hg(II) even when competing cations are present. This is further supported by the observation that the uptake of Hg(II) increases linearly with an increase in sulphur content. Such an observation sheds some light on how the selectivity for a specific pollutant can be achieved by modifying the surface of CNTs and is discussed further in the section on selectivity.

Once again the observation made for the effect CNTs dosage implies that increased quantities of CNTs are required for effective removal. This again has serious cost implications for the practical use of CNTs.

Desorption and Regeneration of Hg(II)

The desorption and regeneration of Hg(II) from the surface of CNTs has not been extensively studied. It was however observed that the desorption of Hg(II) in distilled water and HCl was poor [8,33]. Better and complete desorption was achieved in a solution of 0,5% thiourea in 0,1 M HCl [8,33]. Such an observation has also lead to an understanding of the mechanism of Hg(II) adsorption onto S-MWCNTs which definitely suggests that chemisorptions occurs.

Strategies for Improving the Selectivity of CNTs for Specific Pollutants

As stated earlier the importance of achieving selectivity for a specific pollutant cannot be overemphasized. A review of the information in Table 1 shows clearly that selectivity and competitive adsorption studies are lacking. A number of strategies which can be employed to improve the selectivity of the CNTs for specific pollutants are critically discussed. Where appropriate, the relevant literature in which such studies have been undertaken is cited.

The Introduction of Heteroatoms onto the CNT Surface

In the discussion on the removal of Hg it was noted that S-MWCNTs were selective to the uptake of Hg(II) and the linear increase in Hg(II) uptake with sulphur content. This implied that there was some interaction between the sulphur and the mercury which is expected since mercury has a high affinity for sulphur [43]. This observation therefore implies that the introduction of a heteroatom which can potentially bind a pollutant will improve the selectivity for a specific pollutant.

The introduction of other heteroatoms like nitrogen has also been explored [8]. However, the form in which the element is present seems to play a key role. For instance, the presence of quartenary nitrogen increases the surface charge of the CNTs thereby favouring anion uptake [8]. Conversely, the introduction of heteroatoms like oxygen results in a negative surface charge thereby favouring cation uptake. However, the extent of cation or anion specificity still needs to be explored.

Fe-doping has also been attempted for the extraction of benzonitrile [21]. Although this showed a better adsorption capability than the undoped materials the selectivity for benzonitrile in the presence of other pollutants has not been reported.

Thus, as a whole it seems as though only sulphur doping has been successful at achieving selectivity.

Acid-Base Treatment and Functional Groups

Acid and base treatment of the CNTs is another strategy that could introduce functional groups which could selectively bind specific pollutants. Acid treatment in particular results in the oxidation of CNTs which can bind cations. In Table 1 it was noted that oxidised MWCNTs have been extensively used for cation adsorption. This has been supported by other studies as well [8]. However, the variety of cations that have been adsorbed suggests that selectivity needs to be explored.

Recent studies have shown that base-treated MWCNTs have varying effects on anion and cation uptake depending on the type of heteroatom that is introduced [8]. The selectivity of base-treatment is therefore also uncertain.

The Use of Carbon Nanotubes for Practical Applications in Waste-water Treatment

An extensive amount of research on the ability of CNTs to extract various pollutants has been conducted on a laboratory scale. However, the practical utility of these materials on an industrial scale is still a topic of immense debate. The limited studies conducted on the effect of CNTs dosage on the uptake capacity have suggested that increased amounts of CNTs improves the uptake capacity. This has serious cost implications as it is expensive to synthesis CNTs on a large scale.

A study on mixing small quantities of CNTs in an inert filler such as silicon carbide has been conducted for both industrial effluents containing Cr(VI) and Hg(II) [8]. This study showed that it was possible to adsorb appreciable amounts of Cr(VI) and Hg(II). These materials could also be regenerated and reused over 6 cycles of adsorption and desorption. S-MWCNTs mixed into SiC were highly selective to the uptake of Hg.

However, unfunctionalised CNTs mixed into SiC were not very selective to the uptake of Cr(VI) especially in the presence of competing sulphite ions[8]. This now poses serious challenges for the selective remediation of metal ions.

Conclusion and Future Research Perspectives

In this chapter it has been noted that CNTs in different forms are capable of adsorbing various contaminants to appreciable extents. Such adsorbents even demonstrate a superior adsorption capacity to that of activated carbon. However, there are a few challenges facing the use of these materials for practical industrial applications. The first of these is achieving selectivity of a specific pollutant. Thus far, only selectivity for Hg has been achieved [8,33]

The second issue is the economic issue of synthesizing large scale quantities of CNTs. Although desorption and regeneration can resolve this issue, large scale quantities still need to be synthesized initially. Lastly, the issue of why these materials perform better at adsorption than activated carbon still needs to be resolved.

Based on these issues the following studies are therefore proposed:

More selectivity studies can be explored. For instance, ligands can be attached to CNTs surface to selectively bind a metal ion eg. Diphenylcarbazide which forms complexes with Cr(VI) can be bound to a nanotube surface to selectively extract Cr(VI).

The cost and economic issues can be addressed by forming composites between CNTs and low cost adsorbents by embedding only small quantities of the nanotubes into these adsorbents.

Lastly, more studies on mechanism of uptake by CNTs and activated carbon need to be conducted to work out why CNTs are better adsorbents.

REFERENCES

[1] S. Iijima, *Nature* 354 (1991).

[2] C. Lu and H. Chiu, *Chemical Engineering Science* 61 (2006).

[3] S. Qu, F. Huang, S. Yu, G. Chen and Kong J., *Journal of Hazardous Materials* 168 (2008).

[4] Li, Y.H., Wang, S., Cao, A., Zhao, D., Zang, X., Xu, C., Luan, Z., Ruan, D., Liang, J., Wu, D. and Wei, B. *Chemical Physical Letters*,350 5-6 (2001).

[5] Li, Y.H., Wang, S., Wei, J., Zhang, X., Xu, C., Luan, Z., Wu, D. and Wei, D. *Chemical Physics Letters* 357 3-4 (2002).

[6] Long, R.Q. and Yang, R.T. *J. Am. Che. Soc.*, 123 (2001).

[7] X. Feng., *PhD Thesis,* University of Pittsburgh, 2004.

[8] K. Pillay, *PhD Thesis,* University of the Witwatersrand, 2011.

[9] C.Y. Kuo, *Desalination,* 243, 3, (2009).

[10] B. Pan, Sun, K. and Xing B. *Journal of Soil Sediments*, 10 (2010).

[11] C. Chen and X. Wang, *Ind. Chem. Eng. Res.,* 45 (2007).

[12] M.L. Kandah and M.J. Meunier, *Journal of Hazardous Materials,* 146 (2007).

[13] C. Lu, C. Liu and G.P. Rao, *Journal of Hazardous Materials,*151 (2008).

[14] Y.H. Li, S. Wang, Z. Luan, J. Ding, C. Xu and D. Wu, *Carbon,* 41 (2003).

[15] C.J.M. Chin, M.W. Shih and H.J. Tsai, *Applied Surface Science,* 256 (2010).

[16] G. Sheng, J. Li, D. Shao, J. Hu, C. Chen, Y. Chen and X. Wang, *Journal of Hazardous Materials*, 178, 2010.

[17] H.J. Wang, A.L. Zhou, F. Peng, H. Yu and L.F. Chen, *Materials Science and Engineering A*, 466, 2007.

[18] H. Wang, A. Zhou, F. Peng, H. Yu and J. Yang, *Journal of Hazardous Materials,* 166, 2007.

[19] S. Yang, J. Li, D. Shao, J. Hu and X. Wang, *Journal of Hazardous Materials*, 166, 2009.
[20] G. Sheng, J. Li, D. Shao, J. Hu, C. Chen, Y. Chen and X. Wang, *Journal of Hazardous Materials*, 178, 2010.
[21] A.L. Aguiar, S.B. Fagan, L.B. da Silva, J. Mendez Filho and A.G. Souza Filho. *J. Phys. Chem.,* 114, 2010.
[22] H. Sone, B. Fegetsu, D. Tsukada and M. Endo, *Talanta,* 74, 2008.
[23] K. Pyrzynska and M. Bystrzewski, *Colloids and Surfaces A: Physicochemical and Engineering Aspects*, 362, 2010.
[24] M.A. Salam, M. Mokhtar, S.N. Basahel, S.A. Al-Thalbaiti and A.Y. Obaid, *Journal of Alloys and Compounds* 210 2010.
[25] M.A. Salam and R.C. Burke, *Water, Air, Soil Pollution*, 210, 2010.
[26] K. Pillay, E.M. Cukrowska and N.J. Coville, *Journal of Hazardous Materials*, 166, 2009.
[27] G.P. Rao, C. Lu and F. Su, *Separation and Purification Technology* 58 2007.
[28] F.W. Dixon, *Journal of the American Medical Association.* 93 1929.
[29] J. Franchini, E. Magnani and Mutti, A, *Scandinavion Journal of Work Environmental Health*, 9 1983.
[30] Z.C. Di, J. Ding, X.J. Peng, Y.H. Li, Z.K. Luan and J. Liang, *Chemosphere* 62 2008.
[31] X. Hu, Y. Wu, H., Y.H. Li, Z.K. Luan and J.Liang, *Chemosphere* 62 2006.
[32] B. Tawabini, S. Al-Khali and M. Khaleb, *Water Science and Technology* 61 3 2010.
[33] K. Pillay, E.M. Cukrowska and N.J. Coville, Unpublished Results.

In: Applications of Carbon Nanotubes
Editor: Ajay Kumar Mishra

ISBN: 978-1-62081-918-0

Chapter 15

APPLICATIONS OF CARBON NANOTUBES IN ENVIRONMENTAL ISSUES: CONTROLLING WATER POLLUTION

Jose R. Rangel-Mendez[1*], Nancy V. Perez-Aguilar[1,3] and Paola E. Diaz-Flores[2]

[1]Institute for Scientific and Technological Research of San Luis Potosi (IPICYT), Environmental Sciences Division, San Luis Potosi, Mexico

[2]Universidad Autonoma de San Luis Potosi, Coordinacion Academica Region Altiplano, San Luis Potosi, Mexico

[3]Universidad Autonoma de Coahuila, Facultad de Ciencias Quimicas, Coahuila, Mexico

ABSTRACT

Carbon nanotubes were synthesized for the first time in the 90's and since then these have been used in electrical and electronics devices. In recent years, scientists from all around the world have focused also on carbon nanotubes to control water pollution. Hence, water treatment and purification systems based on carbon nanotubes might emerge in the near future. This chapter describes the most outstanding results obtained for the removal of organic and inorganic pollutants present in aqueous solutions (cadmium, lead and phenol) with different types of carbon nanotubes, such as single wall carbon nanotubes (SWCNT), multiwall carbon nanotubes (MWCNT) and nitrogen-doped multiwall carbon nanotubes (N-MWCNT). The adsorption experiments were carried out with ideal solutions of single and binary compositions of the mentioned pollutants, at different conditions of pH and temperature. Results for single component solutions show that the pH and temperature of the system play an important role in the adsorption capacity of nanotubes. Also, these features can correlate the adsorption capacity when two or more species are in solution (competitive adsorption). On the other hand, the physical-chemical characterization of adsorbents allows knowing their properties as active sites, specific area, pore size distribution, etc., involved in the adsorption process.

* E-mail: rene@ipicyt.edu.mx. Tel: (+52) 444-8342000. Fax: (+52) 444-8342010.

The analysis of these properties may describe possible adsorption mechanisms and kinetics as will be shown throughout the chapter. The results of this research have helped to suggest adsorption mechanisms and to elucidate the relevance of nitrogen atoms in the lattice of carbon nanotubes to adsorption capacity. In addition, results from adsorption kinetic studies are presented, in an effort to determine mass transfer coefficients, which are important to design industrial scale equipment.

1. INTRODUCTION

Water plays a key role in human societies, as the world population grows there is a challenge to supply drinking water that fulfills the established parameters by governmental organizations. As water is becoming increasingly scarce relative to demand, great efforts are necessary to protect the water supply for humans, biodiversity and ecosystems. That is why all countries are encouraged to optimize water consumption and at the same time to prevent its pollution through several actions such as water recycling and reuse, adoption of international standards for water quality, and more investment in purification and wastewater treatment infrastructure [1]. These actions require technological innovation, and nanotechnology can offers significant opportunities for the water sector. This new knowledge would result in using nanoparticles, nanofiltration or other products to improve water quality [2].

Population growth has increased water demand for almost all productive activities such as agriculture, industry, energy and transportation. The wastewater from these activities containing high loads of toxic substances such as toxic metals is frequently discharged into water bodies. These contaminants have the potential for long-range transport, dispersal and deposition. Toxic metals cannot be destroyed or broken down through treatment or environmental degradation, leading to different human health problems such as poisoning and cancer. The international standards for water quality list as priority toxic metals: lead, cadmium, mercury, chromium, nickel, silver, thallium and zinc. These standards are guideline values that are reasonably achievable through practical treatment technologies. However, these values are revised continuously, thus it is necessary to improve current technologies and to develop new ones [3, 4].

In this chapter lead,cadmium and phenol were selected as model inorganic and organic toxic pollutants in water; the World Health Organization recommends a guideline value for the maximum allowable limit of 0.001 mg/ L for total phenol, 0.01 mg/L for lead and 0.003 mg/L for cadmium in drinking water. Lead concentrations in drinking-water are generally below 5 mg/L, although much higher concentrations (above 100 mg/L) have been measured where lead fittings are present. Infants, children up to 6 years of age and pregnant women are most susceptible to its adverse health effects. Lead in aqueous solutions participates in a series of consecutive proton transfers that determine their chemical compounds in solution; lead ion (Pb^{2+}) is stable below pH 5.5, then species $PbOH^{+}$, $Pb(OH)_2$, $Pb(OH)^{3-}$ are consecutively formed as a result of hydrolysis reactions [5]. On the other side, cadmium and its compounds are used in the steel industry, plastics and in batteries. Cadmium is released to the environment in wastewater, and diffuse pollution is caused by contamination from fertilizers and local air pollution. Contamination of drinking-water may also be caused by impurities in the zinc of galvanized pipes and solders, and some metal fittings. Cadmium ion (cadmium) in aqueous solution is stable below pH 7.8, then species $CdOH^+$, $Cd(OH)_2$, $Cd(OH)_3^-$ are consecutively formed as result of hydrolysis reactions [5]. Drinking water purification and wastewater treatment will invariably be required to reduce both health risks and negative environmental impacts due to toxic metals. Different technologies can be employed to remove pollutants as toxic metals: chemical precipitation, coagulation-flocculation, membrane filtration, ion exchange and adsorption.

On the other side, phenol was selected as model organic compound. Phenol and some of its derivatives are present in effluents of many industries such as petrochemicals, petroleum refineries, insecticides and herbicides industries, etc. Exposure to these substances can cause skin and eye burns; comas, convulsions and death can result from overexposure to it. Phenolic derivates are one of the most common organic water pollutants and their presence in natural water can lead further to the formation of substituted compounds during disinfection and oxidation processes. Various treatment technologies available include catalytic wet air oxidation, biological degradation, adsorption and UV-oxidation[6, 7].

Adsorption is a mass transfer operation by which a substance is transferred from the liquid phase to the surface of a solid and becomes bound by physical and/ or chemical interactions; this is used in drinking water treatment as tertiary process, for the removal of color and odor to achieve regulatory requirements. The primary adsorbent used in drinking water treatment is granular activated carbon, which is considered to be the best available technology for removing low-solubility contaminants such as toxic metals in trace concentrations. Some other adsorbents which are capable to remove toxic metals from aqueous solutions are zeolites, clays and polymeric resins [8, 9]. The usefulness of activated carbons derives mainly from its large micropore and mesopore volumes, with a high surface area between 300 and 4000 m^2/g. These also have analogous surface chemical groups of organic compounds which can display acidic, basic and/or neutral characteristics. Some surface acidic functional groups include carboxyl, carbonyl, lactone, hydroxyl and carboxylic anhydride; these specific functional groups are essential for the adsorption of heavy metals due to complexes formed [10, 11]. So far activated carbon has been used as an all-purpose adsorbent, but future applications of adsorption are limited by the availability of new, better and cheaper adsorbents.

1.1. Nanotechnology in Water Treatment

The expansion of scientific knowledge and technological applications is changing the way in which water is used, cleaned and reused, to meet human, economic and environmental needs. Nanotechnology has been emerging rapidly; it has been predicted that it will exert strong influence on the technologies related to supply, use and management of water resources [13]. Nanotechnology is the manipulation of matter for particular applications at the atomic, molecular, or macromolecular levels, using a length scale of approximately one to one hundred nanometers in any dimension [12]. Nanotechnology can develop new materials with enhanced properties and attributes, resulting of larger surface area per unit of volume and quantum effects that occur at the nanometer scale [14]. Characteristics of nanomaterials such as enhanced reactivity, surface area, subsurface transport, and/or sequestration, could benefit environmental remediation by faster and more economical processes than current conventional approaches.

Nanotechnology shows particular promise for water resources, to improve water quality through water treatment or remediation. Key areas are desalinization, water purification, wastewater treatment and monitoring. The first three areas involve the use of nanofiltration technology, nanomaterials and nanoparticles to remove or reduce water contaminants. Monitoring involves the use of new sensor technology coupled with micro and nano-fabrication technology. Nanosensors would be highly accurate and portable for detection of biological and chemical water contaminants, present in very low concentrations [12, 15]. Nanotechnology-based water treatments could replace current chemicals for which existing technologies are inefficient or costly, and could eventually permit human use of heavily polluted and saline water for drinking, sanitation and irrigation [2]. Applications involving dispersive uses of nanomaterials in water have the potential for wide exposures to aquatic life and humans. At this point of time and stage of knowledge, it is unlikely to make any collective judgment about the potential risks of exposure to nanomaterials [12, 14]. Considering this situation, adequate immobilization of nanoparticles before their use in water-related applications could prevent their dispersion, and their probable adverse environmental effects. Nanotechnology-based applications to the water sector are still under fundamental research, and it is unclear when they will be ready for wide-scale use. These new technologies must compete against currently used technologies, looking for cheaper solutions that meet government standards. In addition, it is necessary to overcome factors such as availability of nanomaterials in large scale and lack of knowledge and regulations about their environmental and health impacts [13]. The development of highly efficient nanotechnology-based water treatment and purification systems will occur in the near future. These will offer higher reliability of treatment, enabling reduction of equipment size, more energy saving, simplification and cost reduction compared to conventional technologies.

1.2. Use of Carbon Nanotubes in Water Treatment

Carbon nanotubes are key nanomaterials in nanotechnology. A better performance of the carbon nanotube as adsorbents is expected compared to currently used activated carbons. Carbon nanotubes have been tested as adsorbent of toxic metals in aqueous solution obtaining promising results; most of the studies have been conducted with multiwall carbon nanotubes (MWNT). Early studies were done by Li et al., [16], who used MWNT oxidized with nitric acid to adsorb lead in aqueous solution. The maximum adsorption capacity at room temperature and pH 5 was 17.5 mg/g. Since then, the same researchers have reported the adsorption of lead, copper, and cadmium onto MWNT: the maximum adsorption capacities were 97.1 mg/g for lead, 24.5 mg/g for copper and 10.9 mg/g for cadmium at room temperature, pH 5.0 and metal ion equilibrium concentration of 10 mg/L. It was also found that the adsorption capacities of the MWNT were 3 to 4 times larger than those of powder activated carbon and granular activated carbon [17]. After that, more studies have been performed: copper ions were removed by oxidized MWNT and their maximum adsorption capacity was 27.6 mg/g at pH 5.2 [18]. Also different reagents have been used to modify the nanotubes: hydrogen peroxide (H_2O_2), potassium permanganate ($KMnO_4$) and nitric acid (HNO_3). The oxidized nanotubes removed cadmium with maximum adsorption capacities of 2.6, 5.1 and 11 mg/g, respectively, at the cadmium equilibrium concentration of 4 mg/L [19]. Nitric acid solutions are the most used oxidizing agent to attach oxygen-containing groups onto carbon nanotubes to modify their surface properties for metal ions adsorption. Furthermore, adsorption equilibrium tests at different temperatures were conducted to evaluate the effect on adsorption capacity; it was found that adsorption with carbon nanotubes is an endothermic process [20]. Also, carbon nanotubes synthesized by using different conditions of precursor, catalyst and temperature (xylene-Fe-800°C, benzene-Fe-1150°C, propylene-Ni-750°C or methane-Ni-650°C) were tested to adsorb lead. It was found that those synthesized with methane-Ni-650°C had a higher lead adsorption capacity (82.6 mg/g at pH 5), because their less crystalline structure and higher concentration of carboxylic groups [21].

The adsorption capacity of MWNT for several metal ions has been evaluated at different conditions of ionic strength, pH, temperature, foreign ions, etc. Some of these studies include kinetics analysis of the experimental data by the pseudo-second order reaction model. Published results for maximum nickel adsorption capacity are 9.8 mg/g at 30 °C and pH 6.5 [22], 49.6 mg/g at room temperature and pH 6 [23], 38.46 mg/g at room temperature and pH 7, for MWNT oxidized by sodium hypochlorite [24], and 7.4 mg/g due to the effect of polyacrylic acid (PAA), with a structure like that of natural organic matter [25], 38.46 mg/g at room temperature and pH 7, for MWNT oxidized by sodium hypochlorite [24], and 7.4 mg/g due to the effect of polyacrylic acid (PAA), with a structure like that of natural organic matter [25]. Few studies have been carried out to test the more expensive single wall carbon nanotubes (SWNT) which have higher adsorption capacity than MWNT. Lu and Chiu [26] observed a maximum zinc adsorption capacity of 44 mg/g at room temperature and pH 7 for oxidized SWNT, versus 32.7 mg/g for oxidized MWCNT. The maximum nickel adsorption capacity onto oxidized SWNT is 47.85 mg/g at room temperature and pH 7 versus 38.46 mg/g for oxidized MWNT [27].

Recently, two adsorption studies were conducted with nitrogen-doped carbon nanotubes (N-MWCNTs) to remove metal ions from aqueous solutions. Maximum adsorption capacity of 31 mg/g [28] and 21.6 mg/g at pH 7 [29] were obtained for cadmium and 28.9 mg/g at pH 5 for lead [29].

As mentioned above, organic compounds constitute other type of pollutants usually found in water. For instance chlorobenzenes, which are used as solvents, heat transfer agents, insect repellent deodorants, degreasers and intermediates in dye and pesticide synthesis. Peng et al., [30] removed 1, 2-dichlorobenzene (1, 2-DCB) from water using carbon nanotubes and found that the time needed to achieve equilibrium was 40 min. Moreover, these authors attributed the adsorption of 1, 2-DCB to the interaction with carbonyl type oxygenated groups present on the nanotubes surface. Diaz-Flores et al.,[31] studied the phenol adsorption capacity of N-MWCNTs modified with air (N-MWCNTs5% or 5% weight lost; N-MWCNTs10% or 10% weight lost; etc). The adsorption capacity decreased with the oxidation level as follows: 0.16, 0.14, 0.12 mmol/g for N-MWCNTs5%,N-MWCNTs10% and N-MWCNTS 20%.

In general, carbon nanotubes have shown better adsorption capacities for toxic metal ions in aqueous solution than activated carbons. Their adsorptive features besides a more efficient regeneration, and shorter time to reach equilibrium than activated carbon, have led to a search for an adequate use of CNT in devices as membranes and filters for water purification systems [25, 32-34]. Adsorption processes are designed on equilibrium and kinetics, but this last research topic remains basically unexplored. The overall rate of adsorption of a metal ion on a porous material can be described by two main steps: the external mass transfer from the bulk liquid to the solid surface, followed by mass transport of adsorbate molecules into the particle; this is the most typical of the rate steps in solid adsorbents and is referred to as intraparticle diffusion [35]. Most research studies about adsorption with carbon nanotubes in liquid phase have minimized the role of intraparticle diffusion in kinetics adsorption, assuming that the adsorption reaction is the controlling step. The evaluation of adsorption kinetics is necessary considering the different types of existing carbon nanotubes. The physical structure and the chemical properties should be involved in adsorption kinetics of toxic metals.

2. CARBON NANOTUBES CHARACTERIZATION

The morphology of carbon nanotubes has been observed by electron microscopy; micrographs showed agglomerates of MWCNT and N-MWCNT with a wide distribution of diameters, in the range of 1 to 100 μm. On the other hand, SWCNTs were grouped in 2D nanocrystals with several aligned nanotubes and separated by a distance of approximately 3.354 Å [36]. After nitric acid oxidation carbon nanotubes were exfoliated, probably due to repulsion forces between oxygen-containing surface groups, which reduced the π-conjugation of the graphene sheet, overcoming the van der Waals forces [37]. Proximal dimensions of 500 μm in length and 60 to 90 nm of external diameter were observed for oxidized MWCNTs (ox-MWCNT), while oxidized N-MWCNTs (ox-N-MWCNT) were about 70 to 80 μm in length and 40 to 60 nm of external diameter. The length of a single oxidized SWCNT (ox-SWCNT) was not measured because it was difficult to observe both end caps; their external diameter was in the range of 1 to 3 nm (Fig. 1). Nitric acid oxidation had additional effects such as dissolution of amorphous carbon and catalyst particles, the erosion of the tips in ox-N-MWCNT, the mechanical distortion of the most internal cylindrical planes in ox-MWCNTs (Fig. 1d), probably as a result of decreasing the cohesive energy of the tubes yielding to their partial destruction, as well as pores formation along the sidewall of carbon nanotubes.

On the other hand, X-Ray diffraction patterns showed the reflection of the (002) plane for oxidized carbon nanotubes. This peak corresponds to interlayer spacing between graphene layers, it was slightly larger for the ox-SWCNT (0.355 nm), than for ox-N-MWCNT (0.344 nm) and MWCNT (0.333 nm). The surface defects besides the oxygen-containing groups attached to nanotubes can be viewed as intercalated species in the lattice, as occurs in graphite intercalation compounds, where entire planes of atoms or molecules can be hosted between the carbon layers [36, 38, 39]. Intercalated species on nanotubes lead to expansion of their interplanar as well as their inter-nanotube spacing [38-40].

Moreover, Raman spectra obtained at 514 nm showed the graphite band (G-band) from 1580 to 1590 cm^{-1}, the disorder band (D-band) from 1340 to 1360 cm^{-1}, the second order D-band (D*-band) from 2680 to 2700 cm^{-1} , and the characteristic radial breathing mode (RBM) associated to the diameter of SWCNTs in the range of 130 to 280 cm^{-1}. Diameters of ox-SWCNT were of 0.9 to 1.5 nm, in agreement with values reported by the manufacturer and the literature. The relative intensity between D and G bands (I_D/I_G ratio) is sensitive to the defect density along the sidewall in multiwall carbon nanotubes. An increase of I_D/I_G ratio from 0.365 to 0.514 and 0.769 to 0.836 was determined for ox-MWCNTs and ox-N-MWCNTs, respectively, in relation to pristine nanotubes; therefore sidewall defects were higher for ox-N-MWCNTs than for ox-MWCNTs [41, 42].

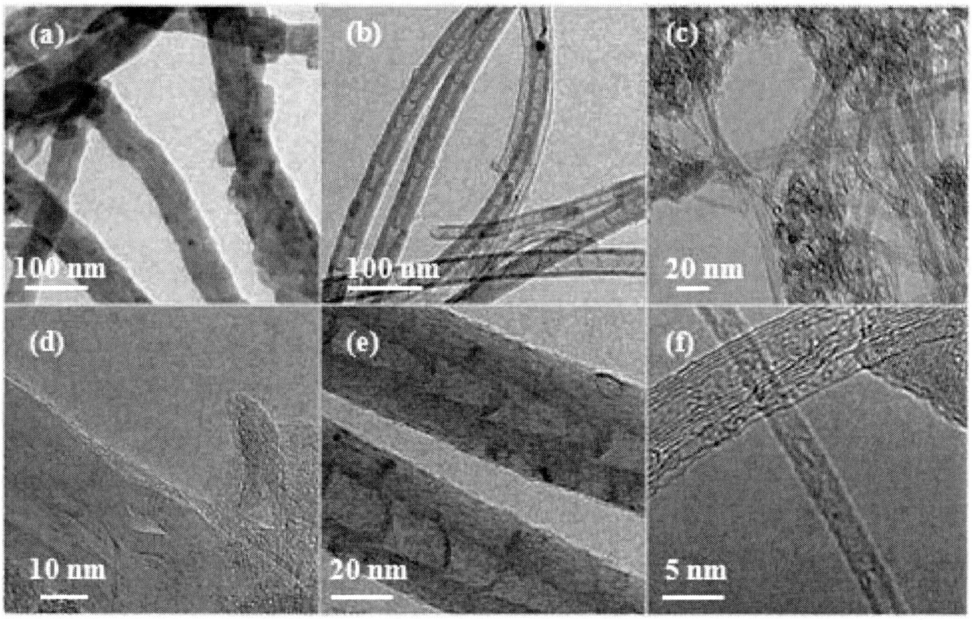

Figure 1. HRTEM images of nitric acid oxidized carbon nanotubes by 5 h at 80 °C; ox-MWCNTs (a, d), ox-N-MWCNTs (b,e), and ox-SWCNTs (c,f).

The relative intensity between the D-band and G-band (I_D/I_G ratio) was 1.098 for pristine nanotubes and 1.2287 for ox-N-MWCNTs. These values indicate a continuously covalent bonding of oxygenated groups along the walls of nanotubes, disrupting the aromatic system of π-electrons [43-45].

Chemical characterization of oxidized carbon nanotubes by elemental analyses showed an important increment in oxygen content for oxidized carbon nanotubes. The acid character and identity of the surface-attached groups measured by titration of total acidic sites (TAS) showed the obtained values of 0.972, 0.246 and 0.155 mmol/g. The slurry pH is related to the concentration of acidic surface functional groups; it was 4.0, 4.6 and 5.3, for ox-SWNT, ox-N-MWCNTs, and ox-MWNT respectively. The main functional group attached to oxidized carbon nanotubes was carboxylic, as identified by FTIR spectroscopy. Comparison between spectra of pristine nanotubes and oxidized N-MWCNT showed the bands at 1444 cm^{-1} [46], 1373 cm^{-1} and 1251 cm^{-1} [47], attributed to vibration of multiwall nanotubes and C-N bonding in ox-N-MWCNTs were overcome by several vibrations in carbonyl and carboxylic functionalities in the range of 1720 to 1250 cm^{-1} [48-50]. A strong band appeared from 3600 cm^{-1} to 3300 cm^{-1} by the stretching of the bonding -OH of carboxylic group in oxidized nanotubes [48, 50, 51]. The nitro group ($-NO_2$) is isoelectronic with the carboxylate ion group and their spectra are very similar, but a weak feature related to nitro group was observed at 1535 cm^{-1} for oxidized nanotubes [49, 50]. Acidic oxygen-containing groups improved wetting of oxidized carbon nanotubes in aqueous solution, and provided nucleophilic sites to which metallic ligands could be adsorbed.

The BET surface area and DFT pore volume of oxidized carbon nanotubes determined by nitrogen physisorption were 812 m^2/g and 0.532 cm^3/g, respectively, for ox-SWCNTs, 74 m^2/g and 0.29 cm^3/g for ox-N-MWCNTs, and finally 38 m^2/g and 0.07 cm^3/g for ox-MWCNTs. A wide polydispersity was observed for multiwall carbon nanotubes, showing maximum pore size diameters around 120 and 150 nm for ox-MWCNTs and ox-N-MWCNTs, respectively. This means that mainly mesopores and macropores between agglomerated nanotubes could contribute to the surface area and the pore volume of these samples, instead of their internal tubular cavity with a proximal diameter of nearly 10 nm as was observed by micrographs (Figure 1).

3. Metals Removal by Carbon Nanotubes

Toxic metals cannot be destroyed or broken down through treatment or environmental degradation, hence it is necessary to develop better technologies to reduce their concentration in water. This study presents the oxidation of N-MWCNTs to improve their performance as adsorbents of toxic metals in water (cadmium and lead).

Adsorption isotherms have been determined at different pH values to evaluate the capacity of ox-N-MWCNT to adsorb cadmium and lead. The increasing adsorption capacity observed for each metal when the pH went up was related to the higher concentration of available carboxylic groups. Data of the adsorption isotherms of lead were correlated with Langmuir model. The maxima adsorption capacities (q_{max}) were 0.083 mmol/g and 0.139 mmol/g at pH values 4 and 5, respectively [29]. The experimental data for adsorption of cadmium were better depicted by the Freundlich model. The values for parameter K were 0.083 mmol-L/g, 0.165 mmol-L/g and 0.192 mmol/L-g at pH 5, 6 and 7, respectively [29]. The mechanism of adsorption of cationic species in aqueous solution can occur by electrostatic interactions and formation of complexes with carboxylic groups, as was explained by [52]. Desorption isotherms carried out at 25 °C showed more than 90% of the mass of both adsorbed metals was desorbed, indicating that metal saturated ox-N-MWCNT can be regenerated and reutilized.

The selectivity of adsorption for lead and cadmium has been evaluated with both metals in solution at pH 5 and 25 °C. At low concentration of both metals, there is enough concentration of acidic sites for adsorption of lead and cadmium, but as the concentration of both metals in solution increased, lead was preferentially adsorbed instead of cadmium.

The difference in hydrated radii (r_H) between lead and cadmium is minimal: $r_{H,Pb}$= 4.01 Å, $r_{H,Cd}$= 4.26 Å [53], and thus both ions can reach the same adsorption sites mainly on the outer surface, since the interlayer spacing of ox-N-MWCNT is not wide enough (3.44 Å as measured by X-ray diffraction) to allow the intraparticle diffusion of these ions. For this reason, the dimension of these contaminants is not a decisive factor in selectivity, although the chemistry of each ion in solution is important.

4. Organic Compunds Removal by Carbon Nanotubes

Phenol and some of its derivatives are one of the most common organic water pollutants which can be removed by adsorption. Phenol in aqueous solutions has been adsorbed on pristine N-MWCNTs and MWCNTs at pH 7 and 10 at 25 °C. Figure 2 shows the adsorption isotherms adjusted by the Langmuir isotherm model. N-MWCNTs showed the maximum adsorption capacity at pH 7, and the adsorption capacity decreased in following order: N-

MWCNTs pH 7>N-MWCNTs pH 10> MWCNTS pH 10> MWCNTS pH 7. Since the pKa value of phenol is 9.99 [52], below this pH phenol is considered a neutral molecule and above this value is found as anionic species (phenolate). The phenol adsorption capacity (C <0.1 mmol / L) of N-MWCNTs decreases when increasing the solution pH from 7 to 10, this is probably due to the electrostatic repulsion between phenol and the adsorbent because the point of zero charge of N-MWCNTs is 6.6. Moreover, when nitrogen is present in the nanotubes lattice, an unpaired of electrons repulse charged molecules, and this is might be why the maximum adsorption capacity of MWCNTs, that do not contain nitrogen, occurred at pH 10. It is important to mention that these carbon nanotubes did not suffer any treatment before adsorption experiment since the main objective was to study the effect of nitrogen on adsorption capacity.

5. Competitive Adsorption on Carbon Nanotubes

Real water systems contain a variety of organic and inorganic species. For this reason, it is important to study competitive adsorption. This section reports some results related to simultaneous adsorption between phenol, cadmium and lead.

Diaz-Flores et. al., [31] carried out adsorption experiment of metal and organic compounds at 25°C and pH 7 onto N-MWCNTs modified with air until to reach 5% in weight lost (N-MWCNTs 5%). The phenol adsorption capacities were higher in comparison with cadmium˙. The adsorption capacities at an equilibrium concentration ≈ 0.4 mmol/L were 0.10 and 0.07 for phenol and cadmium, respectively. This adsorption capacity was estimated employing the Freundlich isotherm, which better adjusted the experimental data.

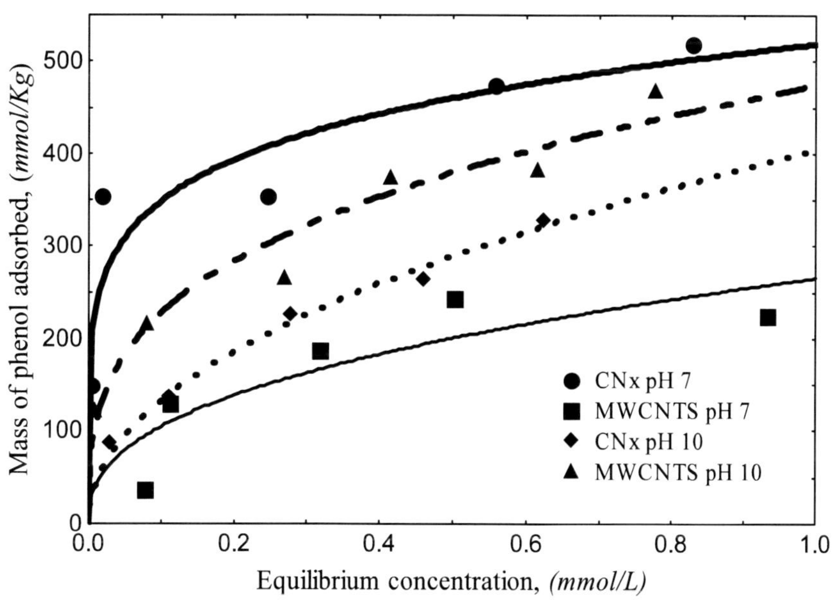

Figure 2. Adsorption isotherm of phenol at different pH solution and onto different carbon nanotubes at 25°C.

This represents almost 30% more adsorption capacity for phenol when compared to cadmium. This was expected since the organic compounds contain aromatic rings in their structure that can be retained by N-MWCNTs tubes due to weak interactions between delocalized electrons and the N-MWCNTs tubes [54], and can be adsorbed by carbonyl groups under the same kind of interaction [55]. In addition, aromatic compounds could be adsorbed onto N-MWCNTs tubes via π-π interactions, whereas heavy metals only anchor on specific sites.

Moreover, Diaz-Flores et. al., [31] developed a study which describes the effect of the adsorption capacity of cadmium when phenol is preadsorbed onto oxidized N-MWCNTs. The individual adsorption capacity of cadmium was 0.083, 0.087, 0.089 mmol/g for N-MWCNTs5%, N-MWCNTs10% and N-MWCNTs20%, respectively; the adsorption capacity increased as the amount of oxygen and specific area raised. The oxygen contents increased as: N-MWCNTS5% (1.7 %) < N-MWCNTs10% (4.4 %) < N-MWCNTs 20% (8.2 %), and the surface area decreased as: N-MWCNTs5% (102 m^2/g); N-MWCNTs10% (58 m^2/g); N-MWCNTs20% (76 m^2/g). Related to this last point, several authors [16, 56] have modified carbon nanotubes using various oxidizing agents such as H_2O_2, $KMnO_4$, and HNO_3 to improve their adsorption capacity towards cadmium, and found that the HNO_3 oxidized material reveals the best adsorption capacity. This modification introduces preferably carboxylic sites that adsorb metals [8]. Diaz-Flores et. al., [31] also showed that the adsorption capacity of phenol is 0.16, 0.14, and 0.12 mmol /g for N-MWCNTs5% N-MWCNTs10% and N-MWCNTs20%. As observed earlier, the phenol adsorption capacity decreased as the oxidation degree increased: 11.9% and 32% for N-MWCNTs10% and N-MWCNTs 20%, respectively. The adsorption of phenol molecules on carbonaceous materials also depends on the surface area of the adsorbent and it is influenced by the presence of oxygenated complexes, as stated by Bansal et al.,[57]. In this study, the presence of a significant amount of oxygen in the material decreased its ability to adsorb phenol [57]. It has also been demonstrated that phenol adsorb onto carbonaceous materials through π-π interactions between the aromatic ring of phenol and carbonyl or quinonic present in the adsorbent [57]. Carbon atoms in the graphitic layers exhibits four valence electrons, three forming σ type bonds and one more forming a π bond with its neighbors. Considering that nitrogen is present in N-MWCNTs tubes and that it has five valences, in the pyridinic site, two of these electrons form σ bonds, two more are in the p_x plane and one more is forming a π bond with its neighbors. For pyrolytic-like bonding, three of the nitrogen valence electrons form σ bonds and the remaining two are donated to the π-bond around the 5-fold ring, which has now six electrons and it is aromatic [58]. It can also be observed that the adsorption capacity of cadmium increased 2.5, 2.9 and 1.4 times for N-MWCNTs5%, N-MWCNTs10% and N-MWCNTs20% when phenol was pre-adsorbed, respectively. The increase of the cadmium adsorption capacity depends on the amount of phenol adsorbed. The same tendency was observed for the adsorption capacity of phenol on N-MWCNTs tubes when cadmium was preadsorbed; it increased almost twice compared with the adsorption capacity of fresh N-MWCNTs tubes (untreated).

A detailed study (Figure 3) shows the competitive adsorption of lead and phenol in aqueous solution at pH 5 onto N-MWCNTs 5% (oxygen content 1.68 %). These experiments were performed at equimolar concentrations of phenol and lead at 25 °C. The results show that the phenol adsorption capacity increased when lead is present in solution. However, the adsorption of lead changes in the presence of phenol. This behavior may be due to the electronic density of phenol that captures the lead atoms and subsequently it is adsorbed on carbon nanotubes: in other words, phenol decreases its electronic density and hence there is less repulsion between this molecule and the surface of carbon nanotubes. These results are consistent and in accordance to the results reported by Diaz-Flores et. al.,[31] who studied the individual and competitive adsorption at equimolar concentration of phenol and cadmium at: the phenol adsorption capacity increases from 0.16 to 0.26 mmol/g when cadmium is present in solution.

6. Kinetic Studies by Three Types of Carbon Nanotubes

Adsorption capacity data and adsorption kinetic parameters are required to design adsorption devices based on carbon nanotubes, and one of the most important parameter is the morphology of carbon nanotubes because it may determine adsorption kinetics.

A standard batch-stirred tank has been used for kinetic experiments; it had 1 L of volume, 0.1 m of internal diameter, and contained 0.5 L of cadmium solution. Four plastic baffles were evenly spaced inside the vessel to prevent vortex formation as well as to reduce the relative motion between liquid and solid particles.

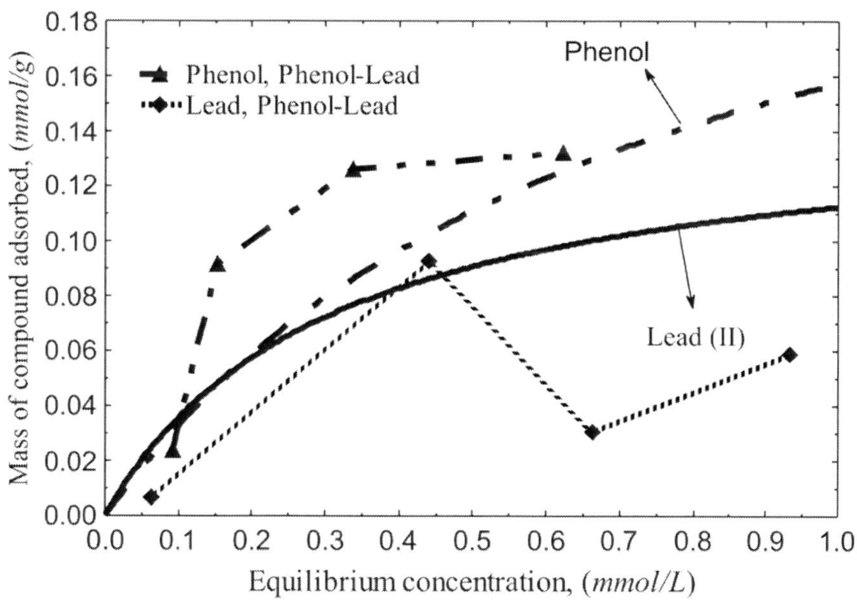

Figure 3. Simultaneous adsorption of lead and phenol at pH 5 and 25°C on pristine N-MWCNTs.

A four-bladed stainless steel impeller of 5 cm of diameter and 1 cm wide, driven by a stainless steel shaft and a variable speed motor was used to mix the cadmium solution containing adsorbent material (Figure 4). These experiments were conducted with 0.2 ± 0.01 g of sample at 25 ± 2 °C, 4.5 ± 0.5 mg/L of cadmium initial concentration, pH 6 ± 0.1, at 200 ± 5 rpm. In due course, the effect of initial cadmium concentration was studied by using 9 ± 1 and 18 ± 1 mg/L of this adsorbate.

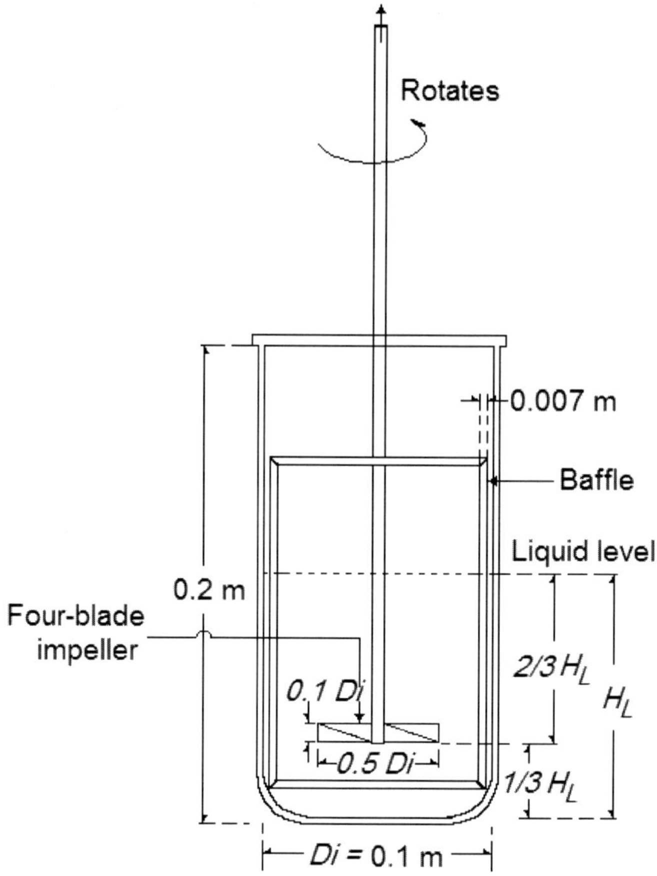

Figure 4. Configuration of stirred vessel for adsorption kinetic experiments.

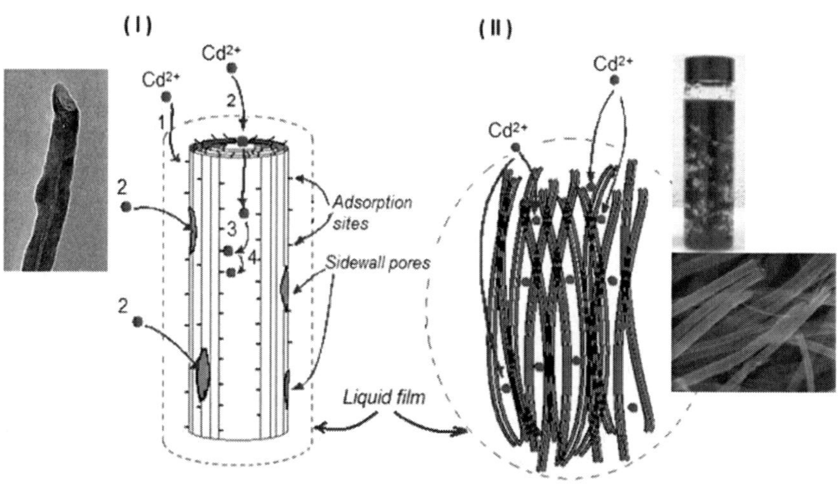

Figure 5. Mass transport processes on cadmium adsorption by carbon nanotubes. (I) Adsorption on a carbon nanotube includes: (1) external mass transport, (2) pore volume diffusion through axial and/or radial direction, (3) adsorption, and (4) surface diffusion; (II) Considering pore volume diffusion (I) in addition with pore volume diffusion between carbon nanotubes.

Samples of 3 mL were withdrawn from the system and immediately filtered with polycarbonate membranes of 0.45 μm of pore diameter. The total volume withdrawn from the vessel was less than 5%. Sampling was more frequent during the first ten minutes.

Cadmium adsorption rate on carbon nanotubes considers different mass transfer processes as illustrated in Figure 5. One or a combination of them can determine the overall adsorption rate.

Equilibrium was reached in proximately 12, 25 and 180 minutes, for ox-N-MWCNT, ox-MWCNT and ox-SWCNT, respectively. The slower adsorption rate observer for ox-SWCNT could be due to cadmium diffusion through spacing between entangled nanotubes, pores along the sidewalls created by oxidation, and internal tubular spaces: open tips were wide enough (0.9 to 1.5 nm) for cadmium ions (hydrated radii of 0.426 nm [53]) to diffuse inside the nanotubes. In contrast, closed tips in some ox-MWCNTs and disrupted internal tubular section in ox-N-MWCNTs could reduce diffusional effects, because adsorption sites along the internal tubular spacing of these carbon nanotubes are not available for cadmium adsorption, consequently the equilibrium was attained faster than for ox-SWCNTs. Based on the previous observations, structural properties and diffusion are closely related, and their impact on the operational cost of processes involving carbon nanotubes should be considered. The following sections include a detailed discussion of the obtained experimental kinetic data analyzed with an external mass diffusion and intraparticle diffusion models.

6.1. Analysis of Adsorption Kinetic Data with an External Mass Transfer Model

The adsorbent particle can be visualized as having a stagnant "film" of liquid of thickness δ surrounding it (Figure 5). The mass transfer resistance of this film is proportional to δ/D_i^0, where the ionic diffusivity of cadmium in water is $D_i^0 = 1.23E\text{-}5\ cm^2/s$ [59, 60]. Controlling the speed of mixing, the film thickness can be reduced and hence accelerate the external mass transfer [35]. Experimental adsorption kinetic data were fitted by the external mass transfer model considering single nanotubes of outer diameter 1.5 nm, 50 nm and 70 nm for ox-SWCNT, ox-N-MWCNT and ox-MWCNT, respectively. Values for density of the particle, pore volume and porosity were also considered in the model. Table 1 presents results for the theoretical mass transfer coefficient ($k_{L,theo}$) and percentage of deviation between the predicted curves and the experimental data.

The external mass transfer model did not adjust well over the entire experimental data for carbon nanotubes. This model predicted a first stage which resulted in an underestimation of the time to reach equilibrium. Also, experimental data depicted a slow second stage suggesting the influence of intraparticle diffusion on cadmium adsorption rate. In addition, theoretical values for k_L were far higher than the experimental ones: around 4000, 17000 and 300 times, for ox-N-MWCNTs, ox-MWCNTs and ox-SWCNTs, respectively.

Such high predicted values could result from the nanometric dimensions of the adsorbents, when the pore diameter (dp) is small enough as in carbon nanotubes and the external diffusion determines the adsorption rate, the mass transfer in the liquid film occurs very fast because due to the high density of easily available adsorption sites outside the nanotubes. In general, the predicted concentration decay curves showed deviations from experimental data higher than 5 % for ox-SWCNTs and ox-N-MWCNT but lower than 5 % for ox-MWCNTs.

The initial cadmium concentration had an influence on k_L ; it was observed as k_L decreased about 20 % and 40 % when C_{A0} increased from 4.2 to 9.3 and 18.5 mg/L.

This suggests that the adsorption rate might occur slower as the initial concentration increases since the external sites are easily saturated and the internal adsorption sites could be available for cadmium just by a diffusion process (Figure 6).

Table 1. External mass-transfer parameters for adsorption kinetic of cadmium onto oxidized carbon nanotubes

Adsorbent	C_{A0} (mg/L)	$k_{L,exp}$ (cm/s)	$k_{L,theo}$ (cm/s)	% D
Effect of different type of carbon nanotubes				
ox-N-MWCNT	4.19	1.3E-5	8.7E-2	8.2
ox-MWCNT	4.33	5.9E-6	9.8E-2	4.1
ox-SWCNT	4.17	1.1E-6	5.0E-4	7.7
Effect of the solution initial concentration, C_{A0}				
ox-N-MWCNT	4.19	1.3E-5	8.7E-2	8.2
ox-N-MWCNT	9.28	1.1E-5	8.0E-2	2.7
ox-N-MWCNT	18.5	7.3E-6	5.2E-2	1.7

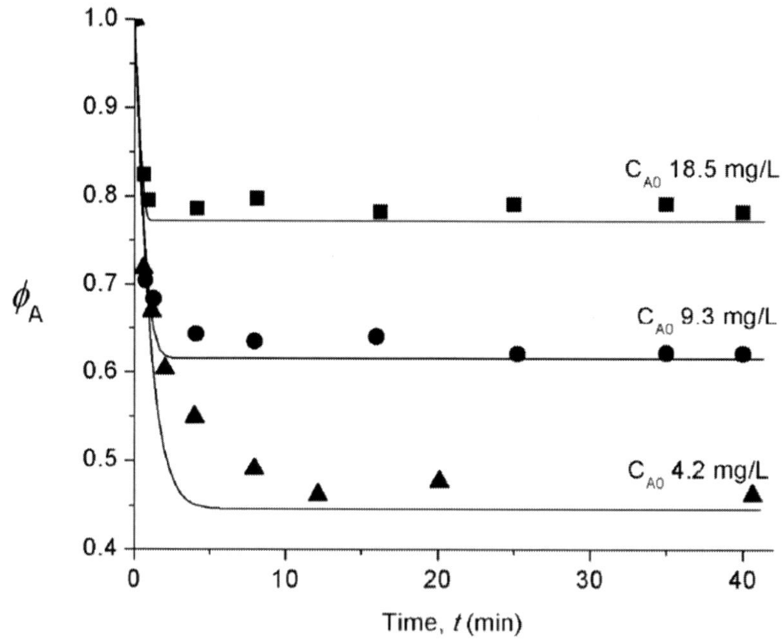

Figure 6. Concentration decay curves for cadmium adsorption onto ox-N-MWCNTs, predicted by the external mass transfer model at C_{A0} 4.2, 9.3 and 18.5 mg/L, pH 6 ± 0.1, 25 °C and 200 rpm. Full symbols and lines represent the experimental data and the predicted curves, respectively.

Experimental k_L showed the highest value for ox-N-MWCNTs, 1.3E-5 cm/s; that is 2.2 and 11.8 times higher than for ox-MWCNTs (5.9E-6 cm/s) and for ox-SWCNTs (1.1E-6 cm/s), respectively, at C_{A0} of 4 mg/L and 200 rpm.

This difference could be due to a larger available surface for ox-N-MWCNTs, near 1E-3 cm^2/g, than 6E-4 and 2.5E-6 cm^2/g for ox-MWCNTs and ox-SWCNTs, correspondingly. The experimental k_L calculated herein were far smaller than 1.5E-2 cm/s, reported by Leyva et al.,[61] for cadmium adsorption onto activated carbon at pH 8 and particle size of 0.4 mm.

These results suggest that intraparticle diffusivity could play an important role in the cadmium adsorption rate onto carbon nanotubes, in contrast with results for macroscopic size adsorbents, as those reported by Allen et al., [62] for cadmium and copper onto peat and lignite, for which external mass transfer controls the adsorption rate.

6.2. Analysis of Adsorption Kinetic Data with an Intraparticle Diffusion Model

The intraparticle diffusion model was implemented assuming two options: cadmium diffusion through single carbon nanotubes, or cadmium diffusion through agglomerated carbon nanotubes, the outer diameter of agglomerated nanoparticles, average of 40 agglomerates, was determined by microscopic analyses. Physical properties of carbon nanotubes considered for this model were outer diameter (d_{out}, μm), pore diameter (d_{pore} , cm), solid density (ρ_s, g/cm^3), pore volume (V_p, cm^3/g), particle density (ρ_p, g/cm^3) and porosity (ε_p) for single carbon nanotubes or agglomerates of carbon nanotubes. More information for these parameters, as well as for the results obtained for the intraparticle diffusion model are reported by Perez-Aguilar et al., [63].

Compared to the external mass transfer model, the intraparticle diffusion model better adjusted the experimental kinetic data, as it was observed from the deviation percentages of about 3% (Figure 7). The sensitivity of the model was tested by using an intraparticle diffusion coefficient, $D_{e,p}$, 0.5 and 2 times higher than the values that best fitted the experimental data.

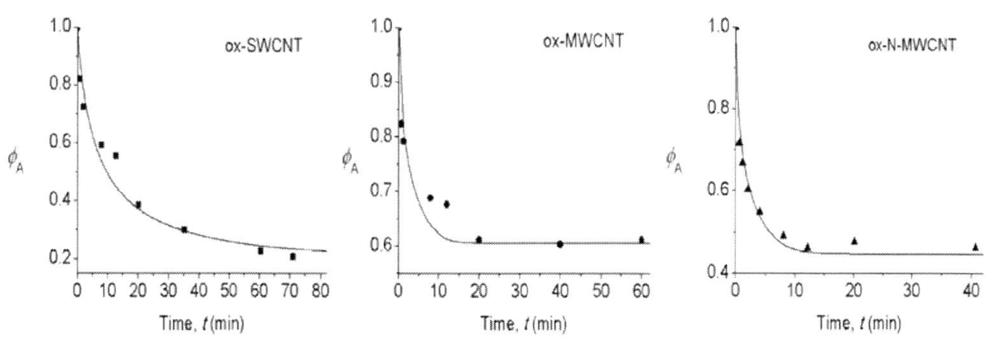

Figure 7. Concentration decay curves predicted by the intraparticle diffusion model, for cadmium adsorption onto ox-SWCNTs; ox-MWCNTs; and ox-N-MWCNTs. Full symbols represent experimental data, and the lines represent the curves predicted by the model considering single nanotubes.

These tested values for $D_{e,p}$ resulted in deviations of 3.5 % and 2.1 %, respectively. The intraparticle diffusivities ($D_{e,p}$) for single nanoparticles were around 4E-9, 1E-9 and 2.4E-11 cm^2/s for ox-N-MWCNT, ox-MWCNT and ox-SWCNT, respectively, all of them smaller than the bulk diffusivity for cadmium in water, 1.23E-5 cm^2/s. Results for $D_{e,p}$ and experimental k_L suggested that intraparticle diffusion could be a key factor in the overall cadmium adsorption rate. Faster speed reduced k_L 70 % and 80 % while $D_{e,p}$ increased 20 % and 5.4 times for ox-MWCNTs and for ox-SWCNTs, respectively.

Comparing the intraparticle diffusion coefficient, $D_{e,p}$ for different carbon nanotubes studied herein, ox-N-MWCNTs had the highest value than ox-MWCNTs and ox-SWCNTs, in correspondence to the pore size trend but contrasting to pore volume and porosity, for which ox-SWCNTs had the highest values. These features in addition to their longer length and probably highest concentration of sidewall pores might explain the slowest transport through ox-SWCNTs compared to that of ox-MWCNTs and ox-N-MWCNTs.

High ratio L/D and flexibility of carbon nanotubes resulted in agglomerates with a different geometry than a cylinder, which might influenced mass transport processes. Predicted concentration decay curves by the intraparticle diffusional model for agglomerates of nanoparticles showed that equilibrium was attained at about 19, 30 and 118 minutes for ox-N-MWCNTs, ox-MWCNTs and ox-SWCNTs, respectively.

Predicted concentration decay curves for both single nanoparticles and agglomerates apparently did not show significant differences. However, the intraparticle diffusivities ($D_{e,p}$) for agglomerated nanotubes were 3.1E-5, 4.5E-5, and 2E-4 cm^2/s for ox-N-MWCNTs, ox-SWCNTs and ox-MWCNTs, respectively. These values for $D_{e,p}$ were 2.5, 3.7 and 16 times higher than the bulk diffusivity for cadmium in water (1.23E-5 cm^2/s), which is not possible because pores limit cadmium transfer to adsorption sites, thus $D_{e,p}$ should be lower than D_i°. Therefore, modeling cadmium adsorption kinetics on agglomerates of nanotubes by using the intraparticle diffusion model may need different assumptions or even a different configuration of the adsorption kinetic experiments.

The cadmium adsorption process onto carbon nanotubes involved both mechanisms: at the initial stage of the adsorption, the transport through the external film controlled the process. However, the intraparticle diffusion rate was slower than the film diffusion, and therefore the overall mass transport process could be mainly controlled by intraparticle diffusion. Both mass transfer mechanisms were related to the modified Sherwood number ($N_{Sh'}$) according to equation 1.

$$N_{Sh'} = \frac{k_L d_p}{D_{e,p}} \qquad (1)$$

The obtained N_{Sh} were 0.0075, 0.028 and 0.35 for ox-SWCNTs, ox-MWCNTs and ox-N-MWCNTs, respectively, or $1/N_{Sh'}$ 133, 36 and 3, correspondingly. This means that intraparticle diffusion was more important than external mass transport in single nanotubes. The observed trend for pore size considering single nanotubes was ox-N-MWCNTs > ox-MWCNTs > ox-SWCNTs, suggesting a high binding energy toward adsorbing molecules in the nanotube interior and/or through the smallest pores of ox-SWCNTs, because closeness of the surrounding sidewalls maximizes the attractive interaction favoring diffusion inside the pores [39]. It is also difficult for the adsorbate to diffuse through the longest and highly entangled ox-SWCNTs to reach internal active sites, resulting in a slower kinetic process than for multiwall carbon nanotubes.

The effective pore volume diffusivity is related to the molecular diffusivity and to the properties of the porous media. Adsorbents of nanometric dimensions would have different adsorption kinetic behavior than those of macroscopic size (ε_p hundred to thousand times bigger than for the carbon nanotubes). This is suggested since diffusivities for single carbon nanotubes obtained by these authors, were smaller than those reported for cadmium adsorption onto bone char, 5.3E-7 cm^2/s at pH 5 [64], and onto activated carbon Filtrasorb 400, 2.5E-6 cm^2/h at pH 8 [61]. However, the equilibrium was attained in 16, 23 and 180 min for single ox-N-MWCNTs, ox-MWCNTs and ox-SWCNTs, respectively, which was far faster than for that reported for activated carbon (around 35 h) and for bone char (more than 10 h). It is to be mentioned that diffusivities, in this case, were determined considering the predicted concentration decay curve by a specific mathematical model that, as already mentioned, has limitations for cylindrical particles of nanometric size and hence the $D_{e,p}$ values reported herein should be taken with care.

Thus, ox-N-MWCNT had the fastest kinetics and ox-SWCNT the slowest, in agreement with trends reported by the external mass transfer model. Moreover, the intraparticle diffusion model includes more physical parameters resulting in a better fitting of the experimental data. Thus, when analysis adsorption kinetics of carbon nanotubes, it is recommended to use an intraparticle diffusion model. Also, it is important to mention the limitations of this last model: this only considers radial diffusion and does not include properties of fluids in nanometric dimensions.

Finally, Perez Aguilar et al.,[63] suggested that the structure of the carbon nanotubes plays a key role in the cadmium adsorption rate, and new studies are needed to determine those factors of nanostructure scale that should be involved in the modeling; for ox-SWCNT and ox-MWCNT, with longer length, diffusion through pores slows the adsorption rate, this effect also occurred in bamboo-type carbon nanotubes (ox- N-MWCNTS), but it was restricted to the tips of the nanotubes. Of the two models used in the study developed by Perez Aguilar et al.,[63], the intraparticle diffusion model provided the best fit to the experimental kinetic data in comparison with the external mass transfer model, mainly in the slow stage. The obtained D_{ep} values were lower than for the bulk diffusivity of cadmium in water only when modeling single nanotubes, suggesting that intraparticle diffusion could be the rate limiting step.

CONCLUSION

The development of devices based on carbon nanotubes for water treatment systems are in the early stages. Considering the results shown in the present chapter, carbon nanotubes are capable to remove inorganic and organic substances present in aqueous solutions, although more adsorbents as activated carbon has better capacity adsorption. However, carbon nanotubes properties can be modified with a chemical treatment to reach higher adsorption capacities and adsorption rates. The modified materials, such as ox-N-MWCNTs might be an option for future applications in adsorption systems, due to their physicochemical features and faster adsorption kinetics in relation with ox-SWCNTs and ox-MWCNTs. Nevertheless, more studies are needed to advance in the knowledge of this exciting subject.

ACKNOWLEDGMENTS

The authors appreciate the economical support from Consejo Nacional de Ciencia y Tecnologia (CONACYT) through Fondos Mixtos CONACYT-State of Puebla (PUE-2004-C02-5), Fondos Mixtos CONACYT-State of San Luis Potosi (FMSLP-2009-C02-106795), Fondos SEP-CONACYT (SEP-2004-C01-45764), Fondos SEP-PROMEP 103.5/09/4244, Fondos SEP-PROMEP 103.5/10/5475 Fondos SEP-PROMEP/103.5/11/4157 and Fondos UASLP FAI-C10-FAI-05-07.35. The authors also thank the National Laboratories LINAN and LANBAMA, and acknowledge Dulce Partida, Guillermo Vidriales and Carmen Rocha for their technical support.

REFERENCES

[1] Forum, W.W., Water for Growth and development. A theme document for the 4th World Water Forum. 2006.

[2] Savage, N. and M.S. Diallo, Nanomaterials and Water Purification: Opportunities and Challenges. *Journal of Nanoparticle Research*, 2005. 7(4): p. 331-342.

[3] USEPA, Introduction to the National Pretreatment Program. , O.o.W. Management., Editor. 1999, U.S. Environmental Protection Agency.: Washington.

[4] Organization, W.H., Guidelines for drinking-water quality: incorporating first addendum, vol 1, recommendations. 2006. 1.

[5] Werner Stumm, J.J.M., Aquatic chemistry. Chemical equilibrium and rates in natural waters. Environmental science and technology. 1996, New York, NY: John Wiley and Sons, Inc.

[6] Busca, G., et al., Technologies for the removal of phenol from fluid streams: A short review of recent developments. *Journal of Hazardous Materials,* 2008. 160(2-3): p. 265-288.

[7] Ahmaruzzaman, M. and D.K. Sharma, Adsorption of phenols from wastewater. *Journal of Colloid and Interface Science*, 2005. 287(1): p. 14-24.

[8] Radovic, L.R., Chemistry and physics of carbon.. Vol. 27. 2001, New York: Marcel Dekker Inc.

[9] Bansal, R.C., Goyal, M. , Activated carbon adsorption. 2005, Boca Raton: CRC Press.

[10] Yin, C.Y., Aroua, M.K., Daud, W.M. , Review of modifications of activated carbon for enhancing contaminants uptakes from aqueous solution. *Separation and Purification Technology* 2007. 52: p. 403-415.

[11] Yang, R.T., Adsorbents: fundamentals and applications. 2003, New Jersey: Wiley.

[12] USEPA, Nanotechnology white paper. S.P.C. U.S., Editor. 2007, US Environmental Protection Agency. Washington.

[13] Programme, W.W.A., The United Nations World Water Development Report 3: Water in a changing world., in 2009, UNESCO: Paris, London.

[14] Renn, O., Roco, M.C., Nanotechnology and the need for risk governance. *Journal of Nanoparticle Research* 2006. 8: p. 153-191.

[15] Tour, J.M., Transition to Organic Materials Science. Passive, Active, and Hybrid Nanotechnologies. *The Journal of Organic Chemistry,* 2007. 72(20): p. 7477-7496.

[16] Li, Y.-H., et al., Lead adsorption on carbon nanotubes. *Chemical Physics Letters,* 2002. 357(3-4): p. 263-266.

[17] Li, Y.-H., et al., Competitive adsorption of Pb2+, Cu2+ and Cadmium ions from aqueous solutions by multiwalled carbon nanotubes. *Carbon*, 2003. 41(14): p. 2787-2792.

[18] Yan-Hui Li, Z.L., Xu Xiao, Xiangwen Zhou, Cailu Xu, Dehai Wu, Bingqing Wei Removal of Cu2+ Ions from Aqueous Solutions by Carbon Nanotubes *Adsorption Science and Technology,* 2003. 21(5): p. 475-485.

[19] Li, Y.-H., et al., Adsorption of cadmium(II) from aqueous solution by surface oxidized carbon nanotubes. *Carbon,* 2003. 41(5): p. 1057-1062.

[20] Li, Y.-H., et al., Adsorption thermodynamic, kinetic and desorption studies of Pb2+ on carbon nanotubes. *Water Research*, 2005. 39(4): p. 605-609.

[21] Li, X.-q. and W.-x. Zhang, Iron Nanoparticles: the Core Shell Structure and Unique Properties for Ni(II) Sequestration. *Langmuir,* 2006. 22(10): p. 4638-4642.

[22] Chen, C. and X. Wang, Adsorption of Ni(II) from Aqueous Solution Using Oxidized Multiwall Carbon Nanotubes. *Industrial and Engineering Chemistry Research,* 2006. 45(26): p. 9144-9149.

[23] Kandah, M.I. and J.-L. Meunier, Removal of nickel ions from water by multi-walled carbon nanotubes. *Journal of Hazardous Materials*, 2007. 146(1-2): p. 283-288.

[24] Lu, C., C. Liu, and G.P. Rao, Comparisons of sorbent cost for the removal of Ni2+ from aqueous solution by carbon nanotubes and granular activated carbon. *Journal of Hazardous Materials*, 2008. 151(1): p. 239-246.

[25] Yang, S., et al., Adsorption of Ni(II) on oxidized multi-walled carbon nanotubes: Effect of contact time, pH, foreign ions and PAA. *Journal of Hazardous Materials,* 2009. 166(1): p. 109-116.

[26] Lu, C. and H. Chiu, Adsorption of zinc(II) from water with purified carbon nanotubes. *Chemical Engineering Science*, 2006. 61(4): p. 1138-1145.

[27] Lu, C. and H. Chiu, Chemical modification of multiwalled carbon nanotubes for sorption of Zn2+ from aqueous solution. *Chemical Engineering Journal*, 2008. 139(3): p. 462-468.

[28] Andrade-Espinosa, G., et al., Acid modified bamboo-type carbon nanotubes and cup-stacked-type carbon nanofibres as adsorbent materials: cadmium removal from aqueous solution. *Journal of Chemical Technology and Biotechnology,* 2009. 84(4): p. 519-524.

[29] Perez-Aguilar, N., et al., Adsorption of cadmium and lead onto oxidized nitrogen-doped multiwall carbon nanotubes in aqueous solution: equilibrium and kinetics. *Journal of Nanoparticle Research*. 12(2): p. 467-480.

[30] Peng, X., et al., Adsorption of 1-2 dichlorobenzene from water to carbon nanotubes. *Chem Phys Letter*, 2003. 276(1): p. 154-158.

[31] Diaz-Flores, P.E., et al., Simultaneous adsorption of Cadmium and phenol on modified N-doped carbon nanotubes: Experimental and DFT studies. *Journal of Colloid and Interface Science*, 2009. 334(2): p. 124-131.
[32] Rao, G.P., C. Lu, and F. Su, Sorption of divalent metal ions from aqueous solution by carbon nanotubes: A review. *Separation and Purification Technology,* 2007. 58(1): p. 224-231.
[33] Stafiej, A. and K. Pyrzynska, Adsorption of heavy metal ions with carbon nanotubes. *Separation and Purification Technology*, 2007. 58(1): p. 49-52.
[34] Wang, H., et al., Mechanism study on adsorption of acidified multiwalled carbon nanotubes to Pb(II). *Journal of Colloid and Interface Science,* 2007. 316(2): p. 277-283.
[35] Cooney, D.O., Adsorption design for wastewater treatment. 1999, Boca Raton Florida: Lewis Publishers.
[36] Dresselhaus, M.S., Dresselhaus, G., Avouris, Ph. , Carbon Nanotubes. Synthesis, structure, properties and applications. 2001, Germany: Springer.
[37] Ago, H., et al., Work Functions and Surface Functional Groups of Multiwall Carbon Nanotubes. *The Journal of Physical Chemistry B,* 1999. 103(38): p. 8116-8121.
[38] L, D., Review of the doping of carbon nanotubes (multiwalled and single-walled). *Carbon*, 2002. 40(10): p. 1751-1764.
[39] Kondratyuk, P. and J.T. Yates, Molecular Views of Physical Adsorption Inside and Outside of Single-Wall Carbon Nanotubes. *Accounts of Chemical Research,* 2007. 40(10): p. 995-1004.
[40] Loiseau, A., Blase, X., Charlier, J., Gadelle, P., Journet, C., Laurent, Ch., Peigney, A., Synthesis methods and growth mechanisms. in Understanding carbon nanotubes. From basics to applications., A. Loiseau, Launois, P., Petit, P., Roche, S., Salvetat, J.P., Editor. 2006, Springer: Netherlands. p. 49 -130.
[41] Liu, Y., C. Pan, and J. Wang, Raman spectra of carbon nanotubes and nanofibers prepared by ethanol flames. *Journal of Materials Science,* 2004. 39(3): p. 1091-1094.
[42] Choi, H.C., et al., Release of N2 from the Carbon Nanotubes via High-Temperature Annealing. *The Journal of Physical Chemistry B*, 2005. 109(5): p. 1683-1688.
[43] Antunes, E.F., et al., Influence of diameter in the Raman spectra of aligned multi-walled carbon nanotubes. *Carbon*, 2007. 45(5): p. 913-921.
[44] Zhang J, Z.H., Qing Q, Yang Y, Li Q, Liu Z, Guo X, Du Z Effect of chemical oxidation on the structure of single-walled carbon nanotubes. *J. Phys. Chem. B* 2003. 107(16): p. 3712-3718.
[45] Dyke, C.A. and J.M. Tour, Overcoming the Insolubility of Carbon Nanotubes through High Degrees of Sidewall Functionalization. *ChemInform,* 2004. 35(23): p. no-no.
[46] Misra, A., et al., FTIR Spectroscopy of Multiwalled Carbon Nanotubes: A Simple Approach to Study the Nitrogen Doping. *Journal of Nanoscience and Nanotechnology*, 2007. 7(6): p. 1820-1823.
[47] Choi, H.C., Bae, S.Y., Park, J., Seo, K., Kim, C., Kim, B., Song, H.J., Shin, H., Experimental and theoretical studies on the structure of N-doped carbon nanotubes: Possibility of intercalated molecular N2. *Applied Physics Letters* 2004. 85: p. 5742-5744.

[48] Pamula, E. and P.G. Rouxhet, Bulk and surface chemical functionalities of type III PAN-based carbon fibres. *Carbon*, 2003. 41(10): p. 1905-1915.

[49] Jerzy, Z., IR spectroscopic investigations of the mechanism of oxidation of carbonaceous films with HNO3 solution. *Carbon*, 1980. 18(4): p. 281-285.

[50] Gomez-Serrano, V., et al., Mass and surface changes of activated carbon treated with nitric acid. Thermal behavior of the samples. *Thermochimica Acta*, 1997. 291(1-2): p. 109-115.

[51] Chen, J.P. and Wu, Acid/Base-Treated Activated Carbons:â€‰ Characterization of Functional Groups and Metal Adsorptive Properties. *Langmuir*, 2004. 20(6): p. 2233-2242.

[52] Wang, S.-G., et al., Removal of lead(II) from aqueous solution by adsorption onto manganese oxide-coated carbon nanotubes. *Separation and Purification Technology,* 2007. 58(1): p. 17-23.

[53] Nightingale, E.R., Phenomenological Theory of Ion Solvation. Effective Radii of Hydrated Ions. *The Journal of Physical Chemistry*, 1959. 63(9): p. 1381-1387.

[54] Woods, L.M., Å.C. BÄƒdescu, and T.L. Reinecke, Adsorption of simple benzene derivatives on carbon nanotubes. *Physical Review B,* 2007. 75(15): p. 155-415.

[55] Tessmer, C.H., R.D. Vidic, and L.J. Uranowski, Impact of Oxygen-Containing Surface Functional Groups on Activated Carbon Adsorption of Phenols. *Environmental Science and Technology,* 1997. 31(7): p. 1872-1878.

[56] Li, Y.H., Zhao, Y. M., Hu, W. B., Ahmad, I., Zhu, Y.Q., Peng, X.J., Carbon nanotubes-the promising adsorbent in wastewater treatment. *Journal of Physics: Conference Series*, 2007. 61 p. 698-702.

[57] Bansal, R.C., Donnet, J.B., Stoeckli, F., Active carbon. 1988, New York: Marcel Dekker Inc,.

[58] Ewels, C.P. and M. Glerup, Nitrogen Doping in Carbon Nanotubes. *Journal of Nanoscience and Nanotechnology*, 2005. 5(9): p. 1345-1363.

[59] Newman, J., Electrochemical Systems. 1973, Englewood Cliffs, NJ: Prentice Hall.

[60] Dobos, D., Handbook for electrochemists in industry and universities. 1994, Amsterdam: Elsevier Scientific.

[61] Leyva-Ramos, R., et al., Intraparticle diffusion of cadmium and zinc ions during adsorption from aqueous solution on activated carbon. *Journal of Chemical Technology and Biotechnology*, 2005. 80(8): p. 924-933.

[62] Allen, S.J., et al., The Adsorption of Pollutants by Peat, Lignite and Activated Chars. *Journal of Chemical Technology and Biotechnology*, 1997. 68(4): p. 442-452.

[63] Perez-Aguilar, N.V., P.E. Diaz-Flores, and J.R. Rangel-Mendez, The adsorption kinetics of cadmium by three different types of carbon nanotubes. *Journal of Colloid and Interface Science.* 364(2): p. 279-287.

[64] Cheung, C.W., et al., Combined Diffusion Model for the Sorption of Cadmium, Copper, and Zinc Ions onto Bone Char. *Environmental Science and Technology,* 2001. 35(7): p. 1511-1522.

INDEX

A

D

E

F

G

H

M

N

O

P

T

U

V

W

X

Y

Z